Principles of
Environmental Science
Inquiry & Applications

FIFTH EDITION

William P. Cunningham
University of Minnesota

Mary Ann Cunningham
Vassar College

 Higher Education

Boston   Burr Ridge, IL   Dubuque, IA   New York   San Francisco   St. Louis
Bangkok   Bogotá   Caracas   Kuala Lumpur   Lisbon   London   Madrid   Mexico City
Milan   Montreal   New Delhi   Santiago   Seoul   Singapore   Sydney   Taipei   Toronto

Higher Education

PRINCIPLES OF ENVIRONMENTAL SCIENCE: INQUIRY & APPLICATIONS, FIFTH EDITION

ISBN 978-0-07-338319-4
MHID 0-07-338319-8

Publisher: *Janice Roerig-Blong*
Executive Editor: *Margaret J. Kemp*
Director of Development: *Kristine Tibbetts*
Senior Developmental Editor: *Rose M. Koos*
Editorial Coordinator: *Ashley Zellmer*
Senior Marketing Manager: *Tami Petsche*
Lead Project Manager: *Peggy J. Selle*
Senior Production Supervisor: *Laura Fuller*
Senior Media Project Manager: *Tammy Juran*
Manager, Creative Services: *Michelle D. Whitaker*
Cover Designer: *Jamie E. O'Neal*
Interior Designer: *John Albert Joran*
(USE) Cover Image: *(front)* ©*Michael Melford/Getty Images; (back inset)* ©*Peter Essick/Aurora/Getty Images*
Senior Photo Research Coordinator: *Lori Hancock*
Photo Research: *LouAnn K. Wilson*
Compositor: *Lachina Publishing Services*
Typeface: *10/12 Times LT*
Printer: *Quebecor World Versailles, KY*

The credits section for this book begins on page 389 and is considered an extension of the copyright page.

Library of Congress Cataloging-in-Publication Data

Cunningham, William P.
  Principles of environmental science : inquiry & applications / William Cunningham, Mary Ann Cunningham.
—5th ed.
      p. cm.
  Includes index.
  ISBN 978-0-07-338319-4 — ISBN 0-07-338319-8 (hard copy : alk. paper) 1. Environmental sciences—
Textbooks. I. Cunningham, Mary Ann. II. Title.
  GE105.C865 2009
  363.7—dc22

                                                                                    2008019443

www.mhhe.com

About the Authors

William P. Cunningham

William P. Cunningham is an emeritus professor at the University of Minnesota. In his 38-year career at the university, he taught a variety of biology courses, including Environmental Science, Conservation Biology, Environmental Health, Environmental Ethics, Plant Physiology, General Biology, and Cell Biology. He is a member of the Academy of Distinguished Teachers, the highest teaching award granted at the University of Minnesota. He was a member of a number of interdisciplinary programs for international students, teachers, and nontraditional students. He also carried out research or taught in Sweden, Norway, Brazil, New Zealand, China, and Indonesia.

Professor Cunningham has participated in a number of governmental and nongovernmental organizations over the past 40 years. He was chair of the Minnesota chapter of the Sierra Club, a member of the Sierra Club national committee on energy policy, vice president of the Friends of the Boundary Waters Canoe Area, chair of the Minnesota governor's task force on energy policy, and a citizen member of the Minnesota Legislative Commission on Energy.

In addition to environmental science textbooks, Cunningham edited three editions of an *Environmental Encyclopedia* published by Thompson-Gale Press. He has also authored or co-authored about 50 scientific articles, mostly in the fields of cell biology and conservation biology as well as several invited chapters or reports in the areas of energy policy and environmental health. His Ph.D. from the University of Texas was in botany.

His hobbies include backpacking, canoe and kayak building (and paddling), birding, hiking, gardening, and traveling. He lives in St. Paul, Minnesota with his wife, Mary. He has three children (one of whom is co-author of this book) and seven grandchildren.

Mary Ann Cunningham

Mary Ann Cunningham is an associate professor of geography at Vassar College, in New York's Hudson Valley. A biogeographer with interests in landscape ecology, geographic information systems (GIS), and remote sensing, she teaches environmental science, natural resource conservation, and land-use planning, as well as GIS and remote sensing. Field research methods, statistical methods, and scientific methods in data analysis are regular components of her teaching. As a scientist and educator, Mary Ann enjoys teaching and conducting research with both science students and non-science liberal arts students. As a geographer, she likes to engage students with the ways their physical surroundings and social context shape their world experience. In addition to teaching at a liberal arts college, she has taught at community colleges and research universities.

Mary Ann has been writing in environmental science for over a decade, and she has been co-author of this book since its first edition. She is also co-author of *Environmental Science* (now in its tenth edition), and an editor of the *Environmental Encyclopedia* (third edition, Thompson-Gale Press). She has published work on pedagogy in cartography, as well as instructional and testing materials in environmental science. With colleagues at Vassar, she has published a GIS lab manual, *Exploring Environmental Science with GIS*, designed to provide students with an easy, inexpensive introduction to spatial and environmental analysis with GIS.

In addition to environmental science, Mary Ann's primary research activities focus on land-cover change, habitat fragmentation, and distributions of bird populations. This work allows her to conduct field studies in the grasslands of the Great Plains as well as in the woodlands of the Hudson Valley. In her spare time she loves to travel, hike, and watch birds.

Mary Ann holds a bachelor's degree from Carleton College, a master's degree from the University of Oregon, and a Ph.D. from the University of Minnesota.

Brief Contents

Table of Contents

v

List of Case Studies

Renewed Passion for Environmental Science

A new energy is invigorating the environmental movement. Analysts once said that environmentalism is dead, but now a diverse, savvy, and passionate movement is taking shape. The need for environmental science education has never been greater as the mounting evidence of environmental threats has become impossible to ignore. Meanwhile, scientists are finding better ways to interpret and explain research results, activists are discovering new approaches for shaping public policy, and the general public is awaking to the importance of clean water and clear air. In the United States, hundreds of colleges, communities, and local governments are working to reduce carbon emissions and to use energy efficiently. More than 400 bills have been passed in 40 states to require renewable energy or to otherwise combat climate change.

Environmental science is truly a global concern. Even people in developing countries are demanding better protection of environmental quality. The Chinese government, for example, responding to thousands of citizen protests, has promised new policies that will promote renewable energy, clean surface waters, and improve air quality. It remains to be seen how well these ambitions will be met, but the dramatic changes in rhetoric, technology, and creativity are remarkable. Most importantly, environmental concern is not just a fringe movement involving efforts to protect and improve our common environment: it is business leaders finding ways to reduce costs by reducing waste, insurance companies concerned about rising sea levels, and inner-city communities trying to lower asthma rates in children. Major changes are occurring across the globe in the quest to save the critical resources that provide life and health to the environment. It's a wonderful time to be studying these issues and to prepare to play a role either as a practitioner or an informed citizen.

What Sets This Book Apart?

A Positive, Balanced Viewpoint

If students are to take the ideas of environmental science to heart, they need positive messages about ways all of us can contribute to a more sustainable world. This book presents the positive developments through introductory **case studies** at the beginning of each chapter, illustrating an important current issue to demonstrate how it relates to practical environmental concerns. Most of these case studies present optimistic examples in which people are working to find solutions to environmental problems. These stories also help to demystify scientific investigation and help students understand how scientists study complex issues. In addition to these introductory stories, case studies and examples of how scientists investigate our environment appear periodically throughout the book to reiterate the practical importance of these issues.

Integrated Approach Emphasizing Sustainability

Environmental problems and their solutions occur at the intersection of natural systems and the human systems that manipulate the natural world. In this book we present an **integrated approach** to physical sciences—biology, ecology, geology, air and water resources—and to human systems that affect nature—food and agriculture, population growth, urbanization, environmental health, resource economics, and policy. Although it is tempting to emphasize purely natural systems, we feel that students can never understand why coral reefs are threatened or why tropical forests are being cut down if they don't know something about the cultural, economic, and political forces that shape our decisions.

Current and Accurate Data

Throughout this book, we present up-to-date tables and graphs with the most current available data. We hope this data will give students an appreciation of the kinds of information available in environmental science. Among the sources we have called upon here are geographic information systems (GIS) data and maps, current census and population data, international news and data sources, and federal data collection agencies. Every chapter in this book has numerous updates that reflect recent events in energy, food, climate, population trends, and other important issues.

Active Learning and Critical Thinking

Learning how scientists approach problems can help students develop habits of independent, orderly, and objective thought. But it takes active involvement to master these skills. *Principles of Environmental Science* integrates numerous learning aids that will encourage students to think for themselves. Data and interpretations aren't presented as immutable truths, but rather as evidence to be examined and tested.

- **Exploring Science** essays promote scientific literacy by demonstrating the methodology scientists use to explore complex environmental questions.

- **What Can You Do?** boxes encourage students to "make a difference" by assuming personal responsibility for environmentally friendly decisions. The text offers many examples of how scientists and citizens have worked to resolve environmental questions, both basic and applied.

- **Data Analysis** exercises conclude each chapter to give students further opportunities to apply skills and put into practice the knowledge they've gained. We pay special attention to graphing techniques in these boxes because data display is such an important part of scientific information delivery. We don't limit this discussion to simple pie charts and line plots, we use this space to demonstrate a variety of ways to display and analyze data.

- **Critical Thinking and Discussion**, a challenging, open-ended set of questions at the end of each chapter, encourages students to think more deeply and independently about issues and principles presented in the chapter. These questions make excellent starting points for discussion sections. They also could be used to practice for essay exams, or might even serve as an essay exam themselves.

What's New in This Edition?

Google Earth™ Placemarks

Throughout this book we've identified interesting and important geographical places that help in understanding environmental issues. Icons in the text identify these places and direct the reader to placemarkers on our web page that take you directly to those places in Google Earth™. You can zoom in for a close view or up to a higher altitude to gain an overall perspective. We believe the exercises we've created around these placemarkers on our website will help students gain a global perspective and will be useful for concept review, class discussion, and lecture enrichment.

Active Learning Exercises

Active Learning exercises encourage students to practice critical thinking skills and apply their understanding of chapter concepts to propose solutions.

Learning Outcomes

Each chapter opens with a list of learning outcomes that will help students organize study priorities. Rather than being imperative requirements, these outcomes have been changed to more friendly questions that lead rather than command.

End-of-Chapter Study Tools

For this edition, we've changed the review questions to practice quizzes to help students prepare for exams. This edition also has a number of new Data Analysis exercises, critical thinking and discussion questions, and conclusions that draw together key ideas in each chapter.

New Chapter Content

- Chapter 1 has been reorganized to engage students more quickly with a major emphasis on environmental problems and progress. A revised presentation of environmental history puts issues in context while a strengthened discussion of critical thinking and sound science helps students analyze information.

- Chapter 2 has improved presentations on systems, nutrients, isotopes, and ecosystems. A new Data Analysis box invites students to explore nutrient flow in a wetland.

- Chapter 3 has a new Exploring Science box on evolution of cichlids in Lake Victoria and a new What Can You Do? box on working locally for ecological diversity. It also has a revised section on human-caused ecological disturbances. A new Data Analysis box on the classic species competition studies of G. F. Gause gives students some historical background and invites them to learn to read graphs.

- Chapter 4 opens with a new case study on successful family planning in Thailand. The chapter goes on to a new discussion of ecological footprints along with updated world population and demographic data. It ends with a new Data Analysis box on communicating with graphs.

- Chapter 5 includes a modified introduction to biodiversity, an updated discussion of the Endangered Species Act, and a new Active Learning box on climate graphs.

- Chapter 6 has a new case study on British Columbia's Great Bear Rainforest, a major new section on world parks and preserves, and a new Exploring Science box on rangeland conservation in New Mexico.

- Chapter 7 has been extensively revised to include a section that emphasizes dramatic changes in food production and hunger in the past 40 years, an individual's relationship to food production, a new discussion of cheap food policies in the U.S., a new section on locavores, and other sustainable activities. It also includes a new Data Analysis exercise and two new Active Learning exercises.

- Chapter 8 opens with a new case study on successful Guinea worm eradication. It has an added section on the role of environmental factors in global disease and a revised section on conservation medicine including recent disease outbreaks. New information about methicillin-resistant Staph A has been added together with a new section on hormesis and epigenetics.

- Chapter 9 is among the most completely updated in the book. It has a new case study on ocean stabilization (geoengineering) as well as a new discussion of data from ice cores in correlation with historic climate shifts. It also includes a new Active Learning box on calculating carbon reductions, an updated section on clean air legislation, and a new Data Analysis exercise on graphing air pollution.

- Chapter 10 opens with a revised case study on saving the Chattahoochee. It also has a revised section on water availability correlating with the drought in the southern United States. A new box on China's South-to-North water diversion project has been added as well as an updated section on water privatization and the conflict over water resources. Also included is a new Exploring Science box on the Gulf "dead zone."

- Chapter 11 opens with a revised case study about the problems associated with coal-bed methane wells. This chapter also provides a brief overview of the flooding in June 2008 that occurred in the Midwest. It ends with a new Data Analysis box on exploring recent earthquakes and evaluating erosion on farmland.

- Chapter 12 has a new emphasis on personal energy use and costs, as well as a new Active Learning box on the costs of driving. It includes a new table on energy use and an expanded discussion of Hubbert's peak and peak oil. The energy-efficient building and design section has been expanded and a new section on biomass fuels has been added to reflect changes in policy and technology.

- Chapter 13 has a new section on landfill methane and an expanded discussion on the export of e-waste to poor countries. This chapter also includes a new section on disposal problems for the 300 billion bottles of water consumed annually worldwide.

- Chapter 14 has a revised introduction to urban environments and economics. It also has a new Active Learning box on microlending.

- Chapter 15 opens with a new case study on greening in China. It contains a new section on the emerging grassroots movement to find solutions to global warming and a new section on sustainability that is tied to the opening story on economic development in China.

Acknowledgements

We express our gratitude to the entire McGraw-Hill book team for their wonderful work in putting together this edition. A special thanks to Janice Roerig-Blong (publisher), Marge Kemp (executive editor), Rose Koos (developmental editor), and Ashley Zellmer (editorial coordinator) who oversaw the developmental stages and made many creative contributions to this book. Cathy Conroy has done a superb job of copy editing, correcting errors, and improving our prose. Lori Hancock and LouAnn Wilson found excellent photos for us. Peggy Selle managed the project through production. Tami Petsche (marketing manager) has supported this project with her enthusiasm and creative ideas.

This text has had the benefit of input from more than 400 researchers, professionals, and instructors who have reviewed this book or our larger text, *Environmental Science: A Global Concern*. These reviewers have helped us keep the text current and focused. We deeply appreciate their many helpful suggestions and comments. Space does not permit inclusion of all the excellent ideas that were provided, but we will continue to do our best to incorporate the ideas that reviewers have given us. In addition, all of us owe a great debt to the many scholars whose work forms the basis of our understanding of environmental science. We stand on the shoulders of giants. If errors persist in spite of our best efforts to root them out, we accept responsibility.

The following individuals provided reviews for this book. We thank them for their suggestions.

Fifth Edition Reviewers

Gary A. Beluzo
Holyoke Community College

Xianfeng Chen
Slippery Rock University of Pennsylvania

Doreen Dewell
Whatcom Community College

Kenneth Engelbrecht
Metropolitan State College of Denver

Colleen Garrity
Arizona State University—Tempe

Barbara A. Giuliano
Chestnut Hill College

Daniel Habib
Queens College

Kerry E. Hartman
Fort Berthold Community College

Tara Jo Holmberg
Northwestern Connecticut Community College

James Hutcherson
Blue Ridge Community College

John M. Lendvay
University of San Francisco

Thomas R. MacDonald
University of San Francisco

John B. McGill
York Technical College

Dr. Kiran P. Misra
Edinboro University of Pennsylvania

Dan Pettus
Columbia College Online Campus

William Roy
University of Illinois—Champaign

Daniel Vogt
University of Washington

Paul Weihe
Central College

Guided Tour

Application-based learning contributes to engaged scientific investigation

Case Studies

All chapters open with a real-world case study to help students appreciate and understand how environmental science impacts lives and how scientists study complex issues.

New! Google Earth™ Placemarks

Google Earth™ interactive satellite imagery gives students a geographic context for global places and topics discussed in the text. Google Earth™ icons indicate when to visit the text's website, where students will find links to locations mentioned in the text, and corresponding exercises that will help them understand environmental topics. Placemark links can be found at www.mhhe.com/cunningham5e.

CASE STUDY

Ocean Fertilization

"Give me a half a tanker of iron and I'll give you an Ice Age." When oceanographer, John Martin, made that remark 25 years ago, most of his audience treated it as a joke, but now some people are taking his idea more seriously. Martin had carried out experiments showing that primary productivity in much of the ocean is limited by nutrient deficiencies. When he added iron to water collected a few hundred kilometers off the Antarctic coast, he found that chlorophyll levels increased by a factor of 10,000. Some subsequent experiments have produced even greater biomass growth (fig. 9.1).

The reason people are so interested in ocean fertilization is that it might offer a way to remove large amounts of carbon dioxide (CO_2) from the air. As we'll discuss in this chapter, a vast majority of climate scientists believe that our emissions of CO_2 and other heat-absorbing gases are now warming the atmosphere and changing our global climate. If these pollutants continue to accumulate at current rates, there could be truly disastrous consequences. Global climate change may well be the most serious threat—outside of nuclear war—that we currently face. We need urgently to reduce our discharges and to find ways to remove these gases from the air.

You'll learn in this chapter that numerous approaches for emissions reductions and carbon sequestration have been suggested. Many steps can be taken using existing technology and would actually save money. Some, like ocean enrichment, have uncertain benefits and potential complications. Critics worry that we might cause new problems in our attempt to solve old ones.

Will algal cells, stimulated by adding extra nutrients to the ocean, eventually die and sink to the bottom, or will they simply decay and release the CO_2 they've absorbed back into the atmosphere? It's possible that growing more algae in the ocean could help restore fish populations depleted by overharvesting. On the other hand, large-scale eutrophication often creates oxygen-depleted "dead zones" in which almost nothing can survive. Sometimes excess nutrients result in blooms of highly toxic algae and dinoflagellates. Many scientists worry that sensitive ecosystems,

such as coral reefs, which already are being stressed by climate change and human intrusions, might be fatally damaged by efforts to change ocean chemistry.

However, at least two commercial companies have already announced plans for large-scale ocean fertilization. The Planktos Corporation of Foster City, California, expects to spread 50 to 100 tons of pulverized iron ore in a 50 to 100 km diameter area of the Pacific about 300 km west of the Galápagos Islands. And the Ocean Nourishment Corporation from New South Wales, Australia, already has approval to dump 500 tons of urea (as a nitrogen source) in the Sulu Sea between the Philippines and Borneo. Ocean Nourishment also has plans for similar fertilization projects in Malaysia, Chile, and the United Arab Emirates. Both these companies hope to sell lucrative carbon offset credits on the global climate exchange as a result of their ocean nutrient enhancement.

In 2007, the International Maritime Organization, the international body that administers the Law of the Sea, declared that all ocean fertilization projects fall under their jurisdiction. They didn't ban these activities outright, but they expressed concern about unintended consequences and unknown ecological effects. Many environmental groups cheered this intervention, but others warn that we need to do everything in our power to combat global warming. What do you think? Is the risk of geoengineering worth its possible benefits? If you were a delegate to the Law of the Sea Convention, what monitoring steps and ecological safeguards would you impose on companies wanting to exploit this technology?

A major part of this chapter will be devoted to air pollutants and global climate change and what we might do about them. First, however, in order to understand our climate system better, we'll look at the factors that normally shape our weather and climate.

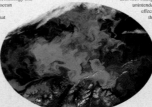

Figure 9.1 A massive plankton bloom fills the ocean off the coast of Norway.
Photo courtesy of NASA.

9.1 The Atmosphere Is a Complex System

We live at the bottom of a virtual ocean of air that extends upward about 500 km (300 mi). In the layer closest to the earth's surface, known as the troposphere, the air moves ceaselessly, flowing, swirling, and continually redistributing heat and moisture from one part of the globe to another. The composition and behavior of the troposphere and other layers control our **weather** (daily temperature and moisture conditions in a place) and our **climate** (long-term weather patterns).

The earth's earliest atmosphere probably consisted mainly of hydrogen and helium. Over billions of years, most of that hydrogen and helium diffused into space. Volcanic emissions added carbon, nitrogen, oxygen, sulfur, and other elements to the atmosphere.

Data Analysis

At the end of each chapter, these exercises give students further opportunities to apply skills and analyze data.

Data Analysis: Graphing Multiple Variables

Is it possible to show relationships between two dependent variables on the same graph? Sometimes that's desirable when you want to make comparisons between them. The graph on page 198 does just that. It's a description of how people perceive different risks. We judge the severity of risks based on how familiar they are and how much control we have over our exposure.

- Take a look at the different risks in Figure 1. Make a list of about 5 that you consider most dangerous. Then list about 5 that you think are least dangerous.

- Do most of your most dangerous activities/items fall into one quadrant? What are the axes of the graph? Do the ideas on the axes help explain why you consider some activities more dangerous than others?

New! Active Learning

Students will be encouraged to practice critical thinking skills and apply their understanding of newly-learned concepts and to propose possible solutions.

Active Learning

Calculating Probabilities

You can calculate the statistical danger of a risky activity by multiplying the probability of danger by the frequency of the activity. For example, in the United States, 1 person in 3 will be injured in a car accident in their lifetime (so the probability of injury is 1 per 3 persons, or $1/3$). In a population of 30 car-riding people, the cumulative risk of injury is 30 people × (1 injury/3 people) = 10 injuries over 30 lifetimes.

1. If the average person takes 50,000 trips in a lifetime and the accident risk is $1/3$ per lifetime, what is the probability of an accident per trip?

2. If you have been riding safely for 20 years, what is the probability of an accident during your next trip?

Answers: 1. Probability of injury per trip = (1 injury/3 lifetimes) × lifetime/50,000 trips) = 1 injury/150,000 trips. 2. 1 in 150,000, statistically, you have the same chance each time.

What Do You Think?

This feature provides challenging environmental studies that offer an opportunity for students to consider contradictory data, special interest and conflicting interpretations within a real scenario.

WHAT DO YOU THINK?

Cultural Choices and the Rate of Population Growth

The good news is that the human population on earth may never double again. The bad news is that it may take a century to stabilize. Meanwhile, the choices we make individually, and collectively through our political, economic, and cultural actions, will determine whether it will be a pleasant, or not so pleasant, hundred years. The choices are already being made in India, where two different states are taking two very different approaches to population growth.

In 1999, having added more than 180 million people in just a decade, India reached a population of 1 billion humans. If current growth rates persist, India will have at least 1.63 billion residents in 2050 and will surpass China as the world's most populous country. How will this country, where more than a quarter of its inhabitants live in abject poverty, feed, house, educate, and employ all those being added each year? A fierce debate is taking place about how to control India's population, with ramifications for the rest of the world as well.

On one side of this issue are those who believe that the best way to reduce the number of children born is poverty eradication and progress for women. Drawing on social justice principles established at the 1994 UN Conference on Population and Development in Cairo, some argue that responsible economic development, a broad-based social welfare system, education and empowerment of women, and high-quality health care—including family planning services—are essential components of population control. Without progress in these areas, they believe, efforts to provide contraceptives or encourage sterilization are futile.

On the other side of this debate are those who contend that, while social progress is an admirable goal, India doesn't have the time or

Its annual growth rate of 1.7 percent suggests that India's population could double in 41 years, becoming the world's most populous country by 2050. Indian states have taken different approaches to slow population growth. Despite its poverty, Kerala has a total fertility rate lower than the United States.

Exploring Science

Current environmental issues exemplify the principles of scientific observation and data-gathering techniques to promote scientific literacy.

Exploring
SCIENCE:
Studying the Gulf Dead Zone

In the 1980s shrimp boat crews noticed that certain locations off the Gulf Coast of Louisiana were emptied of all aquatic life. Since the region supports shrimp, fish, and oyster fisheries worth $250 to $450 million per year, these "dead zones" were important to the economy as well as to the Gulf's ecological systems. In 1985, Nancy Rabelais, a scientist working with Louisiana Universities Marine Consortium, began mapping areas of low oxygen concentrations in the Gulf waters. Her results, published in 1991, showed that vast areas, just above the floor of the Gulf, had oxygen concentration less than 2 parts per million (ppm), a level that eliminated all animal life except primitive worms. Healthy aquatic systems usually have about 10 ppm dissolved oxygen. What caused this hypoxic (oxygen-starved) area to develop?

Rabelais and her team tracked the phenomenon for several years, and it became clear that the dead zone was growing larger over time, that poor shrimp harvests coincided with years when the zone was large, and that the size of the dead zone, which ranges from 5,000 to 20,000 km² (about the size of New Jersey), depended on rainfall and runoff rates from the Mississippi River. Excessive nutrients, mainly nitrogen, from farms and cities far upstream on the Mississippi River, were the suspected culprit.

How did Rabelais and her team know that nutrients were the problem? They noticed that each year, 7–10 days after large spring rains in the agricultural parts of the upper Mississippi watershed, oxygen concentrations in the Gulf drop from 5 ppm to below 2 ppm. These rains are known to wash soil, organic debris, and last year's nitrogen-rich fertilizers from farm fields. The scientists also knew that saltwater ecosystems normally have little available nitrogen, a key nutrient for algae and plant growth. Pulses of agricultural runoff were followed by a pulse of growth of algae and phytoplankton (tiny floating plants). Such a burst of biological activity produces an excess of dead plant cells and

fecal matter that drifts to the seafloor. Shrimp, clams, oysters, and other filter feeders normally consume this debris, but they can't keep up with the sudden flood of material. Instead, decomposing bacteria in the sediment break down the debris, and they consume most of the available dissolved oxygen as well. Putrefying sediments also produce hydrogen sulfide, which further poisons the water near the seafloor.

In well-mixed water bodies, as in the open ocean, oxygen from upper layers of water is frequently mixed into lower water layers. Warm, protected water bodies are often stratified, however, as abundant sunlight keeps the upper layers warmer, and less dense, than lower layers. Denser lower layers cannot mix with upper layers unless strong currents or winds stir the water.

Many enclosed coastal waters, including Chesapeake Bay, Long Island Sound, the Mediterranean Sea, and the Black Sea, tend to be stratified and suffer hypoxic conditions that destroy bottom and near-bottom communities. There are about 200 dead zones around the world, and the number has doubled each decade since dead zones were first observed in the 1970s. The Gulf of Mexico is second in size behind a 100,000 km² dead zone in the Baltic Sea.

Can dead zones recover? Yes. Water is a forgiving medium, and organisms use nitrogen quickly. In 1996 in the Black Sea region,

farmers in collapsing communist economies cut their nitrogen applications by half out of economic necessity: the Black Sea dead zone disappeared, while farmers saw no drop in their crop yields. In the Mississippi watershed, farmers can afford abundant fertilizer, and they fear they can't afford to risk underfertilizing. Because of the great geographic distance between the farm states and the Gulf, Midwestern states have been slow to develop an interest in the dead zone. At the same time, concentrated feedlot production of beef and pork is rapidly increasing, and feedlot runoff is the fastest growing, and least regulated, source of nutrient enrichment in rivers.

In 2001, federal, state, and tribal governments forged an agreement to cut nitrogen inputs by 30 percent and reduce the size of the dead zone to 5,000 km². This agreement represented astonishingly quick research and political response to scientific results, but it doesn't appear to be enough. Computer models suggest that it would take a 40–45 percent reduction in nitrogen to achieve the 5,000 km² goal.

Human activities have increased the flow of nitrogen reaching U.S. coastal waters by four to eight times since the 1950s. Phosphorus, another key nutrient, has tripled. This case study shows how water pollution can connect far-distant places, such as Midwestern farmers and Louisiana shrimpers.

Hypoxic zone

The Mississippi River drains 40 percent of the coterminous United States, including the most heavily farmed states. Nitrogen fertilizer produces a summer "dead zone" in the Gulf of Mexico.

What Can You Do?

Practical ideas that students can employ to make a positive difference in our environment.

What Can You Do?

Reducing Individual CO_2 Emissions

Each of us can take steps to reduce global warming. While individually each change may have only a small impact, collectively they add up. Furthermore, most will save money in the long run and have other environmental and health benefits by reducing air pollution and resource consumption. The savings vary depending on where and how you live, but the following are averages for the United States.

	CO_2 Reduction (Pounds per Year)	Approximate Yearly Savings (U.S. $)
1. Keep your car tires at full pressure, avoid quick starts and stops, and drive within the speed limit.	1,100	$130
2. Carpool, walk, or take the bus once per week.	800	$100
3. Turn off the lights when you leave a room.	300	$10
4. Replace all your incandescent light bulbs with compact fluorescent bulbs.	100 each	$5 each
5. Raise your air conditioning 2°.	400	$20
6. Lower your furnace thermostat 2°.	550	$50
7. When heating or cooling, close doors and windows.	500	$20
8. When heating or cooling aren't needed, open doors and windows.	500	$20
9. Unplug TVs, DVD players, computers, and other instant-on electronics.	250	$20
10. Eat local, seasonal food.	250	variable
11. Take only five-minute showers.	250	$25
12. Take the train instead of flying 500 miles.	300	$100
13. Air dry your clothes.	700	$100
14. Replace your old car, truck, or SUV with one that gets at least 50 mpg.	6,000	$750
15. Defrost your refrigerator and keep coils and door seals clean.	700	$100

Source: Interfaith Power and Light, 2007.

Pedagogical features facilitate student understanding of environmental science

Learning Outcomes

After studying this chapter, you should be able to answer the following questions:

- Why are we concerned about human population growth?
- Will the world's population double again as it did between 1965 and 2000?
- What is the relationship between population growth and environmental impact?
- Why has the human population grown so rapidly since 1800?
- How is human population growth changing in different parts of the world?
- How does population growth change as a society develops?
- What factors slow down or speed up human population growth?

New! Learning Outcomes

Questions at the beginning of each chapter challenge students to find their own answers.

Conclusion

The conclusion summarizes the chapter by highlighting key ideas and relating them to one another.

Conclusion

A few decades ago, we were warned that a human population explosion was about to engulf the world. Exponential population growth was seen as a cause or corollary to nearly every important environmental problem. Some people still warn that the total number of humans might grow to 30 or 40 billion by the end of this century. Birth rates have fallen, however, almost everywhere, and most demographers now believe that we will reach an equilibrium around 9 billion people in about 2050. Some claim that if we promote equality, democracy, human development, and modern family planning techniques, population might even decline to below its current level of 6.7 billion in the next 50 years. How we should carry out family planning and birth control remains a controversial issue. Should we focus on political and economic reforms, and hope that a demographic transition will naturally follow; or should we take more direct action (or any action) to reduce births?

Whether our planet can support 9 billion—or even 6 billion—people on a long-term basis remains a vital question. If all those people try to live at a level of material comfort and affluence now enjoyed by residents of the wealthiest nations, using the old, polluting, inefficient technology that we now employ, the answer is almost certain that even 6 billion people is too many in the long run. If we find more sustainable ways to live, however, it may be that 9 billion people could live happy, comfortable, productive lives. If we don't find new ways to live, we probably face a crisis no matter what happens to our population size. We'll discuss pollution problems, energy sources, and sustainability in subsequent chapters of this book.

Practice Quiz

1. About how many years of human existence passed before the world population reached its first billion? What factors restricted population before that time, and what factors contributed to growth after that point?

2. Describe the pattern of human population growth over the past 200 years. What is the shape of the growth curve (recall chapter 3)?

3. Define *ecological footprint*. Why is it helpful, but why might it also be inaccurate?

4. Why do some economists consider human resources more important than natural resources in determining a country's future?

Practice Quiz

Short-answer questions allow students to check their knowledge of chapter concepts.

Critical Thinking and Discussion Questions

Brief scenarios of everyday occurrences or ideas challenge students to apply what they have learned to their lives.

Critical Thinking and Discussion Questions

Apply the principles you have learned in this chapter to discuss these questions with other students.

1. Suppose that you were head of a family planning agency in India. How would you design a scientific study to determine the effectiveness of different approaches to population stabilization? How would you account for factors such as culture, religion, education, and economics?

2. Why do you suppose that the United Nations gives high, medium, and low projections for future population growth? Why not give a single estimate? What factors would you consider in making these projections?

3. Some demographers claim that the total world population has already begun to slow, while others dispute this claim. How would you recognize a true demographic transition, as opposed to mere random fluctuations in birth and death rates?

4. Why do we usually express crude birth and death rates per thousand people? Why not give the numbers per person or for the entire population?

Relevant photos and instructional art support learning

Numerous high quality photos and realistic illustrations display detailed diagrams, graphs, and real-life situations.

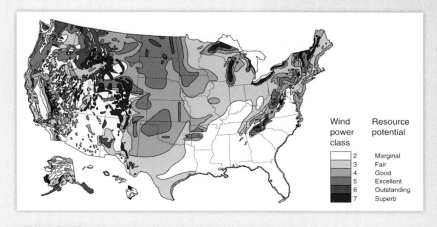

Wind power class | Resource potential
2 | Marginal
3 | Fair
4 | Good
5 | Excellent
6 | Outstanding
7 | Superb

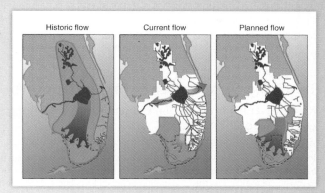

Historic flow | Current flow | Planned flow

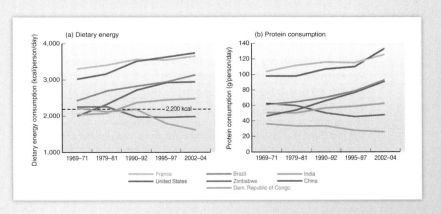

(a) Dietary energy

(b) Protein consumption

2,200 kcal

France
United States
Brazil
Zimbabwe
Dem. Republic of Congo
India
China

Teaching and Learning Supplements

McGraw-Hill offers various tools and technology products to support *Principles of Environmental Science*. Students can order supplemental study materials by contacting their local bookstore or by calling 800-262-4729. Instructors can obtain teaching aids by calling the Customer Service Department at 800-338-3987, visiting the McGraw-Hill website at www.mhhe.com, or by contacting their local McGraw-Hill sales representative.

Teaching Supplements for Instructors

McGraw-Hill's Presentation Center (found at www.mhhe.com/cunningham5e)

Build instructional materials wherever, whenever, and however you want! Presentation Center is an online digital library containing assets such as photos, artwork, animations, PowerPoints, and other media types that can be used to create customized lectures, visually enhanced tests and quizzes, compelling course websites, or attractive printed support materials.

Access to your book, access to all books!
The Presentation Center library includes thousands of assets from many McGraw-Hill titles. This ever-growing resource gives instructors the power to utilize assets specific to an adopted textbook as well as content from all other books in the library.

Nothing could be easier!
Accessed from the instructor side of your textbook's website, Presentation Center's dynamic search engine allows you to explore by discipline, course, textbook chapter, asset type, or keyword. Simply browse, select, and download the files you need to build engaging course materials. All assets are copyright McGraw-Hill Higher Education but can be used by instructors for classroom purposes.

Instructors will find the following digital assets for *Principles of Environmental Science* at Presentation Center:

- **Color Art** Full-color digital files of illustrations in the text can be readily incorporated into lecture presentations, exams, or custom-made classroom materials. These include all of the 3-D realistic art found in this edition, representing some of the most important concepts in environmental science.
- **Photos** Digital files of photographs from the text can be reproduced for multiple classroom uses.
- **Tables** Every table that appears in the text is provided in electronic format.
- **Animations** Ninety-four full-color animations that illustrate many different concepts covered in the study of environmental science are available for use in creating classroom lectures, testing materials, or online course communication. The visual impact of motion will enhance classroom presentations and increase comprehension.
- **Global Base Maps** Twenty-two base maps for all world regions and major subregions are offered in four versions: black-and-white and full-color, both with labels and without labels. These choices allow instructors the flexibility to plan class activities, quizzing opportunities, study tools, and PowerPoint enhancements.
- **PowerPoint Lecture Outlines** Ready-made presentations that combine art and photos and lecture notes are provided for each of the 15 chapters of the text. These outlines can be used as they are or tailored to reflect your preferred lecture topics and sequences.
- **PowerPoint Slides** For instructors who prefer to create their lectures from scratch, all illustrations, photos, and tables are preinserted by chapter into blank PowerPoint slides for convenience.

Earth and Environmental Science DVD by Discovery Channel Education (ISBN: 978-0-07-352541-9; MHID: 0-07-352541-3)

Begin your class with a quick peek at science in action. The exciting NEW DVD by Discovery Channel Education offers 50 short (3–5 minute) videos on topics ranging from conservation to volcanoes. Search by topic and download into your PowerPoint lecture.

McGraw-Hill's Biology Digitized Videos (ISBN: 978-0-07-312155-0; MHID: 0-07-312155-X)

Licensed from some of the highest quality life-science video producers in the world, these brief video clips on DVD range in length from 15 seconds to two minutes and cover all areas of general biology, from cells to ecosystems. Engaging and informative, McGraw-Hill's digitized biology videos will help capture students' interest while illustrating key biological concepts, applications, and processes.

eInstruction

This classroom performance system (CPS) utilizes wireless technology to bring interactivity into the classroom or lecture hall. Instructors and students receive immediate feedback through wireless response pads that are easy to use and engage students. eInstruction can assist instructors by:

- taking attendance
- administering quizzes and tests
- creating a lecture with intermittent questions
- using the CPS grade book to manage lectures and student comprehension
- integrating interactivity into PowerPoint presentations

Contact your local McGraw-Hill sales representative for more information.

Course Delivery Systems

With help from WebCT, Blackboard, and other course management systems, professors can take complete control of their course content. Course cartridges containing website content, online testing, and powerful student tracking features are readily available for use within these platforms.

Learning Supplements for Students

Text-specific Website (http://www.mhhe.com/cunningham5e)

The *Principles of Environmental Science* website provides access to resources including quizzes for each chapter, additional case studies, interactive base maps, Google Earth™ exercises, ecological footprint calculators, and much more.

Field & Laboratory Exercises in Environmental Science, Seventh Edition by Enger and Smith (ISBN: 978-0-07-290913-5; MHID: 0-07-290913-7)

The major objectives of this manual are to provide students with hands-on experiences that are relevant, easy to understand, applicable to the student's life, and presented in an interesting, informative format. Ranging from field and lab experiments to conducting social and personal assessments of the environmental impact of human activities, the manual presents something for everyone, regardless of the budget or facilities of each class. These labs are grouped by categories that can be used in conjunction with any introductory environmental textbook.

Exploring Environmental Science with GIS by Stewart, Cunningham, Schneiderman, and Gold (ISBN: 978-0-07-297564-2; MHID: 0-07-297564-4)

This short book provides exercises for students and instructors who are new to GIS, but are familiar with the Windows operating system. The exercises focus on improving analytical skills, understanding spatial relationships, and understanding the nature and structure of environmental data. Because the software used is distributed free of charge, this text is appropriate for courses and schools that are not yet ready to commit to the expense and time involved in acquiring other GIS packages.

Annual Editions: Environment 08/09 by Sharp (ISBN: 978-0-07-351548-9; MHID: 0-07-351548-5)

This twenty-seventh edition is a compilation of current articles from the best of the public press. The selections explore the global environment, the world's population, energy, the biosphere, natural resources, and pollution.

Taking Sides: Clashing Views on Environmental Issues, Thirteenth Edition by Easton (ISBN: 978-0-07-351444-4; MHID: 0-07-351444-6)

This thirteenth edition presents current controversial issues in a debate-style format designed to stimulate student interest and develop critical thinking skills. Each issue is thoughtfully framed with an issue summary, an issue introduction, and a postscript. An instructor's manual with testing material is available for each volume.

Global Studies: The World at a Glance, Second Edition by Tessema (ISBN: 978-0-07-340408-0; MHID: 0-07-340408-X)

This book features a compilation of up-to-date data and accurate information on some of the important facts about the world we live in. While it is close to impossible to be an expert in all areas, such as a nation's capital, type of government, currency, major languages, population, religions, political structure, climate, economics, etc., this book is intended to assist in understanding these essential facts in order to make useful applications.

Sources: Notable Selections in Environmental Studies, Third Edition by Goldfarb (ISBN: 978-0-07-352758-1; MHID: 0-07-352758-0)

This volume brings together primary source selections of enduring intellectual value—classic articles, book excerpts, and research studies—that have shaped environmental studies and our contemporary understanding of it. The book includes carefully edited selections from the works of the most distinguished environmental observers, past and present. Selections are organized topically around the following major areas of study: energy, environmental degradation, population issues and the environment, human health and the environment, and environment and society.

Student Atlas of Environmental Issues, by Allen (ISBN: 978-0-69-736520-0; MHID: 0-69-736520-4)

This atlas is an invaluable pedagogical tool for exploring the human impact on the air, waters, biosphere, and land in every major world region. This informative resource provides a unique combination of maps and data that help students understand the dimensions of the world's environmental problems and the geographical basis of these problems.

Traditional outrigger canoes and hand lines are still used by villagers on many islands in the southwestern Pacific, but these low-impact fishing methods are being threatened by trawlers, dynamite fishing, and other destructive techniques.

1 Understanding Our Environment

Learning Outcomes

After studying this chapter, you should be able to answer the following questions:

- Describe several of the most important environmental problems facing the world. Are there signs of hope for solving these problems?
- What do we mean by sustainability and sustainable development?
- Why does science support—but rarely prove—particular theories?
- How can scientists know if their research is reliable and important?
- How can critical thinking help us understand environmental issues?
- Explain how we can use graphs and statistics to answer questions in environmental science.
- What are some arguments for conservation or preservation of nature?

> *Today we are faced with a challenge that calls for a shift in our thinking,
> so that humanity stops threatening its life-support system.*
>
> **–Wangari Maathai, winner of 2004 Nobel Peace Prize**

Saving the Reefs of Apo Island

As their outrigger canoes glide gracefully onto Apo Island's beach after an early morning fishing expedition, villagers call to each other to ask how fishing was. "Tunay mabuti!" (very good!) is the cheerful reply. Nearly every canoe has a basketful of fish; enough to feed a family for several days with a surplus to send to the market. Life hasn't always been so good on the island. Thirty years ago, this island, like many others in the Philippines, suffered a catastrophic decline in the seafood that was the mainstay of their diet and livelihood. Rapid population growth coupled with destructive fishing methods such as dynamite or cyanide fishing, small mesh gill nets, deep-sea trawling, and *muroami* (a technique in which fish are chased into nets by pounding on coral with weighted lines) had damaged the reef habitat and exhausted fish stocks.

In 1979, scientists from Silliman University on nearby Negros Island visited Apo to explain how establishing a marine sanctuary could help reverse this decline. The coral reef fringing the island acts as a food source and nursery for many of the marine species sought by fishermen. Protecting that breeding ground, they explained, is the key to preserving a healthy fishery (fig 1.1). The scientists took villagers from Apo to the uninhabited Sumilon Island, where a no-take reserve was teeming with fish.

After much discussion, several families decided to establish a marine sanctuary along a short section of Apo Island shoreline. Initially, the area had high-quality coral but few fish. The participating families took turns watching to make sure that no one trespassed in the no-fishing zone. Within a few years, fish numbers and sizes in the sanctuary increased dramatically, and "spillover" of surplus fish led to higher catches in surrounding areas. In 1985, Apo villagers voted to establish a 500 m (0.3 mi) wide marine sanctuary around the entire island.

Fishing is now allowed in this reserve, but only by low-impact methods such as hand-held lines, bamboo traps, large mesh nets, spearfishing without SCUBA gear, and hand netting. Coral-destroying techniques, such as dynamite, cyanide, trawling, and *muroami* fishing are prohibited. By protecting the reef, villagers are guarding the nursery that forms the base for their entire marine ecosystem. Young fish growing up in the shelter of the coral move out as adults to populate the neighboring waters and yield abundant harvests. Fishermen report that they spend much less time traveling to distant fishing areas now that fish around the island are so much more abundant.

Apo Island's sanctuary is so successful that it has become the inspiration for more than 400 marine preserves throughout the Philippines and many others around the world. Not all are functioning as well as they might, but many, like Apo, have made dramatic progress in restoring abundant fish populations in nearby waters.

The rich marine life and beautiful coral formations in Apo Island's crystal clear water now attract international tourists. Two small hotels and a dive shop provide jobs for island residents. Other villagers take in tourists as boarders or sell food and T-shirts to visitors. The island government collects a diving/snorkeling fee, which has been used to build schools, improve island water supplies, and provide electricity to most of the island's 145 households. Almost all the island men still fish as their main occupation, but the fact that they don't have to go so far or work so hard for the fish they need means they have time for other activities, such as guiding diving tours or helping with household chores.

Higher family incomes now allow most island children to attend high school on Negros. Many continue their education with college or technical programs. Some find jobs elsewhere in the Philippines, and the money they send back home is a big economic boost for Apo families. Others return to their home island as teachers or to start businesses such as restaurants or dive shops. Seeing that they can do something positive to improve their environment and living conditions has empowered villagers to take on self-improvement projects that they may not otherwise have attempted.

Finding ways to live sustainably within the limits of the resource base available to us and without damaging the life-support systems provided by our ecosystem is a pre-eminent challenge of environmental science. Our answers to these challenges must be ecologically sound, economically sustainable, and socially acceptable if they are to succeed in the long term. Sometimes, as this case study shows, actions based on ecological knowledge and local action can spread to have positive effects on a global scale. Economics, policy, planning, and social organization all play vital roles in finding answers to human/environment problems. We'll discuss those disciplines later in this book, but first we'll look at the basic principles of ecological science as a basis for understanding our global environment.

Figure 1.1 Coral reefs are among the most beautiful, species-rich, and productive biological communities on the planet. They serve as the nurseries for many open-water species. At least half the world's reefs are threatened by pollution, global climate change, destructive fishing methods, and other human activities, but they can be protected and restored if we care for them.

1.1 Understanding Our Environment

Understanding the depletion of Apo Island's fishery, and finding solutions to the problem, required an appreciation of many aspects of the environment. Knowledge about population biology, reef ecology, the cultural history of fishing, and even economics of fishing all contribute to understanding environmental resource questions. As you will find throughout this book, the field of environmental science draws on many disciplines to help us understand pressing problems of resource supply, ecosystem stability, and sustainable living. As with the case of Apo Island, environmental science is about identifying and solving problems that affect our lives, and those of future generations.

In this chapter and throughout this book, you will read about many cases in which humans have caused serious environmental problems. You will also read about promising, exciting solutions to many of these problems. Your task as a student of environmental science is to gain an understanding of some of the larger current problems, what some solutions might be, and how you might use knowledge from a variety of disciplines—from biology and chemistry to economics—to develop tomorrow's strategies for more sustainable living on our planet.

We live on a marvelous planet

Before proceeding in our discussion of current dilemmas and how scientists are trying to understand them, we should pause for a moment to consider the extraordinary natural world that we inherited and that we hope to pass on to future generations in as good—or perhaps even better—condition than we found it.

Imagine that you are an astronaut returning to the earth after a long trip to the moon or Mars. What a relief it would be, after experiencing the hostile environment of outer space, to come back to this beautiful, bountiful planet (fig. 1.2). Although there are dangers and difficulties here, we live in a remarkably prolific and hospitable world that is, as far as we know, unique in the universe. Compared with the conditions on other planets in our solar system, temperatures on the earth are mild and relatively constant. Plentiful supplies of clean air, fresh water, and fertile soil are regener-

ated endlessly and spontaneously by biogeochemical cycles and biological communities (discussed in chapters 2 and 3).

Perhaps the most amazing feature of our planet is its rich diversity of life. Millions of beautiful and intriguing species populate the earth and help sustain a habitable environment (fig. 1.3). This vast multitude of life creates complex, interrelated communities where towering trees and huge animals live together with, and depend upon, such tiny life-forms as viruses, bacteria, and fungi. Together, all these organisms make up delightfully diverse, self-sustaining ecosystems, including dense, moist forests; vast, sunny savannas; and richly colorful coral reefs.

From time to time, we should pause to remember that, in spite of the challenges and complications of life on earth, we are incredibly lucky to be here. We should ask ourselves: what is our proper place in nature? What *ought* we do and what *can* we do to protect the irreplaceable habitat that produced and supports us? These are some of the central questions of environmental science.

What is environmental science?

We inhabit two worlds. One is the natural world of plants, animals, soils, air, and water that preceded us by billions of years and of which we are a part. The other is the world of social institutions and artifacts that we create for ourselves using science, technology, and political organization. Both worlds are essential to our lives, but integrating them successfully causes enduring tensions.

Environment (from the French *environner*: to encircle or surround) can be defined as (1) the circumstances and conditions that surround an organism or a group of organisms or (2) the social and cultural conditions that affect an individual or a community. Since humans inhabit the natural world as well as the "built" or technological, social, and cultural world, all constitute important parts of our environment.

Figure 1.2 The life-sustaining ecosystems on which we all depend are unique in the universe, as far as we know.

Figure 1.3 Perhaps the most amazing feature of our planet is its rich diversity of life.

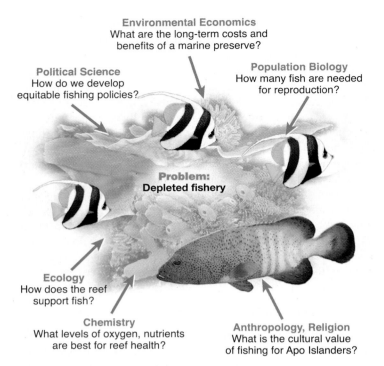

Figure 1.4 Many types of knowledge are needed in environmental science. A few examples are shown here.

Environmental science is the systematic study of our environment and our place in it. Because environmental problems are complex, environmental science draws on many fields of knowledge (fig. 1.4). Sciences such as biology, chemistry, earth science, and geography provide important information. Social sciences and humanities, from political science and economics to art and literature, help us understand how society responds to environmental crises and opportunities. Environmental science is also mission-oriented: it shows that we all have a responsibility to get involved and try to do something about the problems we have created.

The distinguished economist Barbara Ward has pointed out that, for an increasing number of environmental issues, the difficulty is not to identify remedies. Remedies are now well understood; the problem is to make them socially, economically, and politically acceptable. Foresters know how to plant trees, but not how to establish conditions under which villagers in developing countries can manage plantations for themselves. Engineers know how to control pollution, but not how to persuade factories to install the necessary equipment. City planners know how to design urban areas, but not how to make them affordable for the poorest members of society. The solutions to these problems increasingly involve human social systems as well as natural science.

Criteria for environmental literacy have been suggested by the National Environmental Education Advancement Project in Wisconsin. These criteria include awareness and appreciation of the natural and built environment, knowledge of natural systems and ecological concepts, understanding of current environmental

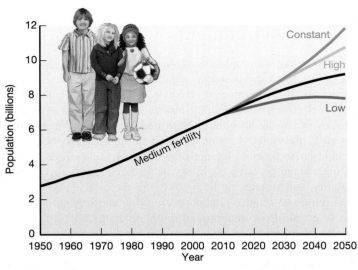

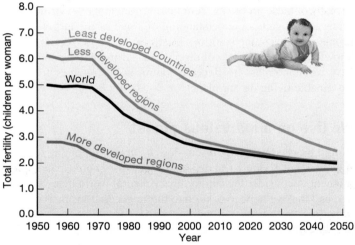

Figure 1.5 Bad news and good news: globally, populations continue to rise, but our rate of growth has plummeted. Some countries are below the replacement rate of about two children per woman.
Source: United Nations Population Program, 2007.

issues, and ability to use analytical and problem-solving skills on environmental issues. These are good goals to keep in mind as you study this book.

1.2 Environmental Problems and Opportunities

A first step in understanding environmental science is to recognize some of the principal problems we face, and some of the recent changes in environmental quality and environmental health. As you read the examples below, think about what factors contribute to these issues—imagine redrawing figure 1.5 for one of these problems. Also think about what steps might be taken to resolve some of these problems.

We face persistent environmental problems

With more than 6.7 billion people on earth, we are adding about 80 million more each year. While demographers report a transition to slower growth rates in most countries, present trends project a population between 8 and 10 billion by 2050 (fig. 1.5). The impact of that many people on our natural resources and ecological systems is a serious concern that complicates many of the other problems we face.

Clean water

Water may well be the most critical resource in the twenty-first century. Already at least 1.1 billion people lack access to safe drinking water, and twice that many don't have adequate sanitation. Polluted water contributes to the death of more than 15 million people every year, most of them children under age 5. About 40 percent of the world population lives in countries where water demands now exceed supplies, and the UN projects that by 2025 as many as three-fourths of us could live under similar conditions.

Food supplies

Over the past century, global food production has more than kept pace with human population growth, but there are worries about whether we will be able to maintain this pace. Soil scientists report that about two-thirds of all agricultural lands show signs of degradation. The biotechnology and intensive farming techniques responsible for much of our recent production gains are too expensive for many poor farmers. Can we find ways to produce the food we need without further environmental degradation? And will that food be distributed equitably? In a world of food surpluses, currently more than 850 million people are chronically undernourished, and at least 60 million face acute food shortages due to bad weather or politics (fig. 1.6).

Energy resources

How we obtain and use energy is likely to play a crucial role in our environmental future. Fossil fuels (oil, coal, and natural gas) presently provide around 80 percent of the energy used in industrialized countries. Supplies of these fuels are diminishing, however, and problems associated with their acquisition and use—air and water pollution, mining damage, shipping accidents, and geopolitics—may limit what we do with remaining reserves. Cleaner, renewable energy resources—solar, wind, geothermal, and biomass power—together with conservation could give us cleaner, less destructive options if we invest in appropriate technology.

Figure 1.6 More than 850 million people are chronically undernourished, and at least 60 million face life-threatening food shortages.

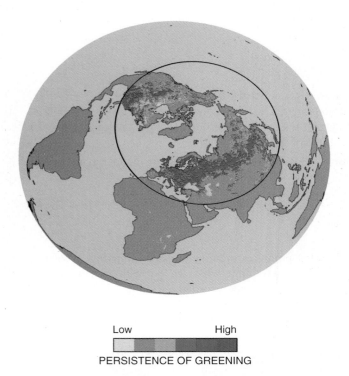

Low   High

PERSISTENCE OF GREENING

Figure 1.7 Satellite images and surface temperature data show that polar regions, especially in Eurasia, are becoming green earlier and staying green longer than ever in recorded history. This appears to be evidence of a changing global climate.
Source: NASA, 2002.

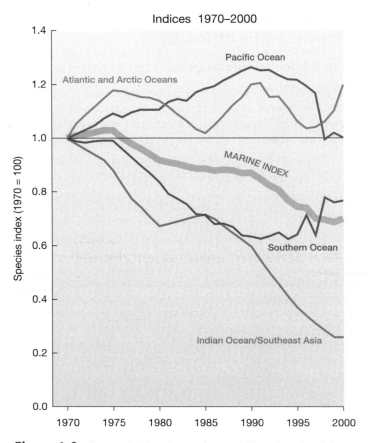

Figure 1.8 Changes in abundance of 217 species of marine fish, birds, mammals, and reptiles, from 1970 to 1999.
Source: World Wide Fund for Nature (WWF), 2000.

Climate change

Burning fossil fuels, making cement, cultivating rice paddies, clearing forests, and other human activities release carbon dioxide and other so-called greenhouse gases, which trap heat in the atmosphere. Over the past 200 years, atmospheric CO_2 concentrations have increased about 30 percent. Climatologists warn that by 2100, if current trends continue, mean global temperatures will probably warm between 1.5° and 6°C (2.7° and 11°F). Global climate change already is affecting a wide variety of biological species (fig. 1.7). Further warming is likely to cause increasingly severe weather events, including droughts in some areas and floods in others. Melting alpine glaciers and snowfields could threaten water supplies on which millions of people depend. Rising sea levels already are flooding low-lying islands and coastal regions, while habitat losses and climate changes are affecting many biological species. Canadian Environment Minister David Anderson said that global climate change is a greater threat than terrorism because it could force hundreds of millions of people from their homes and trigger an economic and social catastrophe.

Air quality

Air quality has worsened dramatically in many areas. Over southern Asia, for example, satellite images recently revealed a 3-km (2-mile)-thick toxic haze of ash, acids, aerosols, dust, and

photochemical products, which regularly covers the entire Indian subcontinent for much of the year. Nobel laureate Paul Crutzen estimates that at least 3 million people die each year from diseases triggered by air pollution. Worldwide, the United Nations estimates, more than 2 billion metric tons of air pollutants (not including carbon dioxide or wind-blown soil) are released each year. Air pollution no longer is merely a local problem. Mercury, polychlorinated biphenyls (PCB), DDT, and other long-lasting pollutants accumulate in arctic ecosystems and native people after being transported by air currents from industrial regions thousands of kilometers to the south. And during certain days, as much as 75 percent of the smog and particulate pollution recorded on the west coast of North America can be traced to Asia.

Biodiversity loss

Biologists report that habitat destruction, overexploitation, pollution, and introduction of exotic organisms are eliminating species at a rate comparable to the great extinction that marked the end of the age of dinosaurs. The UN Environment Programme reports that, over the past century, more than 800 species have disappeared and at least 10,000 species are now considered threatened. This includes about half of all primates and freshwater fish, together with around 10 percent of all plant species. Top predators, including nearly all the big cats in the world, are particularly rare and endangered. A nationwide survey of the United Kingdom in 2004 found that most bird and butterfly populations had declined between 50 and 75 percent over the previous 20 years. At least half of the forests existing before the introduction of agriculture have been cleared, and much of the diverse "old growth" on which many species depend for habitat is rapidly being cut and replaced by secondary growth or monoculture.

Marine resources

As the opening case study for this chapter shows, the ocean is an irreplaceable food resource for many people. More than a billion people in developing countries depend on seafood for their main source of animal protein, but most commercial fisheries around the world are in steep decline (fig. 1.8). According to the World Resources Institute, more than three-quarters of the 441 fish stocks for which information is available are severely depleted or in urgent need of better management. Canadian researchers estimate that 90 percent of all the large predators, including bluefin tuna, marlin, swordfish, sharks, cod, and halibut, have been removed from the ocean.

There are also many signs of hope

The dismal litany of problems facing us seems overwhelming, doesn't it? Is there hope that we can find solutions to these dilemmas? We think so. The example of Apo Island shows that ordinary people can make important progress both in protecting nature and in improving their lives.

Marine resources

Around the world, people who depend on seafood for their livelihood and sustenance are finding that setting aside marine reserves can restore fish populations as well as promote human development (fig. 1.9). Showing that these projects can be ecologically sound, economically sustainable, and socially acceptable on the local scale can lead to wider applications. Marine reserves are being established to protect reproductive areas in California, Hawaii, New Zealand, Great Britain, and many other areas, in addition to the Philippines.

Population and pollution

As you will see in subsequent chapters in this book, progress has been made in many areas in reducing pollution and curbing wasteful resource use. Many cities in Europe and North America, for example, are cleaner and much more livable now than they were a century ago. Population has stabilized in most industrialized countries and even in some very poor countries where social security and democracy have been established. Over the past 25 years, the average number of children born per woman worldwide has decreased from 6.1 to 2.6 (see fig. 1.5). By 2050, the UN Population Division predicts, all developed countries and 75 percent of the developing world will experience a below-replacement fertility rate of 2.1 children per woman. This suggests that the world population will stabilize at about 8.9 billion, rather than the 9.3 billion previously estimated.

Figure 1.9 Fishing has improved dramatically on Apo Island since the reefs have been protected and destructive fishing techniques have been outlawed.

Health

The incidence of life-threatening infectious diseases has been reduced sharply in most countries during the past century, while life expectancies have nearly doubled, on average. Smallpox has been completely eradicated and polio has been vanquished except in a few countries. Since 1990 more than 800 million people have gained access to improved water supplies and modern sanitation. In spite of population growth that added nearly a billion people to the world during the 1990s, the number facing food insecurity and chronic hunger during this period actually declined by about 40 million.

Conservation of forests and nature preserves

Deforestation has slowed in Asia, from more than 8 percent during the 1980s to less than 1 percent in the 1990s. Nature preserves and protected areas have increased dramatically over the past few decades. In 2006, according to UNESCO, there were more than 100,000 parks and nature preserves in the world, representing more than 20 million km^2 (about 7.7 million mi^2), or about 13.5 percent of the world's land area. Ecoregion and habitat protection remains uneven, however, and some areas are protected in name only. Still, this is dramatic progress in biodiversity protection.

Renewable energy

Encouraging progress is being made in a transition to renewable energy sources. The European Union has announced a goal of obtaining 22 percent of its electricity and 12 percent of all energy from renewable sources by 2010. British Prime Minister Tony Blair has laid out ambitious plans to fight global warming by cutting carbon dioxide emissions in Britain by 60 percent through energy conservation and a switch to renewables. If nonpolluting, sustainable energy technology is made available to the world's poorer countries, it may be possible to enhance human development while simultaneously reducing environmental damage.

Figure 1.10 While many of us live in luxury, more than 1.4 billion people lack access to food, housing, clean water, sanitation, education, medical care, and other essentials for a healthy, productive life. Helping them meet their needs is not only humane, it is essential to protect our mutual environment.

Information

The increased speed at which information and technology now flow around the world holds promise that we can continue to find solutions to our environmental dilemmas. Although we continue to face many challenges, collectively we may be able to implement sustainable development that raises living standards for everyone while also reducing our negative environmental impacts.

1.3 Human Dimensions of Environmental Science

Because we live in both the natural and social worlds, and because we and our technology have become such dominant forces on the planet, environmental science must take human institutions and the human condition into account. We live in a world of haves and havenots; a few of us live in increasing luxury, while many others lack the basic necessities for a decent, healthy, productive life. The World Bank estimates that more than 1.4 billion people—about one-fifth of the world's population—live in acute poverty with an income of less than $1 (U.S.) per day. These poorest of the poor generally lack access to an adequate diet, decent housing, basic sanitation, clean water, education, medical care, and other essentials for a humane existence. Seventy percent of those people are women and children. In fact, four out of five people in the world live in what would be considered poverty in the richer countries (fig. 1.10).

Policymakers are becoming aware that eliminating poverty and protecting our common environment are inextricably interlinked because the world's poorest people are both the victims and the agents of environmental degradation. The poorest people are often forced to meet short-term survival needs at the cost of long-term sustainability. Desperate for croplands to feed themselves and their families, many move into virgin forests or cultivate steep, erosion-prone hillsides, where soil nutrients are exhausted after only a few years. Others migrate to the grimy, crowded slums and ramshackle shantytowns that now surround most major cities in the developing world. With no way to dispose of wastes, the residents often foul their environment further and contaminate the air they breathe and the water on which they depend for washing and drinking.

The cycle of poverty, illness, and limited opportunities can become a self-sustaining process that passes from one generation to another. People who are malnourished and ill can't work productively to obtain food, shelter, or medicine for themselves or their children, who also are malnourished and ill. About 250 million children—mostly in Asia and Africa and some as young as 4 years old—are forced to work under appalling conditions weaving carpets, making ceramics and jewelry, or working in the sex trade. Growing up in these conditions leads to educational, psychological, and developmental deficits that condemn these children to perpetuate this cycle.

Faced with immediate survival needs and few options, these unfortunate people often have no choice but to overharvest resources; in doing so, however, they diminish not only their own options but also those of future generations. And in an increasingly interconnected world, the environments and resource bases damaged by poverty and ignorance are directly linked to those on which we depend.

Affluence also has environmental costs

The affluent lifestyle that many of us in the richer countries enjoy consumes an inordinate share of the world's natural resources and produces a shockingly high proportion of pollutants and wastes. The United States, for instance, with less than 5 percent of the total population, consumes about one-quarter of most commercially traded commodities, such as oil, and produces a quarter to

Figure 1.11 "And may we continue to be worthy of consuming a disproportionate share of this planet's resources."
© The New Yorker Collection 1992. Lee Lorenz from cartoonbank.com. All Rights Reserved.

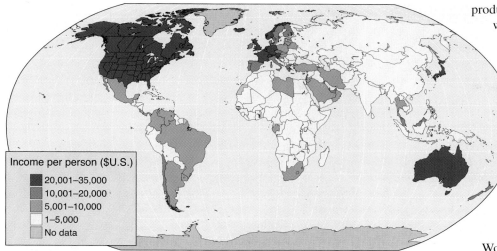

Figure 1.12 About four-fifths of the world's population live in middle or low-income countries.

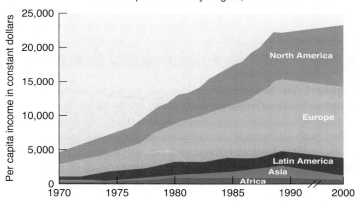

Growing Disparities in Incomes among Regions
Per Capita Income by Region, 1970–2000

Figure 1.13 Although the percentage of the world's population living in poverty has decreased slightly over the past 30 years, the relative gap between the rich and poor nations has increased sharply.
Source: United Nations, 2001.

half of most industrial wastes, such as greenhouse gases, pesticides, and other persistent pollutants.

To get an average American through the day takes about 450 kg (nearly 1,000 lbs) of raw materials, including 18 kg (40 lbs) of fossil fuels, 13 kg (29 lbs) of other minerals, 12 kg (26 lbs) of farm products, 10 kg (22 lbs) of wood and paper, and 450 liters (119 gal) of water. Every year, Americans throw away some 160 million tons of garbage, including 50 million tons of paper, 67 billion cans and bottles, 25 billion styrofoam cups, 18 billion disposable diapers, and 2 billion disposable razors (fig. 1.11).

This profligate resource consumption and waste disposal strains the planet's life-support systems. If everyone in the world tried to live at consumption levels approaching ours, the results would be disastrous. Unless we find ways to curb our desires and produce the things we truly need in less destructive ways, the sustainability of human life on our planet is questionable.

Sustainability is a central theme

Sustainability is a search for ecological stability and human progress that can last over the long term. Of course, neither ecological systems nor human institutions can continue forever. We can work, however, to protect the best aspects of both realms, and to encourage resiliency and adaptability in both of them. World Health Organization Director Gro Harlem Brundtland has defined **sustainable development** as "meeting the needs of the present without compromising the ability of future generations to meet their own needs." In these terms, development means bettering people's lives. Sustainable development, then, means progress in human well-being that we can extend or prolong over many generations, rather than just a few years. To be truly enduring, the benefits of sustainable development must be available to all humans and not just to the members of a privileged group. We will discuss this topic further in chapter 14.

Where do the rich and poor live?

About one-fifth of the world's population lives in the 20 richest countries, where the average per capita income is above $25,000 (U.S.) per year. Most of these countries are in North America or Western Europe, but Japan, Singapore, Australia, New Zealand, the United Arab Emirates, and Israel also fall into this group (fig. 1.12). Almost every country, however, even the richest, such as the United States and Canada, has poor people. No doubt everyone reading this book knows about homeless people or other individuals who lack resources for a safe, productive life. Some 35 million Americans—one-third of them children—live in households without sufficient food.

Eighty percent of the world's population lives in middle- or low-income countries, where nearly everyone is poor by North American standards. More than 3 billion people live in the poorest nations, where the average per capita income is below $620 (U.S.) per year. China and India are the largest of these countries, with a combined population of about 2.3 billion people. Among the 41 other nations in this category, 33 are in sub-Saharan Africa. All the other lowest-income nations, except Haiti, are in Asia. Although poverty levels in countries such as China and Indonesia have fallen in recent years, most countries in sub-Saharan Africa and much of Latin America have made little progress. The destabilizing and impoverishing effects of earlier colonialism continue to play important roles in the ongoing problems of these unfortunate countries. Meanwhile, the relative gap between rich and poor has increased dramatically (fig. 1.13).

Table 1.1	Quality-of-Life Indicators	
	Least-Developed Countries	Most-Developed Countries
GDP/Person[1]	(U.S.)$329	(U.S.)$30,589
Poverty Index[2]	78.1%	~0
Life Expectancy	43.6 years	76.5 years
Adult Literacy	58%	99%
Female Secondary Education	11%	95%
Total Fertility[3]	5.0	1.7
Infant Mortality[4]	97	5
Improved Sanitation	23%	100%
Improved Water	61%	100%
CO_2/capita[5]	0.2 tons	13 tons

[1]Annual gross domestic product

[2]Percent living on less than (U.S.)$2/day

[3]Average births/woman

[4]Per 1,000 live births

[5]Metric tons/yr/person

Source: UNDP Human Development Index, 2006.

The gulf between the richest and poorest nations affects many quality-of-life indicators (table 1.1). The average individual in the highest income countries has an annual income more than 100 times that of those in the lowest-income nations. Because of high infant mortality rates, the average family in the poorest countries has more than four times as many children as those in richer countries. If it were not for immigration, the population in most of the richer countries now would be declining; the total population of poorer countries continues to grow at 2.6 percent per year.

The gulf between rich and poor is even greater at the individual level. The richest 200 people in the world have a combined wealth of $1 trillion. This is more than the total owned by the 3 billion people who make up the poorest half of the world's population.

Indigenous peoples are guardians of much of the world's biodiversity

In both rich and poor countries, native or **indigenous people** are generally the least powerful, most neglected groups in the world. Typically descendants of the original inhabitants of an area taken over by more powerful outsiders, they are distinct from their country's dominant language, culture, religion, and racial communities. Of the world's nearly 6,000 recognized cultures, 5,000 are indigenous ones that account for only about 10 percent of the total world population. In many countries, traditional caste systems, discriminatory laws, economics, or prejudice repress indigenous people. Their unique cultures are disappearing, along with biological diversity, as natural habitats are destroyed to satisfy industrialized world appetites for resources. Traditional ways of life are disrupted further by dominant Western culture sweeping around the globe.

At least half of the world's 6,000 distinct languages are dying because they are no longer taught to children. When the last few

Figure 1.14 Do indigenous people have unique knowledge about nature and inalienable rights to traditional territories?

elders who still speak the language die, so will the culture that was its origin. Lost with those cultures will be a rich repertoire of knowledge about nature and a keen understanding of a particular environment and way of life (fig. 1.14).

Nonetheless, in many places, the 500 million indigenous people who remain in traditional homelands still possess valuable ecological wisdom and remain the guardians of little-disturbed habitats

Highest cultural diversity		Highest biological diversity
Nigeria	Indonesia	Madagascar
Cameroon	New Guinea	South Africa
Australia	Mexico	Malaysia
Congo	China	Cuba
Sudan	Brazil	Peru
Chad	United States	Ecuador
Nepal	Philippines	New Zealand

Figure 1.15 Cultural diversity and biodiversity often go hand in hand. Seven of the countries with the highest cultural diversity in the world are also on the list of "megadiversity" countries with the highest number of unique biological organisms (listed in decreasing order of importance).

Source: Norman Myers, Conservation International and Cultural Survival Inc., 2002.

that are refuges for rare and endangered species and undamaged ecosystems. In his book *The Future of Life*, the eminent ecologist E. O. Wilson argues that the cheapest and most effective way to preserve species is to protect the natural ecosystems in which they now live. Interestingly, just 12 countries account for 60 percent of all human languages (fig. 1.15). Seven of these are also among the "megadiversity" countries that contain more than half of all unique plant and animal species. Conditions that support evolution of many unique species seem to favor development of equally diverse human cultures as well.

Recognizing native land rights and promoting political pluralism can be among the best ways to safeguard ecological processes and endangered species. As the Kuna Indians of Panama say, "Where there are forests, there are native people, and where there are native people, there are forests." A few countries, such as Papua New Guinea, Fiji, Ecuador, Canada, and Australia, acknowledge indigenous title to extensive land areas.

Other countries, unfortunately, ignore the rights of native people. Indonesia, for instance, claims ownership of nearly three-quarters of its forestland and all waters and offshore fishing rights, ignoring the interests of indigenous inhabitants. Similarly, the Philippine government claims possession of all uncultivated land in its territory, while Cameroon and Tanzania recognize no rights at all for forest-dwelling pygmies who represent one of the world's oldest cultures.

1.4 Science Helps Us Understand Our Environment

Because environmental questions are complex, we need orderly methods of examining and understanding them. Environmental science provides such a methodical, orderly way of understanding our environment. In this section, we'll investigate what science is, what the scientific method is, and why that method is important.

What is science? **Science** (from *scire*, "to know" in Latin) is a process for producing empirical knowledge by observing natural phenomena. We develop or test theories (proposed explanations of how a process works) using these observations. "Science" also refers to the cumulative body of knowledge produced by many scientists. Science is valuable because it helps us understand the world and meet practical needs, such as finding new medicines, new energy sources, or new foods. In this section, we'll investigate how and why science follows standard methods.

Science rests on the assumption that the world is knowable and that we can learn about it by careful observation and logical reasoning (table 1.2). For early philosophers of science, this assumption was a radical departure from religious and philosophical approaches. In the Middle Ages the ultimate sources of knowledge about matters such as how crops grow, how diseases spread, or how the stars move, were religious authorities or cultural traditions. While these sources provided many useful insights, there was no way to test their explanations independently and objectively. The benefit of scientific thinking is that it searches for testable evidence: If you suspect a disease spreads through contaminated

Table 1.2	Basic Principles of Science
1.	*Empiricism:* We can learn about the world by careful observation of empirical (real, observable) phenomena; we can expect to understand fundamental processes and natural laws by observation.
2.	*Uniformitarianism:* Basic patterns and processes are uniform across time and space; the forces at work today are the same as those that shaped the world in the past, and they will continue to do so in the future.
3.	*Parsimony:* When two plausible explanations are reasonable, the simpler (more parsimonious) one is preferable. This rule is also known as Ockham's razor, after the English philosopher who proposed it.
4.	*Uncertainty:* Knowledge changes as new evidence appears, and explanations (theories) change with new evidence. Theories based on current evidence should be tested on additional evidence, with the understanding that new data may disprove the best theories.
5.	*Repeatability:* Tests and experiments should be repeatable; if the same results cannot be reproduced, then the conclusions are probably incorrect.
6.	*Proof is elusive:* We rarely expect science to provide absolute proof that a theory is correct, because new evidence may always undermine our current theories. Even evolution, the cornerstone of modern biology, ecology, and other sciences, remains a "theory" because we cannot absolutely prove its operation.
7.	*Testable questions:* To find out whether a theory is correct, it must be tested; we formulate testable statements (hypotheses) to test theories.

water, you can close off access to the water source and observe whether or not the disease stops spreading.

Science depends on skepticism and accuracy

Ideally scientists are skeptical. They are cautious about accepting proposed explanations until there is substantial evidence to support them. Even then, every explanation is considered only provisionally true, because there is always a possibility that some additional evidence may appear to disprove it. Scientists also aim to be methodical and unbiased. Because bias and methodical errors are hard to avoid, scientific tests are subject to review by informed peers, who can evaluate results and conclusions (fig. 1.16). The peer review process is an essential part of ensuring that scientists maintain good standards in study design, data collection, and interpretation of results.

Scientists demand **reproducibility** because they are cautious about accepting conclusions. Making an observation or obtaining a result just once doesn't count for much. You have to produce the same result consistently to be sure that your first outcome wasn't a fluke. Even more important, you must be able to describe the conditions of your study so that someone else can reproduce your findings. Repeating studies or tests is known as **replication**.

Science relies on measurements that are accurate (close to the true value). Accuracy usually requires careful, precise

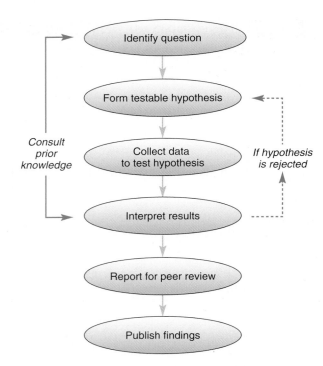

Figure 1.16 Ideally, scientific investigation follows a series of logical, orderly steps to formulate and test hypotheses.

measurement (fig. 1.17). Science also requires tidy record keeping. Inaccurate data and poor records can produce sloppy results and misleading conclusions.

Deductive and inductive reasoning are both useful

Ideally, scientists deduce conclusions from general laws that they know to be true. For example, if we know that massive objects attract each other (because of gravity), then it follows that an apple will fall to the ground when it releases from the tree. This logical reasoning from general to specific is known as **deductive reasoning**. Often, however, we do not know general laws that guide natural systems. We observe, for example, that birds appear and disappear as a year goes by. Through many repeated observations in different places, we can infer that the birds move from place to place. We can develop a general rule that birds migrate seasonally. Reasoning from many observations to produce a general rule is **inductive reasoning**. Although deductive reasoning is more logically sound than inductive reasoning, it only works when our general laws are correct. We often rely on inductive reasoning to understand the world because we have few immutable laws.

Sometimes it is insight, as much as reasoning, that leads us to an answer. Many people fail to recognize the role that insight, creativity, and luck play in research. Some of our most important discoveries were made because the investigators were passionately interested in their topics and pursued hunches that appeared unreasonable to fellow scientists. A good example is Barbara McClintock, the geneticist who discovered that genes in corn can

move and recombine spontaneously. Where other corn geneticists saw random patterns of color and kernel size, McClintock's years of experience in corn breeding and an uncanny ability to recognize patterns, led her to guess that genes could recombine in ways that no one had previously imagined.

The scientific method is an orderly way to examine problems

You may already be using the scientific method without being aware of it. Suppose you have a flashlight that doesn't work. The flashlight has several components (switch, bulb, batteries) that could be faulty. If you change all the components at once, your flashlight might work, but a more methodical series of tests will tell you more about what was wrong with the system—knowledge that may be useful next time you have a faulty flashlight. So you decide to follow the standard scientific steps:

1. *Observe* that your flashlight doesn't light; also, there are three main components of the lighting system (batteries, bulb, and switch).

2. Propose a **hypothesis**, a testable explanation: "The flashlight doesn't work because the batteries are dead."

3. Develop a *test* of the hypothesis and *predict* the result that would indicate your hypothesis was correct: "I will replace the batteries; the light should then turn on."

4. Gather *data* from your test: After you replaced the batteries, did the light turn on?

5. *Interpret* your results: If the light works now, then your hypothesis was right; if not, then you should formulate a new hypothesis, perhaps that the bulb is faulty, and develop a new test for that hypothesis.

Figure 1.17 Making careful, accurate measurements and keeping good records are essential in scientific research.

Figure 1.18 How can you prove who is responsible for environmental contamination, such as the orange ooze in this stream? Careful, repeated measurements and well-formed hypotheses are essential.

In systems more complex than a flashlight, it is almost always easier to prove a hypothesis wrong than to prove it unquestionably true. This is because we usually test our hypotheses with observations, but there is no way to make every possible observation. The philosopher Ludwig Wittgenstein illustrated this problem as follows: Suppose you saw hundreds of swans, and all were white. These observations might lead you to hypothesize that all swans

were white. You could test your hypothesis by viewing thousands of swans, and each observation might support your hypothesis, but you could never be entirely sure that it was correct. On the other hand, if you saw just one black swan, you would know with certainty that your hypothesis was wrong.

As you'll read in later chapters, the elusiveness of absolute proof is a persistent problem in environmental policy and law. You can never absolutely prove that the toxic waste dump up the street is making you sick. The elusiveness of proof often decides environmental liability lawsuits (fig. 1.18).

When an explanation has been supported by a large number of tests, and when a majority of experts have reached a general consensus that it is a reliable description or explanation, we call it a **scientific theory**. Note that scientists' use of this term is very different from the way the public uses it. To many people, a theory is speculative and unsupported by facts. To a scientist, it means just the opposite: While all explanations are tentative and open to revision and correction, an explanation that counts as a scientific theory is supported by an overwhelming body of data and experience, and it is generally accepted by the scientific community, at least for the present.

Understanding probability helps reduce uncertainty

One strategy to improve confidence in the face of uncertainty is to focus on probability. **Probability** is a measure of how likely something is to occur. Usually, probability estimates are based on a set of previous observations or on standard statistical measures. Probability does not tell you what *will* happen, but it tells you what *is likely* to happen. If you hear on the news that you have a 20 percent chance of catching a cold this winter, that means that 20 of every 100 people are likely to catch a cold. This doesn't mean that you will catch one. In fact, it's more likely that you won't catch a cold than that you will. If you hear that 80 out of every 100 people will catch a cold, you still don't know whether you'll get sick, but there's a much higher chance that you will.

Science often involves probability, so it is important to be familiar with the idea. Sometimes probability has to do with random chance: If you flip a coin, you have a random chance of getting heads or tails. Every time you flip, you have the same 50 percent probability of getting heads. The chance of getting ten heads in a row is small (in fact, the chance is 1 in 210, or 1 in 1,024), but on any individual flip, you have exactly the same 50 percent chance, since this is a random test. Sometimes probability is weighted by circumstances: Suppose that about 10 percent of the students in your class earn an A each semester. Your likelihood of being in that 10 percent depends a great deal on how much time you spend studying, how many questions you ask in class, and other factors. Sometimes there is a combination of chance and circumstances: The probability that you will catch a cold this winter depends partly on whether you encounter someone who is sick (largely random chance) and whether you take steps to stay healthy (get enough rest, wash your hands frequently, eat a healthy diet, and so on).

Exploring SCIENCE:

What Are Statistics, and Why Are They Important?

In the Apo Island case study at the beginning of this chapter, islanders observed that the number of fish caught per hour of effort had changed, and they believed that some species had disappeared while others had become smaller. Memory from long experience of fishing allowed them to observe these changes. Often in environmental science, we want to know if a change is happening, or we want to know why change is happening, but we don't have generations of local knowledge to guide us. Also, we may want to document changes in a way that will convince people who are unfamiliar with our study area. How might we examine the state of the problem, in that case? One good strategy is to use statistics. We can use statistics to *describe* the problem, to *compare* trends in one area to those in other areas, in order to find out how unusual our study area is, and to *explain* observed changes, or identify their causes.

Many people find statistics intimidating, but when you take them step by step, a few simple ideas can take you a long way. Here are some examples of ways you can use statistics to examine a problem such as that in Apo Island.

1. *Descriptive statistics help you assess the state of a group.* A good first step to describing the state of the fishery is to count and record the number of fish caught in a day. Of course, on any given day, you might have had bad luck, and different fishers have different skill in fishing. To make a general estimate of the fishing catch, you could calculate the **mean** (average) number of fish per day from a number of people: Count all the fish caught in a day, and divide that count by the number of person-hours spent fishing. Try this for the four hypothetical fishing boats listed in the following table.

Fishing Success

Boat	Number of Fish	Hours
1	10	2
2	15	3
3	5	1
4	20	4
Sum:		

Number of fish per hour = _____

Is this rate of fishing high or low? Suppose you have records telling you that 10 years ago the mean was 12.5 fish per hour spent fishing. If you calculated a mean of 5 fish per hour above, you know your fishing rate has fallen.

2. *Larger samples provide more information.* Fishing records from one village can provide important insights into the state of the local ecosystem, but it is always possible that your village is an *outlier*, or an unusual case. Records from 10 villages, or 50, would tell a much more complete story. In the case of the Philippines, there are hundreds of islands and villages. Recording catches from every one would be incredibly difficult and expensive, but that doesn't mean you cannot make a general statement about fishery declines or growth. You can take a **random sample** from the hundreds of villages. The sample should be random so you don't just pick those nearest the big city (they're convenient to get to, but their fisheries might be especially poor because of pollution and overuse) or just those in one area (that area might be unusual). The larger the sample—50 villages instead of 10—the more confident you can be that you've got a good picture of the general state of fisheries. Often we use a **histogram**, or frequency distribution, to show the number of cases with small and large values (fig. 1). The histogram here is approximately a bell-shaped Gaussian distribution. This pattern is also known as **normal distribution**, because large sets of randomly selected observations tend to have this distribution. In this histogram, our village's average of 5 fish per hour of effort is not unusual, but it is lower than most of the group.

3. *Comparing groups.* Suppose you want to test whether the presence of marine protected areas improves fishing. You could examine this question by comparing villages with protected areas to those without. You could visually compare these groups using a box plot, which shows the mean and range for each group (fig. 2). Here, the line across the middle of a box shows the group mean. Half the observations are in the box, and the "whiskers" above and below the box show the 10th and 90th percentile values (that is, the top and bottom 10 percent). Dots show extreme highs and lows.

The groups in figure 2 look different, but is the difference really meaningful? You can answer this question by estimating the odds that your results could have been achieved entirely by chance. That is, there's a small possibility that if you selected randomly from a single group (say, only villages *without* marine preserves), you could produce the same difference between two groups. You could test the **significance** of the difference, or the probability that the difference could have been produced randomly. To save time, we won't do that now, but in this case, the significance statistic indicates that this probability is very small, about one in 1,000. Even though the ranges overlap, we can conclude that

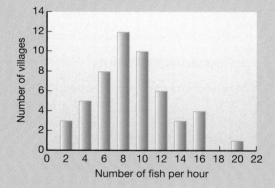

Figure 1 A histogram, or frequency distribution.

this test shows a meaningful difference between villages with and without marine preserves.

4. *Testing explanatory factors*. You can further examine the effect of reef preserves on fishing by plotting the two variables against each other for all your observations (villages). In this case, you consider area of marine preserves the **explanatory variable** (or independent variable), and number of fish per hour is the dependent variable (fig. 3). By convention, we plot the explanatory variable on the X (horizontal) axis. The dependent variable is on the Y (vertical) axis. Following this convention makes it easier for readers to quickly and correctly interpret your graph.

5. *Statistics tell one of many possible stories*. The examples shown here use a small set of hypothetical fishing data. As you read this book you'll encounter many examples of real data sets, describing real and pressing problems and relationships. As you look at numbers and graphs throughout this book, and in newspapers or on television, read them carefully. Think about what the patterns might mean. Also think about what variables really mean—and what they don't tell you. One of the devilish details of representing the world with a few numbers is that numbers can be calculated in many ways, for different purposes. If you want to describe changes in fisheries, do you look at total fish caught? Fish per hour of effort? Changes over 5 years? Over 50 years? All these approaches could yield very different impressions. Asking this sort of question is a central aspect of *critical thinking*—a skill you'll read about in this chapter, and which you'll practice throughout this book. Careful examination of statistics and graphs can also increase your confidence in your own understanding of the issues presented.

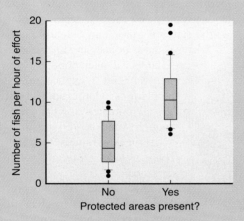

Figure 2 A box plot compares group means (center lines in the boxes) and ranges.

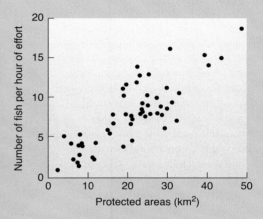

Figure 3 This scatter plot shows a positive relationship: the Y variable increases as the X variable increases.

Figure 1.19 What is an appropriate sample size for an investigation? How many individuals would you need to examine to have a representative sample of a population? These are statistical questions of variability, probability, and accuracy.

Statistics can calculate the probability that your results were random

Statistics can help in experimental design as well as in interpreting data (see Exploring Science, p. 14). Many statistical tests focus on calculating the probability that observed results could have occurred by chance. Often, the degree of confidence we can assign to results depends on sample size as well as the amount of variability between groups (fig. 1.19).

Ecological tests are generally considered significant if there is less than 5 percent probability that the results were achieved by random chance. A probability of less than 1 percent gives still greater confidence in the results.

As you read this book, you will encounter many statistics, including many measures of probability. When you see these numbers, stop and think: Is the probability high enough to worry about? How high is it compared to other risks or chances you've read about? What are the conditions that make probability higher or lower?

Experimental design can reduce bias

The study of fishing success (above) is an example of an observational experiment, one in which you observe natural events and interpret a causal relationship between the variables. This kind of study is also called a **natural experiment**, one that involves observation of events that have already happened. Many scientists depend on natural experiments: A geologist, for instance, might want to study mountain building, or an ecologist might want to learn about how species coevolve, but neither scientist can spend millions of years watching the process happen. Similarly, a toxicologist cannot give people a disease just to see how lethal it is.

Other scientists can use **manipulative experiments**, in which conditions are deliberately altered, and all other variables are held constant. Most manipulative experiments are done in the laboratory, where conditions can be carefully controlled. Suppose you were interested in studying whether lawn chemicals contributed to deformities in tadpoles. You might keep two groups of tadpoles in fish tanks, and expose one to chemicals. In the lab, you could ensure that both tanks had identical temperatures, light, food, and oxygen. By comparing a treatment (exposed) group and a control (unexposed) group, you have also made this a **controlled study**.

Often, there is a risk of experimenter bias. Suppose the researcher sees a tadpole with a small nub that looks like it might become an extra leg. Whether she calls this nub a deformity might depend on whether she knows that the tadpole is in the treatment group or the control group. To avoid this bias, **blind experiments** are often used, in which the researcher doesn't know which group is treated until after the data have been analyzed. In health studies, such as tests of new drugs, **double-blind experiments** are used, in which neither the subject (who receives a drug or a placebo) nor the researcher knows who is in the treatment group and who is in the control group.

In each of these studies there is one **dependent variable** and one, or perhaps more, **independent variables**. The dependent variable, also known as a response variable, is affected by the independent variables. In a graph, the dependent variable is on the vertical (Y) axis, by convention. Independent variables are rarely really independent (they may be affected by the same environmental conditions as the dependent variable, for example). Many people prefer to call them **explanatory variables**, because we hope they will explain differences in the dependent variable.

Science is a cumulative process

The scientific method outlined in figure 1.16 is the process used to carry out individual studies. Larger-scale accumulation of scientific knowledge involves cooperation and contributions from countless people. Good science is rarely carried out by a single individual working in isolation. Instead, a community of scientists

collaborates in a cumulative, self-correcting process. You often hear about big breakthroughs and dramatic discoveries that change our understanding overnight, but in reality these changes are usually the culmination of the labor of many people, each working on different aspects of a common problem, each adding small insights to solve a problem. Ideas and information are exchanged, debated, tested, and re-tested to arrive at **scientific consensus**, or general agreement among informed scholars.

The idea of consensus is important. For those not deeply involved in a subject, the multitude of contradictory results can be bewildering: Are coral reefs declining, and does it matter? Is climate changing, and how much? Among those who have performed and read many studies, there tends to emerge a general agreement about the state of a problem. Scientific consensus now holds that many coral reefs are in danger, though opinions vary on how severe the problem is. Consensus is that global climates are changing, though models differ somewhat on how rapidly they will change under different policy scenarios.

Sometimes new ideas emerge that cause major shifts in scientific consensus. These great changes in explanatory frameworks were termed **paradigm shifts** by Thomas Kuhn (1967), who studied revolutions in scientific thought. According to Kuhn, paradigm shifts occur when a majority of scientists accept that the old explanation no longer explains new observations very well. For example, two centuries ago, geologists explained many earth features in terms of Noah's flood. The best scientists held that the flood created beaches well above modern sea level, scattered boulders erratically across the landscape, and gouged enormous valleys where there is no water now (fig. 1.20). Then the Swiss glaciologist Louis Agassiz and others suggested that the earth had once been much colder and that glaciers had covered large areas.

Periodic ice ages better explained changing sea levels, boulders transported far from their source rock, and the great, gouged valleys. This new idea completely altered the way geologists explained their subject. Similarly, the idea of tectonic plate movement, in which continents shift slowly around the earth's surface, revolutionized the ways geologists, biogeographers, ecologists, and others explained the development of the earth and its life-forms.

Figure 1.20 Paradigm shifts change the ways we explain our world. Geologists now attribute Yosemite's valleys to glaciers, where once they believed Noah's flood carved its walls.

What is sound science?

Environmental science often deals with questions that are emotionally or politically charged. Scientific studies of climate change may be threatening to companies that sell coal; studies of health costs of pesticides might worry people who use or sell these chemicals. When controversy surrounds science, claims about **sound science** and accusations of "junk science" often arise. What do these terms mean, and how can you evaluate who is right?

When you hear arguments about whose science is valid, you need to remember the basic principles of science: Are the disputed studies reproducible? Are conclusions drawn with caution and skepticism? Are samples large and random? Are conclusions supported by a majority of scholars who have studied the problem and know the field?

Often media figures on television or radio will take a position contrary to the scientific majority. A contrarian position gains them publicity and political allies (and sometimes money). This strategy has been especially popular around large issues such as climate change. For decades now, almost all climate scientists have agreed that human activities, such as fossil fuel burning and land clearing, are contributing to climate change. But it is always possible to find a contrarian scientist who is happy to contradict the majority of evidence. Especially when political favors, publicity, or money are involved, it is always possible to find "expert" witnesses who will testify on opposite sides of a case.

Scientific uncertainty is frequently invoked as a reason to postpone or reverse policies that a vast majority of informed scientists consider to be prudent. In questions of chemical safety, energy conservation, climate change, or air pollution control, opponents of change may charge that the evidence doesn't constitute absolute proof, so that no action needs to be taken. You will see examples of this in chapter 8 (environmental health), chapter 9 (air quality and climate), and elsewhere in this book.

Similarly, disputes over evolution often hinge on the concept of theories in science. Opponents of teaching evolution in public schools often charge that because scientists call evolution a theory, evolution is just a matter of conjecture. This is a deliberate confusion of terminology. The theory of evolution is supported by overwhelming amounts of evidence, but we still call it a theory because scientists prefer to be cautious about proof.

If you see claims of sound science and junk science, how can you evaluate them? How can you identify bogus analysis that is dressed up in quasi-scientific jargon but that has no objectivity? This is such an important question that astronomer Carl Sagan has proposed a "Baloney Detection Kit" (table 1.3) to help you out.

Is environmental science the same as environmentalism?

Environmental science is the use of scientific methods to study processes and systems in the environment in which we live. Environmentalism is working to influence attitudes and policies that affect our environment. The two are often separate goals. Petroleum geologists, for example, are environmental scientists who study geology in order to find oil resources. They work to increase access to resources, to benefit their company, perhaps to benefit our standard of living, but usually not to protect the environment in which they work and live. Other environmental scientists are also environmentalists. Many ecologists actively work to defend the ecosystems they study. Many environmental scien-

Table 1.3	Questions for Baloney Detection
1.	How reliable are the sources of this claim? Is there reason to believe that they might have an agenda to pursue in this case?
2.	Have the claims been verified by other sources? What data are presented in support of this opinion?
3.	What position does the majority of the scientific community hold in this issue?
4.	How does this claim fit with what we know about how the world works? Is this a reasonable assertion or does it contradict established theories?
5.	Are the arguments balanced and logical? Have proponents of a particular position considered alternate points of view or only selected supportive evidence for their particular beliefs?
6.	What do you know about the sources of funding for a particular position? Are they financed by groups with partisan goals?
7.	Where was evidence for competing theories published? Has it undergone impartial peer review or is it only in proprietary publication?

Source: Carl Sagan

tists work in the public interest, to serve public health, for example, without necessarily being interested in nature or other species.

Whether we use science to pursue public health, economic success, environmental quality, or other goals depends on issues outside of science. Many of these issues have to do with worldviews and ethics, which are discussed below.

1.5 Critical Thinking

Perhaps the most valuable skill you can learn in any of your classes is the ability to think clearly, creatively, and purposefully. Much of the most important information in environmental science is hotly disputed. Evidence can vary, depending on how, when, and by whom it was gathered. **Critical thinking** is a term we use to describe logical, orderly, analytical assessment of ideas, evidence, and arguments. Developing this skill is essential for the course you are taking now. Critical thinking is also an extremely important skill for your life in general. You can use it when you evaluate the claims of a car salesman, a credit card offer, or the campaign promises of a political candidate.

An important complication is that distinguished authorities can vehemently disagree about many important topics. Disagreements may be based on contradictory data, on different interpretations of the same data, or on different priorities. One expert might consider economic health the overriding priority; another might prioritize environmental quality. (Still another expert might consider both priorities to be closely related.) You have already read about ways that you can evaluate contradictory claims by remembering the principles of how science should work. You can also examine the validity of contradictory claims by practicing critical thinking. In this section, we'll examine some of the different aspects of critical thinking.

Critical thinking helps us analyze information

A number of skills, attitudes, and approaches can help us evaluate information and make decisions. **Critical thinking** asks, "What am I trying to accomplish here, and how will I know when I've succeeded?" **Analytical thinking** asks, "How can I break this problem down into its constituent parts?" **Creative thinking** asks, "How might I approach this problem in new and inventive ways?" **Logical thinking** asks, "How can orderly, deductive reasoning help me think clearly?" **Reflective thinking** asks, "What does it all mean?"

While much of science is based on analytical and logical thinking, critical thinking adds elements of contextual sensitivity and empathy that can be very helpful in your study of environmental science. It challenges us to examine theories, facts, and options in a systematic, purposeful, and responsible manner. It shares many methods and approaches with other methods of reasoning but adds some important skills, attitudes, and dispositions. Furthermore, it challenges us to plan methodically and to assess the process of thinking as well as the implications of our decisions. Thinking critically can help us discover hidden ideas and meanings, develop strategies for evaluating reasons and conclusions in arguments, recognize the differences between facts and values, and avoid jumping to conclusions (table 1.4).

Notice that many critical thinking processes are self-reflective and self-correcting. This form of thinking is sometimes called "thinking about thinking." It is not critical in the sense of finding fault, but it makes a conscious, active, disciplined effort to be aware of hidden motives and assumptions (fig. 1.21); to uncover bias; and to recognize the reliability or unreliability of sources.

While critical thinking shares many of the orderly, systematic approaches of formal logic, it also invokes traits such as empathy, sensitivity, courage, and humility. Formulating intelligent opinions about some of the complex issues you will encounter in environmental science requires more than simple logic. Developing these attitudes and skills is not easy or simple. It takes practice. You have to develop your mental faculties just as you need to train for a sport. Intellectual integrity, modesty, fairness, compassion, and fortitude are not traits you use only occasionally. They must be cultivated until they become a part of your normal way of thinking.

What do you need to think critically?

We all use critical or reflective thinking at times. Suppose a television commercial tells you that a new breakfast cereal is tasty and good for you. You may be suspicious and ask yourself a few questions: What do they mean by *good*? Good for whom or what? Does *tasty* simply mean more sugar and salt? Might the sources of this information have other motives besides your health and happiness? You probably practice critical analysis regularly.

Here are some steps you can use in critical thinking:

1. *Identify and evaluate premises and conclusions in an argument.* What is the basis for the claims made here? What evidence is presented to support these claims, and what conclusions are drawn from this evidence? If the premises and evidence are correct, does it follow that the conclusions are necessarily true?

Table 1.4	Steps in Critical Thinking
1.	What is the purpose of my thinking?
2.	What precise question am I trying to answer?
3.	Within what point of view am I thinking?
4.	What information am I using?
5.	How am I interpreting that information?
6.	What concepts or ideas are central to my thinking?
7.	What conclusions am I aiming toward?
8.	What am I taking for granted; what assumptions am I making?
9.	If I accept the conclusions, what are the implications?
10.	What would the consequences be if I put my thoughts into action?

Source: R. Paul, National Council for Critical Thinking.

Figure 1.21 Critical thinking evaluates premises, contradictions, and assumptions. Was this sign, in the middle of a popular beach near Chicago, the only way to reduce human exposure to bacteria? What other strategies might there be? Why was this one chosen? Who might be affected?

2. *Acknowledge and clarify uncertainties, vagueness, equivocation, and contradictions.* Do the terms used have more than one meaning? If so, are all participants in the argument using the same meanings? Is ambiguity or equivocation deliberate? Can all the claims be true simultaneously?

3. *Distinguish between facts and values.* Can claims be tested? (If so, these are statements of fact and should be verifiable by gathering evidence.) Are claims made about the worth or lack of worth of something? (If so, these are value statements or opinions and probably cannot be verified objectively.)

4. *Recognize and assess assumptions.* Given the backgrounds and views of the protagonists, what underlying reasons might there be for the premises, evidence, or conclusions presented?

Does anyone have an "axe to grind" or a personal agenda in this issue? What does s/he think you know, need, want, or believe? Is there a hidden message based on race, gender, ethnicity, economics, or some belief system that distorts this discussion?

5. *Distinguish source reliability or unreliability.* What qualifies the experts on this issue? What special knowledge or information do they have? What evidence do they present? How can we determine whether the information offered is accurate, true, or even plausible?

6. *Recognize and understand conceptual frameworks.* What are the basic beliefs, attitudes, and values that this person, group, or society holds? What dominating philosophy or ethics control their outlooks and actions? How do these beliefs and values affect the way people view themselves and the world around them? If there are conflicting or contradictory beliefs and values, how can these differences be resolved?

Critical thinking helps you learn environmental science

In this book, you will have many opportunities to practice critical thinking skills. Every chapter includes many facts, figures, opinions, and theories. Are all of them true? No, probably not. They were the best information available when this text was written, but much in environmental science is in a state of flux. Data change constantly, as does our interpretation of data. Do the ideas presented here give a complete picture of the state of our environment? Unfortunately, they probably don't. No matter how comprehensive our discussion is of this complex, diverse subject, it can never capture everything worth knowing, nor can it reveal all possible points of view.

When reading this text, try to distinguish between statements of fact and opinion. Ask yourself if the premises support the conclusions drawn from them. Although we have tried to be fair and even-handed in presenting controversies, our personal biases and values—some of which we may not even recognize—affect how we see issues and present arguments. Watch for cases in which you need to think for yourself, and use your critical and reflective thinking skills to uncover the truth.

You'll find more on critical thinking, as well as some useful tips on how to study effectively, on our web page at www.mhhe.com/cunningham5e.

1.6 A Brief History of Conservation and Environmental Thought

Although some early societies had negative impacts on their surroundings, others lived in relative harmony with nature. In modern times, however, growing human populations and the power of our technology have heightened concern about what we are doing to our environment.

How have our ideas about our environment evolved in response to these great changes? We can divide conservation history and environmental activism into at least four distinct stages: (1) pragmatic resource conservation, (2) moral and aesthetic nature preservation, (3) a growing concern about health and ecological damage caused by pollution, and (4) global environmental citizenship. These stages aren't necessarily mutually exclusive. Parts of each persist today in the environmental movement, and one person may embrace them all simultaneously.

Nature protection has historic roots

Recognizing human misuse of nature is not unique to modern times. Plato complained in the fourth century B.C. that Greece once was blessed with fertile soil and clothed with abundant forests of fine trees. After the trees were cut to build houses and ships, however, heavy rains washed the soil into the sea, leaving only a rocky "skeleton of a body wasted by disease." Springs and rivers dried up, while farming became all but impossible. Despite these early observations, most modern environmental ideas developed in response to resource depletion associated with more recent agricultural and industrial revolutions.

Some of the earliest recorded scientific studies of environmental damage were carried out in the eighteenth century by French or British colonial administrators, many of whom were trained scientists and who observed rapid soil loss and drying wells that resulted from intensive colonial production of sugar and other commodities. Some of these colonial administrators considered responsible environmental stewardship as an aesthetic and moral priority, as well as an economic necessity. These early conservationists observed and understood the connections between deforestation, soil erosion, and local climate change. The pioneering British plant physiologist Stephen Hales, for instance, suggested that conserving green plants preserves rainfall. His ideas were put into practice in 1764 on the Caribbean island of Tobago, where about 20 percent of the land was marked as "reserved in wood for rains."

Pierre Poivre, an early French governor of Mauritius, an island in the Indian Ocean, was appalled at the environmental and social devastation caused by destruction of wildlife (such as the flightless dodo) and the felling of ebony forests on the island by early European settlers. In 1769 Poivre ordered that one-quarter of the island be preserved in forests, particularly on steep mountain slopes and along waterways. Mauritius remains a model for balancing nature and human needs. Its forest reserves shelter a larger percentage of its original flora and fauna than most other human-occupied islands.

Resource waste triggered pragmatic resource conservation

Many historians consider the publication of *Man and Nature* in 1864 by geographer George Perkins Marsh as the wellspring of environmental protection in North America. Marsh, who also was a lawyer, politician, and diplomat, traveled widely around the Mediterranean as part of his diplomatic duties in Turkey and Italy. He read widely in the classics (including Plato) and

(a) President Teddy Roosevelt

(b) Gifford Pinchot

(c) John Muir

(d) Aldo Leopold

Figure 1.22 Some early pioneers of the American conservation movement. President Teddy Roosevelt (a) and his main advisor Gifford Pinchot (b) emphasized pragmatic resource conservation, while John Muir (c) and Aldo Leopold (d) focused on ethical and aesthetic relationships.

personally observed the damage caused by excessive grazing by goats and sheep and by the deforestation of steep hillsides. Alarmed by the wanton destruction and profligate waste of resources still occurring on the American frontier in his lifetime, he warned of its ecological consequences. Largely because of his book, national forest reserves were established in the United States in 1873 to protect dwindling timber supplies and endangered watersheds.

Among those influenced by Marsh's warnings were U.S. President Theodore Roosevelt and his chief conservation adviser, Gifford Pinchot (fig. 1.22 a and b). In 1905 Roosevelt, who was the leader of the populist, progressive movement, moved forest management out of the corruption-filled Interior Department into the Department of Agriculture. Pinchot, who was the first American-born professional forester, became the first chief of the new Forest Service. He put resource management on an honest, rational, and scientific basis for the first time in American history. Together with naturalists and activists such as John Muir, Roosevelt and Pinchot established the framework of the national forest, park, and wildlife refuge system. They passed game protection laws and tried to stop some of the most flagrant abuses of the public domain. In 1908 Pinchot organized and chaired the White House Conference on Natural Resources, perhaps the most prestigious and influential environmental meeting ever held in the United States. Pinchot also

was governor of Pennsylvania and founding head of the Tennessee Valley Authority, which provided inexpensive power to the southeastern United States.

The basis of Roosevelt's and Pinchot's policies was pragmatic **utilitarian conservation**. They argued that the forests should be saved "not because they are beautiful or because they shelter wild creatures of the wilderness, but only to provide homes and jobs for people." Resources should be used "for the greatest good, for the greatest number, for the longest time." "There has been a fundamental misconception," Pinchot wrote, "that conservation means nothing but husbanding of resources for future generations. Nothing could be further from the truth. The first principle of conservation is development and use of the natural resources now existing on this continent for the benefit of the people who live here now. There may be just as much waste in neglecting the development and use of certain natural resources as there is in their destruction." This pragmatic approach still can be seen in the multiple-use policies of the U.S. Forest Service.

Ethical and aesthetic concerns inspired the preservation movement

John Muir (fig. 1.22c), amateur geologist, popular author, and first president of the Sierra Club, strenuously opposed Pinchot's utilitarian policies. Muir argued that nature deserves to exist for its own sake, regardless of its usefulness to us. Aesthetic and spiritual values formed the core of his philosophy of nature protection. This outlook prioritizes **preservation** because it emphasizes the fundamental right of other organisms—and nature as a whole—to exist and to pursue their own interests (fig. 1.23). Muir wrote,

Figure 1.23 A conservationist might say this forest was valuable as a supplier of useful resources, including timber and fresh water. A preservationist might argue that this ecosystem is important for its own sake. Many people are sympathetic with both outlooks.

"The world, we are told, was made for man. A presumption that is totally unsupported by the facts. . . . Nature's object in making animals and plants might possibly be first of all the happiness of each one of them. . . . Why ought man to value himself as more than an infinitely small unit of the one great unit of creation?"

Muir, who was an early explorer and interpreter of California's Sierra Nevada range, fought long and hard for establishment of Yosemite and Kings Canyon National Parks. The National Park Service, established in 1916, was first headed by Muir's disciple, Stephen Mather, and has always been oriented toward preservation of nature rather than consumptive uses. Muir's preservationist ideas have often been at odds with Pinchot's utilitarian approach. One of Muir and Pinchot's biggest battles was over the damming of Hetch Hetchy Valley in Yosemite. Muir regarded flooding the valley a sacrilege against nature. Pinchot, who championed publicly owned utilities, viewed the dam as a way to free San Francisco residents from the clutches of greedy water and power monopolies.

In 1935, pioneering wildlife ecologist Aldo Leopold (fig. 1.22d) bought a small, worn-out farm in central Wisconsin. A dilapidated chicken shack, the only remaining building, was remodeled into a rustic cabin. Working together with his children, Leopold planted thousands of trees in a practical experiment in restoring the health and beauty of the land. "Conservation," he wrote, "is the positive exercise of skill and insight, not merely a negative exercise of abstinence or caution." The shack became a writing refuge and became the main focus of *A Sand County Almanac*, a much beloved collection of essays about our relation with nature. In it, Leopold wrote, "We abuse land because we regard it as a commodity belonging to us. When we see land as a community to which we belong, we may begin to use it with love and respect." Together with Bob Marshall and two others, Leopold was a founder of the Wilderness Society.

Rising pollution levels led to the modern environmental movement

The undesirable effects of pollution probably have been recognized as long as people have been building smoky fires. In 1723 the acrid coal smoke in London was so severe that King Edward I threatened to hang anyone who burned coal in the city. In 1661 the English diarist John Evelyn complained about the noxious air pollution caused by coal fires and factories and suggested that sweet-smelling trees be planted to purify city air. Increasingly dangerous smog attacks in Britain led, in 1880, to formation of a national Fog and Smoke Committee to combat this problem. But nearly a century later, London's air (like that of many cities) was still bad. In 1952 an especially bad episode turned midday skies dark and may have caused 12,000 deaths. This event was extreme, but noxious air was common in many large cities.

The tremendous expansion of chemical industries during and after World War II added a new set of concerns to the environmental agenda. *Silent Spring*, written by Rachel Carson (fig. 1.24a) and published in 1962, awakened the public to the threats of pollution and toxic chemicals to humans as well as other species. The movement she engendered might be called **modern environmentalism** because its concerns extended to include both natural resources and environmental pollution.

Two other pioneers of this movement were activist David Brower and scientist Barry Commoner (fig. 1.24b and c). Brower, as executive director of the Sierra Club, Friends of the Earth, and Earth Island Institute, introduced many of the techniques of environmental lobbying and activism, including litigation, testifying at regulatory hearings, book and calendar publishing, and use of mass media for publicity campaigns. Commoner, who was trained as a molecular biologist, has been a leader in analyzing the links between science, technology, and society. Both activism and research remain hallmarks of the modern environmental movement.

Under the leadership of a number of other brilliant and dedicated activists and scientists, the environmental agenda was expanded in the 1970s, to most of the issues addressed in this textbook, such as human population growth, atomic weapons testing and atomic power, fossil fuel extraction and use, recycling, air and water pollution, and wilderness protection. Environmentalism has become well established in the public agenda since the first national Earth Day in 1970. A majority of Americans now consider themselves environmentalists, although there is considerable variation in what that term means.

Figure 1.24 Among many distinguished environmental leaders in modern times, Rachel Carson (a), David Brower (b), Barry Commoner (c), and Wangari Maathai (d) stand out for their dedication, innovation, and bravery.

(a) Rachel Carson

(b) David Brower

(c) Barry Commoner

(d) Wangari Maathai

Environmental quality is tied to social progress

Many people today believe that the roots of the environmental movement are elitist—promoting the interests of a wealthy minority, who can afford to vacation in wilderness. In fact, most environmental leaders have seen social justice and environmental equity as closely linked. Gifford Pinchot, Teddy Roosevelt, and John Muir all strove to keep nature accessible to everyone, at a time when public lands, forests, and waterways were increasingly controlled by a few wealthy individuals and private corporations. The idea of national parks, one of our principal strategies for nature conservation, is to provide public access to natural beauty and outdoor recreation. Aldo Leopold, a founder of the Wilderness Society, promoted ideas of land stewardship among farmers, fishers, and hunters. Robert Marshall, also a founder of the Wilderness Society, campaigned all his life for social and economic justice for low-income groups. Both Rachel Carson and Barry Commoner were principally interested in environmental health—an issue that is especially urgent for low-income, minority, and inner-city residents. Many of these individuals grew up in working class families, so their sympathy with social causes is not surprising.

Increasingly, environmental activists are linking environmental quality and social progress on a global scale. One of the core concepts of modern environmental thought is **sustainable development**, the idea that economic improvement for the world's poorest populations is possible without devastating the environment. This idea became widely publicized after the Earth Summit, a United Nations meeting held in Rio de Janeiro, Brazil, in 1992. The Rio meeting was a pivotal event because it brought together many diverse groups. Environmentalists and politicians from wealthy countries, indigenous people and workers struggling for rights and land, and government representatives from developing countries all came together and became more aware of their common needs.

Some of today's leading environmental thinkers come from developing nations, where poverty and environmental degradation together plague hundreds of millions of people. Dr. Wangari Maathai of Kenya is a notable example. In 1977, Dr. Maathai (fig. 1.24d) founded the Green Belt Movement in her native Kenya as a way of both organizing poor rural women and restoring their environment. Beginning at a small, local scale, this organization has grown to more than 600 grassroots networks across Kenya. They have planted more than 30 million trees while mobilizing communities for self-determination, justice, equity, poverty reduction, and environmental conservation. Dr. Maathai was elected to the Kenyan Parliament and served as Assistant Minister for Environment and Natural Resources. Her leadership has helped bring democracy and good government to her country. In 2004, she received the Nobel Peace Prize for her work, the first time a Nobel has been awarded for environmental action. In her acceptance speech, she said, "Working together, we have proven that sustainable development is possible; that reforestation of degraded land is possible; and that exemplary governance is possible when ordinary citizens are informed, sensitized, mobilized and involved in direct action for their environment."

Photographs of the earth from space (see fig. 1.2) provide a powerful icon for the fourth wave of ecological concern, which might be called **global environmentalism**. Such photos remind us how small, fragile, beautiful, and rare our home planet is. We all share an environment at this global scale. As Ambassador Adlai Stevenson noted, in his 1965 farewell address to the United Nations, we now need to worry about the life-support systems of the planet as a whole. He went on to say in this speech, "We cannot maintain it half fortunate, half miserable, half confident, half despairing, half slave to the ancient enemies of mankind and half free in a liberation of resources undreamed of until this day. No craft, no crew, can travel with such vast contradictions. On their resolution depends the security of us all."

Conclusion

Environmental science gives us useful tools and ideas for understanding both environmental problems and new solutions to those problems. We face many severe and persistent problems, including human population growth, contaminated water and air, climate change, and biodiversity losses. We can also see many encouraging examples of progress. Human population growth rates have slowed, the extent of habitat preserves has expanded greatly in recent years, we have promising new energy options, and in many regions we have made improvements in air and water quality.

Both poverty and affluence are linked to environmental degradation. Impoverished populations, desperate for cropland, often clear forests and eliminate wildlife habitat. Lacking water treatment, they contaminate water supplies. Illness from contaminated water often leads to a cycle of environmental degradation, illness, and poverty. Affluence also has great environmental costs. Wealthy populations can afford to consume or degrade extraordinary amounts of resources, including energy, water, paper, food, and soil. Differences between the rich and poor can be measured in terms of life expectancy, infant mortality, and other measures of well-being, in addition to income levels.

Science helps us analyze and resolve these problems because it provides an orderly, methodical approach to examining problems. Ideally, scientists are skeptical about evidence and cautious about conclusions. The scientific method provides an approach to doing this: it involves developing a hypothesis and collecting data to test it. Analyzing data often involves statistics, which provide some simple approaches to evaluating and comparing observations.

Critical thinking also provides orderly steps for analyzing the assumptions and logic of arguments. Critical thinking is an essential skill, both in school and in life.

Environmental thought has evolved in response to environmental deterioration. Conservation focuses on maintaining usable resources; preservation focuses on maintaining nature for its own sake. Throughout history, these ideas have been closely tied to concerns for social equity, for the rights of low-income people to have access to resources and to a healthy environment. In recent years these twin concerns have expanded to recognize the possibilities of change in developing countries, and the global interconnections of environmental and social concerns.

1. Describe how fishing has changed at Apo Island, and the direct and indirect effects on people's lives.

2. What are some basic assumptions of science?

3. Distinguish between a hypothesis and a theory.

4. Describe the steps in the scientific method.

5. What is probability? Give an example.

6. In a graph, which axis represents the independent variable? The dependent variable?

7. What's the first step in critical thinking according to table 1.4?

8. Distinguish between utilitarian conservation and biocentric preservation. Name two environmental leaders associated with each of these philosophies.

9. Why do some experts regard water as the most critical natural resource for the twenty-first century?

10. Where in figure 1.7 do the largest areas of persistence of greening occur? What is persistence of greening?

11. Describe some signs of hope in overcoming global environmental problems.

12. What is the link between poverty and environmental quality?

13. Define *sustainability* and *sustainable development*.

Critical Thinking and Discussion Questions

Apply the principles you have learned in this chapter to discuss these questions with other students.

1. How do you think the example of Apo Island's marine sanctuary meets the criteria of being scientifically sound, economically sustainable, and socially acceptable? If you were studying this situation, what information would you look for to support or refute this conclusion?

2. The analytical approaches of science are suitable for answering many questions. Are there some questions that science cannot answer? Why or why not?

3. Many social theorists argue that there is no such thing as objective truth or an impartial observer. If you were researching a controversial topic—perhaps whether marine reserves improve fishing and lead to economic development—what steps would you take to try to maintain objectivity and impartiality?

4. Does the world have enough resources for 8 or 10 billion people to live decent, secure, happy, fulfilling lives? What do those terms mean to you? Try to imagine what they mean to others in our global village.

5. Suppose you wanted to study the environmental impacts of a rich versus a poor country. What factors would you examine, and how would you compare them?

Data Analysis: Working with Graphs

Do you find it easier to evaluate trends in a graph or in a table? Many people find the visual presentation of a graph quicker to read than a table. But reading graphs takes practice, and it takes some patience. This exercise asks you to examine different kinds of graphs, as preparation for reading many others that will follow in this book. Solidifying this skill is extremely important for studying environmental science—and many other topics.

Examining Relationships

Many graphs show the relationship between two variables. Usually we want to know whether changes in one variable are associated with changes in the other. For example, has per capita income gone up or down as time has passed (fig. 1)? Or does fishing tend to be better or worse where there are more marine preserves (fig. 2)?

These two graphs have different styles because they show slightly different kinds of information. In figure 1, lines connect the data points to show that they represent changing values for a place. This graph, of course, shows several places simultaneously.

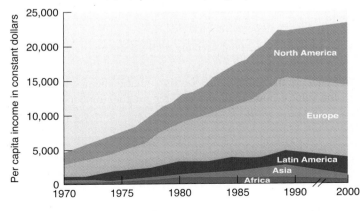

Figure 1 A line graph shows how the *y* variable changes as the *x* variable increases. What are *x* and *y* here?

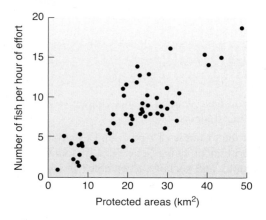

Figure 2 A scatter plot shows the relationship of *x* and *y* for many observations. Here each observation is a village.

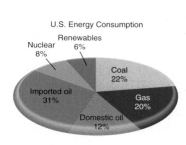

Figure 3 A pie chart shows proportions.

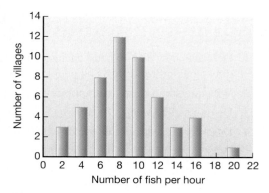

Figure 4 A bar graph shows values for classes or groups.

Here, the question the graph answers is *whether there is a change* in income, and how great the change is. In figure 2, on the other hand, the question is *whether there is a relationship* between *x* and *y*.

Some graphs ask *what is the relative distribution* of something? A pie chart is a common, easy-to-read method of showing the sizes of portions of a whole, for example, percentages of total energy consumption (fig. 3). Bar charts also show relative amounts of something. For example, figure 4 shows relative amounts of villages with small and large fish catches. Note that we often use the words "graph" and "chart" interchangeably.

Elements of a Graph

The first step in reading a graph is to look at its elements. By convention, we put the dependent variable on the vertical (Y) axis and the independent (or explanatory) variable on the horizontal (X) axis. Axes are labeled, and tic marks show values along the axes.

Whenever you look at a graph, you should ask yourself a few questions to be sure you understand it. Answer the following questions for figure 1:

1. What is the subject of the graph?
2. What is the independent variable?
3. What is the dependent variable?
4. Does the independent variable really change independently of the other variable?
5. What are the units on the X-axis? What is the magnitude of the values (how small and how big are the numbers)?

6. What are the units on the Y-axis? What is the magnitude of the values?
7. When there are multiple lines, what is the difference in magnitude between the highest and lowest lines (here, in dollars) at the left end of the graph? What is the difference at the right end? Has the difference gotten bigger or smaller?
8. What do the trends in the graph mean? What factors might explain the changes you see?

Answer these questions again, this time using figure 2. Instead of question 7, since there are no lines on this scatter plot, you should ask yourself if the dots form a fairly linear pattern or a loose cloud with no direction. If they form a linear pattern, does the line go upward? Downward? Is it flat? What would those different patterns mean?

9. For review, describe the following:
   - a scatter plot
   - a bar chart
   - a pie chart
   - a line graph
   - a dependent variable
   - the X-axis

For Additional Help in Studying This Chapter, please visit our website at www.mhhe.com/cunningham5e. You will find practice quizzes, key terms, answers to end of chapter questions, additional case studies, an extensive reading list, and Google Earth™ mapping quizzes.

Arcata, California, built an artificial marsh as a low-cost, ecologically based treatment system for sewage effluent.

2 Environmental Systems:
Connections, Cycles, Flows, and Feedback Loops

Learning Outcomes

After studying this chapter, you should be able to answer the following questions:

- What are systems, and how do feedback loops regulate them?
- Chemical bonds hold atoms and molecules together. What would our world be like if there were no chemical bonds, or if they were so strong they never broke apart?
- Ecologists say there is no "away" to throw things, and that everything in the universe tends to slow down and fall apart. What do they mean?
- All living things—except for some unusual organisms living in extreme conditions— depend on the sun as their ultimate energy source. Explain.
- What qualities make water so unique and essential for life as we know it?
- Why are big, fierce animals rare?
- How and why do elements such as carbon, nitrogen, phosphate, and sulfur cycle through ecosystems?

Most institutions demand unqualified faith; but the institution of science makes skepticism a virtue.

–Robert King Merton

A Natural System for Wastewater Treatment

Arcata, a small town in northwestern California's redwood country, sits at the north end of Humbolt Bay, about 450 km (280 mi) north of San Francisco. Logging, fishing, and farming, long the economic engines of the area, are now giving way to tourism, education, and diversified industries. Many of the 17,000 residents of Arcata choose to live there because of the outstanding environmental quality and recreational opportunities of the area. Until the early 1970s, however, Arcata's sewage system discharged unchlorinated sewage effluent directly into the nearly enclosed bay. Algal blooms discolored the water, while fishing and swimming appeal declined.

In 1974 California enacted a policy prohibiting discharge of untreated wastewater into bays and estuaries. State planners proposed a large and expensive regional sewage treatment plant for the entire Humbolt Bay area. The plan was for large interceptor sewers to encircle the bay with a major underwater pipeline crossing the main navigation channel. Effluent from the proposed plant was to be released offshore into an area of shifting sea bottom in heavy winter storms. The cost and disruption required by this project prompted city officials to look for alternatives.

Ecologists from Humbolt State University pointed out that natural processes could be harnessed to solve Arcata's wastewater problems at a fraction of the cost and disturbance of a conventional treatment plant. After several years of study and experimentation, the city received approval to build a constructed wetland for wastewater treatment. This solved two problems at the same time. Arcata's waterfront was blighted by an abandoned lumbermill pond, channelized sloughs, marginal pasture land, and an abandoned sanitary land-fill. Building a new wetland on this degraded area would beautify it while also solving the sewage treatment problem (fig. 2.1).

Today, Arcata's waterfront has been transformed into 40 ha (about 100 acres) of freshwater and saltwater marshes, brackish ponds, tidal sloughs, and estuaries. This diverse habitat supports a wide variety plants and animals, and is now a wildlife sanctuary and a major tourist attraction. As a home or rest stop for over 200 bird species, the wetland is regarded as one of the best birding sites along the Pacific North Coast. In 1987, the city was awarded a prize by the Ford Foundation for innovative local government projects that included $100,000 to build an interpretive center in the marsh. The wetland is now an outstanding place for outdoor education, scientific study, and recreation.

Arcata's constructed wetland is also an unqualified success in wastewater treatment. After primary clarification to remove grit and sediment, wastewater passes through several oxidation ponds that remove about half the organic material (measured by biological oxygen demand) and suspended solids. Sludge captured by the clarifiers goes to digesters that generate methane gas, which is burned to provide heat that speeds the digestion process. Effluent from the oxidation ponds passes through treatment marshes, where sunlight and oxygen kill pathogens while aquatic plants and animals remove remaining organic material along with nitrogen, phosphorus, and other plant nutrients. The effluent is treated with chlorine gas to kill any residual pathogens. Finally, the treated water trickles through several large enhancement marshes that allow chlorine to evaporate and complete pollutant removal.

After 20 years of successful operation, Arcata's constructed wetland has inspired many other communities to find ecological solutions to their problems. Arcata Bay now produces more than half the oysters grown in California. The city also operates a wastewater aquaculture project where salmon, trout, and other fish species are raised in a mixture of wastewater effluent and seawater. Economically, the project is a success, providing excellent water quality treatment far below the cost of conventional systems. And the innovative approach of working with nature rather than against it has made Arcata an important ecotourist destination. Around the world, hundreds of communities have followed Arcata's model and now use constructed wetlands to eliminate up to 98 percent of the pollutants from their wastewater.

We'll discuss water quality and pollution control further in chapter 10. First, however, we'll look, in this chapter, at some of the ecological relationships that determine energy flows and nutrient recycling in biological communities, such as those in Arcata's marshes and wetlands. We'll study how those communities function as systems, as well as the feedback loops that regulate those systems. Understanding these ecological principles will help us find natural solutions to other environmental problems we face.

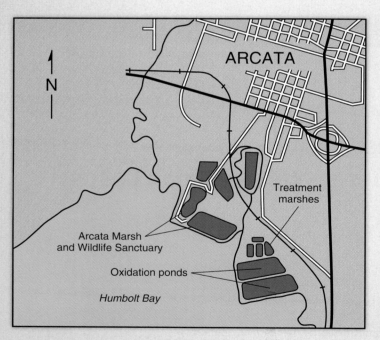

Figure 2.1 Arcata's constructed wetlands remove pathogens, pollutants, and nutrients from municipal wastewater using native biological communities and ecological processes in a setting that also serves as a wildlife sanctuary and tourist attraction.
Source: Environmental Protection Agency.

2.1 | Systems Describe Interactions

Arcata's wetland is remarkable because it is a balanced, functioning system—one that approximates a natural ecosystem. **Systems** are a central idea in environmental science: a system is a network of interdependent components and processes, with materials and energy flowing from one component of the system to another. For example, "ecosystem" is probably a familiar term for you. This simple word represents a complex assemblage of animals, plants, and their environment, through which materials and energy move.

The idea of systems is useful because it helps us organize our thoughts about the inconceivably complex phenomena around us. For example, an ecosystem might consist of countless animals, plants, and their physical surroundings. (You are a system consisting of millions of cells, complex organs, and innumerable bits of energy and matter that move through you.) Keeping track of all the elements and their relationships in an ecosystem would probably be an impossible task. But if we step back and think about them in terms of plants, herbivores, carnivores, and decomposers, then we can start to comprehend how it works (fig. 2.2).

We can use some general terms to describe the components of a system. A simple system consists of state variables (also called compartments), which store resources such as energy, matter, or water; and flows, or the pathways by which those resources move from one state variable to another. In figure 2.2, the plant and animals represent state variables. The plant represents many different plant types, all of which are things that store solar energy and create carbohydrates from carbon, water, and sunlight. The rabbit represents many kinds of herbivores, all of which consume plants, then store energy, water, and carbohydrates until they are used, transformed, or consumed by a carnivore. We can describe the flows in terms of herbivory, predation, or photosynthesis, all processes that transfer energy and matter from one state variable to another.

It may seem cold and analytical to describe a rabbit or a flower as a state variable, but it is also helpful to do so. When we start discussing natural complexity in the simple terms of systems, we can identify common characteristics. Understanding these characteristics can help us diagnose disturbances or changes in the system: for example, if rabbits become too numerous, herbivory can become too rapid for plants to sustain. Overgrazing can lead to widespread collapse of this system. Let's examine some of the common characteristics we can find in systems.

Systems can be described in terms of their characteristics

Open systems are those that receive inputs from their surroundings and produce outputs that leave the system. Almost all natural systems are open systems. In principle, a **closed system** exchanges no energy or matter with its surroundings, but these are rare. Often we think of pseudo-closed systems, those that

Figure 2.2 A system can be described in very simple terms.

exchange only a little energy but no matter with their surroundings. **Throughput** is a term we can use to describe the energy and matter that flow into, through, and out of a system. Larger throughput might expand the size of state variables. For example, you can consider your household economy in terms of throughput. If you get more income, you have the option of enlarging your state variables (bank account, car, television, . . .). Usually an increase in income is also associated with an increase in outflow (the money spent on that new car and TV). In terms of a wetland like Arcata's, a larger system would generally tolerate, or require, larger throughput. A large wetland can absorb and process more nutrients and energy than a small wetland. If inputs rose dramatically, however, plant growth and decomposition might not be able to keep up with the rapid change.

Arcata's constructed wetland is an open system, which receives nutrients (biological waste) from the city's sewer system. Those nutrients are processed by plants and decomposers and transformed into vegetation, which feeds birds and fish. To some extent, an increase in nutrients can support an increase in plant growth, which allows greater nutrient uptake, which supports greater plant growth. . . This increasing response, where an increase in the state variable leads to further increases in the same variable, is called a **positive feedback** (fig. 2.3). When positive feedbacks accelerate out of control, the system can become unstable and even collapse. A term for this process in wetlands is eutrophication (chapter 10), the runaway growth of plants, in response to excessive nutrients, which leads to biological collapse (at least temporarily) of a wetland system. In contrast, a **negative feedback** has a dampening effect: too many fish in a pond leads to food scarcity, which leads to fish mortality

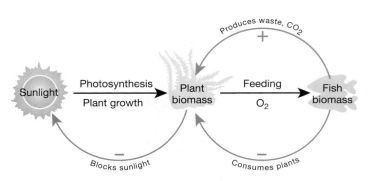

Figure 2.3 Positive feedbacks (*red*) and negative feedbacks (*blue*) can increase or decrease the size of state variables (or compartments) in a system.

(fig. 2.3). Your body is a system with many active negative feedback mechanisms: for example, if you exercise, you become hot, so your skin sweats to keep you from overheating.

Systems may exhibit stability

Negative feedback loops tend to maintain stability in a system. We often think of systems exhibiting **homeostasis**, or a tendency to remain more or less stable and unchanging. Your body temperature stays remarkably constant despite dramatic changes in your environment and your activity levels. Changing by just a few degrees is extremely unusual—just think how little temperature change is involved in a serious fever. Natural systems such as wetlands tend to be stable, too, at least on the scale of years or decades. Over long periods of time, they may change quite a bit, though. Systems might also fluctuate: the population of a species in a wetland might rise and fall repeatedly, or in a cycle, and that cycle might be part of the system's normal functioning.

Disturbances, events that destabilize or change the system, might cause these population fluctuations. Summer drought in a wetland, for example, will periodically reduce some populations and increase others. Disturbances might also be large enough to cause more dramatic and persistent change. A wetland might get a huge flush of nutrient inputs, causing organisms to die and causing the ecosystem to collapse. A large storm, or a change in onshore drainage networks, might raise water levels, killing the emergent plants that need air and sunlight to grow. Sometimes ecosystems show **resilience**, or an ability to return to their previous condition after disturbance. Sometimes ecosystems undergo a **state shift**, where conditions do not return to "normal." For example, if an earthquake were to lower the ground level, Arcata's wetland might become permanently flooded. It might turn into a bay instead of a marsh. Or if climate change caused a substantial decrease in rainfall, the wetland might dry up and turn into a meadow. State shifts are a concern in many situations: when a forest is logged, does it return to its original conditions? Or does it transform into something entirely new? That depends on many factors.

Emergent properties are another interesting aspect of systems. Sometimes a system is more than the sum of its parts. A wetland has complex varieties of living organisms, colorful dragonflies, intricate webs of relationships that can reinforce each other through their complexity. This system has larger influences

Figure 2.4 Emergent properties of systems, including beautiful sights and sounds, make them exciting to study.

beyond its borders—helping to stabilize local temperatures and humidity, producing biodiversity that spills over into surrounding areas, or supporting migrant waterfowl. A wetland also has beautiful sights and sounds that may be irrelevant to its functioning as a system, but that we appreciate nonetheless (fig. 2.4). In a similar way, you are a system made up of component parts, but you have many emergent properties, including your ability to think, share ideas, build relationships, sing, and dance.

2.2 Elements of Life

What exactly are the materials that flow through a system like the Arcata wetland? If a principal concern in such a wetland system is the control of nutrients, such as NO_3 and PO_4, as well as maintaining O_2 levels, just what exactly are these elements, and why are they important? What does "O_2" or "NO_3" mean? In this section we will examine matter and the elements and compounds on which all life depends. In the sections that follow, you'll study how organisms use those elements and compounds to capture and store solar energy, and how materials cycle through global systems, as well as ecosystems.

In a sense, every organism is a chemical factory that captures matter and energy from its environment and transforms them into structures and processes that make life possible. To understand how these processes work, we will begin with some of the fundamental properties of matter and energy.

Matter is made of atoms, molecules, and compounds

Everything that takes up space and has mass is **matter**. Matter exists in three distinct states—solid, liquid, and gas—due to differences in the arrangement of its constitutive particles. Water, for example, can exist as ice (solid), as liquid water, or as water vapor (gas).

Under ordinary circumstances, matter is neither created nor destroyed but rather is recycled over and over again. Some of the molecules that make up your body probably contain atoms that once made up the body of a dinosaur and most certainly were part of many smaller prehistoric organisms, as chemical elements are used and reused by living organisms. Matter is transformed and combined in different ways, but it doesn't disappear; everything goes somewhere. These statements paraphrase the physical principle of **conservation of matter**.

How does this principle apply to human relationships with the biosphere? Particularly in affluent societies, we use natural resources to produce an incredible amount of "disposable" consumer goods. If everything goes somewhere, where do the things we dispose of go after the garbage truck leaves? As the sheer amount of "disposed-of stuff" increases, we are having greater problems finding places to put it. Ultimately, there is no "away" where we can throw things we don't want any more.

Matter consists of **elements**, which are substances that cannot be broken down into simpler forms by ordinary chemical reactions. Each of the 115 known elements (92 natural, plus 23 created under special conditions) has distinct chemical characteristics. Just four elements—oxygen, carbon, hydrogen, and nitrogen—are responsible for more than 96 percent of the mass of most living organisms.

All elements are composed of **atoms**, which are the smallest particles that exhibit the characteristics of the element. Atoms are composed of positively charged protons, negatively charged electrons, and electrically neutral neutrons. Protons and neutrons, which have approximately the same mass, are clustered in the nucleus in the center of the atom (fig. 2.5). Electrons, which are tiny in comparison to the other particles, orbit the nucleus at the speed of light.

Each element has a characteristic number of protons per atom, called its **atomic number**. The number of neutrons in different atoms of the same element can vary slightly. Thus, the atomic mass, which is the sum of the protons and neutrons in each nucleus, also can vary. We call forms of a single element that differ in atomic mass **isotopes**. For example, hydrogen, the lightest element, normally has only one proton (and no neutrons) in its nucleus. A small percentage of hydrogen atoms have one proton and one neutron. We call this isotope deuterium (^{2}H). An even smaller

percentage of natural hydrogen called tritium (^{3}H) has one proton plus two neutrons. The heavy form of nitrogen (^{15}N) has one more neutron in its nucleus than does the more common ^{14}N. Both these nitrogen isotopes are stable but some isotopes are unstable—that is, they may spontaneously emit electromagnetic energy, or subatomic particles, or both. Radioactive waste, and nuclear energy, result from unstable isotopes of elements such as uranium and plutonium.

Why should you know about isotopes? Although you might not often think of it this way, both stable and unstable types are in the news every day. Unstable, radioactive isotopes are distributed in the environment by both radioactive waste (potentially from nuclear power plants) and nuclear bombs. Every time you hear debates about nuclear power plants—which might produce the energy that powers the lights you are using right now—or about nuclear weapons in international politics, the core issue is radioactive isotopes. Understanding why they are dangerous, because emitted particles damage living cells, is important as you decide your opinions about these policy issues.

Stable isotopes—those that do not change mass by losing neutrons—are important because they help us understand climate history and many other environmental processes. This is because light-weight isotopes move differently than heavier ones. For example, oxygen occurs as both ^{16}O (a lighter isotope) and ^{18}O (a heavier isotope). Some water molecules (H_2O) contain the light-weight oxygen isotopes. These light-weight molecules evaporate, and turn into rain or snowfall, *slightly* more easily than heavier water molecules, especially in cool weather. Consequently, ice stored in the Antarctic during cool periods in the earth's history will contain slightly higher proportions of light-isotope water, with ^{16}O. Ice from warm periods contains relatively higher proportions of ^{18}O. By examining the proportions of isotopes in ancient ice layers, climate scientists can measure the earth's temperature from hundreds of thousands of years ago.

So while atoms, elements, and isotopes may seem a little arcane, they actually are fundamental to understanding climate change, nuclear weapons, and energy policy—all issues you can read about in the news almost any day of the week.

Electric charges keep atoms together

Atoms frequently gain or lose electrons, acquiring a negative or positive electrical charge. Charged atoms (or combinations of atoms) are called **ions**. Negatively charged ions (with one or more extra electrons) are *anions*. Positively charged ions are *cations*. A sodium (Na) atom, for example, can give up an electron to become a sodium ion (Na^+). Chlorine (Cl) readily gains electrons, forming chlorine ions (Cl^-).

Atoms often join to form **compounds**, or substances composed of different kinds of atoms (fig. 2.6). A pair or group of atoms that can exist as a single unit is known as a **molecule**. Some elements commonly occur as molecules, such as molecular oxygen (O_2) or molecular nitrogen (N_2), and some compounds can exist as molecules, such as glucose ($C_6H_{12}O_6$). In contrast to these molecules, sodium chloride (NaCl, table salt) is a compound that cannot exist as a single pair of atoms. Instead it occurs in a large

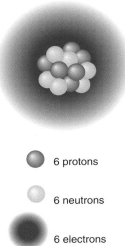

Figure 2.5 As difficult as it may be to imagine when you look at a solid object, all matter is composed of tiny, moving particles, separated by space and held together by energy. This model represents a carbon-12 atom, with a nucleus containing six protons and six neutrons; the six electrons are represented as a fuzzy cloud of potential locations, rather than as individual particles.

6 protons

6 neutrons

6 electrons

mass of Na and Cl atoms or as two ions, Na$^+$ and Cl$^-$, in solution. Most molecules consist of only a few atoms. Others, such as proteins and nucleic acids, can include millions or even billions of atoms.

When ions with opposite charges form a compound, the electrical attraction holding them together is an *ionic bond*. Sometimes atoms form bonds by *sharing* electrons. For example, two hydrogen atoms can bond by sharing a single electron—it orbits the two hydrogen nuclei equally and holds the atoms together. Such electron-sharing bonds are known as *covalent bonds*. Carbon (C) can form covalent bonds simultaneously with four other atoms, so carbon can create complex structures such as sugars and proteins. Atoms in covalent bonds do not always share electrons evenly. An important example in environmental science is the covalent bonds in water (H_2O). The oxygen atom attracts the shared electrons more strongly than do the two hydrogen atoms. Consequently, the hydrogen portion of the molecule has a slight positive charge, while the oxygen has a slight negative charge. These charges create a mild attraction between water molecules, so that water tends to be somewhat cohesive. This fact helps explain some of the remarkable properties of water (see Exploring Science, p. 31).

When an atom gives up one or more electrons, we say it is *oxidized* (because it is very often oxygen that takes the electron, as bonds are formed with this very common and highly reactive element). When an atom gains electrons, we say it is *reduced*. Chemical reactions necessary for life involve oxidation and reduction: Oxidation of sugar and starch molecules, for example, is an important part of how you gain energy from food.

Breaking bonds requires energy, while forming bonds generally releases energy. Burning wood in your fireplace breaks up large molecules, such as cellulose, and forms many smaller ones, such as carbon dioxide and water. The net result is a release of energy (heat). Generally, some energy input (activation energy) is needed to initiate these reactions. In your fireplace, a match might provide the needed activation energy. In your car, a spark from the battery provides activation energy to initiate the oxidation (burning) of gasoline.

Acids and bases release reactive H$^+$ and OH$^-$

Substances that readily give up hydrogen ions in water are known as **acids**. Hydrochloric acid, for example, dissociates in water to form H$^+$ and Cl$^-$ ions. In later chapters, you may read about acid rain (which has an abundance of H$^+$ ions), acid mine drainage, and many other environmental problems involving acids. In general, acids cause environmental damage because the H$^+$ ions react readily with living tissues (such as your skin or tissues of fish larvae) and with nonliving substances (such as the limestone on buildings, which erodes under acid rain).

Substances that readily bond with H$^+$ ions are called **bases** or alkaline substances. Sodium hydroxide (NaOH), for example, releases hydroxide ions (OH$^-$) that bond with H$^+$ ions in water. Bases can be highly reactive, so they also cause significant environmental problems. Acids and bases can also be essential to living things: The acids in your stomach help you digest food,

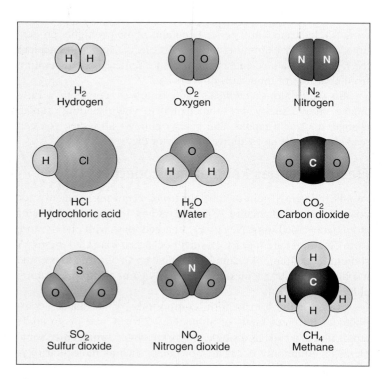

Figure 2.6 These common molecules, with atoms held together by covalent bonds, are important components of the atmosphere or important pollutants.

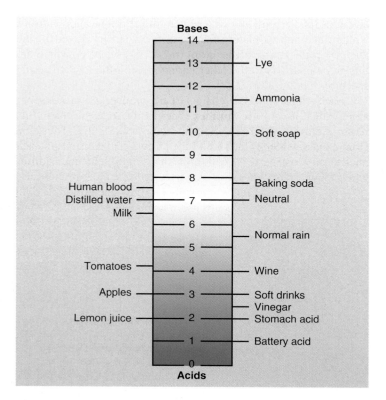

Figure 2.7 The pH scale. The numbers represent the negative logarithm of the hydrogen ion concentration in water. Alkaline (basic) solutions have a pH greater than 7. Acids (pH less than 7) have high concentrations of reactive H$^+$ ions.

Exploring
SCIENCE:

A "Water Planet"

If travelers from another solar system were to visit our lovely, cool, blue planet, they might call it Aqua rather than Terra because of its outstanding feature: the abundance of streams, rivers, lakes, and oceans of liquid water. Our planet is the only place we know where water exists as a liquid in any appreciable quantity. Water covers nearly three-fourths of the earth's surface and moves around constantly through evaporation, precipitation, and runoff that distribute nutrients, replenish freshwater supplies, and shape the land. Water makes up 60 to 70 percent of the weight of most living organisms. It fills cells, giving form and support to tissues. Among water's unique, almost magical qualities, are the following:

1. Water molecules are polar, that is, they have a slight positive charge on one side and a slight negative charge on the other side. Therefore, water readily dissolves polar or ionic substances, including sugars and nutrients, and carries materials to and from cells.
2. Water is the only inorganic liquid that occurs in nature under normal conditions at temperatures suitable for life. Most substances exist as either a solid or a gas, with only a very narrow liquid temperature range. Organisms synthesize organic compounds such as oils and alcohols that remain liquid at ambient temperatures and are therefore extremely valuable to life, but the original and predominant liquid in nature is water.

3. Water molecules are cohesive, tending to stick together tenaciously. You have experienced this property if you have ever done a belly flop off a diving board. Water has the highest surface tension of any common, natural liquid. Water also adheres to surfaces. As a result, water is subject to *capillary action*: it can be drawn into small channels. Without capillary action, movement of water and nutrients into groundwater reservoirs and through living organisms might not be possible.
4. Water is unique in that it expands when it crystallizes. Most substances shrink as they change from liquid to solid. Ice floats because it is less dense than liquid water. When temperatures fall below freezing, the surface layers of lakes, rivers, and oceans cool faster and freeze before deeper water. Floating ice then insulates underlying layers, keeping most water bodies liquid (and aquatic organisms alive) throughout the winter in most places. Without this feature, many aquatic systems would freeze solid in winter.
5. Water has a high heat of vaporization, using a great deal of heat to convert from liquid to vapor. Consequently, evaporating water is an effective way for organisms to shed excess heat. Many animals pant or sweat to moisten evaporative cooling surfaces. Why do you feel less comfortable on a hot, humid day than on a hot, dry day? Because the water vapor–

laden air inhibits the rate of evaporation from your skin, thereby impairing your ability to shed heat.
6. Water also has a high specific heat; that is, a great deal of heat is absorbed before it changes temperature. The slow response of water to temperature change helps moderate global temperatures, keeping the environment warm in winter and cool in summer. This effect is especially noticeable near the ocean, but it is important globally.

All these properties make water a unique and vitally important component of the ecological cycles that move materials and energy and make life on earth possible.

Surface tension is demonstrated by the resistance of a water surface to penetration, as when it is walked upon by a water strider.

for example, and acids in soil help make nutrients available to growing plants.

We describe acids and bases in terms of **pH**, the negative logarithm of its concentration of H^+ ions (fig. 2.7). Acids have a pH below 7; bases have a pH greater than 7. A solution of exactly pH 7 is "neutral." Because the pH scale is logarithmic, pH 6 represents *ten times* more hydrogen ions in solution than pH 7.

A solution can be neutralized by adding buffers, or substances that accept or release hydrogen ions. In the environment, for example, alkaline rock can buffer acidic precipitation, decreasing its acidity. Lakes with acidic bedrock, such as granite, are especially vulnerable to acid rain because they have little buffering capacity.

Organic compounds have a carbon backbone

Organisms use some elements in abundance, others in trace amounts, and others not at all. Certain vital substances are concentrated within cells, while others are actively excluded. Car-

bon is a particularly important element because chains and rings of carbon atoms form the skeletons of **organic compounds**, the material of which biomolecules, and therefore living organisms, are made.

The four major categories of organic compounds in living things ("bio-organic compounds") are lipids, carbohydrates, proteins, and nucleic acids. Lipids (including fats and oils) store energy for cells, and they provide the core of cell membranes and other structures. Many hormones are also lipids. Lipids do not readily dissolve in water, and their structure is a chain of carbon atoms with attached hydrogen atoms. This structure makes them part of the family of hydrocarbons (fig. 2.8a). Carbohydrates (including sugars, starches, and cellulose) also store energy and provide structure to cells. Like lipids, carbohydrates have a basic structure of carbon atoms, but hydroxyl (OH) groups replace half the hydrogen atoms in their basic structure, and they usually consist of long chains of simple sugars. Glucose (fig. 2.8b) is an example of a very simple sugar.

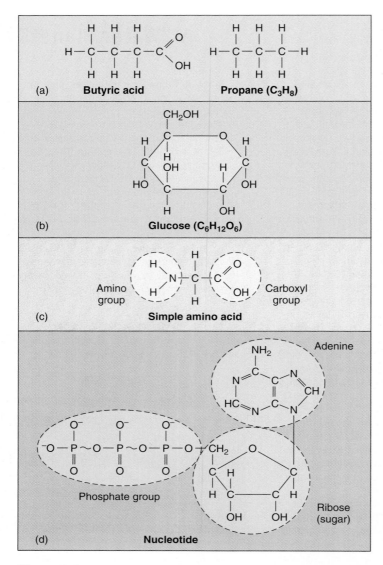

Figure 2.8 The four major groups of biologically important organic molecules are based on repeating subunits of these carbon-based structures. Basic structures are shown for (a) butyric acid (a building block of lipids) and a hydrocarbon, (b) a simple carbohydrate, (c) a protein, and (d) a nucleic acid.

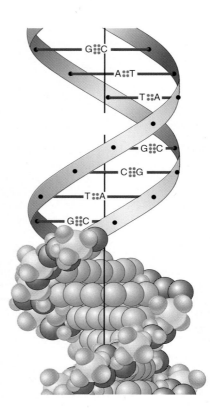

Figure 2.9 A composite molecular model of DNA. The lower part shows individual atoms, while the upper part has been simplified to show the strands of the double helix held together by hydrogen bonds (small dots) between matching nucleotides (A, T, G, and C). A complete DNA molecule contains millions of nucleotides and carries genetic information for many specific, inheritable traits.

Proteins are composed of chains of subunits called amino acids (fig. 2.8c). Folded into complex three-dimensional shapes, proteins provide structure to cells and are used for countless cell functions. Enzymes, such as those that release energy from lipids and carbohydrates, are proteins. Proteins also help identify disease-causing microbes, make muscles move, transport oxygen to cells, and regulate cell activity.

Nucleotides are complex molecules made of a five-carbon sugar (ribose or deoxyribose), one or more phosphate groups, and an organic nitrogen-containing base (fig. 2.8d). Nucleotides are extremely important as signaling molecules (they carry information between cells, tissues, and organs) and as sources of energy within cells. They also form long chains called *ribonucleic acid* (RNA)

or **deoxyribonucleic acid (DNA)** that are essential for storing and expressing genetic information. Only four kinds of nucleotides (adenine, guanine, cytosine, and thyamine) occur in DNA, but DNA contains millions of these molecules arranged in very specific sequences. These sequences of nucleotides provide genetic information, or instructions, for cells. These instructions direct the growth and development of an organism. They also direct the formation of proteins or other compounds, such as those in melanin, a pigment that protects your skin from sunlight.

Long chains of DNA bind together to form a double helix (a two-stranded spiral, fig. 2.9). These chains replicate themselves when cells divide, so that as you grow, your DNA is reproduced in all your cells, from blood cells to hair cells. Since every individual (except identical twins) has a distinctive DNA pattern, DNA has proven useful in identifying individuals in forensics. Since DNA includes records of ancestors' DNA, it has allowed scientists to establish relationships, for example that dinosaurs were more closely related to birds than to lizards. Understanding DNA's structure has also allowed us to combine genes, for example inserting disease resistant traits into food crops (chapter 7). See the related story "Bar-Coding Life" at www.mhhe.com/cunningham5e.

Cells are the fundamental units of life

All living organisms are composed of **cells**, minute compartments within which the processes of life are carried out (fig. 2.10). Microscopic organisms such as bacteria, some algae, and protozoa are composed of single cells. Most higher organisms are multicellular, usually with many different cell varieties. Your body, for instance, is composed of several trillion cells of about two hundred distinct types. Every cell is surrounded by a thin but dynamic membrane of lipid and protein that receives information about the exterior world and regulates the flow of materials between the cell and its environment. Inside, cells are subdivided into tiny organelles and subcellular particles that provide the machinery for life. Some of these organelles store and release energy. Others manage and distribute information. Still others create the internal structure that gives the cell its shape and allows it to fulfill its role.

Nitrogen and phosphorus are key nutrients

In the opening case study about the Arcata wetland, one of the chief concerns is managing nitrogen and phosphorus. These elements are essential nutrients for plant growth, and both were abundant in the raw sewage that Arcata released in the bay. Before the constructed wetland was built, algal blooms (explosive growth of floating algae) discolored the water and reduced fish populations. An important function of the wetland is to allow aquatic plants to take up nitrogen and phosphorus before it reaches the open bay.

Notice that in figure 2.8 there are only a few different elements that are especially important. Carbon (C) is captured from air by green plants, and oxygen (O) and hydrogen (H) derive from air or water. The most important additional elements are nitrogen (N) and phosphorus (P), which are essential parts of the complex proteins, lipids, sugars, and nucleic acids that keep you alive. Of course your cells use many other elements, but these are the most abundant. You derive all these elements by consuming molecules produced by green plants. Plants, however, must extract these elements from their environment. Low levels of N and P often limit growth in ecosystems where they are scarce. Abundance of N and P can cause runaway growth, however, destabilizing an ecosystem. If you fertilize your plants, these two nutrients, along with potassium (K), are probably the main ingredients in your fertilizer. These elements often occur in the form of nitrate (NO_3), ammonium (NH_4), and phosphate (PO_4). Later in this chapter you will read more about how C, H_2O, N, and P move through our environment.

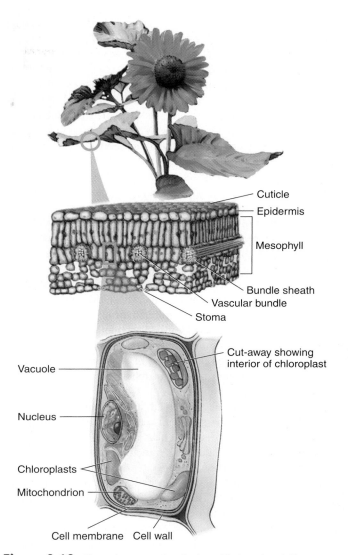

Figure 2.10 Plant tissues and a single cell's interior. Cell components include a cellulose cell wall; a nucleus; a large, empty vacuole; and several chloroplasts, which carry out photosynthesis.

2.3 Energy

If matter is the material of which things are made, energy provides the force to hold structures together, tear them apart, and move them from one place to another. In this section we will look at some fundamental characteristics of these components of our world.

Energy occurs in different types and qualities

Energy is the ability to do work such as moving matter over a distance or causing a heat transfer between two objects at different temperatures. Energy can take many different forms. Heat, light, electricity, and chemical energy are examples that we all experience. The energy contained in moving objects is called **kinetic energy**. A rock rolling down a hill, the wind blowing through the trees, water flowing over a dam (fig. 2.11), or electrons speeding

Figure 2.11 Water stored behind this dam represents potential energy. Water flowing over the dam has kinetic energy, some of which is converted to heat.

around the nucleus of an atom are all examples of kinetic energy. **Potential energy** is stored energy that is available for use. A rock poised at the top of a hill and water stored behind a dam are examples of potential energy. **Chemical energy** stored in the food that you eat and the gasoline that you put into your car are also examples of potential energy that can be released to do useful work. Energy is often measured in units of heat (calories) or work (joules). One joule (J) is the work done when one kilogram is accelerated at one meter per second per second. One calorie is the amount of energy needed to heat one gram of pure water one degree Celsius. A calorie can also be measured as 4.184 J.

Heat describes the energy that can be transferred between objects of different temperature. When a substance absorbs heat, the kinetic energy of its molecules increases, or it may change state: a solid may become a liquid, or a liquid may become a gas. We sense change in heat content as change in temperature (unless the substance changes state).

An object can have a high heat content but a low temperature, such as a lake that freezes slowly in the fall. Other objects, like a burning match, have a high temperature but little heat content. Heat storage in lakes and oceans is essential to moderating climates and maintaining biological communities. Heat absorbed in changing states is also critical. As you will read in chapter 10, evaporation and condensation of water in the atmosphere helps distribute heat around the globe.

Energy that is diffused, dispersed, and low in temperature is considered low-quality energy because it is difficult to gather and use for productive purposes. The heat stored in the oceans, for instance, is immense but hard to capture and use, so it is low quality. Conversely, energy that is intense, concentrated, and high in temperature is high-quality energy because of its usefulness in carrying out work. The intense flames of a very hot fire or high-voltage electrical energy are examples of high-quality forms that are valuable to humans. Many of our alternative energy sources (such as wind) are diffuse compared to the higher-quality, more concentrated chemical energy in oil, coal, or gas.

Thermodynamics describes the conservation and degradation of energy

Atoms and molecules cycle endlessly through organisms and their environment, but energy flows in a one-way path. A constant supply of energy—nearly all of it from the sun—is needed to keep biological processes running. Energy can be used repeatedly as it flows through the system, and it can be stored temporarily in the chemical bonds of organic molecules, but eventually it is released and dissipated.

The study of thermodynamics deals with how energy is transferred in natural processes. More specifically, it deals with the rates of flow and the transformation of energy from one form or quality to another. Thermodynamics is a complex, quantitative discipline, but you don't need a great deal of math to understand some of the broad principles that shape our world and our lives.

The **first law of thermodynamics** states that energy is *conserved;* that is, it is neither created nor destroyed under normal conditions. Energy may be transformed, for example, from the energy in a chemical bond to heat energy, but the total amount does not change.

The **second law of thermodynamics** states that, with each successive energy transfer or transformation in a system, less energy is available to do work. That is, energy is degraded to lower-quality forms, or it dissipates and is lost, as it is used. When you drive a car, for example, the chemical energy of the gas is degraded to kinetic energy and heat, which dissipates, eventually, to space. The second law recognizes that disorder, or **entropy**, tends to increase in all natural systems. Consequently, there is always less *useful* energy available when you finish a process than there was before you started. Because of this loss, everything in the universe tends to fall apart, slow down, and get more disorganized.

How does the second law of thermodynamics apply to organisms and biological systems? Organisms are highly organized, both structurally and metabolically. Constant care and maintenance is required to keep up this organization, and a continual supply of energy is required to maintain these processes. Every time some energy is used by a cell to do work, some of that energy is dissipated or lost as heat. If cellular energy supplies are interrupted or depleted, the result—sooner or later—is death.

2.4 Energy for Life

Where does the energy needed by living organisms come from? How is it captured and transferred among organisms? For nearly all life on earth, the sun is the ultimate energy source, and the sun's energy is captured by green plants. Green plants are often called **primary producers** because they create carbohydrates and other compounds using just sunlight, air, and water.

There are organisms that get energy in other ways, and these are interesting because they are exceptions to the normal rule. Deep in the earth's crust, or deep on the ocean floor, and in hot springs such as those in Yellowstone National Park, we can find extremophiles, organisms that gain their energy from **chemosynthesis**, or extracting energy from inorganic chemical compounds such as hydrogen sulfide (H_2S). Until 30 years ago, we knew almost nothing about these organisms and their ecosystems. Recent deep-sea explorations have shown that an abundance of astonishingly varied and abundant life occurs hundreds of meters deep on the ocean floor. These ecosystems cluster around thermal vents. Thermal vents are cracks where boiling-hot water, heated by magma in the earth's crust, escapes from the ocean floor. Here, microorganisms grow by oxidizing hydrogen sulfide; bacteria support an ecosystem that includes blind shrimp, giant tube worms, crabs, clams, and other organisms (fig. 2.12).

These fascinating systems are distinctive because we have discovered them so recently. They are also interesting because of their contrast to the incredible profusion of photosynthesis-based life we enjoy here at the earth's surface.

Green plants get energy from the sun

Our sun is a star, a fiery ball of exploding hydrogen gas. Its thermonuclear reactions emit powerful forms of radiation, including potentially deadly ultraviolet and nuclear radiation (fig. 2.13), yet

Figure 2.12 A colony of tube worms and mussels clusters over a cool, deep-sea methane seep in the Gulf of Mexico.

life here is nurtured by, and dependent upon, this searing energy source. Solar energy is essential to life for two main reasons.

First, the sun provides warmth. Most organisms survive within a relatively narrow temperature range. In fact, each species has its own range of temperatures within which it can function normally. At high temperatures (above 40°C), most biomolecules begin to break down or become distorted and nonfunctional. At low temperatures (near 0°C), some chemical reactions of metabolism occur too slowly to enable organisms to grow and reproduce. Other planets in our solar system are either too hot or too cold to

support life as we know it. The earth's water and atmosphere help to moderate, maintain, and distribute the sun's heat.

Second, nearly all organisms on the earth's surface depend on solar radiation for life-sustaining energy, which is captured by green plants, algae, and some bacteria in a process called **photosynthesis**. Photosynthesis converts radiant energy into useful, high-quality chemical energy in the bonds that hold together organic molecules.

How much of the available solar energy is actually used by organisms? The amount of incoming solar radiation is enormous, about 1,372 watts/m^2 at the top of the atmosphere (1 watt = 1 J per second). However, more than half of the incoming sunlight may be reflected or absorbed by atmospheric clouds, dust, and gases. In particular, harmful, short wavelengths are filtered out by gases (such as ozone) in the upper atmosphere; thus, the atmosphere is a valuable shield, protecting life-forms from harmful doses of ultraviolet and other forms of radiation. Even with these energy reductions, however, the sun provides much more energy than biological systems can harness, and more than enough for all our energy needs if technology could enable us to tap it efficiently.

Of the solar radiation that does reach the earth's surface, about 10 percent is ultraviolet, 45 percent is visible, and 45 percent is infrared. Most of that energy is absorbed by land or water or is reflected into space by water, snow, and land surfaces. (Seen from outer space, Earth shines about as brightly as Venus.)

Of the energy that reaches the earth's surface, photosynthesis uses only certain wavelengths, mainly red and blue light. (Most plants reflect green wavelengths, so that is the color they appear to us.) Half of the energy plants absorb is used in evaporating water. In the end, only 1 to 2 percent of the sunlight falling on plants is available for photosynthesis. This small percentage represents the energy base for virtually all life in the biosphere!

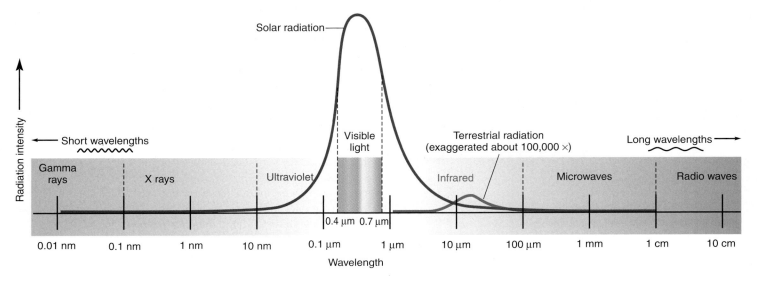

Figure 2.13 The electromagnetic spectrum. Our eyes are sensitive to visible light wavelengths, which make up nearly half the energy that reaches the earth's surface (represented by the area under the "solar radiation" curve). Photosynthesizing plants use the most abundant solar wavelengths (light and infrared). The earth reemits lower-energy, longer wavelengths (shown by the "terrestrial radiation" curve), mainly the infrared part of the spectrum.

How does photosynthesis capture energy?

Photosynthesis occurs in tiny organelles called chloroplasts that reside within plant cells (see fig. 2.10). The most important key to this process is chlorophyll, a unique green molecule that can absorb light energy and use it to create high-energy chemical bonds in compounds that serve as the fuel for all subsequent cellular metabolism. Chlorophyll doesn't do this important job all alone, however. It is assisted by a large group of other lipid, sugar, protein, and nucleotide molecules. Together these components carry out two interconnected cyclic sets of reactions (fig. 2.14).

Photosynthesis begins with a series of steps called light-dependent reactions: These occur only while the chloroplast is receiving light. Enzymes split water molecules and release molecular oxygen (O_2). This is the source of nearly all the oxygen in the atmosphere on which all animals, including you, depend for life. The light-dependent reactions also create mobile, high-energy molecules (adenosine triphosphate, or ATP, and nicotinamide adenine dinucleotide phosphate, or NADPH), which provide energy for the next set of processes, the light-independent reactions. As their name implies, these reactions do not use light directly. Here, enzymes extract energy from ATP and NADPH to add carbon atoms (from carbon dioxide) to simple sugar molecules, such as glucose. These molecules provide the building blocks for larger, more complex organic molecules.

In most temperate-zone plants, photosynthesis can be summarized in the following equation:

$$6H_2O + 6CO_2 + \text{solar energy} \xrightarrow[\text{chlorophyll}]{} C_6H_{12}O_6 \text{ (sugar)} + 6O_2$$

We read this equation as "water plus carbon dioxide plus energy produces sugar plus oxygen." The reason the equation uses six water and six carbon dioxide molecules is that it takes six carbon atoms to make the sugar product. If you look closely, you will see that all the atoms in the reactants balance with those in the products. This is an example of conservation of matter.

You might wonder how making a simple sugar benefits the plant. The answer is that glucose is an energy-rich compound that serves as the central, primary fuel for all metabolic processes of cells. The energy in its chemical bonds—the ones created by photosynthesis—can be released by other enzymes and used to make other molecules (lipids, proteins, nucleic acids, or other carbohydrates), or it can drive kinetic processes such as movement of ions across membranes, transmission of messages, changes in cellular shape or structure, or movement of the cell itself in some cases. This process of releasing chemical energy, called **cellular respiration**, involves splitting carbon and hydrogen atoms from the sugar molecule and recombining them with oxygen to recreate carbon dioxide and water. The net chemical reaction, then, is the reverse of photosynthesis:

$$C_6H_{12}O_6 + 6O_2 \longrightarrow 6H_2O + 6CO_2 + \text{released energy}$$

Note that in photosynthesis, energy is *captured*, while in respiration, energy is *released*. Similarly, photosynthesis *uses* water and carbon dioxide to *produce* sugar and oxygen, while respiration does just the opposite. In both sets of reactions, energy is stored temporarily in chemical bonds, which constitute a kind of energy currency for the cell. Plants carry out both photosynthesis and respiration, but during the day, if light, water, and CO_2 are available, they have a net production of O_2 and carbohydrates.

We animals don't have chlorophyll and can't carry out photosynthetic food production. We do have the components for cellular respiration, however. In fact, this is how we get all our energy for life. We eat plants—or other animals that have eaten plants—and break down the organic molecules in our food through cellular respiration to obtain energy (fig. 2.15). In the process, we also consume oxygen and release carbon dioxide, thus completing the cycle of photosynthesis and respiration. Later in this chapter we will see how these feeding relationships work.

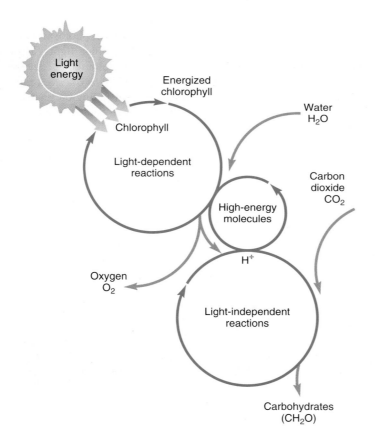

Figure 2.14 Photosynthesis involves a series of reactions in which chlorophyll captures light energy and forms high-energy molecules, ATP and NADPH. Light-independent reactions then use energy from ATP and NADPH to fix carbon (from air) in organic molecules.

2.5 From Species to Ecosystems

While many biologists study life at the cellular and molecular level, ecologists study interactions at the species, population, biotic community, or ecosystem level. In Latin, *species* literally

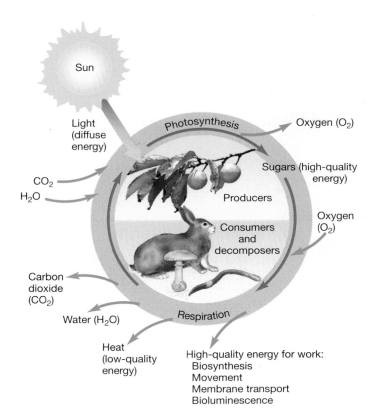

Figure 2.15 Energy exchange in ecosystems. Plants use sunlight, water, and carbon dioxide to produce sugars and other organic molecules. Consumers use oxygen and break down sugars during cellular respiration. Plants also carry out respiration, but during the day, if light, water, and CO_2 are available, they have a net production of O_2 and carbohydrates.

means *kind*. In biology, **species** refers to all organisms of the same kind that are genetically similar enough to breed in nature and produce live, fertile offspring. There are several qualifications and some important exceptions to this definition of species (especially among bacteria and plants), but for our purposes this is a useful working definition.

Organisms occur in populations, communities, and ecosystems

A **population** consists of all the members of a species living in a given area at the same time. Chapter 4 deals further with population growth and dynamics. All of the populations of organisms living and interacting in a particular area make up a **biological community**. What populations make up the biological community of which you are a part? The population sign marking your city limits announces only the number of humans who live there, disregarding the other populations of animals, plants, fungi, and microorganisms that are part of the biological community within the city's boundaries. Characteristics of biological communities are discussed in more detail in chapter 3.

An ecological system, or **ecosystem**, is composed of a biological community and its physical environment. Arcata's constructed wetlands, described in the opening case study for this chapter, is a good example of an ecosystem. Its environment includes abiotic factors (nonliving components), such as climate, water, minerals, and sunlight, as well as biotic factors, such as organisms, their products (secretions, wastes, and remains), and effects in a given area. It is useful to think about the biological community and its environment together, because energy and matter flow through both. Understanding how those flows work is a major theme in ecology.

Food chains, food webs, and trophic levels link species

Photosynthesis (and rarely chemosynthesis) is the base of all ecosystems. Organisms that produce organic material by photosynthesis, mainly green plants and algae, are therefore known as **producers**. One of the most important properties of an ecosystem is its **productivity**, the amount of **biomass** (biological material) produced in a given area during a given period of time. Photosynthesis is described as *primary productivity* because it is the basis for almost all other growth in an ecosystem. Manufacture of biomass by organisms that eat plants is termed *secondary productivity*. A given ecosystem may have very high total productivity, but if decomposers decompose organic material as rapidly as it is formed, the *net primary productivity* will be low. Remote sensing from aircraft or satellites allows us to measure productivity in large systems (see Exploring Science, p. 39).

In ecosystems, some consumers feed on a single species, but most consumers have multiple food sources. Similarly, some species are prey to a single kind of predator, but many species in an ecosystem are beset by several types of predators and parasites. In this way, individual food chains become interconnected to form a **food web**. Figure 2.16 shows feeding relationships among some of the larger organisms in a woodland and lake community. If we were to add all the insects, worms, and microscopic organisms that belong in this picture, however, we would have overwhelming complexity. Perhaps you can imagine the challenge ecologists face in trying to quantify and interpret the precise matter and energy transfers that occur in a natural ecosystem!

An organism's feeding status in an ecosystem can be expressed as its **trophic level** (from the Greek *trophe*, food). In our first example, the corn plant is at the producer level; it transforms solar energy into chemical energy, producing food molecules. Other organisms in the ecosystem are **consumers** of the chemical energy harnessed by the producers. An organism that eats producers is a primary consumer. An organism that eats primary consumers is a secondary consumer, which may, in turn, be eaten by a tertiary consumer, and so on. Most terrestrial food chains are relatively short (seeds ⟶ mouse ⟶ owl), but aquatic food chains may be quite long (microscopic algae ⟶ copepod ⟶ minnow ⟶ crayfish ⟶ bass ⟶ osprey). The length of a food chain also may reflect the physical characteristics of a

Figure 2.16 Each time an organism feeds, it becomes a link in a food chain. In an ecosystem, food chains become interconnected when predators feed on more than one kind of prey, thus forming a food web. The arrows in this diagram indicate the direction in which matter and energy are transferred through feeding relationships.

particular ecosystem. A harsh arctic landscape generally has a much shorter food chain than a temperate or tropical one.

Organisms can be identified both by the trophic level at which they feed and by the *kinds* of food they eat (fig. 2.17). **Herbivores** are plant eaters, **carnivores** are flesh eaters, and **omnivores** eat both plant and animal matter. What are humans? We are natural omnivores, by history and by habit.

One of the most important trophic levels is occupied by the many kinds of organisms that remove and recycle the dead bodies and waste products of others. **Scavengers** such as crows, jackals, and vultures clean up dead carcasses of larger animals. **Detritivores** such as ants and beetles consume litter, debris, and dung, while **decomposer** organisms such as fungi and bacteria complete the final breakdown and recycling of organic materials. It could be argued that these microorganisms are second in importance only to producers, because without their activity nutrients would remain locked up in the organic compounds of dead organisms and discarded body wastes, rather than being made available to successive generations of organisms.

Ecological pyramids describe trophic levels

If we arrange the organisms in a food chain according to trophic levels, they often form a pyramid with a broad base representing primary producers and only a few individuals in the highest trophic levels. This pyramid arrangement is especially true if we look at the energy content of an ecosystem (fig. 2.18). True to the second principle of thermodynamics, less food energy is available to the top trophic level than is available to preceding levels. For

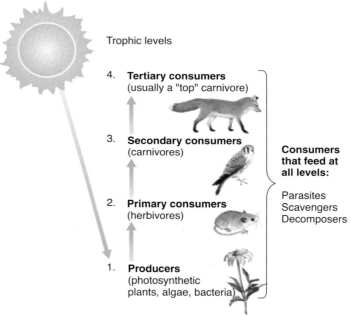

Trophic levels

4. **Tertiary consumers**
(usually a "top" carnivore)

3. **Secondary consumers**
(carnivores)

2. **Primary consumers**
(herbivores)

1. **Producers**
(photosynthetic plants, algae, bacteria)

Consumers that feed at all levels:

Parasites
Scavengers
Decomposers

Figure 2.17 Trophic levels describe an organism's position in a food chain. Some consumers feed on all trophic levels.

example, it takes a huge number of plants to support a modest colony of grazers such as prairie dogs. Several colonies of prairie dogs, in turn, might be required to feed a single coyote. And a very large top carnivore like a tiger may need a home range of hundreds of square kilometers to survive.

Exploring
SCIENCE:
Remote Sensing, Photosynthesis, and Material Cycles

Measuring primary productivity is important for understanding individual plants and local environments. Understanding the rates of primary productivity is also key to understanding global processes, such as material cycling, and biological activity:

- In global carbon cycles, how much carbon is stored by plants, how quickly is it stored, and how does carbon storage compare in contrasting environments, such as the Arctic and the tropics?
- How does this carbon storage affect global climates (chapter 9)?
- In global nutrient cycles, how much nitrogen and phosphorus wash offshore, and where?

How can environmental scientists measure primary production (photosynthesis) at a global scale? In a small, relatively closed ecosystem, such as a pond, ecologists can collect and analyze samples of all trophic levels. But that method is impossible for large ecosystems, especially for oceans, which cover 70 percent of the earth's surface. One of the newest methods of quantifying biological productivity involves remote sensing, or using data collected from satellite sensors that observe the energy reflected from the earth's surface.

As you have read in this chapter, chlorophyll in green plants *absorbs* red and blue wavelengths of light and *reflects* green wavelengths. Your eye receives, or senses, these green wavelengths. A white-sand beach, on the other hand, reflects approximately equal amounts of all light wavelengths that reach it from the sun, so it looks white (and bright!) to your eye. In a similar way, different surfaces of the earth reflect characteristic wavelengths. Snow-covered surfaces reflect light wavelengths; dark-green forests with abundant chlorophyll-rich leaves—and ocean surfaces rich in photosynthetic algae and plants—reflect greens and near-infrared wavelengths. Dry, brown forests with little active

chlorophyll reflect more red and less infrared energy than do dark-green forests (fig. 1).

To detect land cover patterns on the earth's surface, we can put a sensor on a satellite that orbits the earth. As the satellite travels, the sensor receives and transmits to earth a series of "snapshots." One of the best known earth-imaging satellites, *Landsat 7*, produces images that cover an area 185 km (115 mi) wide, and each pixel represents an area of just 30 × 30 m on the ground. *Landsat* orbits approximately from pole to pole, so as the earth spins below the satellite, it captures images of the entire surface every 16 days. Another satellite, *SeaWIFS*, was designed mainly for monitoring biological activity in oceans (fig. 2). *SeaWIFS* follows a path similar to *Landsat*'s but it revisits each point on the earth every day and produces images with a pixel resolution of just over 1 km.

Since satellites detect a much greater range of wavelengths than our eyes can, they are able to monitor and map chlorophyll abundance. In

oceans, this is a useful measure of ecosystem health, as well as carbon dioxide uptake. By quantifying and mapping primary production in oceans, climatologists are working to estimate the role of ocean ecosystems in moderating climate change: for example, they can estimate the extent of biomass production in the cold, oxygen-rich waters of the North Atlantic (fig. 2). Oceanographers can also detect near-shore areas where nutrients washing off the land surface fertilize marine ecosystems and stimulate high productivity, such as near the mouth of the Amazon or Mississippi River. Monitoring and mapping these patterns helps us estimate human impacts on nutrient flows from land to sea.

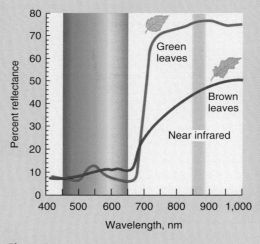

Figure 1 Energy wavelengths reflected by green and brown leaves.

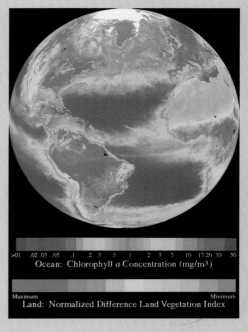

Figure 2 *SeaWiFS* image showing chlorophyll abundance in oceans and plant growth on land (normalized difference vegetation index).

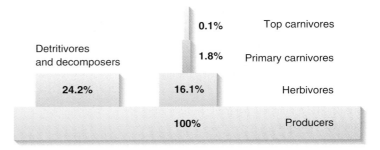

Figure 2.18 An energy pyramid. Numbers show the percentage of energy remaining at each level. Detritivores and decomposers feed at every level but are shown attached to the producer bar because this level provides most of their energy.

Source: Data from Howard T. Odum, "Trophic Structure and Productivity of Silver Springs, Florida" in *Ecological Monographs*, 27:55–112, 1957, Ecological Society of America.

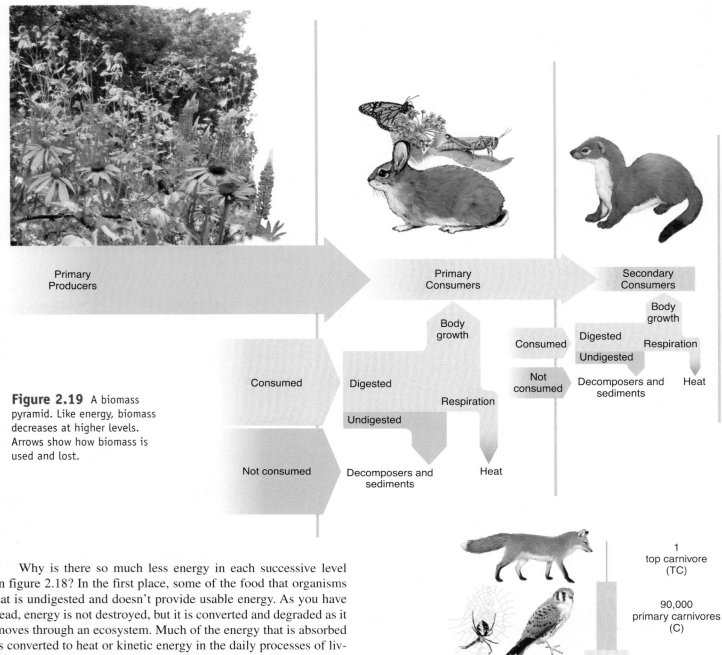

Figure 2.19 A biomass pyramid. Like energy, biomass decreases at higher levels. Arrows show how biomass is used and lost.

Primary Producers

Primary Consumers

Secondary Consumers

Consumed

Digested

Body growth

Respiration

Undigested

Not consumed

Decomposers and sediments

Heat

Consumed

Digested

Body growth

Respiration

Undigested

Not consumed

Decomposers and sediments

Heat

Why is there so much less energy in each successive level in figure 2.18? In the first place, some of the food that organisms eat is undigested and doesn't provide usable energy. As you have read, energy is not destroyed, but it is converted and degraded as it moves through an ecosystem. Much of the energy that is absorbed is converted to heat or kinetic energy in the daily processes of living, and thus isn't stored as biomass that can be eaten.

Furthermore, predators don't operate at 100 percent efficiency. If there were enough foxes to catch all the rabbits available in the summer when the supply is abundant, there would be too many foxes in the middle of the winter when rabbits are scarce. A general rule of thumb is that only about 10 percent of the energy in one consumer level is represented in the next higher level (fig. 2.19). The amount of energy available is often expressed in biomass. For example, it generally takes about 100 kg of clover to make 10 kg of rabbit and 10 kg of rabbit to make 1 kg of fox.

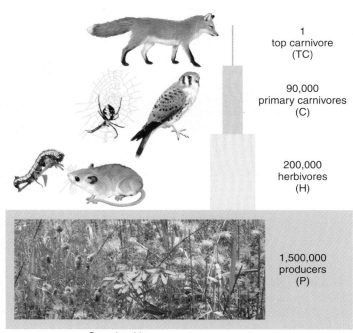

1
top carnivore
(TC)

90,000
primary carnivores
(C)

200,000
herbivores
(H)

1,500,000
producers
(P)

Grassland in summer

Figure 2.20 A numbers pyramid. Millions of producers may support just one top carnivore.

The total number of organisms and the total amount of biomass in each successive trophic level of an ecosystem also may form pyramids (fig. 2.20) similar to those describing energy content. Top predators, such as tigers, grizzly bears, and great white sharks, have always been rare because they require huge territories. It takes many individuals at lower trophic levels to support one of these large carnivores. Furthermore, the characteristics that make them superb predators also make them a threat to us. We have eliminated many of the large fierce animals in the world either directly or by removing their prey and habitat.

The relationship between biomass and numbers is not as dependable as energy, however. The biomass pyramid, for instance, can be inverted by periodic fluctuations in producer populations (for example, low plant and algal biomass present during winter in temperate aquatic ecosystems). The numbers pyramid also can be inverted. One coyote can support numerous tapeworms, for example, or one large tree can support thousands of caterpillars.

2.6 Biogeochemical Cycles and Life Processes

The elements and compounds that sustain us are cycled endlessly through living things and through the environment. As the great naturalist John Muir said, "When one tugs at a single thing in nature, he finds it attached to the rest of the world." On a global scale, this movement is referred to as biogeochemical cycling. Substances can move quickly or slowly: you might store carbon for hours or days, while carbon is stored in the earth for millions of years. When human activity alters flow rates or storage times in these natural cycles, overwhelming the environment's ability to process them, these materials can become pollutants. Sulfur, nitrogen, carbon dioxide, and phosphorus are some of the more serious examples of these. Global-scale movement of energy is discussed in detail in chapter 9. Here we will explore some of the paths involved in cycling several important elements: water, carbon, nitrogen, sulfur, and phosphorus.

The hydrologic cycle

The path of water through our environment is perhaps the most familiar material cycle, and it is discussed in greater detail in chapter 10 (fig. 2.21). Most of the earth's water is stored in the oceans, but solar energy continually evaporates this water, and winds

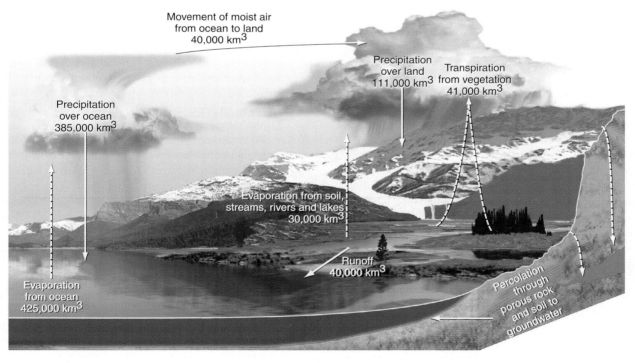

Figure 2.21 The hydrologic cycle. Most exchange occurs with evaporation from oceans and precipitation back to oceans. About one-tenth of water evaporated from oceans falls over land, is recycled through terrestrial systems, and eventually drains back to oceans in rivers.

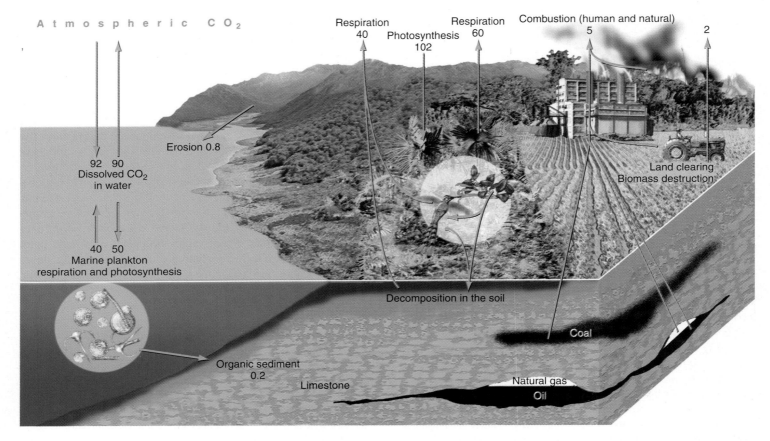

Figure 2.22 The carbon cycle. Numbers indicate approximate exchange of carbon in gigatons (Gt) per year. Natural exchanges are balanced, but human sources produce a net increase of CO_2 in the atmosphere.

distribute water vapor around the globe. Water that condenses over land surfaces, in the form of rain, snow, or fog, supports all terrestrial (land-based) ecosystems. Living organisms emit the moisture they have consumed through respiration and perspiration. Eventually this moisture reenters the atmosphere or enters lakes and streams, from which it ultimately returns to the ocean again.

As it moves through living things and through the atmosphere, water is responsible for metabolic processes within cells, for maintaining the flows of key nutrients through ecosystems, and for global-scale distribution of heat and energy (chapter 9). Water performs countless services because of its unusual properties (see Exploring Science, p. 31). Water is so important that, when astronomers look for signs of life on distant planets, traces of water are the key evidence they seek.

The carbon cycle

Carbon serves a dual purpose for organisms: (1) it is a structural component of organic molecules, and (2) chemical bonds in carbon compounds provide metabolic energy. The **carbon cycle** begins with photosynthetic organisms taking up carbon dioxide (CO_2) (fig. 2.22). This is called carbon-fixation because carbon is changed from gaseous CO_2 to less mobile organic molecules. Once a carbon atom is incorporated into organic compounds, its path to recycling may be very quick or extremely slow. Imagine

for a moment what happens to a simple sugar molecule you swallow in a glass of fruit juice. The sugar molecule is absorbed into your bloodstream, where it is made available to your cells for cellular respiration or the production of more complex biomolecules. If it is used in respiration, you may exhale the same carbon atom as CO_2 in an hour or less, and a plant could take up that exhaled CO_2 the same afternoon.

Alternatively, your body may use that sugar molecule to make larger organic molecules that become part of your cellular structure. The carbon atoms in the sugar molecule could remain a part of your body until it decays after death. Similarly, carbon in the wood of a thousand-year-old tree will be released only when fungi and bacteria digest the wood and release carbon dioxide as a by-product of their respiration.

Sometimes recycling takes a very long time. Coal and oil are the compressed, chemically altered remains of plants and microorganisms that lived millions of years ago. Their carbon atoms (and hydrogen, oxygen, nitrogen, sulfur, etc.) are not released until the coal and oil are burned. Enormous amounts of carbon also are locked up as calcium carbonate ($CaCO_3$), used to build shells and skeletons of marine organisms from tiny protozoans to corals. The world's extensive surface limestone deposits are biologically formed calcium carbonate from ancient oceans, exposed by geological events. The carbon in limestone has been locked away for millennia, which is probably the fate of carbon currently being deposited in ocean sediments. Eventually, even the deep ocean deposits are recycled as they are drawn into deep molten layers and released via volcanic activity. Geologists estimate that every carbon atom on the earth has made about 30 such round trips over the past 4 billion years.

Materials that store carbon, including geologic formations and standing forests, are known as carbon sinks. When carbon is released from these sinks, as when we burn fossil fuels and inject CO_2 into the atmosphere, or when we clear extensive forests, natural recycling systems may not be able to keep up. This is the root of the global warming problem, discussed in chapter 9. Alternatively, extra atmospheric CO_2 could support faster plant growth, speeding some of the recycling processes.

The nitrogen cycle

Organisms cannot exist without amino acids, peptides, and proteins, all of which are organic molecules that contain nitrogen. Nitrogen is therefore an extremely important nutrient for living things. (Nitrogen is a primary component of many household and agricultural fertilizers.) Even though nitrogen makes up about 78 percent of the air around us, plants cannot use N_2, the stable diatomic (two-atom) molecule in the air.

Plants acquire nitrogen through an extremely complex **nitrogen cycle** (fig. 2.23). The key to this cycle is nitrogen-fixing bacteria (including some blue-green algae or cyanobacteria). These organisms have a highly specialized ability to "fix" nitrogen, or combine gaseous N_2 with hydrogen to make ammonia (NH_3).

Other bacteria then combine ammonia with oxygen to form nitrites (NO_2^-). Another group of bacteria converts nitrites to nitrates (NO_3^-), which green plants can absorb and use. After plant cells absorb nitrates, the nitrates are reduced to ammonium (NH_4^+), which cells use to build amino acids that become the building blocks for peptides and proteins.

Members of the bean family (legumes) and a few other kinds of plants are especially useful in agriculture because nitrogen-fixing bacteria actually live in their root tissues (fig. 2.24). Legumes and their associated bacteria add nitrogen to the soil, so interplanting and rotating legumes with crops, such as corn, that use but cannot replace soil nitrates are beneficial farming practices that take practical advantage of this relationship.

Nitrogen reenters the environment in several ways. The most obvious path is through the death of organisms. Fungi and bacteria decompose dead organisms, releasing ammonia and ammonium ions, which then are available for nitrate formation. Organisms don't have to die to donate proteins to the environment, however. Plants shed their leaves, needles, flowers, fruits, and cones; animals shed hair, feathers, skin, exoskeletons, pupal cases, and silk. Animals also produce excrement and urinary wastes that contain nitrogenous compounds. Urine is especially high in nitrogen

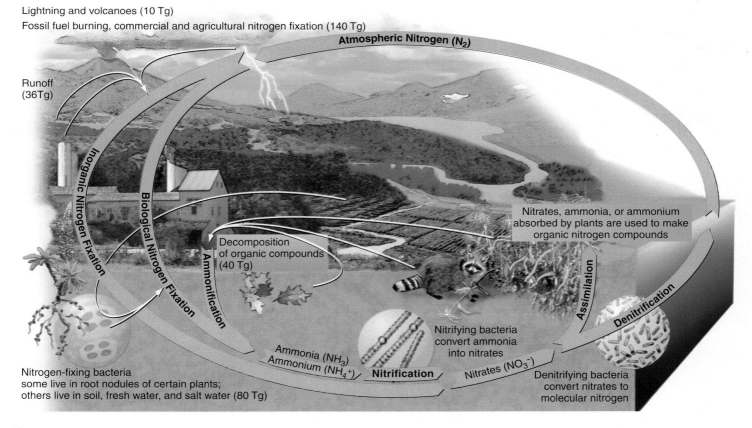

Lightning and volcanoes (10 Tg)
Fossil fuel burning, commercial and agricultural nitrogen fixation (140 Tg)
Atmospheric Nitrogen (N_2)
Runoff (36 Tg)
Inorganic Nitrogen Fixation
Biological Nitrogen Fixation
Ammonification
Decomposition of organic compounds (40 Tg)
Nitrates, ammonia, or ammonium absorbed by plants are used to make organic nitrogen compounds
Assimilation
Denitrification
Ammonia (NH_3) Ammonium (NH_4^+)
Nitrifying bacteria convert ammonia into nitrates
Nitrification
Nitrates (NO_3^-)
Nitrogen-fixing bacteria some live in root nodules of certain plants; others live in soil, fresh water, and salt water (80 Tg)
Denitrifying bacteria convert nitrates to molecular nitrogen

Figure 2.23 The nitrogen cycle. Human sources of nitrogen fixation (conversion of molecular nitrogen to ammonia or ammonium) are now about 50 percent greater than natural sources. Bacteria convert ammonia to nitrates, which plants use to create organic nitrogen. Eventually, nitrogen is stored in sediments or converted back to molecular nitrogen (1 Tg = 10^{12} g).

because it contains the detoxified wastes of protein metabolism. All of these by-products of living organisms decompose, replenishing soil fertility (see related Case Study, "Why Trees Need Salmon," at www.mhhe.com/cunningham5e).

How does nitrogen reenter the atmosphere, completing the cycle? Denitrifying bacteria break down nitrates into N_2 and nitrous oxide (N_2O), gases that return to the atmosphere; thus, denitrifying bacteria compete with plant roots for available nitrates. However, denitrification occurs mainly in waterlogged soils that have low oxygen availability and a large amount of decomposable organic matter. These are suitable growing conditions for many wild plant species in swamps and marshes, but not for most cultivated crop species, except for rice, a domesticated wetland grass.

In recent years, humans have profoundly altered the nitrogen cycle. By using synthetic fertilizers, cultivating nitrogen-fixing crops, and burning fossil fuels, we now convert more nitrogen to ammonia and nitrates than all natural land processes combined. This excess nitrogen input causes algal blooms and excess plant growth in water bodies, called eutrophication, which we will discuss in more detail in chapter 10. Excess nitrogen also causes serious loss of soil nutrients such as calcium and potassium; acidification of rivers and lakes; and rising atmospheric concentrations of nitrous oxide, a greenhouse gas. It also encourages the spread of weeds into areas such as prairies, where native plants are adapted to nitrogen-poor environments.

Figure 2.24 Nitrogen molecules (N_2) are converted to useable forms in the bumps (nodules) on the roots of this bean plant. Each nodule is a mass of root tissue containing many bacteria that help convert nitrogen in the soil to a form that the bean plant can assimilate and use to manufacture amino acids.

The phosphorus cycle

Minerals become available to organisms after they are released from rocks. Two mineral cycles of particular significance to organisms are phosphorus and sulfur. Why do you suppose phosphorus is a primary ingredient in fertilizers? At the cellular level, energy-rich phosphorus-containing compounds are primary participants in energy-transfer reactions, as discussed earlier.

The amount of available phosphorus in an environment, therefore, can dramatically affect productivity. Abundant phosphorus stimulates lush plant and algal growth, making phosphorus a major contributor to water pollution.

The phosphorus cycle begins when phosphorus compounds leach from rocks and minerals over long periods of time (fig. 2.25). Because phosphorous has no atmospheric form, it is usually transported in water. Producer organisms take in inorganic phosphorus, incorporate it into organic molecules, and then pass it on to consumers. Phosphorus returns to the environment by decomposition. An important aspect of the phosphorus cycle is the very long time it takes for phosphorus atoms to pass through it. Deep ocean sediments are significant phosphorus sinks of extreme longevity. Phosphate ores that now are mined to make detergents and inorganic fertilizers represent exposed ocean sediments that are millennia old.

Figure 2.25 The phosphorus cycle. Natural movement of phosphorus is slight, involving recycling within ecosystems and some erosion and sedimentation of phosphorus-bearing rock. Use of phosphate (PO_4^{-3}) fertilizers and cleaning agents increases phosphorus in aquatic systems, causing eutrophication. Units are teragrams (Tg) phosphorus per year.

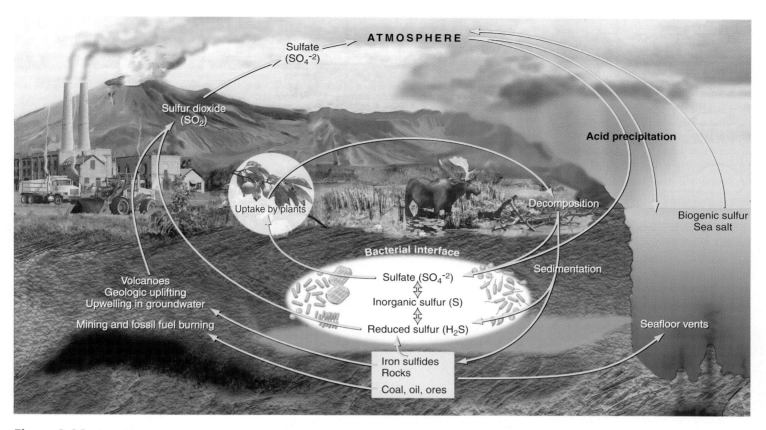

Figure 2.26 The sulfur cycle. Sulfur is present mainly in rocks, soil, and water. It cycles through ecosystems when it is taken in by organisms. Combustion of fossil fuels causes increased levels of atmospheric sulfur compounds, which create problems related to acid precipitation.

You could think of our present use of phosphates, which are washed out into the river systems and eventually the oceans, as an accelerated mobilization of phosphorus from source to sink. Aquatic ecosystems often are dramatically affected in the process because excess phosphates can stimulate explosive growth of algae and photosynthetic bacteria populations (algae blooms), upsetting ecosystem stability (see related Case Study, "The Environmental Chemistry of Phosphorus," at www.mhhe.com/cunningham5e). Can you think of ways we could reduce the amount of phosphorus we put into our environment?

The sulfur cycle

Sulfur plays a vital role in organisms, especially as a minor but essential component of proteins. Sulfur compounds are important determinants of the acidity of rainfall, surface water, and soil. In addition, sulfur in particles and tiny airborne droplets may act as critical regulators of global climate. Most of the earth's sulfur is tied up underground in rocks and minerals, such as iron disulfide (pyrite) and calcium sulfate (gypsum). Weathering, emissions from deep seafloor vents, and volcanic eruptions release this inorganic sulfur into the air and water (fig. 2.26).

The sulfur cycle is complicated by the large number of oxidation states the element can assume, producing hydrogen sulfide (H_2S), sulfur dioxide (SO_2), sulfate ion (SO_4^{2-}), and others. Inor-

ganic processes are responsible for many of these transformations, but living organisms, especially bacteria, also sequester sulfur in biogenic deposits or release it into the environment. Which of the several kinds of sulfur bacteria prevails in any given situation depends on oxygen concentrations, pH level, and light level.

Human activities also release large quantities of sulfur, primarily through burning fossil fuels. Total yearly anthropogenic sulfur emissions rival those of natural processes, and acid rain (caused by sulfuric acid produced as a result of fossil fuel use) is a serious problem in many areas (see chapter 9). Sulfur dioxide and sulfate aerosols cause human health problems, damage buildings and vegetation, and reduce visibility. They also absorb ultraviolet (UV) radiation and create cloud cover that cools cities and may be offsetting greenhouse effects of rising CO_2 concentrations.

Interestingly, the biogenic sulfur emissions of oceanic phytoplankton may play a role in global climate regulation. When ocean water is warm, tiny, single-celled organisms release dimethylsulfide (DMS), which is oxidized to SO_2 and then SO_4^{2-} in the atmosphere. Acting as cloud droplet condensation nuclei, these sulfate aerosols increase the earth's albedo (reflectivity) and cool the earth. As ocean temperatures drop because less sunlight gets through, phytoplankton activity decreases, DMS production falls, and clouds disappear. Thus, DMS, which may account for half of all biogenic sulfur emissions, could be a feedback mechanism that keeps temperature within a suitable range for all life.

Conclusion

The movement of matter and energy through systems, such as the Arcata wetland, maintains the world's living environments. Matter consists of atoms, which make up molecules or compounds. Among the principal substances we consider in ecosystems are water, carbon, nitrogen, phosphorus, and sulfur. Nitrogen and phosphorus, especially, are key nutrients for living things. Energy also moves through systems. The laws of thermodynamics tell us that energy is neither created nor destroyed (first law), but it is degraded and dissipates as it moves through ecosystems (second law). For example, the chemical energy in food molecules is concentrated potential energy that degrades to less concentrated forms such as heat or kinetic energy as we use it. Matter, similarly, is neither created nor destroyed; it is reused continually.

Primary producers provide the energy and matter in an ecosystem. Nearly all ecosystems rely on green plants, which pho-tosynthesize and create organic compounds that store energy, nutrients, and carbon. These processes are what the city of Arcata employed to take up nutrients produced in sewage effluent. Excessive amounts of nutrients can create a positive feedback in plant or algae reproduction and growth, and positive feedbacks can destabilize a system. Negative feedbacks tend to maintain system stability. Cellular respiration is the reverse of photosynthesis: this is how organisms extract energy and nutrients from organic molecules.

Primary producers support smaller numbers of consumers in an ecosystem. Thus, Arcata's wetland supports hundreds of bird, fish, and insect species. Top level predators are normally rare because large numbers of organisms at lower trophic levels are needed to support them. We can think about this pyramid structure of trophic levels in terms of energy, biomass, or numbers of individuals. We can also understand these organisms as components of a system, through which carbon, water, and nutrients move.

Practice Quiz

1. What two problems did Arcata, California, solve with its constructed wetland?
2. What are systems and how do feedback loops regulate them?
3. Your body contains vast numbers of carbon atoms. How is it possible that some of these carbons may have been part of the body of a prehistoric creature?
4. List six unique properties of water. Describe, briefly, how each of these properties makes water essential to life as we know it.
5. What is DNA, and why is it important?
6. The oceans store a vast amount of heat, but this huge reservoir of energy is of little use to humans. Explain the difference between high-quality and low-quality energy.
7. In the biosphere, matter follows circular pathways, while energy flows in a linear fashion. Explain.
8. Which wavelengths do our eyes respond to, and why? (Refer to fig. 2.13.) About how long are short ultraviolet wavelengths to microwave lengths?
9. Where do extremophiles live? How do they get the energy they need for survival?
10. Ecosystems require energy to function. From where does this energy come? Where does it go?
11. How do green plants capture energy, and what do they do with it?
12. Define the terms *species*, *population*, and *biological community*.
13. Why are big fierce animals rare?
14. Most ecosystems can be visualized as a pyramid with many organisms in the lowest trophic levels and only a few individuals at the top. Give an example of an inverted numbers pyramid.
15. What is the ratio of human-caused carbon releases into the atmosphere shown in figure 2.22 compared to the amount released by terrestrial respiration?

Critical Thinking and Discussion Questions

Apply the principles you have learned in this chapter to discuss these questions with other students.

1. Ecosystems are often defined as a matter of convenience because we can't study everything at once. How would you describe the characteristics and boundaries of the ecosystem in which you live? In what respects is your ecosystem an open one?
2. Think of some practical examples of increasing entropy in everyday life. Is a messy room really evidence of thermodynamics at work, or merely personal preference?
3. Some chemical bonds are weak and have a very short half-life (fractions of a second, in some cases); others are strong and stable, lasting for years or even centuries. What would our world be like if all chemical bonds were either very weak or extremely strong?
4. If you had to design a research project to evaluate the relative biomass of producers and consumers in an ecosystem, what would you measure? (*Note:* This could be a natural system or a human-made one.)
5. Understanding storage compartments is essential to understanding material cycles, such as the carbon cycle. If you look around your backyard, how many carbon storage compartments are there? Which ones are the biggest? Which ones are the longest lasting?

As you have read, movements of nitrogen and phosphorus are among the most important considerations in many wetland systems, because high levels of these nutrients can cause excessive algae and bacteria growth. This is a topic of great interest, and many excellent scientific studies have been done to examine how nutrients move in a wetland, and in other ecosystems. Taking a little time to examine these nutrient cycles in detail will draw on your knowledge of atoms, compounds, systems, cycles, and other ideas in this chapter. Understanding nutrient cycling will also help you in later chapters of this book.

One excellent overview was produced by the Environmental Protection Agency. Go to this website and download a PDF document of the study: http://www.epa.gov/waterscience/criteria/nutrient/guidance/wetlands/index.html. If you prefer, you can also look at just one chapter at a time on this website.

Find chapter 2, An Overview of Wetland Science, and answer the following questions:

1. Look first at the page numbered 15 (section 2.1). How many nutrients are discussed in this chapter? Why are iron (Fe), aluminum (Al), and calcium of interest in understanding phosphorus?

2. Now look at the next page, with the photograph. What are the sources from which nutrients enter a wetland? Think of at least three ways in which human activities can increase these sources.

3. Proceed to page 27, figure 2.5. Study the online figure and fill in the boxes on figure 1 on this page. How many differ-

ent forms of nitrogen, or compounds containing nitrogen, are there? List them.

4. What is meant by "organic N"? (Think about what the term *organic* means in this chapter.) In what form does the plant take up N? How does N return from a living plant to the floor of the wetland?

5. Two important processes in the nitrogen cycle are nitrification (making nitrate, NO_3) and denitrification (breaking up nitrate). One of these processes *increases* the number of O atoms per N; the other *decreases* the number of O atoms per N. Which one decreases the number of O atoms?

6. If you look at the figure 2.6, page 29, you will see that phosphorus compounds are described mainly in terms of whether the P atoms are in an organic (carbon-based) molecule. Which form does the plant take up: dissolved *organic* phosphorus (DOP), or dissolved *inorganic* phosphorus (DIP)? Is a phosphate molecule (PO_4) organic or inorganic? Should it be taken up easily by plants?

This EPA document contains a great deal of additional information. Peruse it briefly to see what other questions scientists have in studying wetland systems.

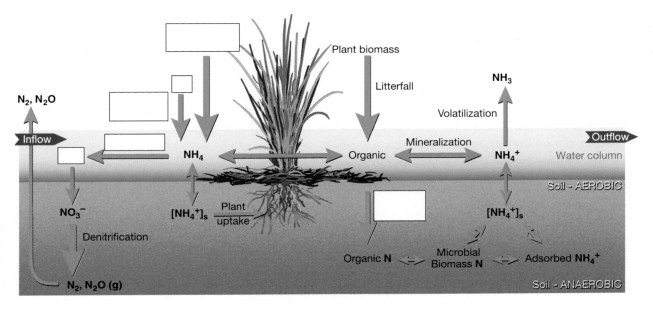

Figure 1 A detailed schematic diagram of the nitrogen cycle in a wetland. Study the online original to fill in the boxes.
Source: EPA Nutrient Criteria Technical Guidance Manual www.epa.gov/waterscience/criteria/nutrient/guidance/.

Students labor for science and future generations, weeding experimental grassland plots at Cedar Creek Natural History Area in Minnesota.

 **3** # Evolution, Species Interactions, and Biological Communities

Learning Outcomes

After studying this chapter, you should be able to answer the following questions:

- How does species diversity arise?
- Why do species live in different locations?
- How do interactions among species affect their fates and that of communities?
- If a species has unlimited growth potential, why doesn't it fill the earth?
- What special properties does a community of species have, and why are they important?
- What is the relationship between species diversity and community stability?
- What is disturbance, and how does it affect communities?

When I view all beings not as special creations, but as lineal descendents of some few beings which have lived long before the first bed of the Cambrian system was deposited, they seem to me to become ennobled.

–Charles Darwin

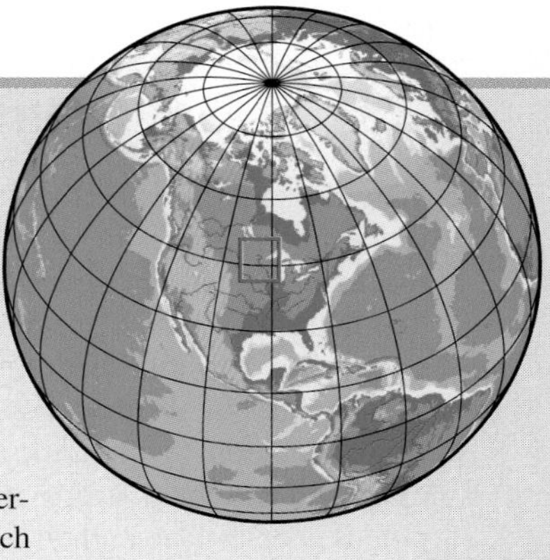

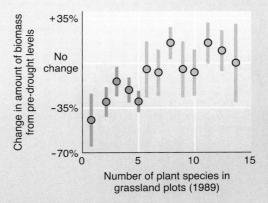

Species Diversity Promotes Community Recovery

Does diversity matter? Ecologists long believed that a diversity of species in a community is not just beautiful and fascinating, but also functionally important. If disturbed, do communities recover quickly when they contain many species? Are they more stable, in other words? This question, raised by ecologists in the 1950s, suggests a strong reason to preserve the earth's biodiversity, but not until 1994 did someone present data that helped answer it. In a long-term study of prairie plant communities, ecologist Dave Tilman and his colleagues planted dozens of experimental plots at Cedar Creek Natural History Area, each plot containing different numbers of species (fig. 3.1). In 1988, central Minnesota experienced its hottest, driest summer in 50 years. As Tilman's team watched their plots recover from the drought, they measured growth in the plots by carefully clipping, counting, and weighing every plant. By 1992, it was clear that plots with five or fewer species were recovering slowly, while higher-diversity plots had reached or exceeded their pre-drought productivity.

Because the question of diversity and recovery was controversial, the strength of Tilman's results depended on duplicated plots (replicates) and a long period of study. If he had used only one plot for each level of species diversity, chance alone might have depressed productivity. Another question in any experiment is how well experimental results represent other situations and other samples. Tilman's team approximated an answer to this question by reporting a range of values around the average recovery rates, in this case the "standard error." This standard statistical measure and its related value, confidence interval, show the range within which nearly all means should fall if someone else did the same experiment. In this way we describe how confident we are that the results can be applied to other situations.

Later experiments at Cedar Creek showed that species-rich plots recovered more fully after drought because some species were severely harmed by drought, others went dormant but didn't die, and still others continued growing but at a slow rate. When rain returned, surviving plants revived filling the plots with green leaves and stems. The chance that a species-poor plot contained quickly recovering species was lower than for a species-rich plot. Consequently, low-diversity plots were more likely to attain their former productivity slowly, if at all. This explanation was bolstered by another experimental discovery that nitrogen—a limiting factor in the sterile, sandy soils of the plots—was used up more completely in species-rich plots than in species-poor plots (fig. 3.2).

The experimental data from Cedar Creek prairies stimulated tests in other ecosystems. Is resilience after disturbance always evident in species-rich plant communities, such as rainforests of Central and South America? (Preliminary results show that it depends on how big and severe the disturbance is.) Do species-poor plant communities, such as dune grasslands and salt marshes, recover only slowly following disturbance? Actually, some simple plant communities appear quite stable where disturbances like wind, waves, and storms are frequent.

What are the implications of Tilman's research for agricultural fields populated by a single crop plant? One important application may be in biofuels production. Ethanol (a biomass fuel) made from starch or cellulose may be an attractive alternative to petroleum-based fuels (chapter 12). Tilman's study suggests that, given variable weather over several years, a diverse biomass crop can achieve an economical net energy yield while providing greater environmental benefits than does a single-species crop. The general lesson is that keeping a diversity of species as a backdrop to human existence may serve us best in the long run; and if species diversity is maintained, nature may be more resilient than we imagine.

Figure 3.1 Experimental grassland plots at Cedar Creek.

Figure 3.2 By 1992, plots with five or fewer species were still below pre-drought productivity, while higher-diversity plots had recovered. (Bars show standard error, or variability, around the means.)
Source: Tilman & Dowling (1994) *Nature* 367:363–365.

3.1 Evolution Leads to Diversity

Why do some species live in one place but not another? A more important question to environmental scientists is, what are the mechanisms that promote the great variety of species on earth and that determine which species will survive in one environment but not another? In this section you will come to understand (1) concepts behind the theory of speciation by means of natural selection and adaptation (evolution); (2) the characteristics of species that make some of them weedy and others endangered; and (3) the limitations species face in their environments and implications for their survival. First we'll start with the basics: How do species arise?

Natural selection and adaptation modify species

How does a polar bear stand the long, sunless, super-cold arctic winter? How does the saguaro cactus survive blistering temperatures and extreme dryness of the desert? We commonly say that each species is *adapted* to the environment where it lives, but what does that mean? **Adaptation**, the acquisition of traits that allow a species to survive in its environment, is one of the most important concepts in biology.

We use the term *adapt* in two ways. An individual organism can respond immediately to a changing environment in a process called acclimation. If you keep a houseplant indoors all winter and then put it out in full sunlight in the spring, the leaves become damaged. If the damage isn't severe, your plant may grow new leaves with thicker cuticles and denser pigments that block the sun's rays. However, the change isn't permanent. After another winter inside, it will still get sun-scald in the following spring. The leaf changes are not permanent and cannot be passed on to offspring, or even carried over from the previous year. Although the capacity to acclimate is inherited, houseplants in each generation must develop their own protective leaf epidermis.

Another type of adaptation affects populations consisting of many individuals. Genetic traits are passed from generation to generation and allow a species to live more successfully in its environment (fig. 3.3). This process of adaptation to environment is explained by the theory of evolution. It was simultaneously proposed by Charles Darwin (1809–1882) based on his studies in the Galápagos Islands and elsewhere, and by Alfred Russel Wallace (1823–1913), who investigated the species of the Malay Archipelago. The basic idea of evolution is that species change over generations because individuals compete for scarce resources. Better competitors in a population survive—they have greater reproductive potential or fitness—and their offspring inherit the beneficial traits. Over generations, those traits become common in a population. The process of better-selected individuals passing their traits to the next generation is called **natural selection**. The traits are encoded in a species' DNA, but from where does the original DNA coding come, which then gives some individuals greater fitness? Every organism has a dizzying array of genetic diversity in its DNA. It has been demonstrated in experiments and by observing natural populations that changes to the DNA coding sequence of individuals occurs, and that the changed sequences are inherited by offspring. Exposure to ionizing radiation and toxic materials, and random recombination and mistakes in replication of DNA strands during reproduction are the main causes of genetic mutations. Sometimes a single mutation has a large effect, but evolutionary change is mostly brought about by many mutations accumulating over time. Only mutations in reproductive cells (gametes) matter; body cell changes—cancers, for example—are not inherited. Most mutations have no effect on fitness, and many actually have a negative effect. During the course of a species' life span—a million or more years—some mutations are thought to have given those individuals an advantage under the **selection pressures** of their environment at that time. The result is a species population that differs from those of numerous preceding generations.

All species live within limits

Environmental factors exert selection pressure and influence the fitness of individuals and their offspring. For this reason, species are limited in where they can live. Limitations include the following: (1) physiological stress due to inappropriate levels of some critical environmental factor, such as moisture, light, temperature, pH, or specific nutrients; (2) competition with other species; (3) predation, including parasitism and disease; and (4) luck. In some cases, the individuals of a population that survive environmental catastrophes or find their way to a new habitat, where they start a new population, may simply be lucky rather than more fit than their contemporaries.

An organism's physiology and behavior allow it to survive only in certain environments. Temperature, moisture level, nutrient supply, soil and water chemistry, living space, and other environmental factors must be at appropriate levels for organisms to persist. In 1840, the chemist Justus von Liebig proposed that the single factor in shortest supply relative to demand is the **critical factor** determining where a species lives. The giant saguaro cactus (*Carnegiea gigantea*), which grows in the dry, hot Sonoran desert of southern Arizona and northern Mexico, offers an example

Figure 3.3 Giraffes don't have long necks because they stretch to reach tree-top leaves, but those giraffes that happened to have longer necks got more food and had more offspring, so the trait became fixed in the population.

Figure 3.4 Saguaro cacti inhabit the Sonoran desert, and their northern limit is partly controlled by a critical environmental factor—freezing temperatures. Local topography allows some individuals to push that boundary, such that temperature alone doesn't explain the exact edge of the cactus's northern range.

(fig. 3.4). Saguaros are extremely sensitive to freezing temperatures. A single winter night with temperatures below freezing for 12 or more hours kills growing tips on the branches, preventing further development. Thus the northern edge of the saguaro's range corresponds to a zone where freezing temperatures last less than half a day at any time.

Ecologist Victor Shelford (1877–1968) later expanded Liebig's principle by stating that each environmental factor has both minimum and maximum levels, called **tolerance limits**, beyond which a particular species cannot survive or is unable to reproduce (fig. 3.5). The single factor closest to these survival limits, Shelford postulated, is the critical factor that limits where a particular organism can live. At one time, ecologists tried to identify unique factors limiting the growth of every plant and animal

population. We now know that several factors working together, even in a clear-cut case like the saguaro, usually determine a species' distribution. If you have ever explored the rocky coasts of New England or the Pacific Northwest, you have probably noticed that mussels and barnacles grow thickly in the intertidal zone, the place between high and low tides. No one factor decides this pattern. Instead, the distribution of these animals is determined by a combination of temperature extremes, drying time between tides, salt concentrations, competitors, and food availability.

In some species, tolerance limits affect the distribution of young differently than adults. The desert pupfish, for instance, lives in small, isolated populations in warm springs in the northern Sonoran desert. Adult pupfish can survive temperatures between 0° and 42°C (a remarkably high temperature for a fish) and tolerate an equally wide range of salt concentrations. Eggs and juvenile fish, however, can survive only between 20° and 36°C and are killed by high salt levels. Reproduction, therefore, is limited to a small part of the range of the adult fish.

Sometimes the requirements and tolerances of species are useful **indicators** of specific environmental characteristics. The presence or absence of such species indicates something about the community and the ecosystem as a whole. Lichens and eastern white pine, for example, are indicators of air pollution because they are extremely sensitive to sulfur dioxide and ozone, respectively. Bull thistle and many other plant weeds grow on disturbed soil but are not eaten by cattle; therefore, a vigorous population of bull thistle or certain other plants in a pasture indicates it is being overgrazed. Similarly, anglers know that trout species require cool, clean, well-oxygenated water; the presence or absence of trout is used as an indicator of good water quality.

The ecological niche is a species' role and environment

Habitat describes the place or set of environmental conditions in which a particular organism lives. A more functional term,

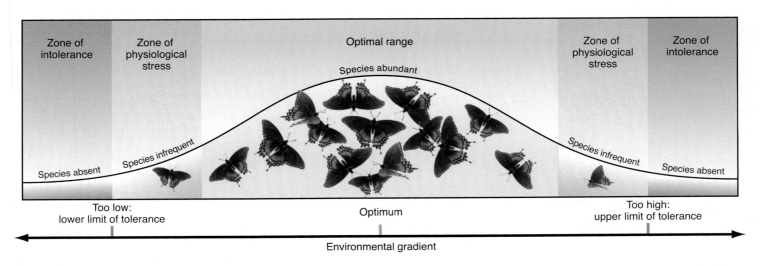

Figure 3.5 Tolerance limits affect species populations. For every environmental factor, there is an optimal range where a species is most abundant. As an organism moves away from locations with optimal conditions, the harder life becomes. At some point along the environmental gradient for that factor, individuals become physiologically stressed and few, if any, survive those conditions.

Figure 3.6 Each of the species in this African savanna has its own ecological niche that determines where and how it lives.

ecological niche, describes both the role played by a species in a biological community and the total set of environmental factors that determine a species distribution. The concept of niche was first defined in 1927 by the British ecologist Charles Elton (1900–1991). To Elton, each species had a role in a community of species, and the niche defined its way of obtaining food, the relationships it had with other species, and the services it provided to its community. Thirty years later, the American limnologist G. E. Hutchinson (1903–1991) proposed a more biophysical definition of niche. Every species, he pointed out, exists within a range of physical and chemical conditions (temperature, light levels, acidity, humidity, salinity, etc.) and also biological interactions (predators and prey present, defenses, nutritional resources available, etc.). The niche is more complex than the idea of a critical factor (fig. 3.6). A graph of a species niche would be multidimensional, with many factors being simultaneously displayed.

For generalists, like rats or cockroaches, the ecological niche is broad. In other words, generalists have a wide range of tolerance for many environmental factors. For others, such as the giant panda (*Ailuropoda melanoleuca*), only a narrow ecological niche exists (fig. 3.7). Bamboo is low in nutrients, but provides 95 percent of a

Figure 3.7 The giant panda eats almost nothing but bamboo. It has a digestive system like a carnivore, but it is a vegetarian. Periodic die-offs of bamboo, combined with destruction of habitat, cause pandas to starve, with the last large die-off in the 1970s.

panda's diet, requiring it to spend as much as 16 hours a day eating. There are virtually no competitors for bamboo, except other pandas, yet the species is endangered, primarily due to shrinking habitat. Giant pandas, like many species on earth, are habitat specialists. Specialists have more exacting habitat requirements, tend to have lower reproductive rates, and care for their young longer. They may

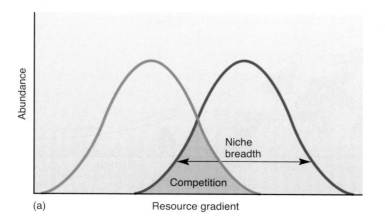

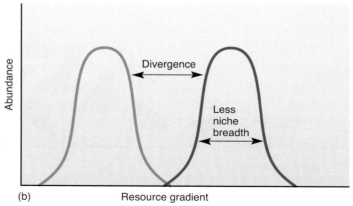

Figure 3.8 Competition causes resource partitioning and niche specialization. (a) Where niches of two species overlap along a resource gradient, competition occurs (shaded area). Individuals in this part of the niche have less success producing young. (b) Over time the traits of the populations diverge, leading to specialization, narrower niche breadth, and less competition between species.

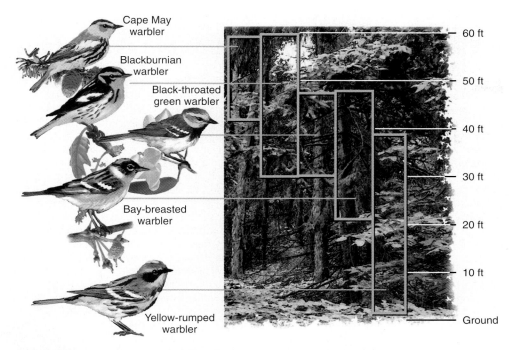

Figure 3.9 Several species of insect-eating wood warblers occupy the same forests in eastern North America. The competitive exclusion principle predicts that the warblers should partition the resource—insect food—in order to reduce competition. And in fact, the warblers feed in different parts of the forest.
Source: Original observations by R. H. MacArthur (1958) *Ecology* 39:599–619.

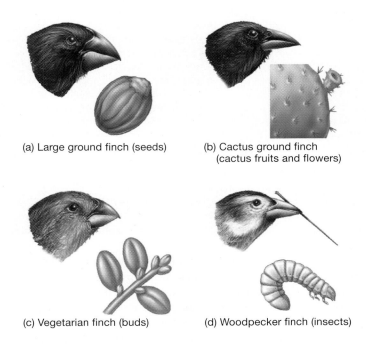

(a) Large ground finch (seeds)

(b) Cactus ground finch (cactus fruits and flowers)

(c) Vegetarian finch (buds)

(d) Woodpecker finch (insects)

Figure 3.10 Four of the 13 species of finches in the Galápagos with their preferred foods. Distinctive beaks and behaviors arose from the ancestral, pointy-beaked warbler finch (*Certhidia olivacea*, not shown) in response to different foods available on the islands. Recent research on the finches' DNA suggests that beak shape changed first, followed by divergence in beak size.

be less resilient in response to environmental change. Plants can also be habitat specialists—for instance, **endemic** plant species exist on serpentine and other unusual types of rock outcrops, and nowhere else.

Over time, niches change as species develop new strategies to exploit resources. Species of greater intelligence or complex social structures, such as elephants, chimpanzees, and dolphins, learn from their social group how to behave and can invent new ways of doing things when presented with novel opportunities or challenges. In effect, they alter their ecological niche by passing on cultural behavior from one generation to the next. Most organisms, however, are restricted to their niche by their genetically determined bodies and instinctive behaviors. When two such species compete for limited resources, one eventually gains the larger share, while the other finds different habitat, dies out, or experiences a change in its behavior or physiology so that competition is minimized.

The idea that "complete competitors cannot coexist" was proposed by the Russian biologist G. F. Gause (1910–1986) to explain why mathematical models of species competition always ended with one species disappearing. The **competitive exclusion principle**, as it is called, states that no two species can occupy the same ecological niche for long. The one that is more efficient in using available resources will exclude the other. We call this process of niche evolution **resource partitioning** (fig. 3.8). Partitioning can allow several species to utilize different parts of the same resource and coexist within a single habitat (fig. 3.9). Species can specialize in time, too. Swallows and insectivorous bats both catch insects, but some insect species are active during the day and others at night, providing noncompetitive feeding opportunities for day-active swallows and night-active bats. The competitive exclusion principle does not explain all situations, however. For example, many similar plant species coexisted in the Cedar Creek experimental plots. Do they avoid competition in ways we cannot observe, or are resources so plentiful that no competition need occur?

Speciation maintains species diversity

As an interbreeding species population becomes better adapted to its ecological niche, its genetic heritage (including mutations passed from parents to offspring) gives it the potential to change further as circumstances dictate. In the case of Galápagos finches made famous by Charles Darwin, evidence from body shape, behavior, and genetics leads to the idea that modern Galápagos finches look, behave, and bear DNA related to an original seed-eating finch species (fig. 3.10). It is proposed to have blown to the islands from the mainland where a similar species still exists. Today there are 13 distinct species on the islands that differ markedly in appearance, food preferences, and habitat. Fruit eaters have thick, parrot-like bills; seed eaters have heavy, crushing bills; insect eaters have thin, probing beaks to catch their prey. One

Exploring
SCIENCE:

The Cichlids of Lake Victoria

If you visit your local pet store, chances are you'll see some cichlids (*Haplochromis* sp.). These small colorful, prolific fish come in a wide variety of colors and shapes from many parts of the world. The greatest cichlid diversity on earth—and probably the greatest vertebrate diversity anywhere—is found in the three great African rift lakes: Victoria, Malawi, and Tanganyika. Together, these lakes once had about 1,000 types of cichlids—more than all the fish species in Europe and North America combined. All these cichlids apparently evolved from a few ancestral varieties in the 15,000 years or so since the lakes were formed by splitting of the continental crust. This is one of the fastest and most extensive examples of vertebrate speciation known.

We believe that one of the factors that allowed cichlids to evolve so quickly is that they found few competitors or predators and

a multitude of ecological roles to play in these new lakes. There are mud biters, algae scrapers, leaf chewers, snail crushers, zooplankton eaters, prawn predators, and fish feeders. Because they live in different habitats in the lakes, are active at different times of the day, and have developed different body sizes and shapes to feed on specialized prey, the cichlids have been ecologically isolated for long enough to evolve into an amazing variety of species. Cichlids are a good example of radiative evolution.

Unfortunately, a well-meaning but disastrous fish-stocking experiment has wiped out at least half the cichlid species in these lakes in just a few decades and set off a series of changes that are upsetting important ecological relationships. Lake Victoria, which lies between Kenya, Tanzania, and Uganda, has been particularly hard hit. Cichlids once made up 80 percent of the animal biomass in the lake

and were the base for a thriving local fishery, supplying much-needed protein for local people. Management agencies regarded the bony little cichlids as "trash fish," and decided in the 1960s to introduce Nile perch (*Lates niloticus*), a voracious, exotic predator that can weigh up to 100 kg (220 lbs) and grow up to 2 m long. The perch, they believed, would support a lucrative commercial export trade.

The perch gobbled up the cichlids so quickly that, by 1980, two-thirds of the haplochromine species in the lake were extinct.

Although there are still lots of fish in the lake, 80 percent of the biomass is now made up of perch, which are too large and powerful for the small boats, papyrus nets, and woven baskets traditionally used to harvest cichlids. International fishing companies now use large power boats and nylon nets to harvest great schools of perch, which are filleted, frozen, and shipped to markets in Europe and the Middle East. Because the perch are oily, local fishers can't sun dry them as they once did the cichlids. Instead, they are cooked or smoked over wood fires for local consumption. Forests are being denuded for firewood, and protein malnutrition is common in a region that exports 200,000 tons of fish each year.

Perhaps worst of all, Lake Victoria, which covers an area the size of Switzerland, is dying. Algae blooms clot the surface, oxygen levels have fallen alarmingly, and thick layers of soft silt are filling-in shallow bays. Untreated sewage, chemical pollution, and farm runoff are the immediate causes of these deleterious changes, but destabilization of the natural community plays a role as well. The swarms of cichlids that once ate algae and rotting detritus were the lake's self-cleaning system. Eliminating them threatens the long-term ability of the lake to support any useful aquatic life.

As this example shows, species and ecological diversity are important. Misguided management and development schemes that destroyed native species in Lake Victoria have resulted in an ecosystem that no longer supports the natural community or the local people dependent on it. It's difficult to see how we could replace the variety of species and the ecological roles they played, which evolution provided for free.

For more information, see
Stiassny, M. L. J., and A. Meyer. 1999. Cichlids of the rift lakes. *Scientific American* 280(2): 64–69.

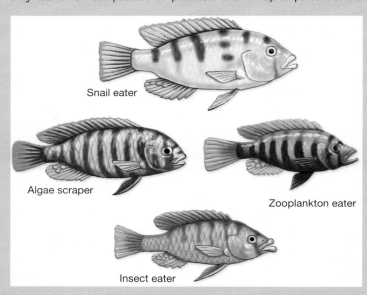

Cichlid fishes of Lake Victoria. More than 300 species have evolved from an original ancestor to take advantage of different food sources and habitats.

Snail eater

Algae scraper

Zooplankton eater

Insect eater

of the most unusual species is the woodpecker finch, which pecks at tree bark for hidden insects. Lacking the woodpecker's long tongue, the finch uses a cactus spine as a tool to extract bugs.

The development of a new species is called **speciation**. No one has observed a new species springing into being, but in some organisms, especially plants, it is inferred to occur somewhat frequently. Nevertheless, given the evidence, speciation is a reasonable proposal for how species arise. Speciation may be relatively rapid on millennial timescales (punctuated equilibrium). For example, after a long period of stability, a new species may arise from parents following the appearance of a new food source, predator, or competitor, or a change in climate. One mechanism

of speciation is **geographic isolation**. This is termed **allopatric speciation**—species arise in non-overlapping geographic locations. The original Galápagos finches were separated from the rest of the population on the mainland, could no longer share genetic material, and became reproductively isolated.

The barriers that divide subpopulations are not always physical. For example, two virtually identical tree frogs (*Hyla versicolor, H. chrysoscelis*) live in similar habitats of eastern North America but have different mating calls. This is an example of behavioral isolation. It also happens that one species has twice the chromosomes of the other. This example of **sympatric speciation** takes place in the same location as the ancestor species. Fern spe-

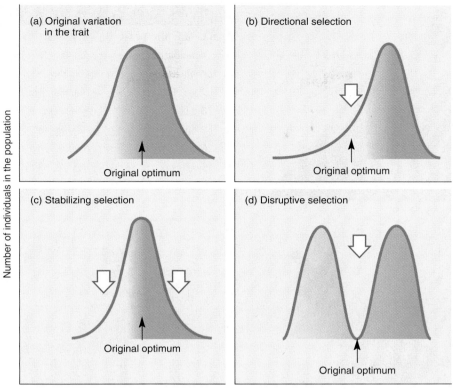

Figure 3.11 A species trait, such as beak shape, changes in response to selection pressure. (a) The original variation is acted on by selection pressure (arrows) that (b) shifts the characteristics of that trait in one direction, (c) or to an intermediate condition. (d) Disruptive selection moves characteristics to the extremes of the trait. Which selection type plausibly resulted in two distinct beak shapes among Galápagos finches—narrow in tree finches versus stout in ground finches?

cies and other plants seem prone to sympatric speciation by doubling or quadrupling the chromosome number of their ancestors.

Once isolation is imposed, the two populations begin to diverge in genetics and physical characteristics. Genetic drift ensures that DNA of two formerly joined populations eventually diverges; in several generations, traits are lost from a population during the natural course of reproduction. Under more extreme circumstances, a die-off of most members of an isolated population strips much of the variation in traits from the survivors. The cheetah experienced a genetic bottleneck about 10,000 years ago and exists today as virtually identical individuals.

In isolation, selection pressures shape physical, behavioral, and genetic characteristics of individuals, causing population traits to shift over time (fig. 3.11). From an original range of characteristics, the shift can be toward an extreme of the trait (directional selection), it can narrow the range of a trait (stabilizing selection), or it can cause traits to diverge to the extremes (disruptive selection). Directional selection is implied by increased pesticide resistance in German cockroaches (*Blattella germanica*). Apparently some individuals can make an enzyme that detoxifies pesticides. Individual cockroaches that lack this characteristic are dying out, and as a result, populations of cockroaches with pesticide resistance are developing.

A small population in a new location—island, mountaintop, unique habitat—encounters new environmental conditions that favor some individuals over others (fig. 3.12). The physical and behavioral traits these individuals have are passed to the next generation, and the frequency of the trait shifts in the population. Where a species may have existed but has died out, others arise and contribute to the incredible variety of life-forms seen in nature. The fossil record is one of ever-increasing species diversity, despite several catastrophes, which were recorded in different geological strata and which wiped out a large proportion of the earth's species each time.

Taxonomy describes relationships among species

Taxonomy is the study of types of organisms and their relationships. With it you can trace how organisms have descended from common ancestors. Taxonomic relationships among species are displayed like a family tree. Botanists, ecologists, and other scientists often use the most specific levels of the tree, genus and species, to compose **binomials**. Also called scientific or Latin names, they identify and describe species using Latin, or Latinized nouns and adjectives, or names of people or places. Scientists

1. Single population

2. Geographically isolated populations

Figure 3.12 Geographic isolation is a mechanism in allopatric speciation. In cool, moist glacial periods, Arizona was forested and red squirrels traveled and interbred freely. As the climate warmed and dried, desert replaced forests on the plains, while cooler mountaintops remained wooded and the only place in Arizona that red squirrels could survive. When mountaintop red squirrel populations became reproductively isolated from one another, they began to develop different characteristics.

communicate about species using these scientific names instead of common names (e.g., buttercup or bluebell), to avoid confusion. A common name can refer to any number of species in different places, and a single species might have many common names. The bionomial *Pinus resinosa,* on the other hand, always is the same tree, whether you call it a red pine, Norway pine, or just pine.

Taxonomy also helps organize specimens and subjects in museum collections and research. You are *Homo sapiens* (human) and eat chips made of *Zea mays* (corn or maize). Both are members of two well-known kingdoms (table 3.1). Scientists, however, recognize six kingdoms (fig. 3.13): animals, plants, fungi (molds and mushrooms), protists (algae, protozoans, slime molds), bacteria (or eubacteria), and archaebacteria (ancient, single-celled organisms that live in harsh environments, such as hot springs). Within these kingdoms are millions of different species, which you will learn more about in chapter 5.

Table 3.1	Taxonomy of Two Common Species	
Taxonomic Level	**Humans**	**Corn**
Kingdom	Animalia	Plantae
Phylum	Chordata	Anthophyta
Class	Mammalia	Monocotyledons
Order	Primates	Commenales
Family	Hominidae	Poaceae
Genus	*Homo*	*Zea*
Species	*Homo sapiens*	*Zea mays*
Subspecies	*H. sapiens sapiens*	*Zea mays mays*

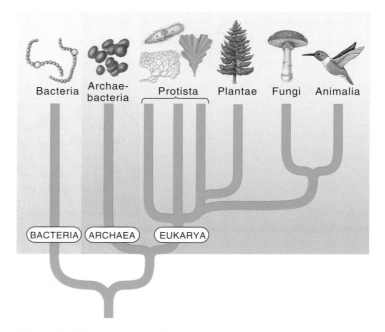

Figure 3.13 The six great kingdoms representing all life on earth. The kingdoms are grouped in domains indicating common origins.

3.2 Species Interactions Shape Communities of Species

We have learned that adaptation to one's environment, determination of ecological niche, and even speciation is affected not just by bodily limits and behavior, but also by competition and predation. Don't despair. Not all biological interactions are antagonistic, and many, in fact, involve cooperation or at least benign interactions and tolerance. In some cases, different organisms depend on each other to acquire resources. Now we will look at the interactions within and between species that affect their success and shape biological communities.

Competition leads to resource allocation

Competition is a type of antagonistic relationship within a biological community. Organisms compete for resources that are in

Figure 3.14 In this tangled Indonesian rainforest, space and light are at a premium. Plants growing beneath the forest canopy have adaptations that help them secure these limited resources. The ferns and bromeliads seen here are epiphytes; they find space and get closer to the sun by perching on limbs and tree trunks. Strangler figs start out as epiphytes, but send roots down to the forest floor and, once contact is made, put on a growth spurt that kills the supporting tree. These are just some of the adaptations to life in the dark jungle.

limited supply: energy and matter in usable forms, living space, and specific sites to carry out life's activities. Plants compete for growing space to develop root and shoot systems so that they can absorb and process sunlight, water, and nutrients (fig. 3.14). Animals compete for living, nesting, and feeding sites, and also for mates. Competition among members of the same species is called **intraspecific competition**, whereas competition between members of different species is called **interspecific competition**. Recall the competitive exclusion principle as it applies to interspecific competition. Competition shapes a species population and biological community by causing individuals and species to shift their focus from one segment of a resource type to another. Thus, warblers all competing with each other for insect food in New England tend to specialize on different areas of the forest's trees, reducing or avoiding competition. Since the 1950s there have been hundreds of interspecific competition studies in natural populations. In general, scientists assume it does occur, but not always, and in some groups—carnivores and plants—it has little effect.

In intraspecific competition, members of the same species compete directly with each other for resources. Several avenues exist to reduce competition in a species population. First, the young of the year disperse. Even plants practice dispersal; seeds are carried by wind, water, and passing animals to less crowded conditions away from the parent plants. Second, by exhibiting strong territoriality, many animals force their offspring or trespassing adults out of their vicinity. In this way territorial species, which include bears, songbirds, ungulates, and even fish, minimize competition between individuals and generations. A third way to reduce intraspecific com-

Figure 3.15 Insect herbivores are predators as much as are lions and tigers. In fact, insects consume the vast majority of biomass in the world. Complex patterns of predation and defense have often evolved between insect predators and their plant prey.

petition is resource partitioning between generations. The adults and juveniles of these species occupy different ecological niches. For instance, monarch caterpillars munch on milkweed leaves, while metamorphosed butterflies lap nectar. Crabs begin as floating larvae and do not compete with bottom-dwelling adult crabs.

We think of competition among animals as a battle for resources—"nature red in tooth and claw" is the phrase. In fact, many animals avoid fighting if possible, or confront one another with noise and predictable movements. Bighorn sheep and many other ungulates, for example, engage in ritualized combat, with the weaker animal knowing instinctively when to back off. It's worse to be injured than to lose. Instead, competition often is simply about getting to food or habitat first, or being able to use it more efficiently. As we discussed, each species has tolerance limits for nonbiological (abiotic) factors. Studies often show that, when two species compete, the one living in the center of its tolerance limits for a range of resources has an advantage and, more often than not, prevails in competition with another species living outside its optimal environmental conditions.

Predation affects species relationships

All organisms need food to live. Producers make their own food, while consumers eat organic matter created by other organisms. As we saw in chapter 2, photosynthetic plants and algae are the producers in most communities. Consumers include herbivores, carnivores, omnivores, scavengers, detritivores, and decomposers. You may think only carnivores are predators, but ecologically a predator is any organism that feeds directly on another living organism, whether or not this kills the prey (fig. 3.15). Herbivores, carnivores, and omnivores, which feed on live prey, are predators, but scavengers, detritivores, and decomposers, which feed on dead things, are not. In this sense, parasites (organisms that feed on a host organism or steal resources from it without necessarily killing it) and even pathogens (disease-causing organisms) can be considered predator organisms. Herbivory is the type of predation practiced by grazing and browsing animals on plants.

Predation is a powerful but complex influence on species populations in communities. It affects (1) all stages in the life cycles of predator and prey species; (2) many specialized food-obtaining mechanisms; and (3) the evolutionary adjustments in behavior and body characteristics that help prey escape being eaten, and predators more efficiently catch their prey. Predation also interacts with competition. In **predator-mediated competition**, a superior competitor in a habitat builds up a larger population than its competing species; predators take note and increase their hunting pressure on the superior species, reducing its abundance and allowing the weaker competitor to increase its numbers. To test this idea, scientists remove predators from communities of competing species. Often the superior competitors eliminated other species from the habitat. In a classic example, the ochre starfish (*Pisaster ochraceus*) was removed from Pacific tidal zones and its main prey, the common mussel (*Mytilus californicus*), exploded in numbers and crowded out other intertidal species.

Knowing how predators affect prey populations has direct application to human needs, such as pest control in cropland. The cyclamen mite (*Phytonemus pallidus*), for example, is a pest of California strawberry crops. Its damage to strawberry leaves is reduced by predatory mites (*Typhlodromus* and *Neoseiulus*), which arrive naturally or are introduced into fields. Pesticide spraying to control the cyclamen mite can actually increase the infestation because it also kills the beneficial predatory mites.

Predatory relationships may change as the life stage of an organism changes. In marine ecosystems, crustaceans, mollusks, and worms release eggs directly into the water where they and hatchling larvae join the floating plankton community (fig. 3.16). Planktonic animals eat each other and are food for larger carni-

Figure 3.17 Poison arrow frogs of the family Dendrobatidae display striking patterns and brilliant colors that alert potential predators to the extremely toxic secretions on their skin. Indigenous people in Latin America use the toxin to arm blowgun darts.

vores, including fish. As prey species mature, their predators change. Barnacle larvae are planktonic and are eaten by small fish, but as adults their hard shells protect them from fish, but not starfish and predatory snails. Predators often switch prey in the course of their lives. Carnivorous adult frogs usually begin their lives as herbivorous tadpoles. Predators also switch prey when it becomes rare, or something else becomes abundant. Many predators have morphologies and behaviors that make them highly adaptable to a changing prey base, but some, like the polar bear, are highly specialized in their prey preferences.

Some adaptations help avoid predation

Predator-prey relationships exert selection pressures that favor evolutionary adaptation. In this world, predators become more efficient at searching and feeding, and prey become more effective at escape and avoidance. Toxic chemicals, body armor, extraordinary speed, and the ability to hide are a few strategies organisms use to protect themselves. Plants have thick bark, spines, thorns, or distasteful and even harmful chemicals in tissues—poison ivy and stinging nettle are examples. Arthropods, amphibians, snakes, and some mammals produce noxious odors or poisonous secretions that cause other species to leave them alone. Animal prey are adept at hiding, fleeing, or fighting back. On the Serengeti Plain of East Africa, the swift Thomson's gazelle and even swifter cheetah are engaged in an arms race of speed, endurance, and quick reactions. The gazelle escapes often because the cheetah lacks stamina, but the cheetah accelerates from 0 to 72 kph in 2 seconds, giving it the edge in a surprise attack. The response of predator to prey and vice versa, over tens of thousands of years, produces physical and behavioral changes in a process known as **coevolution**. Coevolution can be mutually beneficial: many plants and pollinators have forms and behaviors that benefit each other. A classic case is that of fruit bats, which pollinate and disperse seeds of fruit-bearing tropical plants.

Often species with chemical defenses display distinct coloration and patterns to warn away enemies (fig. 3.17). In a neat evolutionary twist, certain species that are harmless resemble

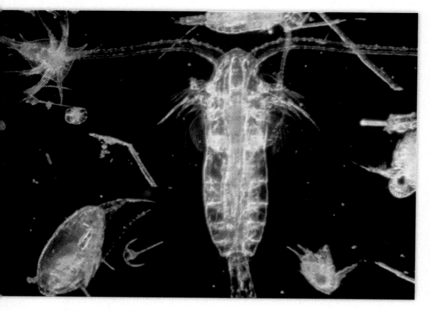

Figure 3.16 Microscopic plants and animals form the basic levels of many aquatic food chains and account for a large percentage of total world biomass. Many oceanic plankton are larval forms that have habitats and feeding relationships very different from their adult forms.

(a)

(b)

Figure 3.18 In a clear example of Batesian mimicry, (a) a dangerous wasp carries bold yellow and black bands that (b) are imitated by the much rarer but harmless longhorn beetle (family Cerambycidae). The beetle even behaves like a wasp, tricking predators which have experienced wasps into avoiding the beetle.

poisonous or distasteful ones, gaining protection against predators who remember a bad experience with the actual toxic organism. This is called **Batesian mimicry**, after the English naturalist H. W. Bates (1825–1892), a traveling companion of Alfred Wallace. Many wasps, for example, have bold patterns of black and yellow stripes to warn off potential predators (fig. 3.18a). The much rarer longhorn beetle has no stinger but looks and acts much like a wasp, tricking predators into avoiding it (fig. 3.18b). The distasteful monarch and benign viceroy butterflies are a classic case of Batesian mimicry. Another form of mimicry, **Müllerian mimicry** (after the biologist Fritz Müller) involves two unpalatable or dangerous species who look alike. When predators learn to avoid either species, both benefit. Species also display forms, colors, and patterns that help avoid being discovered. Insects that look exactly like dead leaves or twigs are among the most remarkable examples (fig. 3.19). Unfortunately for prey, predators also use camouflage to conceal themselves as they lie in wait for their next meal.

Symbiosis: Intimate relations among species

In contrast to predation and competition, some interactions between organisms can be nonantagonistic, even beneficial (table 3.2). In such relationships, called **symbiosis**, two or more species live intimately together, with their fates linked. Symbiotic relationships often enhance the survival of one or both partners. In lichens, a fungus and a photosynthetic partner (either an alga or a cyanobacterium) combine tissues to mutual benefit (fig. 3.20a). This association is called **mutualism**. Some ecologists believe that cooperative, mutualistic relationships may be more important in evolution than commonly thought (fig. 3.20b). Survival of the fittest may also mean survival of organisms that can live together.

Symbiotic relationships often entail some degree of coevolution of the partners, shaping—at least in part—their structural and behavioral characteristics. This mutualistic coadaptation is evident between swollen thorn acacias (*Acacia collinsii*) and the ants (*Pseudomyrmex ferruginea*) that tend them in Central and South America. Acacia ant colonies live inside the swollen thorns on the acacia tree branches. Ants feed on nectar that is produced in glands at the leaf bases and also eat special

Figure 3.19 This walking stick is highly camouflaged to blend in with the forest floor. Natural selection and evolution have created this remarkable shape and color.

Table 3.2	Types of Species Interactions	
Interaction between Two Species	**Effect on First Species**	**Effect on Second Species**
Mutualism	+	+
Commensalism	+	0
Parasitism	+	−
Predation	+	−
Competition	±	±

(+beneficial; −harmful; 0 neutral; ±varies)

protein-rich structures that are produced on leaflet tips. The acacias thus provide shelter and food for the ants. Although they spend energy to provide these services, the trees are not harmed by the ants. What do the acacias get in return? Ants aggressively defend their territories, driving away herbivorous insects that want to feed on the acacias. Ants also trim away vegetation that grows around the tree, reducing competition by other plants for water and nutrients. You can see how mutualism is structuring the biological

(a) Symbiosis

(b) Mutualism

(c) Commensalism

Figure 3.20 Symbiotic relationships. (a) Lichens represent an obligatory mutualism between a fungus and alga or cyanobacterium. (b) Mutualism between a parasite-eating red-billed oxpecker and parasite-infested impala. (c) Commensalism between a tropical tree and free-loading bromeliad.

community in the vicinity of acacias harboring ants, just as competition or predation shapes communities.

Mutualistic relationships can develop quickly. In 2005 the Harvard entomologist E. O. Wilson pieced together evidence to explain a 500-year-old agricultural mystery in the oldest Spanish settlement in the New World, Hispaniola. Using historical accounts and modern research, Dr. Wilson reasoned that mutualism developed between the tropical fire ant (*Solenopsis geminata*), native to the Americas, and a sap-sucking insect that was probably introduced from the Canary Islands in 1516 on a shipment of plantains. The plantains were planted, the sap-suckers were distributed across Hispaniola, and in 1518 a great die-off of crops occurred. Apparently the native fire ants discovered the foreign sap-sucking insects, consumed their excretions of sugar and protein, and protected them from predators, thus allowing the introduced insect population to explode. The Spanish assumed the fire ants caused the agricultural blight, but a little ecological knowledge would have led them to the real culprit.

Commensalism is a type of symbiosis in which one member clearly benefits and the other apparently is neither benefited nor harmed. Many mosses, bromeliads, and other plants growing on trees in the moist tropics are considered commensals (fig. 3.20c). These epiphytes are watered by rain and obtain nutrients from leaf litter and falling dust, and often they neither help nor hurt the trees on which they grow. Robins and sparrows that inhabit suburban yards are commensals with humans. **Parasitism**, a form of predation, may also be considered symbiosis because of the dependency of the parasite on its host.

Keystone species: Influence all out of proportion

A **keystone species** plays a critical role in a biological community that is out of proportion to its abundance. Originally, keystone species were thought to be top predators—lions, wolves, tigers—which limited herbivore abundance and reduced the herbivory of plants. Scientists now recognize that less-conspicuous species also play keystone roles. Tropical figs, for example, bear fruit year-

round at a low but steady rate. If figs are removed from a forest, many fruit-eating animals (frugivores) would starve in the dry season when fruit of other species is scarce. In turn, the disappearance of frugivores would affect plants that depend on them for pollination and seed dispersal. It is clear that the effect of a keystone species on communities often ripples across trophic levels.

Keystone functions have been documented for vegetation-clearing elephants, the predatory ochre sea star, and frog-eating salamanders in coastal North Carolina. Even microorganisms can play keystone roles. In many temperate forest ecosystems, groups of fungi that are associated with tree roots (mycorrhizae) facilitate the uptake of essential minerals. When fungi are absent, trees grow poorly or not at all. Overall, keystone species seem to be more common in aquatic habitats than in terrestrial ones.

The role of keystone species can be difficult to untangle from other species interactions. Off the northern Pacific coast, a giant brown alga (*Macrocystis pyrifera*) forms dense "kelp forests," which shelter fish and shellfish species from predators, allowing them to become established in the community. It turns out, however, that sea otters eat sea urchins living in the kelp forests; when sea otters are absent, the urchins graze on and eliminate kelp forests (fig. 3.21). To complicate things, around 1990, killer whales began preying on otters because of the dwindling stocks of seals and sea lions, thereby creating a cascade of effects. Is the kelp, otter, or orca the keystone here? Whatever the case, keystone species exert their influence by changing competitive relationships. In some communities, perhaps we should call it a "keystone set" of organisms. The functional redundancy seen in the Cedar Creek experimental plots, for instance, suggests that a diversity of species—no single keystone—ensures that essential ecological functions will continue without much change.

3.3 The Growth of Species Populations

Many biological organisms can produce unbelievable numbers of offspring if environmental conditions are right. Consider the

Figure 3.21 Sea otters protect kelp forests in the northern Pacific Ocean by eating sea urchins that would otherwise destroy the kelp. But the otters are being eaten by killer whales. Which is the keystone in this community—or is there a keystone set of organisms?

common housefly (*Musca domestica*). Each female fly lays 120 eggs (assume half female) in a generation. In 56 days those eggs become mature adults, able to reproduce. In one year, with seven generations of flies being born and reproducing, that original fly would be the proud parent of 5.6 trillion offspring. If this rate of reproduction continued for ten years, the entire earth would be covered in several meters of housefly bodies. Luckily housefly reproduction, as for most organisms, is constrained in a variety of ways—scarcity of resources, competition, predation, disease, accident. The housefly merely demonstrates the remarkable amplification—the **biotic potential**—of unrestrained biological reproduction. Population dynamics describes these changes in the number of organisms in a population over time.

Growth without limits is exponential

As you learned in chapter 2, a population consists of all the members of a single species living in a specific area at the same time. The growth of the housefly population just described is **exponential**, having no limit and possessing a distinctive shape when graphed over time. An exponential growth rate (increase in numbers per unit of time) is expressed as a constant fraction, or exponent, which is used as a multiplier of the existing population. The mathematical formula for exponential growth is:

$$\frac{dN}{dt} = rN$$

Here "*d*" means "change" so the change in numbers of individuals (*dN*) per change in time (*dt*) equals the rate of growth (*r*) times the number of individuals in the population (*N*). The *r* term (intrinsic capacity for increase) is a fraction representing the average individual contribution to population growth. If *r* is positive, the population is increasing. If *r* is negative, the population is shrinking. If *r* is zero, there is no change, and *dN/dt* = 0.

A graph of exponential population growth is described as a **J curve** (fig. 3.22) because of its shape. As you can see, the number of individuals added to a population at the beginning of an exponential growth curve can be rather small. But within a very short time, the numbers begin to increase quickly because a fixed percentage leads to a much larger increase as the population size grows.

The exponential growth equation is a very simple model; it is an idealized, simple description of the real world. The same

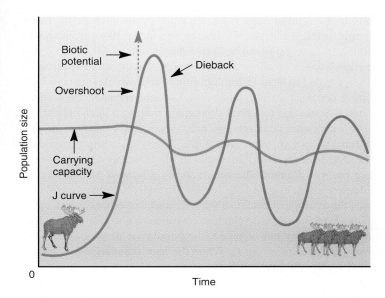

Figure 3.22 J curve, or exponential growth curve, with overshoot of carrying capacity. Exponential growth in an unrestrained population (*left side of curve*) leads to a population crash and oscillations below former levels. After the overshoot, carrying capacity may be reduced because of damage to the resources of the habitat. Moose on Isle Royale in Lake Superior may be exhibiting this growth pattern in response to their changing environment.

equation is used to calculate growth in your bank account due to compounded interest rates; achieving the potential of your savings depends on you never making a withdrawal. Just as species populations lose individuals and experience reduced biotic potential, not all of your dollars will survive to a ripe old age and contribute fully to your future cash position.

Carrying capacity relates growth to its limits

In the real world there are limits to growth. Around 1970, ecologists developed the concept of **carrying capacity** to mean the number or biomass of animals that can be supported (without harvest) in a certain area of habitat. The concept is now used more generally to suggest a limit of sustainability that an environment has in relation to the size of a species population. Carrying capacity is helpful in understanding the population dynamics of some species, perhaps even humans.

When a population overshoots or exceeds the carrying capacity of its environment, resources become limited and death rates rise. If deaths exceed births, the growth rate becomes negative and the population may suddenly decrease, a change called a population crash or dieback (fig. 3.22). Populations may oscillate from high to low levels around the habitat's carrying capacity, which may be lowered if the habitat is damaged. Moose and other browsers or grazers sometimes overgraze their food plants such that future populations of herbivores in the same habitat find less preferred food to sustain them, at least until the habitat recovers. Some species go through predictable cycles if simple factors are involved, such as the seasonal light- and temperature-dependent bloom of algae in a lake. Cycles can be irregular if complex environmental and biotic relationships exist. Irregular cycles include migratory locust outbreaks in the Sahara, or tent caterpillars in temperate forests—these represent irruptive population growth. Population dynamics are also affected by the emigration of organisms from an overcrowded habitat, or immigration of individuals into new habitat, as occurred in 2005 when owls suddenly invaded the northern United States due to a food shortage in their Canadian habitat.

Sometimes predator and prey populations oscillate in synchrony with each other (fig. 3.23). This classic study employed the 200-year record of furs sold at Hudson Bay Company trading posts in Canada (the figure shows a portion of that record). Charles Elton, responsible for the diversity-stability hypothesis, was the ecological detective who showed that numbers of Canada lynx (*Lynx canadensis*) fluctuate on about a ten-year cycle which mirrors, slightly out of phase, the population peaks of snowshoe hares (*Lepus americanus*). When the hare population is high and food is plentiful, life for the lynx is very good, it reproduces well, and populations grow. Eventually, declining food supplies shrink hare populations. For a while the lynx benefits because starving hares are easier to catch than healthy ones. As hares become scarce, however, so do lynx. When hares are at their lowest levels, their food supply recovers and the whole cycle starts over again. This predator-prey oscillation is described mathematically in the Lotka-Volterra model, after the scientists who developed it.

Feedback produces logistic growth

Not all biological populations cycle through exponential overshoot and catastrophic dieback. Many species are regulated by both internal and external factors and come into equilibrium with their environmental resources while maintaining relatively stable popu-

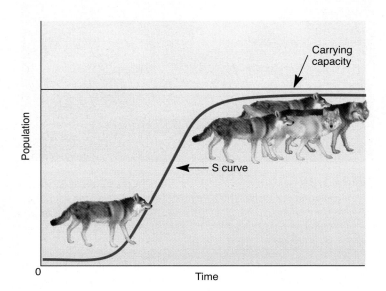

Figure 3.24 S curve, or logistic growth curve, describes a population's changing numbers over time in response to feedback from the environment or its own population density. Over the long run, a conservative and predictable population dynamic may win the race over an exponential population dynamic. Species with this growth pattern tend to be *K*-selected.

lation sizes. When resources are unlimited, they may even grow exponentially, but this growth slows as the carrying capacity of the environment is approached. This population dynamic is called **logistic growth** because of its constantly changing rate.

Mathematically, this growth pattern is described by the following equation, which adds a feedback term for carrying capacity (*K*) to the exponential growth equation:

$$\frac{dN}{dt} = rN\left(1 - \frac{N}{K}\right)$$

The logistic growth equation says that the change in numbers over time (*dN/dt*) equals the exponential growth rate (*rN*) times the portion of the carrying capacity (*K*) not already taken by the current population size (*N*). The term $(1 - N/K)$ establishes the relationship between population size at any given time step and the number of individuals the environment can support. Depending on whether *N* is less than or greater than *K*, the rate of growth will be positive or negative (see Active Learning).

The logistic growth curve has a different shape than the exponential growth curve. It is a sigmoidal-shaped, or **S curve** (fig. 3.24). It describes a population that decreases if its numbers exceed the carrying capacity of the environment.

Population growth rates are affected by external and internal factors. External factors are habitat quality, food availability, and interactions with other organisms. As populations grow, food becomes scarcer and competition for resources more intense. With a larger population, there is an increased risk that disease or parasites will spread, or that predators will be attracted to the area. Some organisms become physiologically stressed when conditions are crowded, but other internal factors of maturity, body size, and hormonal status may cause them to reduce their

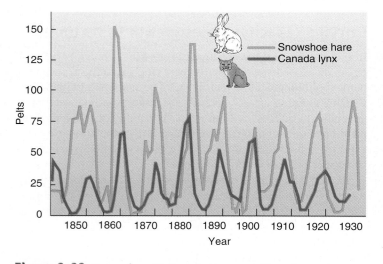

Figure 3.23 Ten-year oscillations in the population of snowshoe hare and lynx in Canada suggest a close linkage of predator and prey, but may not tell the whole story. These data are based on the number of pelts received by the Hudson Bay Company each year, meaning fur-traders were unwitting accomplices in later scientific research.
Source: Data from D. A. MacLulich. Fluctuations in the Numbers of the Varying Hare (*Lepus americanus*). Toronto: University of Toronto Press, 1937, reprinted 1974.

Table 3.3	Reproductive Strategies	
r-Selected Species	**K-Selected Species**
1. Short life	1. Long life
2. Rapid growth	2. Slower growth
3. Early maturity	3. Late maturity
4. Many, small offspring	4. Few, large offspring
5. Little parental care and protection	5. High parental care or protection
6. Little investment in individual offspring	6. High investment in individual offspring
7. Adapted to unstable environment	7. Adapted to stable environment
8. Pioneers, colonizers	8. Later stages of succession
9. Niche generalists	9. Niche specialists
10. Prey	10. Predators
11. Regulated mainly by intrinsic factors	11. Regulated mainly by extrinsic factors
12. Low trophic level	12. High trophic level

reproductive output. Overcrowded house mice (>1,600/m³), for instance, average 5.1 baby mice per litter, while uncrowded house mice (<34/m³) produce 6.2 babies per litter. All these factors are **density-dependent**, meaning as population size increases, the effect intensifies. With density-independent factors, a population is affected no matter what its size. Drought, an early killing frost, flooding, landslide, or habitat destruction by people—all increase mortality rates regardless of the population size. Density-independent limits to population are often nonbiological, capricious acts of nature.

Species respond to limits differently: *r*- and *K*-selected species

The story of the race between the hare and the tortoise has parallels to the way that species deal with limiting factors in their environment. Some organisms, such as dandelions and barnacles, depend on a high rate of reproduction and growth (*rN*) to secure a place in the environment. These organisms are called ***r*-selected species** because they employ a high reproductive rate (*r*) to overcome the high mortality of virtually ignored offspring. These species may even overshoot carrying capacity and experience population crashes, but as long as vast quantities of young are produced, a few will survive. Other organisms that reproduce more conservatively—longer generation times, late sexual maturity, fewer young—are referred to as ***K*-selected species**, because their growth slows as the carrying capacity (*K*) of their environment is approached.

Many species blend exponential (*r*-selected) and logistic (*K*-selected) growth characteristics. Still, it's useful to contrast the advantages and disadvantages of organisms at the extremes of the continuum. It also helps if we view differences in terms of "strategies" of adaptation and the "logic" of different reproductive modes (table 3.3).

Organisms with *r*-selected, or exponential, growth patterns tend to occupy low trophic levels in their ecosystems (chapter 2) or they are successional pioneers. Niche generalists occupy disturbed or new environments, grow rapidly, mature early, and produce many offspring with excellent dispersal abilities. As individual parents, they do little to care for their offspring or protect them from predation. They invest their energy in producing huge numbers of young and count on some surviving to adulthood (fig. 3.25).

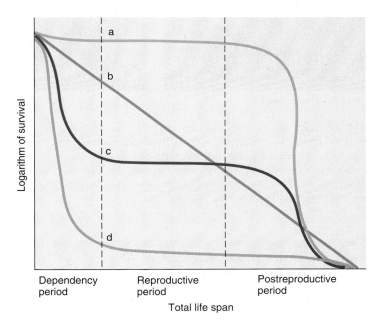

Figure 3.25 Four basic types of survivorship curves for organisms with different life histories. Curve (a) represents organisms such as humans or elephants, which tend to live out the full physiological life span if they survive early growth. Curve (b) represents organisms such as sea gulls, which have a fairly constant mortality at all age levels. Curve (c) represents such organisms as antelope, which have high mortality rates in early and late life. Curve (d) represents such organisms as clams and redwood trees, which have a high mortality rate early in life but live a full life if they reach adulthood.

A female clam, for example, can release up to 1 million eggs in her lifetime. The vast majority of young clams die before reaching maturity, but if even a few survive, the species will continue.

(a) Survive to old age

(b) Die randomly

(c) High infant mortality

(d) Long adult life span

Figure 3.26 Which of these appears to be *r*- and which *K*-selected in their reproductive strategy? (a) Most elephants live long lives, but reproduce relatively late. (b) Seagulls die at all ages, mostly from accidents. (c) Antelope have high infant mortality, but survive well after that. (d) Seedlings of redwood trees die off in huge numbers, but the oldest redwood is at least 2,200 years old.

Many marine invertebrates, parasites, insects, rodents, and annual plants follow this reproductive strategy. Also included in this group are most invasive and pioneer organisms, weeds, pests, and nuisance species.

So-called *K*-selected organisms are usually larger, live long lives, mature slowly, produce few offspring in each generation, and have few natural predators. Elephants, for example, are not reproductively mature until they are 18 to 20 years old. In youth and adolescence, a young elephant belongs to an extended family that cares for it, protects it, and teaches it how to behave. A female elephant normally conceives only once every 4 or 5 years. The gestation period is about 18 months; thus, an elephant herd doesn't produce many babies in any year. Since elephants have few enemies and live a long life (60 or 70 years), this low reproductive rate produces enough elephants to keep the population stable, given good environmental conditions and no poachers.

When you consider the species you recognize from around the world, can you pigeonhole them into categories of *r*- or *K*- selected species (fig. 3.26)? What strategies seem to be oper-

ating for ants, bald eagles, cheetahs, clams, giraffes, pandas, or sharks? An important question in the back of your mind might be: Where do people fit? Are we like wolves and elephants or more similar to clams and rabbits in our population growth strategy?

3.4 Properties of Communities Depend on Species Diversity

No species is an island. It always lives with other species in a biological community in a particular environment. You've seen how interactions among species affect biological communities. In this section, you will learn how species interact to create the fundamental properties of biological communities and ecosystems: (1) diversity and abundance; (2) community structure and patchiness; and (3) complexity, resilience, productivity, and stability.

Diversity and abundance

In the opening case study, diversity resulted in ecological resilience, but what is diversity exactly? **Diversity** is the number of different species per unit area. The number of herbaceous plant species per square meter of African savanna, the number of bird species in Costa Rica, the insect species on earth (a big number!)—all describe diversity. Diversity is important because it indicates the variety of ecological niches and genetic variation in a community. **Abundance** refers to the number of individuals of a species in an area. It is expressed by density of individuals of either a single species or multiple species: the number of diatoms floating in 1 m³ of sea water, the number of sparrows per hectare of urban habitat. Diversity and abundance are often related. Communities with high diversity often have few individuals of any one species. Most communities contain a few common species and many rarer ones.

As a general rule, diversity is greatest at the equator and drops toward the poles. For many species, the abundance of individuals tends to increase along the same gradient. Individual mosquitoes are abundant in the Arctic, but there are few insect species overall. In the tropics, hundreds of thousands of insect species have developed bizarre forms and behaviors, but at any one location only a few individuals of a species are found. The same pattern is seen in trees of high-latitude boreal forests versus tropical rainforests, and in bird populations. In Greenland there reside 56 species of breeding birds, while in Colombia, with one-fifth the area, there are 1,395. Why are there so many more species in Colombia? Climate and history play important roles. Greenland has a harsh climate and a short, cool growing season that restricts the biological activity that can take place. Limited energy overwhelms all other factors and prevents specialization and the development of ecological niches—hence, fewer species. Furthermore, because glaciers

What Can You Do?

Working Locally for Ecological Diversity

You might think that diversity and complexity of ecological systems are too large or too abstract for you to have any influence. But you can contribute to a complex, resilient, and interesting ecosystem, whether you live in the inner city, a suburb, or a rural area.

- Keep your cat indoors. Our lovable domestic cats are also very successful predators. Migratory birds, especially those nesting on the ground, have not evolved defenses against these predators.

- Plant a butterfly garden. Use native plants that support a diverse insect population. Native trees with berries or fruit also support birds. (Be sure to avoid non-native invasive species.) Allow structural diversity (open areas, shrubs, and trees) to support a range of species.

- Join a local environmental organization. Often, the best way to be effective is to concentrate your efforts close to home. City parks and neighborhoods support ecological communities, as do farming and rural areas. Join an organization working to maintain ecosystem health; start by looking for environmental clubs at your school, park organizations, a local Audubon chapter, or a local Nature Conservancy branch.

- Take walks. The best way to learn about ecological systems in your area is to take walks and practice observing your environment. Go with friends and try to identify some of the species and trophic relationships in your area.

- Live in town. Suburban sprawl consumes wildlife habitat and reduces ecosystem complexity by removing many specialized plants and animals. Replacing forests and grasslands with lawns and streets is the surest way to simplify, or eliminate, ecosystems.

covered Greenland until about 10,000 years ago, new species have had little time to develop.

Many areas in the tropics, by contrast, have abundant rainfall and warm temperatures year-round, so that ecosystems there are highly productive. The year-round availability of food, moisture, and warmth supports an exuberance of life and allows a high

degree of specialization in physical shape and behavior. Many niches exist in small areas, with associated high species diversity. Coral reefs are similarly stable, productive, and conducive to proliferation of diverse and exotic life-forms. An enormous abundance of brightly colored and fantastically shaped fishes, corals, sponges, and arthropods live in the reef community. Glaciers never covered tropical forests or coral reefs, although the past climate of the tropics was different from today's.

Species patterns create community structure

Just like jig-saw puzzle pieces, boundaries of species populations and communities form patterns that fit together: (1) individuals and species are spaced throughout communities in different ways; (2) the communities themselves are arranged over a large geographic area or landscape; and (3) communities have relatively uniform interiors ("cores") and also "edges" that meet. **Community** (or ecological) **structure** refers to these patterns of spatial distribution of individuals, species, and communities.

Individuals in communities are distributed in various ways

Even in a relatively uniform environment, individuals of a species population are distributed randomly, arranged in uniform patterns, or clustered together. In randomly distributed populations, individuals live wherever resources are available and chance events allowed (fig. 3.27a). Uniform patterns arise from the physical environment also, but more often are caused by competition. For example, penguins or seabirds compete fiercely for nesting sites in their colonies. Each nest tends to be just out of reach of neighbors sitting on their own nests. Constant squabbling produces a highly regular pattern (fig. 3.27b). Plants also compete, producing a uniform pattern. Sagebrush releases toxins from roots and fallen leaves, which inhibit the growth of competitors and create a circle of bare ground around each bush. Neighbors grow up to the limit of this chemical barrier, and regular spacing results.

Other species cluster together for protection, mutual assistance, reproduction, or access to an environmental resource. Ocean and freshwater fish form dense schools, increasing their chances of detecting and escaping predators (fig. 3.27c). Meanwhile many

(a) Random   (b) Uniform   (c) Clustered

Figure 3.27 Distribution of members of a population in a given space can be (a) random, (b) uniform, or (c) clustered. These patterns in the distribution of individuals give structure to communities.

Figure 3.28 Vertical layering of plants and animals in a tropical rainforest is a type of community structure.

Figure 3.29 Ecological edges are called ecotones. In this scene an ecotone exists between the lake and wetland, and another marks the boundary between wetland and forest. The strip of forest in the center of the photo is all edge—sunny, hotter conditions on both sides penetrate to its center. The forest strip could be a travel route, or corridor, however, for forest animals crossing the wetland.

predators—whether wolves or humans—hunt in packs. When blackbirds flock in a cornfield, or baboons troop across the African savanna, their bands help them evade predators and find food efficiently. Plants also cluster for protection in harsh environments. You often see groves of wind-sheared evergreens at mountain treelines or behind foredunes at seashores. These treelines protect the plants from wind damage, and incidentally shelter other animals and plants.

Individuals can also be distributed vertically in a community. Forests, for instance, have many layers, each with different environmental conditions and combinations of species. Distinct communities of plants, animals, and microbes live in the treetops, at mid-canopy level, and near the ground. This vertical layering is best developed in tropical rainforests (fig. 3.28). Aquatic communities also are often stratified into layers formed by species responding to varying light levels, temperature, salinity, nutrient supply, and pressure.

Communities are distributed in patterns across a landscape

If you have flown in an airplane, you may have noticed that the land consists of patches of different colors and shapes. Some appear long and narrow (hedgerows or rivers), while others are rectangular (pastures and cropfields), or green and lumpy in summer (forests). Each patch represents a biological community with its own set of species and environmental conditions. The name for this pattern of community distribution across a landscape is patchiness or habitat patchiness. The largest patches contain **core habitat**, a mostly uniform environment big enough to support nearly all the plants and animals that are typically found in that community. For example, the largest patches of old coniferous forest in the Pacific Northwest contain the most persistent populations of northern spotted owl, a threatened species. In smaller patches the owl fares poorly, possibly due to competition with the barred owl. A single pair of northern spotted owls may require over 1,000 hectares of core habitat to survive.

Outside habitat cores, species encounter different habitat in what ecologists call the **ecotone**, or border between two communities (fig. 3.29). Ecotones are often rich in species because individuals from both environments occupy the boundary area. Species that prefer ecotones and use the resources of both environments are called edge species. The white-tailed deer is one, browsing in fields but hiding in forests. Sometimes a community edge seems sharp and distinct, but other times one habitat type grades gradually into another.

Where communities meet, the environmental conditions blend and the species and microclimate of one community can penetrate the other. Called **edge effects**, the penetrating influences may extend hundreds of meters into an adjacent community. Grassland makes an adjacent forest edge sunnier, drier, hotter, and more susceptible to storm damage than the center of the forest. Generalist grassland species, including weeds and predators, move into the forest and negatively affect forest species. Depending on how far the edge effects penetrate, the shape of a community patch makes a big difference (fig. 3.30). In a narrow, irregularly shaped patch, far-reaching edge effects would leave no core habitat. In a similar-sized square patch, however, interior species would still find core habitat.

Many game animals (deer, pheasant, etc.) are most plentiful in ecotones and adapt well to human disturbance. With this knowledge, in the 1930s, North American game managers began to boost game populations by cutting openings in forests and planting trees and shrubs in grasslands. Since 1980, however, wildlife biologists recognized that creating edge effects everywhere is harmful to a region's biota, and that core habitat can benefit edge species by serving as refuges. In this view of habitat management, large community patches are preserved, connected to each other, and linked to smaller patches (see fig. 6.27). You can learn about bird life, core habitat, and edge effects by reading "Where Have All the Songbirds Gone?" at www.mhhe.com/cunningham5e.

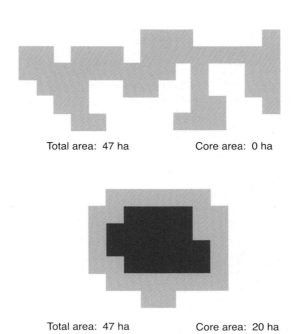

Total area: 47 ha   Core area: 0 ha

Total area: 47 ha   Core area: 20 ha

Figure 3.30 Shape can be as important as size in small preserves. While these areas are similar in size, no place in the top figure is far enough from the edge to have characteristics of core habitat, while the bottom patch has a significant core.

Table 3.4	Community Complexity in the Antarctic Ocean
Type of Function	**Members of Functional Group**
Top ocean predator	Sperm and killer whales, leopard and elephant seals
Aerial predator	Albatross, skuas
Other ocean predator	Weddell and Ross seals, king penguin, pelagic fish
Krill/plankton-feeder	Minke, humpback, fin, blue, and sei whales
Ocean herbivore	Krill, zooplankton (many species)
Ocean-bottom predator	Many species of octopods and bottom-feeding fish
Ocean-bottom herbivore	Many species of echinoderms, crustaceans, mollusks
Photosynthesizer	Many species of phytoplankton and algae

Community properties emerge from diversity and structure

As the opening case study for this chapter shows, more-diverse communities can recover faster from disturbance. But there's more to resilience than diversity and structure, and that is community complexity. **Complexity** refers to the number of trophic levels in a community and to the number of species at each of those trophic levels. A diverse community may not be complex if all of its species cluster in a few trophic levels and form a simple food chain. Take one key level or species away, and the community may unravel.

A complex, interconnected community might have many trophic levels and groups of species performing the same functions (table 3.4). In the Antarctic, the sun powers the system, but detritus that consists of dead organisms and animal excrement is a key energy source. Whales and leopard seals are top predators, while krill (tiny crustaceans) play a key energy-transfer role. In tropical rainforests and many other communities, herbivores form guilds based on the specialized ways they feed on plants. There may be fruit-eaters, leaf-nibblers, root-borers, seed-gnawers, and sap-suckers—each guild is composed of species of different sizes, shapes, and even biological kingdoms, but they feed in similar ways.

In 1955, Robert MacArthur (1930–1972), then a graduate student at Yale, proposed that the more complexity a community possesses, the more resilient it is when disturbance strikes. He reasoned that, if many different species occupy each trophic level, some can fill in if others are stressed or eliminated by external forces. The whole community has **resilience** and either resists or recovers quickly from disturbance, as witnessed in the Cedar Creek experiment.

In complex ecosystems, like kelp forests or intertidal zones, keystone species confuse the picture. For example, cutting all the fig trees from a tropical forest may destroy pollinators and fruit dispersers. We might replant the figs, but could we bring back the

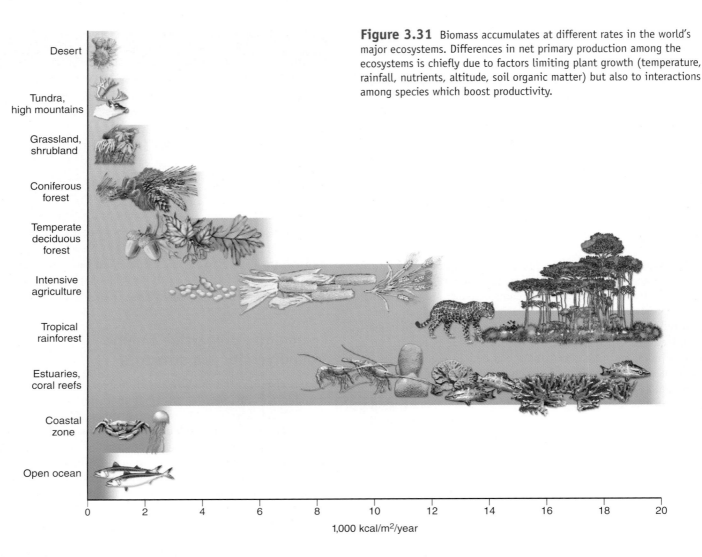

Figure 3.31 Biomass accumulates at different rates in the world's major ecosystems. Differences in net primary production among the ecosystems is chiefly due to factors limiting plant growth (temperature, rainfall, nutrients, altitude, soil organic matter) but also to interactions among species which boost productivity.

Desert

Tundra,
high mountains

Grassland,
shrubland

Coniferous
forest

Temperate
deciduous
forest

Intensive
agriculture

Tropical
rainforest

Estuaries,
coral reefs

Coastal
zone

Open ocean

0   2   4   6   8   10   12   14   16   18   20

1,000 kcal/m^2/year

entire web of relationships that previously existed? In such cases, complexity seems to make a community less resilient, rather than more. While the concept that diversity and complexity lead to resilience is not perfect, it helps explain many observations about communities and urges us to preserve communities in as diverse and complex a condition as we can.

When we say something is productive, we mean it produces a lot of something. Communities produce biomass by converting solar energy into chemical energy that is stored in living (or once-living) organisms. **Primary productivity**, a community's annual output of biomass or energy, is expressed as units of biomass or energy per unit area per year. Since cellular respiration in organisms uses much of that energy, a more useful term is **net primary productivity**, or the amount of biomass stored after respiration. Productivity depends on light levels, temperature, moisture, and nutrient availability. Major ecosystems differ in how fast they produce biomass (fig. 3.31). Tropical forests, coral reefs, and the bays and drowned valleys where rivers meet the ocean (estuaries) have high productivity because of abundant resource supply. In deserts, a lack of water limits photosynthesis, and productivity is low. On the arctic tundra or on high mountains, low temperatures

inhibit plant growth and productivity. In the open ocean, a lack of nutrients reduces the ability of algae to make use of plentiful sunshine and water.

Even the most photosynthetically active ecosystems capture only a small percentage of the available sunlight and use it to make energy-rich compounds. In a temperate-climate oak forest, leaves absorb only about half the available light on a midsummer day. Of this absorbed energy, 99 percent is used in respiration and in evaporating water to cool leaves. A large oak tree can transpire (evaporate) several thousand liters of water on a warm, dry, sunny day, while making only a few kilograms of sugars and other energy-rich organic compounds.

Stability is a complex concept. When we say a community or ecosystem is stable, we mean it resists changes despite disturbance, springs back resiliently after disturbance, and supports the same species in about the same numbers as before the disturbance. Yet sometimes only one of these happens. When the Great Plains experienced devastating droughts in the 1930s and early 1940s, overall productivity declined, some species populations practically vanished, and others held their own but at lower abundance. When the rains returned, the plant communities were different, especially because before, during, and after the drought millions of cattle overgrazed the range. Today's Great Plains grasslands are very different from the grasslands George Armstrong Custer viewed from his horse in the late 1800s. Yet it remains grassland, produces forage, and feeds cattle. Are the Great Plains grasslands stable because they are still relatively productive, or unstable because species diversity and abundance changed since the 1800s? If the range was grazed according to ecological principles, would that restore the original species diversity and abundance, and raise productivity? Thinking about stability in this way allows us to figure out how best to use resources, rather than to simply say that diversity equals stability—which may be true or not, depending on your definition of stability.

3.5 Communities Are Dynamic and Change Over Time

If fire sweeps through a biological community, it's destroyed, right? Not necessarily. Fire may be good for that community. Up until now, we've focused on the day-to-day interactions of organisms with their environments, set in a context of adaptation and selection. In this section, we'll step back and look at more dynamic aspects of communities and how they change over time.

The nature of communities is debated

For several decades starting in the early 1900s, ecologists in North America and Europe argued about the basic nature of communities. It doesn't make interesting party conversation, but those discussions affected how we study and understand communities, view the changes taking place within them, and ultimately use them. Both J. E. B. Warming (1841–1924) in Denmark and Henry Chandler Cowles (1869–1939) in the United States came up with the idea that communities develop in a sequence of stages, starting either from bare rock or after a severe disturbance. They worked in sand dunes and watched the changes as plants first took root in bare sand and, with further development, created forest. This example represents constant change, not stability. In sand dunes, the community that developed last and lasted the longest was called the **climax community**.

The idea of climax community was first championed by the biogeographer F. E. Clements (1874–1945). He viewed the process as a relay—species replace each other in predictable groups and in a fixed, regular order. He argued that every landscape has a characteristic climax community, determined mainly by climate. If left undisturbed, this community would mature to a characteristic set of organisms, each

performing its optimal functions. A climax community to Clements represented the maximum complexity and stability that was possible. He and others made the analogy that the development of a climax community resembled the maturation of an organism. Both communities and organisms, they argued, began simply and primitively, maturing until a highly integrated, complex community developed.

This organismal theory of community was opposed by Clements' contemporary, H. A. Gleason (1882–1975), who saw community history as an unpredictable process. He argued that species are individualistic, each establishing in an environment according to its ability to colonize, tolerate the environmental conditions, and reproduce there. This idea allows for myriad temporary associations of plants and animals to form, fall apart, and reconstitute in slightly different forms, depending on environmental conditions and the species in the neighborhood. Imagine a time-lapse movie of a busy airport terminal. Passengers come and go; groups form and dissipate. Patterns and assemblages that seem significant may not mean much a year later. Gleason suggested that we think ecosystems are uniform and stable only because our lifetimes are too short and our geographic scope too limited to understand their actual dynamic nature.

Ecological succession describes a history of community development

In any landscape, you can read the history of biological communities. That history is revealed by the process of ecological succession. During succession, organisms occupy a site and change the environmental conditions. In **primary succession** land that is bare of soil—a sandbar, mudslide, rock face, volcanic flow—is colonized by living organisms where none lived before (fig. 3.32). When an existing community is disturbed, a new one develops from the biological legacy of the old in a process

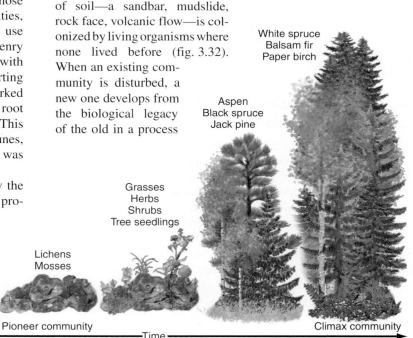

Figure 3.32 Primary succession in boreal forest involves five stages (*left to right*). Bare rock is colonized by lichens and mosses which trap moisture and build soil for grasses, shrubs, and eventually trees. Natural, periodic outbreaks of insects kill or weaken trees. Wildfires often follow, initiating secondary succession. Aspens spring back from roots, jack pine cones open and spread seeds when heated by wildfires, and a new forest begins.

called **secondary succession**. In both kinds of succession, organisms change the environment by modifying soil, light levels, food supplies, and microclimate. This change permits new species to colonize and eventually replace the previous species, a process known as ecological development or facilitation.

In primary succession on land, the first colonists are hardy **pioneer species**, often microbes, mosses, and lichens that can withstand a harsh environment with few resources. When they die, the bodies of pioneer species create patches of organic matter. Organics and other debris accumulate in pockets and crevices, creating soil where seeds lodge and grow. As succession proceeds, the community becomes more diverse and interspecies competition arises. Pioneers disappear as the environment favors new colonizers that have competitive abilities more suited to the new environment.

You can see secondary succession all around you, in abandoned farm fields, in clear-cut forests, and in disturbed suburbs and lots. Soil and possibly plant roots and seeds are present. Because soil lacks vegetation, plants that live one or two years (annuals and biennials) do well. Their light seeds travel far on the wind, and their seedlings tolerate full sun and extreme heat. When they die, they lay down organic material that improves the soil's fertility and shelters other seedlings. Soon long-lived and deep-rooted perennial grasses, herbs, shrubs, and trees take hold, building up the soil's organic matter and increasing its ability to store moisture. Forest species that cannot survive bare, dry, sunny ground eventually find ample food, a diverse community structure, and shelter from drying winds and low humidity.

Generalists figure prominently in early succession. Over thousands of years, however, competition should decrease as niches proliferate and specialists arise. In theory, long periods of community development lead to greater community complexity, high nutrient conservation and recycling, stable productivity, and great resistance to disturbance—an ideal state to be in when the slings and arrows of misfortune arrive.

Appropriate disturbances can benefit communities

Disturbances are plentiful on earth: landslides, mudslides, hailstorms, earthquakes, hurricanes, tornadoes, tidal waves, wildfires, and volcanoes, to name just the obvious. A **disturbance** is any force that disrupts the established patterns of species diversity and abundance, community structure, or community properties. The drought that reduced plant productivity in the opening case study for this chapter is a good example. Animals can cause disturbance. African elephants rip out small trees, trample shrubs, and tear down tree limbs as they forage and move about, opening up forest communities and creating savannas.

People also cause disturbances in many ways. Aboriginal people set fires, introduced new species, harvested resources, or changed communities in many places. Sometimes, those disturbances had major ecological effects. Archaeological evidence suggests that when humans colonized islands, such as New Zealand, Hawaii, and Madagascar, large-scale extinction of many

Figure 3.33 The barrens community at Kingston Plains in Michigan's eastern Upper Peninsula was created when clear-cutting of a dense white pine forest coincided with several devastating fires set by people. Note the stumps from trees cut over 100 years ago.

animal species followed. Although there is debate about whether something similar happened in the Americas at the end of the last ice age, about the same time that humans arrived, many of the large mammal species disappeared. In some cases, the landscapes created by aboriginal people may have been maintained for so long that we assume that those conditions always existed there. In eastern North America, for example, fires set by native people maintained grassy, open savannas that early explorers assumed were the natural state of the landscape.

The disturbances caused by modern technological societies are often much more obvious and irreversible. You undoubtedly have seen scars from road-building, mining, clear-cut logging, or other disruptive activities. Obviously, once a landscape has been turned into a highway, a parking lot, or a huge hole in the ground, it will be a long time—if ever—before it returns to its former condition. But sometimes even if the disturbance seems relatively slight, it can have profound effects.

Consider the Kingston Plains in Michigan's Upper Peninsula, for example. Clear-cut logging at the end of the nineteenth century removed the white pine forest that once grew there. Repeated burning by pioneers, who hoped to establish farms, removed nutrients from the sandy soil and changed ecological conditions so that more than a century later, the forest still hasn't regenerated (fig. 3.33). Given extensive changes by either humans or nature, it may take centuries for a site to return to its predisturbance state, and if climate or other conditions change in the meantime, it may never recover.

Ecologists often find that disturbance benefits many species, much as predation does, because it sets back supreme competitors and allows less-competitive species to persist. In northern temperate forests, maples (especially sugar maple) are more prolific seeders and more shade tolerant at different stages of growth than nearly any other tree species. Given decades of succession, maples outcompete other trees for a place in the forest canopy. Most species of oak, hickory, and other light-requiring trees diminish in abundance, as do species of forest herbs. The dense shade of maples basically starves other species for light. When windstorms, tornadoes, wildfires, or ice storms hit a maple forest, trees are toppled, branches broken, and light again reaches the forest floor and stimulates seedlings of oaks and hickories, as well as forest herbs. Breaking the grip of a supercompetitor is the helpful role disturbances often play. The 1988 Yellowstone fires knocked back lodgepole pine (*Pinus contorta*) that had expanded its acreage in the park (fig. 3.34). In a couple of years a greater variety of plant species grew where previously there was mostly pine. Wildlife responded vigorously to the greater diversity.

Some landscapes never reach a stable climax in a traditional sense because they are characterized by periodic disturbance and are made up of **disturbance-adapted species** that survive fires underground, or resist the flames, and then reseed quickly after fires. Grasslands, the chaparral scrubland of California and the Mediterranean region, savannas, and some kinds of coniferous forests are shaped and maintained by periodic fires that have long been a part of their history. In fact, many of the dominant plant species in these communities need fire to suppress competitors, to prepare the ground for seeds to germinate, or to pop open cones or split thick seed coats and release seeds. Without fire, community structure would be quite different.

People taking an organismal view of such communities believe that disturbance is harmful. In the early 1900s this view merged with the desire to protect timber supplies from ubiquitous wildfires, and to store water behind dams while also controlling floods. Fire suppression and flood control became the central policies in American natural resource management (along with predator control) for most of the twentieth century. Recently, new concepts about natural disturbances are entering land management discussions and bringing change to land management policies. Grasslands and some forests are now considered "fire-adapted" and fires are allowed to burn in them if weather conditions are appropriate. Floods also are seen as crucial for maintaining floodplain and river health. Policymakers and managers increasingly consider ecological information when deciding on new dams and levee construction projects. You can read more about this in chapter 6 and in the related stories "Prescribed Fires in New Mexico" and "Fires in the Boundary Waters Canoe Area Wilderness" at www.mhhe.com/cunningham5e.

From another view, disturbance resets the successional clock that always operates in every community. Even though all seems chaotic after a disturbance, it may be that preserving species diversity by allowing in natural disturbances (or judiciously applied human disturbances) actually ensures stability over the long run, just as diverse prairies managed with fire recover after drought.

Conclusion

Evolution is one of the key organizing principles of biology. It explains how species diversity originates, and how organisms are able to live in highly specialized ecological niches. Natural selection, in which beneficial traits are passed from survivors in one generation to their progeny, is the mechanism by which evolution occurs. Species interactions—competition, predation, symbiosis, and coevolution—are important factors in natural selection. The unique set of organisms and environmental conditions in an ecological community give rise to important properties, such as productivity, abundance, diversity, structure, complexity, connectedness, resilience, and succession. Human introduction of new species as well as removal of existing ones can cause profound changes in biological communities and can compromise the life-supporting ecological services on which we all depend. Understanding these community ecology principles is a vital step in becoming an educated environmental citizen.

Figure 3.34 In 1988 (the driest year on record) lightning-ignited wildfires burned 1.4 million acres in the Greater Yellowstone area. Before the fires, 80 percent of Yellowstone National Park was dominated by lodgepole pine forest like this one. After the fires, some forests became savannas, while others grew into dense thickets. Overall, fires increased variety in the region and promoted species that need more light than is available in forests.

Practice Quiz

1. Explain how tolerance limits to environmental factors determine distribution of a highly specialized species such as the saguaro cactus.

2. Productivity, diversity, complexity, resilience, and structure are exhibited to some extent by all communities and ecosystems. Describe how these characteristics apply to the ecosystem in which you live.

3. Define *selective pressure* and describe one example that has affected species where you live.

4. Define *keystone species* and explain their importance in community structure and function.

5. The most intense interactions often occur between individuals of the same species. What concept discussed in this chapter can be used to explain this phenomenon?

6. Explain how predators affect the adaptations of their prey.

7. Competition for a limited quantity of resources occurs in all ecosystems. This competition can be interspecific or intraspecific. Explain some of the ways an organism might deal with these different types of competition.

8. Describe the process of succession that occurs after a forest fire destroys an existing biological community. Why may periodic fire be beneficial to a community?

9. Which world ecosystems are most productive in terms of biomass (fig. 3.31)? Which are least productive? What units are used in this figure to quantify biomass accumulation?

10. Discuss the dangers posed to existing community members when new species are introduced into ecosystems.

Critical Thinking and Discussion Questions

Apply the principles you have learned in this chapter to discuss these questions with other students.

1. The concepts of natural selection and evolution are central to how most biologists understand and interpret the world, yet the theory of evolution appears to question the beliefs of many religious groups. Why do you think this theory is so widely accepted in science? Does it necessarily question religious beliefs?

2. What is the difference between saying that a duck has webbed feet because it needs them to swim and saying that a duck is able to swim because it has webbed feet?

3. Given the information in this chapter, do you suspect scientists of having an ulterior motive in coming up with the idea of speciation through natural selection and adaptation? How would you communicate these scientific ideas to a nonscientist?

4. The concept of keystone species is controversial among ecologists because most organisms are highly interdependent. If each trophic level in a community depends on all the others, how can we say that one is most important? Choose a familiar ecosystem and decide whether it has a keystone species or keystone set.

5. Some scientists look at the boundary between two biological communities and see a sharp dividing line. Others looking at the same boundary see a gradual transition with much intermixing of species and many interactions between communities. Is there a difference in education or personality that might cause one scientist to see clarity, and another complexity in the same landscape?

6. Ecologists debate whether biological communities have self-sustaining, self-regulating characteristics or are highly variable, accidental assemblages of individually acting species. What outlook or worldview might lead scientists to favor one or the other of these theories?

Data Analysis: Species Competition

In a classic experiment on competition between species for a common food source, the Russian microbiologist G. F. Gause grew populations of different species of ciliated protozoans separately and together in an artificial culture medium. He counted the number of cells of each species and plotted the total volume of each population. The organisms were *Paramecium caudatum* and its close relative, *Paramecium aurelia*. He plotted the aggregate volume of cells rather than the total number in each population because *P. caudatum* is much larger than *P. aurelia* (this size difference allowed him to distinguish between them in a mixed culture). The graphs in this box show the experimental results. As we mentioned earlier in the text, this was one of the first experimental demonstrations of the principle of competitive exclusion. After studying these graphs, answer the following questions.

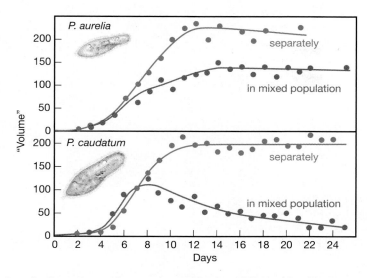

Growth of two *paramecium* species separately and in combination.
Source: Gause, Georgyi Frantsevitch. 1934 *The Struggle for Existence*. Dover Publications, 1971, reprint of original text.

1. How do you read these graphs? What is shown in the top and bottom panels?

2. How did the total volume of the two species compare after 14 days of separate growth?

3. If *P. caudatum* is roughly twice as large as *P. aurelia*, how did the total number of cells compare after 14 days of separate growth?

4. How did the total volume of the two species compare after 24 days of growth in a mixed population?

5. Which of the two species is the more successful competitor in this experiment?

6. Does the larger species always win in competition for food? Why not?

For Additional Help in Studying This Chapter, please visit our website at www.mhhe.com/cunningham5e. You will find practice quizzes, key terms, answers to end of chapter questions, additional case studies, an extensive reading list, and Google Earth™ mapping quizzes.

Thailand's highly successful family planning program combines humor and education with economic development.

Human Populations

Learning Outcomes

After studying this chapter, you should be able to answer the following questions:

- Why are we concerned about human population growth?
- Will the world's population double again as it did between 1965 and 2000?
- What is the relationship between population growth and environmental impact?
- Why has the human population grown so rapidly since 1800?
- How is human population growth changing in different parts of the world?
- How does population growth change as a society develops?
- What factors slow down or speed up human population growth?

For every complex problem there is an answer that is clear, simple, and wrong.

–H. L. Mencken

Family Planning in Thailand: A Success Story

Down a narrow lane off Bangkok's busy Sukhumvit Road, is a most unusual café. Called Cabbages and Condoms, it's not only highly rated for its spicy Thai food, but it's also the only restaurant in the world dedicated to birth control. In an adjoining gift shop, baskets of condoms stand next to decorative handicrafts of the northern hill tribes. Piles of T-shirts carry messages, such as, "A condom a day keeps the doctor away," and "Our food is guaranteed not to cause pregnancy." Both businesses are run by the Population and Community Development Association (PDA), Thailand's largest and most influential nongovernmental organization.

The PDA was founded in 1974 by Mechai Viravaidya, a genial and fun-loving former Thai Minister of Health, who is a genius at public relations and human motivation (fig. 4.1). While traveling around Thailand in the early 1970s, Mechai recognized that rapid population growth—particularly in poor rural areas—was an obstacle to community development. Rather than lecture people about their behavior, Mechai decided to use humor to promote family planning. PDA workers handed out condoms at theaters and traffic jams, anywhere a crowd gathered. They challenged governmental officials to condom balloon-blowing contests, and taught youngsters Mechai's condom song: "Too Many Children Make You Poor." The PDA even pays farmers to paint birth control ads on the sides of their water buffalo.

This campaign has been extremely successful at making birth control and family planning, which once had been taboo topics in polite society, into something familiar and unembarrassing. Although condoms—now commonly called "mechais" in Thailand—are the trademark of PDA, other contraceptives, such as pills, spermicidal foam, and IUDs, are promoted as well. Thailand was one of the first countries to allow the use of the injectable contraceptive DMPA, and remains a major user. Free non-scalpel vasectomies are available on the king's birthday. Sterilization has become the most widely used form of contraception in the country. The campaign to encourage condom use has also been helpful in combating AIDS.

In 1974, when PDA started, Thailand's growth rate was 3.2 percent per year. In just 15 years, contraceptive use among married couples increased from 15 to 70 percent, and the growth rate had dropped to 1.6 percent, one of the most dramatic birth rate declines ever recorded. Now Thailand's growth rate is 0.7 percent, or nearly the same as the United States. The fertility rate (or average number of children per woman) decreased from 7 in 1974 to 1.7 in 2006. The PDA is credited with the fact that Thailand's population is 20 million less than it would have been if it had followed its former trajectory.

In addition to Mechai's creative genius and flair for showmanship, there are several reasons for this success story. Thai people love humor and are more egalitarian than most developing countries. Thai spouses share in decisions regarding children, family life, and contraception. The government recognizes the need for family planning and is willing to work with volunteer organizations, such as the PDA. And Buddhism, the religion of 95 percent of Thais, promotes family planning.

The PDA hasn't limited itself to family planning and condom distribution. It has expanded into a variety of economic development projects. Microlending provides money for a couple of pigs, or a bicycle, or a small supply of goods to sell at the market. Thousands of water-storage jars and cement rainwater-catchment basins have been distributed. Larger scale community development grants include road building, rural electrification, and irrigation projects. Mechai believes that human development and economic security are keys to successful population programs.

This case study introduces several important themes of this chapter. What might be the effects of exponential growth in human populations? How might we manage fertility and population growth? And what are the links between poverty, birth rates, and our common environment? Keep in mind, as you read this chapter, that resource limits aren't simply a matter of total number of people on the planet, they also depend on consumption levels and the types of technology used to produce the things we use.

Figure 4.1 Mechai Viravaidya (*right*) is joined by Peter Piot, Executive Director of UNAIDS, in passing out free condoms on family planning and AIDS awareness day in Bangkok.

4.1 Past and Current Population Growth Are Very Different

Every second, on average, four or five children are born, somewhere on the earth. In that same second, two other people die. This difference between births and deaths means a net gain of roughly 2.3 more humans per second in the world's population. The U.S. Census Bureau estimates the mid-2008 world population to be about 6.7 billion people and growing at 1.15 percent per year. This means we are adding roughly 75 million more people per year. Humans are now one of the most numerous vertebrate species on the earth. We also are more widely distributed and manifestly have a greater global environmental impact than any other species. For the families to whom these children are born, each birth may be a joyous and long-awaited event (fig. 4.2). But is a continuing increase in humans good for the planet in the long run?

Many people worry that overpopulation will cause—or perhaps already is causing—resource depletion and environmental degradation that threaten the ecological life-support systems on which we all depend. These fears often lead to demands for immediate, worldwide birth-control programs to reduce fertility rates and to eventually stabilize or even shrink the total number of humans.

Others believe that human ingenuity, technology, and enterprise can expand the world's carrying capacity and allow us to overcome any problems we encounter. From this perspective, more people may be beneficial, rather than disastrous. A larger population means a larger workforce, more geniuses, and more ideas about what to do. Along with every new mouth comes a pair of hands. Proponents of this worldview argue that continued economic and technological growth can both feed the world's billions and enrich everyone enough to end the population explosion voluntarily.

Still another perspective on this subject derives from social justice concerns. According to this worldview, resources are sufficient for everyone. Current shortages are only signs of greed, waste, and oppression. The root cause of environmental degradation, in this view, is inequitable distribution of wealth and power rather than merely population size. Fostering democracy, empowering women and minorities, and improving the standard of living of the world's poorest people are what is really needed. A narrow focus on population growth fosters racism and blames the poor for their problems, while ignoring the high resource consumption of richer nations.

Whether human populations will continue to grow at present rates and what that growth would imply for environmental quality and human life are among the most central and pressing questions in environmental science. In this chapter, we will look at some causes of population growth, as well as at how populations are measured and described. Family planning and birth control are essential for stabilizing populations. The number of children a couple decides to have and the methods they use to regulate fertility, however, are strongly influenced by culture, religion, politics, and economics, as well as basic biological and medical considerations. We will examine how some of these factors influence human demographics.

Human populations grew slowly until recently

For most of our history, humans were not very numerous, compared with other species. Studies of hunting and gathering societies suggest that the total world population was probably only a few million people before the invention of agriculture and the domestication of animals around 10,000 years ago. The agricultural revolution produced a larger and more secure food supply and allowed the human population to grow, reaching perhaps 50 million people by 5000 B.C. For thousands of years, the number of humans increased very slowly. Archaeological evidence and historical descriptions suggest that only about 300 million people were living at the time of Christ (table 4.1).

As you can see in figure 4.3, human populations began to increase rapidly after about A.D. 1600. Many factors contributed to this rapid growth. Increased sailing and navigating skills stimulated commerce and communication among nations. Agricultural

Figure 4.2 The number of children a family chooses to have is determined by many factors, including access to a variety of birth control methods.

Table 4.1	World Population Growth and Doubling Times	
Date	**Population**	**Doubling Time**
5000 B.C.	50 million	?
800 B.C.	100 million	4,200 years
200 B.C.	200 million	600 years
A.D. 1200	400 million	1,400 years
A.D. 1700	800 million	500 years
A.D. 1900	1,600 million	200 years
A.D. 1965	3,200 million	65 years
A.D. 2000	6,100 million	51 years
A.D. 2050 (estimate)	8,920 million	215 years

Source: United Nations Population Division.

similar to those described in chapter 3? As you will see later in this chapter, there is some evidence that population growth already is slowing, but whether we will reach equilibrium soon enough and at a size that can be sustained over the long term remains a difficult but important question.

4.2 Perspectives on Population Growth

As with many topics in environmental science, people have widely differing opinions about population and resources. Some believe that population growth is the ultimate cause of poverty and environmental degradation. Others argue that poverty, environmental degradation, and overpopulation are all merely symptoms of deeper social and political factors. The worldview we choose to believe will profoundly affect our approach to population issues. In this section, we will examine some of the major figures and their arguments in this debate.

Does environment or culture control human population growth?

Since the time of the Industrial Revolution, when the world population began growing rapidly, individuals have argued about the causes and consequences of population growth. In 1798 Thomas Malthus (1766–1834) wrote *An Essay on the Principle of Population*, changing the way European leaders thought about population growth. Malthus marshaled evidence to show that populations tended to increase at an exponential, or compound, rate while food production either remained stable or increased only slowly. Eventually human populations would outstrip their food supply and collapse into starvation, crime, and misery. He converted most economists of the day from believing that high fertility increased

developments, better sources of power, and better health care and hygiene also played a role. We are now in an exponential, or J curve, pattern of growth, described in chapter 3.

It took all of human history to reach 1 billion people in 1804 but little more than 150 years to reach 3 billion in 1960. It took us only 12 years—from 1987 to 1999—to add the sixth billion. Another way to look at population growth is that the number of humans tripled during the twentieth century. Will it do so again in the twenty-first century? If it does, will we overshoot our environment's carrying capacity and experience a catastrophic dieback

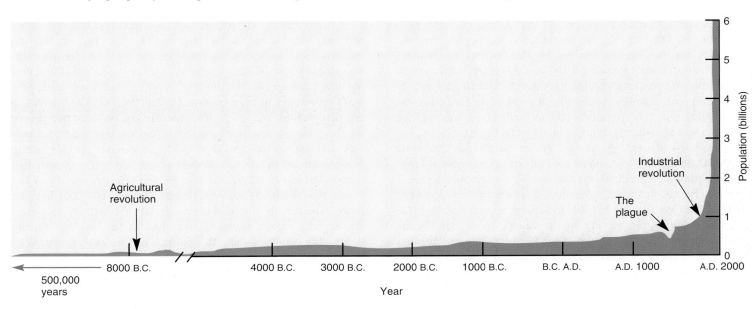

Figure 4.3 Human population levels throughout history. It is clear from the J-shaped growth curve that human population is growing exponentially. When will the growth curve assume an S shape and population growth level off? Choices made in the next 50 years will decide.

gross domestic output to believing that per capita output actually fell with rapidly rising population.

In Malthusian terms, growing human populations stop growing when disease or famine kills many, or when constraining social conditions compel others to reduce their birth rates—late marriage, insufficient resources, celibacy, and "moral restraint." Several decades later, the economist Karl Marx (1818–1883) presented an opposing view, that population growth resulted from poverty, resource depletion, pollution, and other social ills. Slowing population growth, said Marx, required that people be treated justly, and that exploitation and oppression be eliminated from social arrangements.

Both Marx and Malthus developed their theories about human population growth when understanding of the world, technology, and society were much different than they are today. Some believe that we are approaching, or may have surpassed, the earth's carrying capacity. Joel Cohen, a mathematical biologist at Rockefeller University, reviewed published estimates of the maximum human population size the planet can sustain. The estimates, spanning 300 years of thinking, converged on a median value of 10–12 billion. We are more than 6.7 billion strong today, and growing, an alarming prospect for some. Cornell University entomologist David Pimental, for example, has said: "By 2100, if current trends continue, twelve billion miserable humans will suffer a difficult life on Earth." In this view, birth control should be our top priority.

Technology increases carrying capacity for humans

Optimists argue that Malthus was wrong in his predictions of famine and disaster 200 years ago because he failed to account for scientific and technical progress. In fact, food supplies have increased faster than population growth since Malthus' time. For example, according to the UN FAO Statistics Division, each person on the planet averaged 2,435 calories of food per day in 1970, while in 2000 the caloric intake reached 2,807 calories. Even poorer, developing countries saw a rise, from an average of 2,135 calories per day in 1970 to 2,679 in 2000. In that same period the world population went from 3.7 to more than 6 billion people. Certainly terrible famines have stricken different locations in the past 200 years, but they were caused more by politics and economics than by lack of resources or population size. Whether the world can continue to feed its growing population remains to be seen, but technological advances have vastly increased human carrying capacity so far (chapter 7).

The burst of world population growth that began 200 years ago was stimulated by scientific and industrial revolutions. Progress in agricultural productivity, engineering, information technology, commerce, medicine, sanitation, and other achievements of modern life have made it possible to support approximately 1,000 times as many people per unit area as was possible 10,000 years ago. Economist Stephen Moore of the Cato Institute in Washington, D.C., regards this achievement as "a real tribute to human ingenuity and our ability to innovate." There is no reason, he argues, to think that our ability to find technological solutions to our problems will diminish in the future.

Much of our agricultural growth and rising standard of living in the past 200 years, however, has been based on easily acquired natural resources, especially cheap, abundant fossil fuels. Whether rising prices of fossil fuels will constrain that production and result in a crisis in food production and distribution, or in some other critical factor in human society, concerns many people.

However, technology can be a double-edged sword. Our environmental effects aren't just a matter of sheer population size; they also depend on what kinds of resources we use and how we use them. This concept is summarized as the **I = PAT** formula. It says that our environmental impacts (I) are the product of our population size (P) times affluence (A) and the technology (T) used to produce the goods and services we consume. A single American living an affluent lifestyle that depends on high levels of energy and material consumption, and that produces excessive amounts of pollution, probably has a greater environmental impact than a whole village of Asian or African farmers. Ideally, Americans will begin to use nonpolluting, renewable energy and material sources. Better yet, Americans will extend the benefits of environmentally friendly technology to those villages of Asians and Africans so everyone can enjoy the benefits of a better standard of living without degrading their environment.

One way to estimate our environmental impacts is to express our consumption choices into the equivalent amount of land required to produce goods and services. This gives us a single number, called our **ecological footprint**, which estimates the relative amount of bioproductive land required to support each of us. The calculations are imperfect (see What Do You Think? p. 79), but this does give us a way to compare different lifestyle effects. The average resident of the United States, for instance, lives at a level of consumption that requires 9.7 ha of bioproductive land, while the average Indonesian has an ecological footprint of only 1.1 ha.

Population growth could bring benefits

Think of the gigantic economic engine that China is becoming as it continues to industrialize and its population becomes more affluent. More people mean larger markets, more workers, and efficiencies of scale in mass production of goods. Moreover, adding people boosts human ingenuity and intelligence that will create new resources by finding new materials and discovering new ways of doing things. Economist Julian Simon (1932–1998), a champion of this rosy view of human history, believed that people are the "ultimate resource" and that no evidence suggests that pollution, crime, unemployment, crowding, the loss of species, or any other resource limitations will worsen with population growth. In a famous bet in 1980, Simon challenged Paul Ehrlich, author of *The Population Bomb*, to pick five commodities that would become more expensive by the end of the decade. Ehrlich chose metals that actually became cheaper, and he lost the bet. Leaders of many developing countries share this outlook and insist that, instead of being obsessed with population growth, we should focus on the inordinate consumption of the world's resources by people in richer countries. For his part, Ehrlich did not really want to bet about metals, but about renewable resources or certain critical measures of environmental health. He and Simon were negotiating a second bet just before Simon's death.

Calculating Your Ecological Footprint

Can the earth sustain our current lifestyles? Will there be adequate natural resources for future generations? These questions are among the most important in environmental science today. We depend on nature for food, water, energy, oxygen, waste disposal, and other life-support systems. Sustainability implies that we cannot turn our resources into waste faster than nature can recycle that waste and replenish the supplies on which we depend. It also recognizes that degrading ecological systems ultimately threatens everyone's well-being. Although we may be able to overspend nature's budget temporarily, future generations will have to pay the debts we leave them. Living sustainably means meeting our own vital needs without compromising the ability of future generations to meet their own needs.

How can we evaluate and illustrate our ecological impacts? Redefining Progress, a nongovernmental environmental organization, has developed a measure called the **ecological footprint** to compute the demands placed on nature by individuals and nations. A simple questionnaire of 16 items gives a rough estimate of your personal footprint. A more complex assessment of 60 categories including primary commodities (such as milk, wood, or metal ores), as well as the manufactured products derived from them, gives a measure of national consumption patterns.

According to Redefining Progress, the average world citizen has an ecological footprint equivalent to 2.3 hectares (5.6 acres), while the biologically productive land available is only 1.9 hectares (ha) per person. How can this be? The answer is that we're using nonrenewable resources (such as fossil fuels) to support a lifestyle beyond the productive capacity of our environment. It's like living by borrowing on your credit cards. You can do it for a while, but eventually you have to pay off the deficit. The unbalance is far more pronounced in some of the richer countries. The average resident of the United States, for example, lives at a consumption level that requires 9.7 ha of bioproductive land. A dramatic comparison of consumption levels versus population size is shown in figure 1. If everyone in the world were to adopt a North American lifestyle, we'd need about four more planets to support us all. You can check out your own ecological footprint by going to www.redefiningprogress.org/.

Like any model, an ecological footprint gives a useful description of a system. Also like any model, it is built on a number of assumptions: (1) various measures of resource consumption and waste flows can be converted into the biologically productive area required to maintain them; (2) different kinds of resource use and dissimilar types of productive land can be standardized into roughly equivalent areas; (3) because these areas stand for mutually exclusive uses, they can be added up to a total—a total representing humanity's demand—that can be compared to the total world area of bioproductive land. The model also implies that our world has a fixed supply of resources that can't be expanded. Part of the power of this metaphor is that we all can visualize a specific area of land and imagine it being divided into smaller and smaller parcels as our demands increase. But this perspective doesn't take into account technological progress. For example, since 1950, world food production has increased about fourfold. Some of this growth has come from expansion of croplands, but most has come from technological advances such as irrigation, fertilizer use, and higher-yielding crop varieties. Whether this level of production is sustainable is another question, but this progress shows that land area isn't always an absolute limit. Similarly, switching to renewable energy sources such as wind and solar power would make a huge impact on estimates of our ecological footprint. Notice that in figure 2 energy consumption makes up about half of the calculated footprint.

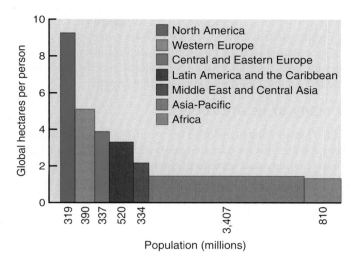

Figure 1 Ecological footprint by region in 2001. The height of each bar is proportional to each region's average footprint per person, the width of the bar is proportional to its population, and the area of the bar is proportional to the region's total ecological footprint.

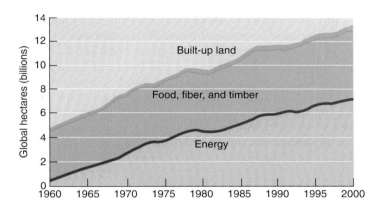

Figure 2 Humanity's ecological footprint grew by about 160 percent from 1961 to 2001, significantly faster than population, which doubled over the same period.

Source: WWF, 2004.

What do you think? Does analyzing our ecological footprint inspire you to correct our mistakes, or does it make sustainability seem an impossible goal? If we in the richer nations have the technology and political power to exploit a larger share of resources, do we have a right to do so, or do we have an ethical responsibility to restrain our consumption? And what about future generations? Do we have an obligation to leave resources for them, or can we assume they'll make technological discoveries to solve their own problems if resources become scarce? You'll find that many of the environmental issues we discuss in this book aren't simply a matter of needing more scientific data. Ethical considerations and intergenerational justice often are just as important as having more facts.

4.3 Many Factors Determine Population Growth

Demography (derived from the Greek words *demos* [people] and *graphein* [to write or to measure]) encompasses vital statistics about people, such as births, deaths, and where they live, as well as total population size. In this section, we will investigate ways to measure and describe human populations and discuss demographic factors that contribute to population growth.

How many of us are there?

The U.S. Census Bureau estimate of over 6.7 billion people in the world in 2009 is only an educated guess. Even in this age of information technology and communication, counting the number of people in the world is like shooting at a moving target. People continue to be born and to die. Furthermore, some countries have never even taken a census, and those that have been done may not be accurate. Governments may overstate or understate their populations to make their countries appear larger and more important or smaller and more stable than they really are. Individuals, especially if they are homeless, refugees, or illegal aliens, may not want to be counted or identified.

We really live in two very different demographic worlds. One of these worlds is poor, young, and growing rapidly, while the other is rich, old, and shrinking in population size. The poorer world is occupied by the vast majority of people who live in the less-developed countries of Africa, Asia, and Latin America. These countries represent 80 percent of the world population but will contribute more than 90 percent of all projected future growth (fig. 4.4). The richer world is made up of North America, Western Europe, Japan, Australia, and New Zealand. The average age in richer countries is 40, and life expectancy of their residents will exceed 90 by 2050. With many couples choosing to have either one or no children, the populations of these countries are expected to decline significantly over the next century.

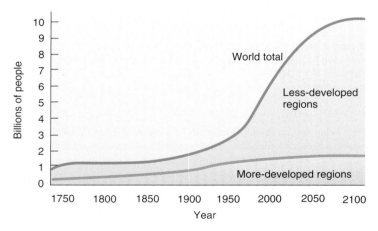

Figure 4.4 Since about 1950, less-developed regions of the planet have experienced exponential growth while growth in more-developed regions has slowed. More than 90 percent of future population growth will occur in less-developed countries.

Table 4.2	The World's Largest Countries		
2008		**2050**	
Country	Population (millions)	Country	Population (millions)
China	1,330	India	1,628
India	1,148	China	1,437
United States	304	United States	420
Indonesia	238	Indonesia	308
Brazil	192	Pakistan	295
Pakistan	168	Brazil	260
Bangladesh	153	Nigeria	258
Russia	140	Bangladesh	231
Nigeria	138	Congo, Dem. Rep. of	183
Japan	127	Ethiopia	170

Source: Data from the U.S. Census Bureau, 2008.

The highest population growth rates occur in a few "hot spots" in the developing world, such as sub-Saharan Africa and the Middle East, where economics, politics, religion, and civil unrest keep birth rates high and contraceptive use low. In Chad and the Democratic Republic of Congo, for example, annual population growth is above 3.2 percent. Less than 10 percent of all couples use any form of birth control, women average more than seven children each, and nearly half the population is less than 15 years old. Even faster growth rates occur in Oman and Palestine, where the population doubling time is only 18 years.

Some countries in the developing world are growing so fast that they will reach immense population sizes by the middle of the twenty-first century (table 4.2). While China was the most populous country throughout the twentieth century, India is expected to pass China in the twenty-first century. Nigeria, which had only 33 million residents in 1950, is forecast to have 258 million in 2050. Ethiopia, with about 18 million people 50 years ago, is likely to grow almost tenfold over a century. In many of these countries, rapid population growth poses huge challenges to food supplies and stability. Bangladesh, about the size of Iowa, is already overcrowded at 153 million people. If rising sea levels flood one-third of the country by 2050, as some climatologists predict, adding another 100 million people will make the situation impossible.

On the other hand, shrinking populations characterize some richer countries. Japan, which has 128 million residents now, is expected to shrink to about 100 million by 2050. Europe, which now makes up about 12 percent of the world population, will constitute less than 7 percent in 50 years, if current trends continue. Even the United States and Canada would have nearly stable populations if immigration were stopped. For the moment, the U.S. population continues to grow. In October 2006, the U.S. population reached 300 million.

It isn't only wealthy countries that have declining populations. Russia, for instance, is now declining by nearly 1 million people per year as death rates have soared and birth rates have

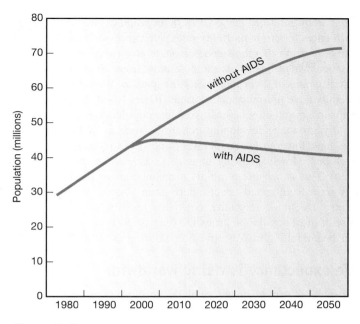

Figure 4.5 Projected population of south Africa with and without AIDS.
Data source: UN Population Division, 2006.

plummeted. A collapsing economy, hyperinflation, crime, corruption, and despair have demoralized the population. Horrific pollution levels left from the Soviet era, coupled with poor nutrition and health care, have resulted in high levels of genetic abnormalities, infertility, and infant mortality. Abortions are twice as common as live births. Death rates, especially among adult men, have risen dramatically. According to some demographers, the life expectancy of Russian males fell by 10 years after 1990; in 2005 it began creeping back up, to 59 years. By 2050 Russia is expected to have a smaller population than Vietnam, the Philippines, and the Democratic Republic of Congo. Bulgaria will lose a third of its population, simply due to declining birth rates.

The situation is even worse in many African countries, where AIDS and other communicable diseases are killing people at a terrible rate. In Zimbabwe, Botswana, Zambia, and Namibia, for example, up to 39 percent of the adult population have AIDS or are HIV positive. Health officials predict that more than two-thirds of the 15-year-olds now living in Botswana will die of AIDS before age 50. Without AIDS, the average life expectancy is estimated to be 69.7 years. Now, with AIDS, Botswana's life expectancy has dropped to only 31.6 years. The populations of many African countries are now falling because of this terrible disease (fig. 4.5). Altogether, Africa's population is expected to be nearly 200 million lower in 2050 than it would have been without AIDS.

The world population density map in appendix 2 on p. 377 shows human population distribution around the world. Notice the high densities supported by fertile river valleys of the Nile, Ganges, Yellow, Yangtze, and Rhine Rivers and the well-watered coastal plains of India, China, and Europe. Historic factors, such as technology diffusion and geopolitical power, also play a role in geographic distribution.

Fertility varies among cultures and at different times

Fecundity is the physical ability to reproduce, while fertility is the actual production of offspring. Those without children may be fecund but not fertile. The most accessible demographic statistic of fertility is usually the **crude birth rate**, the number of births in a year per thousand persons. It is statistically "crude" in the sense that it is not adjusted for population characteristics, such as the number of women of reproductive age.

The **total fertility rate** is the number of children born to an average woman in a population during her entire reproductive life. Upper-class women in seventeenth- and eighteenth-century Europe, whose babies were given to wet nurses immediately after birth, often had 25 or 30 pregnancies. The highest recorded total fertility rates for working-class people are among some Anabaptist agricultural groups in North America, who have averaged up to 12 children per woman. In most tribal or traditional societies, food shortages, health problems, and cultural practices limit total fertility to about six or seven children per woman, even without modern methods of birth control.

Zero population growth (ZPG) occurs when births plus immigration in a population just equal deaths plus emigration. It takes several generations of replacement-level fertility (in which people just replace themselves) to reach ZPG. Where infant mortality rates are high, the replacement level may be 5 or more children per couple. In the more highly developed countries, however, this rate is usually about 2.1 children per couple because some people are infertile, have children who do not survive, or choose not to have children.

Fertility rates have declined dramatically in every region of the world except Africa over the past 50 years (fig. 4.6). In the 1960s,

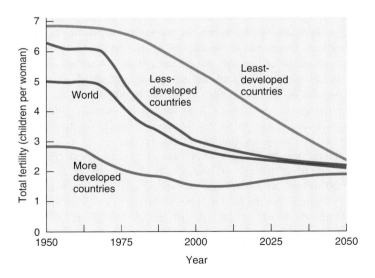

Figure 4.6 Average total fertility rates for less-developed countries fell by more than half over the past 50 years, mostly due to China's one-child policy. By 2050 even the least-developed countries should approach the replacement rate of 2.1 children per reproductively fertile woman.
Source: Data from UN Population Division, *World Population Prospects*, 1996, and Population Reference Bureau, 2004.

Figure 4.7 China's one-child-per-family policy has been remarkably successful in reducing birth rates.

total fertility rates above 6 were common in many countries. The average family in Mexico in 1975, for instance, had 7 children. By 2005, however, the average Mexican woman had only 2.6 children. According to the World Health Organization, 61 of the world's 190 countries are now at or below a **replacement rate** of 2.1 children per couple. The greatest fertility reduction has been in Southeast Asia, where rates have fallen by more than half, primarily in the past few decades. Contrary to what many demographers expected, some of the poorest countries in the world have been remarkably successful in lowering growth rates. Bangladesh, for instance, reduced its fertility rate from 6.9 in 1980 to only 3.0 children per woman in 2005.

China's one-child-per-family policy decreased the fertility rate from 6 in 1970 to 1.8 in 1990 (fig. 4.7). This policy, however, along with a strong preference for sons, has sometimes resulted in abortions, forced sterilizations, and even infanticide. It also has been so successful that the government is now worried about a shortage of workers. By mid century, at least one-third of the Chinese population will be retirees, if current trends continue.

Although the number of boys and girls normally should be fairly balanced in a population, 20 years of female infanticides and sex-based abortions in China have created ratios as high as 140 boys to 100 girls in some regions. This shortage of women has created a flourishing trade in abducted brides. A crackdown in 2002 released 110,000 women who had been kidnapped and sold into marriage, but it's thought that the total number abducted is much higher.

While the world as a whole still has an average fertility rate of 2.6, growth rates are now lower than at any time since World War II. If fertility declines like those in Bangladesh and China were to occur everywhere in the world, our total population could begin to decline by the end of the twenty-first century.

Mortality offsets births

A traveler to a foreign country once asked a local resident, "What's the death rate around here?" "Oh, the same as anywhere," was the reply, "about one per person." In demographics, however, **crude death rates** (or crude mortality rates) are expressed in terms of the number of deaths per thousand persons in any given year. Countries in Africa where health care and sanitation are limited may have mortality rates of 20 or more per 1,000 people. Wealthier countries generally have mortality rates around 10 per 1,000. The number of deaths in a population is sensitive to the population's age structure. Rapidly growing, developing countries, such as Belize or Costa Rica, have lower crude death rates (4 per 1,000) than do the more-developed, slowly growing countries, such as Denmark (12 per 1,000). This is because a rapidly growing country has proportionately more youths and fewer elderly than a more slowly growing country. Declining mortality, not rising fertility, was the primary cause of most population growth in the past 300 years. Crude death rates began falling in Western Europe during the late 1700s.

Life expectancy is rising worldwide

Life span is the oldest age to which a species is known to survive. Although there are many claims in ancient literature of kings living for a thousand years or more, the oldest age that can be certified by written records was that of Jeanne Louise Calment of Arles, France, who was 122 years old at her death in 1997. While modern medicine has made it possible for many of us to survive much longer than our ancestors, it doesn't appear that the maximum life span has increased much at all. Apparently, cells in our bodies have a limited ability to repair damage and produce new components. Sooner or later they simply wear out, and we fall victim to disease, degeneration, accidents, or senility.

Life expectancy is the average age that a newborn infant can expect to attain in any given society. It is another way of expressing the average age at death. For most of human history, life expectancy in most societies probably has been 35 to 40 years. This doesn't mean that no one lived past age 40 but, rather, that many people died at earlier ages (mostly early childhood), which balanced out those who managed to live longer.

The twentieth century saw a global transformation in human health unmatched in history. This revolution can be seen in the dramatic increases in life expectancy in most places (table 4.3). Worldwide, the average life expectancy rose from about 40 to 67.2 years over the past 100 years. The greatest progress was in developing countries. For example, in 1900, the average Indian man could

Table 4.3	Life Expectancy at Birth for Selected Countries in 1900 and 2007			
	1900		**2007**	
Country	Males	Females	Males	Females
India	23	23	63	66
Russia	31	33	59	73
United States	46	48	76	81
Sweden	57	60	79	83
Japan	42	44	79	86

Source: Data from Population Reference Bureau, 2007.

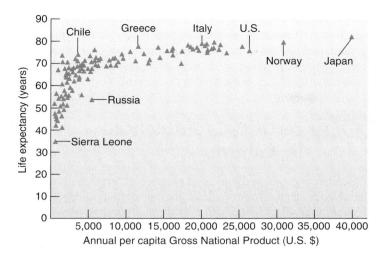

Figure 4.8 Up to about $4,000 (U.S.), rising income increases life expectancy. However, in some poorer countries people live as long as in rich countries. In Chile, for example, people live 76 years on average, while in the United States with over four times the per capita gross national income, people live 78 years.

Sources: United Nations (UN) Population Division, World Population Prospects, 1950–2050 (The 1996 Revision). The United Nations, New York, 1996. Population Reference Bureau, 2005, World Population Data Sheet.

that of countries with ten times its income level. Longer lives were due primarily to better nutrition, improved sanitation, clean water, and education, rather than to miracle drugs or high-tech medicine. While the gains were not as great for the already industrialized countries, residents of the United States, Sweden, and Japan, for example, now live about half-again as long as they did at the beginning of the twentieth century, and they can expect to enjoy much of that life in relatively good health. The Disability Adjusted Life Years (DALYs, a measure of disease burden that combines premature death with loss of healthy life resulting from illness or disability) that someone living in Japan can expect is now 74.5 years, compared with only 64.5 DALYs two decades ago.

As figure 4.8 shows, annual income and life expectancy are strongly correlated up to about $4,000 (U.S.) per person. Beyond that level—which is generally enough for adequate food, shelter, and sanitation for most people—life expectancies level out at about 75 years for men and 85 for women.

Large discrepancies in how the benefits of modernization and social investment are distributed within countries are revealed in differential longevities of various groups. The greatest life expectancy reported anywhere in the United States is for women in New Jersey, who live to an average age of 91. By contrast, Native American men on the Pine Ridge Indian Reservation in neighboring South Dakota live, on average, only to age 45. Only a few countries in Africa have a lower life expectancy. The Pine Ridge Reservation is the poorest area in America, with an unemployment rate near 75 percent and high rates of poverty, alcoholism, drug use, and alienation. Similarly, African-American men in

expect to live less than 23 years, while the average woman would reach just over 23 years. By 2005, although India had an annual per capita income of less than $440 (U.S.), the average life expectancy for both men and women had nearly tripled and was very close to

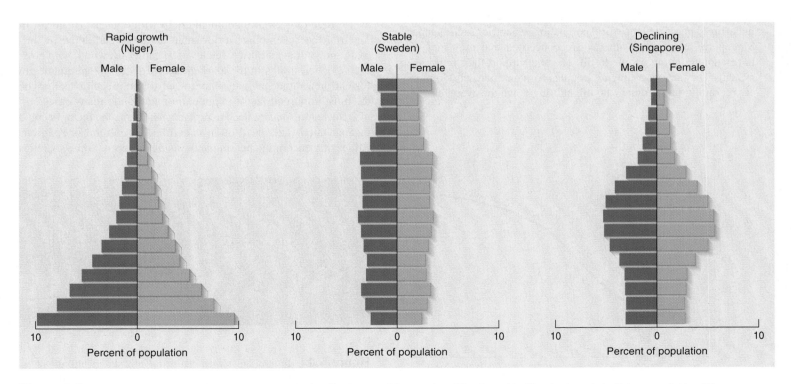

Figure 4.9 The shape of each age class histogram is distinctive for a rapidly growing (Niger), stable (Sweden), and declining population (Singapore). Horizontal bars represent the percentage of the country's population in consecutive age classes (0–5 yrs, 6–10 yrs, etc.)
Source: U.S. Census Bureau, 2003.

Washington, D.C., live, on average, only 57.9 years, which is less than the life expectancy in Lesotho or Swaziland.

Living longer has profound social implications

A population growing rapidly by natural increase has more young people than does a stationary population. One way to show these differences is to graph age classes in a histogram, as figure 4.9 shows. In Niger, which is growing at a rate of 3.5 percent per year, 47.8 percent of the population is in the prereproductive category (below age 15). Even if total fertility rates fell abruptly, the total number of births, and the population size, would continue to grow for some years as these young people entered reproductive age. This phenomenon is called population momentum.

By contrast, a country with a relatively stable population will have nearly the same number in most cohorts. Notice that females outnumber males in Sweden's oldest group because of differences in longevity between sexes. A rapidly declining population, such as Singapore's, can have a pronounced bulge in middle-age cohorts as fewer children are born than in their parents' generation.

Both rapidly growing countries and slowly growing countries can have a problem with their **dependency ratio**, or the number of nonworking compared with working individuals in a population. In Niger, for example, each working person supports a high number of children. In the United States, by contrast, a declining working population is now supporting an ever larger number of retired persons.

This changing age structure and shifting dependency ratio are occurring worldwide (fig. 4.10). In 1950, there were only 130 million people in the world over 65 years old. In 2000, more than 420 million had reached this age. By 2150, those over 65 might make up 25 percent of the world population. Countries such as Japan, Singapore, and Taiwan already are concerned that they may not have enough young people to fill jobs and support their retirement system. They are encouraging couples to have more children and are recruiting immigrants who might bring down the average age of the population. A group of business leaders, including former Federal Reserve Chairman Alan Greenspan, have called for increased immigration into the United States to fill jobs, prevent inflation, and boost economic productivity. Indeed, immigration accounted for the great majority of U.S. population growth after 2000.

4.4 Fertility Is Influenced by Culture

A number of social and economic pressures affect decisions about family size, which in turn affects the population at large. In this section, we will examine both positive and negative pressures on reproduction.

People want children for many reasons

Factors that increase people's desires to have babies are called **pronatalist pressures**. Raising a family may be the most enjoyable and rewarding part of many people's lives. Children can be a source of pleasure, pride, and comfort. They may be the only source of support for elderly parents in countries without a social security system. Where infant mortality rates are high, couples may need to have many children to ensure that at least a few will survive to take care of them when they are old. Where there is little opportunity for upward mobility, children give status in society, express parental creativity, and provide a sense of continuity and accomplishment otherwise missing from life. Often children are valuable to the family not only for future income but even more as a source of current income and help with household chores. In much of the developing world, small children tend domestic animals and younger siblings, fetch water, gather firewood, help grow crops, or sell things in the marketplace (fig. 4.11). Parental desire for children rather than an unmet need for contraceptives may be the most important factor in population growth in many cases.

Society also has a need to replace members who die or become incapacitated. This need often is codified in cultural or religious values that encourage bearing and raising children. Some societies

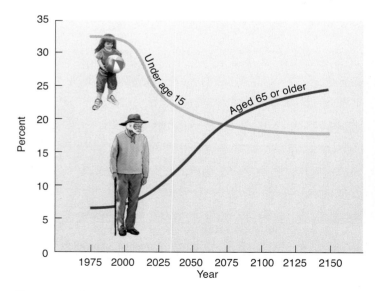

Figure 4.10 By the mid-twenty-first century, children under age 15 will make up a smaller percentage of world population, while people over age 65 will contribute a larger and larger share of the population.

(a)   (b)

Figure 4.11 In rural areas with little mechanized agriculture (a) children are needed to tend livestock, care for younger children, and help parents with household chores. Where agriculture is mechanized (b) rural families view children just as urban families do—helpful, but not critical to survival. This affects the decision about how many children to have.

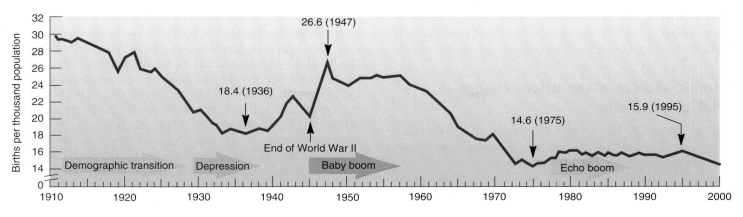

Figure 4.12 Birth rates in the United States, 1910–2000. The falling birth rate from 1910 to 1929 represents a demographic transition from an agricultural to an industrial society. The baby boom following World War II lasted from 1945 to 1965. A much smaller "echo boom" occurred around 1980 when the baby boomers started to reproduce.

Source: Data from Population Reference Bureau and U.S. Bureau of the Census.

look upon families with few or no children with pity or contempt. The idea of deliberately controlling fertility may be shocking, even taboo. Women who are pregnant or have small children have special status and protection. Boys frequently are more valued than girls because they carry on the family name and are expected to support their parents in old age. Couples may have more children than they really want in an attempt to produce a son.

Male pride often is linked to having as many children as possible. In Niger and Cameroon, for example, men, on average, want 12.6 and 11.2 children, respectively. Women in these countries want only 5 or 6. Even though a woman might desire fewer children, however, she may have few choices and little control over her own fertility. In many societies, a woman has no status outside of her role as wife and mother. Yet without children, she may have no source of support in her old age.

Education and income affect the desire for children

In more highly developed countries, many pressures tend to reduce fertility. Higher education and personal freedom for women often result in decisions to limit childbearing. A desire to spend time and money on other goods and activities offsets the desire to have children. When women have opportunities to earn a salary, they are less likely to stay home and have many children. Not only do many women find the challenge and variety of a career attractive, but the money that they earn outside the home becomes an important part of the family budget. Thus, education and socioeconomic status are usually inversely related to fertility in richer countries. In some developing countries, however, fertility initially increases as educational levels and socioeconomic status rise. With higher income, families are better able to afford the children they want. More money also means that women are healthier and therefore better able to conceive and carry a child to term. It may be a generation before this unmet desire for children abates.

In less-developed countries, where feeding and clothing children can be a minimal expense, adding one more child to a family usually doesn't cost much. By contrast, raising a child in

a developed country can cost hundreds of thousands of dollars by the time the child finishes school and is independent. Under these circumstances, parents are more likely to choose to have one or two children on whom they can concentrate their time, energy, and financial resources.

Figure 4.12 shows U.S. birth rates between 1910 and 2000. As you can see, birth rates fell and rose in an irregular pattern. The period between 1910 and 1930 was a time of industrialization and urbanization. Women were getting more education than ever before and entering the workforce in large numbers. The Great Depression in the 1930s made it economically difficult for families to have children, and birth rates were low. The birth rate increased at the beginning of World War II (as it often does in wartime). For reasons that are unclear, a higher percentage of boys are usually born during war years.

A "baby boom" followed World War II, as couples were reunited and new families started. During this time, the government encouraged women to leave their wartime jobs and stay home. A high birth rate persisted through the times of prosperity and optimism of the 1950s but began to fall in the 1960s. Part of this decline was caused by the small number of babies born in the 1930s, which resulted in fewer young adults to give birth in the 1960s. Part was due to changed perceptions of the ideal family size. While in the 1950s women typically wanted four children or more, the norm dropped to one or two (or no) children in the 1970s. A small "echo boom" occurred in the 1980s, as baby boomers began to have babies, but changing economics and attitudes seem to have permanently altered our view of ideal family size in the United States.

4.5 A Demographic Transition Can Lead to Stable Population Size

In 1945 demographer Frank Notestein pointed out that a typical pattern of falling death rates and birth rates due to improved living conditions usually accompanies economic development. He called this pattern the **demographic transition** from high birth and death rates to lower birth and death rates. Figure 4.13 shows an idealized model of a demographic transition. This model is often used to explain connections between population growth and economic development.

Economic and social conditions change mortality and births

Stage I in figure 4.13 represents the conditions in a premodern society. Food shortages, malnutrition, lack of sanitation and medicine, accidents, and other hazards generally keep death rates in such a society around 30 per 1,000 people. Birth rates are correspondingly high to keep population densities relatively constant. Economic development in Stage II brings better jobs, medical care, sanitation, and a generally improved standard of living, and death rates often fall very rapidly. Birth rates may actually rise at first as more money and better nutrition allow people to have the children they always wanted. In a couple of generations, however, birth rates fall as people see that all their children are more likely to survive and that the whole family benefits from concentrating more resources on fewer children. Note that populations grow rapidly during Stage III when death rates have already fallen but birth rates remain high. Depending on how long it takes to complete the transition, the population may go through one or more rounds of doubling before coming into balance again.

The Demographic Transition Model

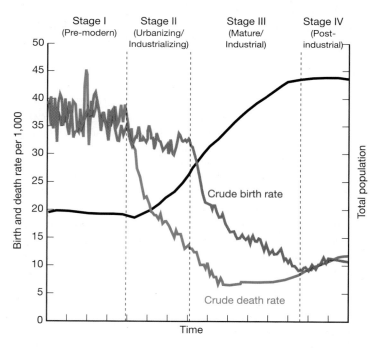

Figure 4.13 Theoretical birth, death, and population growth rates in a demographic transition accompanying economic and social development. In a predevelopment society, birth and death rates are both high, and total population remains relatively stable. During development, death rates tend to fall first, followed in a generation or two by falling birth rates. Total population grows rapidly until both birth and death rates stabilize in a fully developed society.

Stage IV represents conditions in developed countries, where the transition is complete and both birth rates and death rates are low, often a third or less than those in the predevelopment era. The population comes into a new equilibrium in this phase, but at a much larger size than before. Most of the countries of northern and western Europe went through a demographic transition in the nineteenth or early twentieth century similar to the curves shown in figure 4.13. In countries such as Italy, where fertility levels have fallen below replacement rates, there are now fewer births than deaths, and the total population curve has started to decline.

A huge challenge facing countries in the final stage of the demographic transition is the imbalance between people in their most productive years and people who are retired or in their declining years. The debate in the U.S. Congress about how to fund the social security system is due to the fact that, when this program was established, the United States was in the middle of the demographic transition, with a large number of young people relative to older people. In ten to fifteen years, that situation will change, with many more elderly people living longer, and fewer younger workers to support them.

Many of the most rapidly growing countries in the world, such as Kenya, Yemen, Libya, and Jordan, now are in Stage III of this demographic transition. Their death rates have fallen close to the rates of the fully developed countries, but birth rates have not fallen correspondingly. In fact, both their birth rates and total population are higher than those in most European countries when industrialization began 300 years ago. The large disparity between birth and death rates means that many developing countries now are growing at 3 to 4 percent per year. Such high growth rates in developing countries could boost total world population to over 9 billion by the end of the twenty-first century. This raises what may be the two most important questions in this entire chapter: why are birth rates not yet falling in these countries, and what can be done about it?

Many countries are in a demographic transition

Some demographers claim that a demographic transition already is in progress in most developing nations. They believe that problems in taking censuses and a normal lag between falling death and birth rates may hide this for a time but that the world population should stabilize sometime in the twenty-first century. Some evidence supports this view. As mentioned earlier in this chapter, fertility rates have fallen dramatically nearly everywhere in the world over the past half century.

Some countries have had remarkable success in population control. In Thailand, Indonesia, and Colombia, for instance, total fertility dropped by more than half in 20 years. Morocco, the Dominican Republic, Jamaica, Peru, and Mexico all have seen fertility rates fall between 30 and 40 percent in a single generation. Surprisingly, one of the most successful family planning programs in recent years has been in Iran.

The following factors help stabilize populations:

- Growing prosperity and social reforms that accompany development reduce the need and desire for large families in most countries.

Figure 4.14 Do we reduce pressure on the environment by achieving population control in developing countries, or by limiting resource use in developed countries? It depends on whom you ask.
Used with permission of the Asian Cultural Forum on Development.

- Technology is available to bring advances to the developing world much more rapidly than was the case a century ago, and the rate of technology exchange is much faster than it was when Europe and North America were developing.

- Less-developed countries have historic patterns to follow. They can benefit from the mistakes of more-developed countries and chart a course to stability relatively quickly.

- Modern communications (especially television) have caused a revolution of rising expectations that act as stimuli to spur change and development.

Two ways to complete the demographic transition

The Indian states of Kerala and Andra Pradesh exemplify two approaches to population growth (see What Do You Think? p. 88). In Kerala, providing a fair share of social benefits to everyone was the key to completing the demographic transition. This social justice strategy assumes that the world has enough resources for everyone, but inequitable social and economic systems cause maldistributions of those resources. Hunger, poverty, violence, environmental degradation, and overpopulation are symptoms of a lack of justice, rather than a lack of resources. Although overpopulation exacerbates other problems, a focus on overpopulation alone encourages racism and hatred of the poor. Proponents of the social justice view argue that people in the richer nations should recognize their excess consumption levels and the impact that this consumption has on others (fig. 4.14).

The leaders of Andra Pradesh, on the other hand, adopted a strategy of aggressively promoting birth control, rather than social justice. This strategy assumes that better access to birth control is the key to reducing populations, and that social justice is an unrealistic goal.

These two strategies are very different, but both aim to help impoverished countries caught in the "demographic trap," in which they can't escape the transition. Their populations are growing so rapidly that human demands exceed the sustainable yield of local forests, grasslands, croplands, and water resources. The resulting resource shortages, environmental deterioration, economic decline, and political instability may prevent these countries from ever completing modernization. Their populations may continue to grow until catastrophe intervenes.

Many people believe that the only way to break out of the demographic trap is to reduce population growth immediately and drastically by whatever means are necessary, as occurred in Andra Pradesh. They argue strongly for birth control education and bold national policies to encourage lower birth rates. Some agree with Malthus that helping the poor will simply increase their reproductive success and further threaten the resources on which we all depend. Author Garret Hardin described this view as lifeboat ethics:

> Each rich nation amounts to a lifeboat full of comparatively rich people. The poor of the world are in other much more crowded lifeboats. Continuously, so to speak, the poor fall out of their lifeboats and swim for a while, hoping to be admitted to a rich lifeboat, or in some other way to benefit from the goodies on board. . . . We cannot

Cultural Choices and the Rate of Population Growth

The good news is that the human population on earth may never double again. The bad news is that it may take a century to stabilize. Meanwhile, the choices we make individually, and collectively through our political, economic, and cultural actions, will determine whether it will be a pleasant, or not so pleasant, hundred years. The choices are already being made in India, where two different states are taking two very different approaches to population growth.

In 1999, having added more than 180 million people in just a decade, India reached a population of 1 billion humans. If current growth rates persist, India will have at least 1.63 billion residents in 2050 and will surpass China as the world's most populous country. How will this country, where more than a quarter of its inhabitants live in abject poverty, feed, house, educate, and employ all those being added each year? A fierce debate is taking place about how to control India's population, with ramifications for the rest of the world as well.

On one side of this issue are those who believe that the best way to reduce the number of children born is poverty eradication and progress for women. Drawing on social justice principles established at the 1994 UN Conference on Population and Development in Cairo, some argue that responsible economic development, a broad-based social welfare system, education and empowerment of women, and high-quality health care—including family planning services—are essential components of population control. Without progress in these areas, they believe, efforts to provide contraceptives or encourage sterilization are futile.

On the other side of this debate are those who contend that, while social progress is an admirable goal, India doesn't have the time or resources to wait for an indirect approach to population control. The government must push aggressively, they argue, to reduce births now or the population will be so huge and its use of resources so great that only mass starvation, class war, crime, and disease will be able to bring it down to a manageable size.

Unable to reach a consensus on population policy, the Indian government decided in 2000 to let each state approach the problem in its own way. Some states have chosen to focus on social justice, while others have adopted more direct, interventionist policies.

The model for the social justice approach is the southern state of Kerala, which achieved population stabilization in the mid-1980s, the first Indian state to do so. Although still one of the poorest places in the world economically, Kerala's fertility rate is comparable to that of many industrialized nations, including the United States. Both women and men have a nearly 100 percent literacy rate and share affordable and accessible health care, family planning, and educational opportunities; therefore, women have only the number of children they want, usually two. The Kerala experience suggests that increased wealth isn't a prerequisite for zero population growth.

Taking a far different path to birth reduction is the nearby state of Andra Pradesh, which reached a stable growth rate in 2001. Boasting the most dramatic fertility decline of any large Indian state, Andra Pradesh has focused on targeted, strongly enforced sterilization programs. The poor are encouraged—some would say coerced—to be sterilized after having only one or two children. The incentives include cash payments. You might receive 500 rupees—equivalent to $11 (U.S.) or a month's

Its annual growth rate of 1.7 percent suggests that India's population could double in 41 years, becoming the world's most populous country by 2050. Indian states have taken different approaches to slow population growth. Despite its poverty, Kerala has a total fertility rate lower than the United States.

wages for an illiterate farm worker—if you agree to have "the operation." In addition, participants are eligible for better housing, land, wells, and subsidized loans.

The pressure to be sterilized is overwhelmingly directed at women, for whom the procedure is major abdominal surgery. Sterilizations often are done by animal husbandry staff and carried out in government sterilization camps. This practice raises troubling memories of the 1970s for many people, when then-Prime Minister Indira Gandhi suspended democracy and instituted a program of forced sterilization of poor people. There were reports at the time of people being rounded up like livestock and castrated or neutered against their will.

While many feminists and academics regard Andra Pradesh's policies as appallingly intrusive and coercive to women and the poor, the state has successfully reduced population growth. By contrast, the hugely populous northern states of Uttar Pradesh and Bihar have increased their growth rates slightly over the past two decades to a current rate above 2.5 percent per year. How will they slow this exponential growth, and what might be the social and environmental costs of not doing so?

What do you think? Do the methods of population control in India offer insights and models for other countries? Why or why not? Are there ethical principles that apply to these solutions? Which approach would you favor if you were advising other countries?

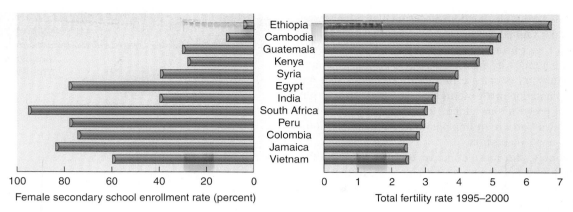

Figure 4.15 Total fertility declines as women's education increases.
Source: Data from Worldwatch Institute, 2003.

risk the safety of all the passengers by helping others in need. What happens if you share space in a lifeboat? The boat is swamped and everyone drowns. Complete justice, complete catastrophe.

Improving women's lives helps reduce birth rates

Kerala's leaders believed that helping women was key to achieving population stability. The 1994 International Conference on Population and Development in Cairo, Egypt, signaled this approach to population issues. A broad consensus reached by the 180 participating countries agreed that responsible economic development, education and empowerment of women, and high-quality health care (including family planning services) must be accessible to everyone if population growth is to be slowed. Child survival is one of the most critical factors in stabilizing population. When infant and child mortality rates are high, as they are in much of the developing world, parents tend to have high numbers of children to ensure that some will survive to adulthood. There has never been a sustained drop in birth rates that was not first preceded by a sustained drop in infant and child mortality.

Increasing family income does not always translate into better welfare for children, since men in many cultures control most financial assets. As the UN Conference in Cairo noted, the best way to improve child survival often is to ensure the rights of mothers. Opportunities for women's education, for instance, as well as land reform, political rights, opportunities to earn an independent income, and improved health status of women often are better indicators of family welfare than is rising gross national product (fig. 4.15).

4.6 Family Planning Gives Us Choices

Family planning allows couples to determine the number and spacing of their children. It doesn't necessarily mean fewer children—people could use family planning to have the maximum number of children possible—but it does imply that the parents will control

their reproductive lives and make rational, conscious decisions about how many children they will have and when those children will be born, rather than leaving it to chance. As the desire for smaller families becomes more common, birth control often becomes an essential part of family planning. In this context, **birth control** usually means any method used to reduce births, including celibacy, delayed marriage, contraception, methods that prevent embryo implantation, and induced abortions.

Humans have always regulated their fertility

The high human birth rate of the last two centuries is not the norm compared to previous millennia of human existence. Evidence suggests that people in every culture and every historic period used a variety of techniques to control population size. Studies of hunting and gathering people, such as the !Kung, or San, of the Kalahari Desert in southwest Africa, indicate that our early ancestors had stable population densities, not because they killed each other or starved to death regularly but because they controlled fertility.

For instance, San women breast-feed children for three or four years. When calories are limited, lactation depletes body fat stores and suppresses ovulation. Coupled with taboos against intercourse while breast-feeding, this is an effective way of spacing children. (However, breast-feeding among well-nourished women in modern societies doesn't necessarily suppress ovulation or prevent conception.) Other ancient techniques to control population size include celibacy, folk medicines, abortion, and infanticide. We may find some or all of these techniques unpleasant or morally unacceptable, but we shouldn't assume that other people are too ignorant or too primitive to make decisions about fertility.

Today there are many options

Modern medicine gives us many more options for controlling fertility than were available to our ancestors. The major categories of birth control techniques include (1) avoidance of sex during fertile periods (for example, celibacy or the use of changes in body temperature or cervical mucus to judge when ovulation will occur), (2) mechanical barriers that prevent contact between sperm and egg (for example, condoms, spermicides, diaphragms, cervical caps, and vaginal sponges), (3) surgical methods that prevent release of sperm or egg (for example, tubal ligations in females and vasectomies in males), (4) hormone-like chemicals that prevent maturation or release of sperm or eggs or that prevent embryo implantation in the uterus (for example, estrogen plus progesterone, or progesterone alone, for females; gossypol for males), (5) physical barriers to implantation (for example, intrauterine devices), and (6) abortion.

Not surprisingly, the most effective birth control methods are also the ones most commonly used (table 4.4). In the United States, the majority of women younger than 30 who eventually want to become pregnant use the Pill. Most women over 35, with their child-bearing years behind them, choose sterilization. Male condom use is more effective than the remaining techniques in the table, and increases in effectiveness when used with a spermicide. Only 2 to 6 women in a hundred become pregnant in a year using this combination method. Condoms have the added advantage of protecting partners against sexually transmitted diseases, including AIDS, if they are made of latex and used correctly. That may partly explain why their use in the United States went from 3.5 million users in 1980 to 8 million in 2000. Condoms are an ancient birth control method; the Egyptians used them some 3,000 years ago.

More than 100 new contraceptive methods are now being studied, and some appear to have great promise. Nearly all are biologically based (e.g., hormonal), rather than mechanical (e.g.,

Table 4.4	Some Birth Control Methods and Pregnancy Prevention Rates
Method	**Number of Women in 100 Who Become Pregnant**
Sterilization (male, female)	<1
IUD	<1
Oral contraceptive (the Pill)	1–2
Hormones (implant, patch, injection, etc.)	1–2
Male condom	11
Sponge and spermicide	14–28
Female condom (e.g., cervical cap)	15–23
Diaphragm together with spermicide	17
Abstinence during fertile periods	20
Morning-after-pill (e.g., Preven)	20
Spermicide alone	20–50
Actively seeking pregnancy	85

Source: U.S. Food and Drug Administration, *Birth Control Guide,* 2003 Revision.

condom, IUD). Recently, the U.S. Food and Drug Administration approved five new birth control products. Four of these use various methods to administer female hormones that prevent pregnancy. Other methods are years away from use, but take a new direction entirely. Vaccines for women are being developed that will prepare the immune system to reject the hormone chorionic gonadotropin, which maintains the uterine lining and allows egg implant, or that will cause an immune reaction against sperm. Injections for men are focused on reducing sperm production, and have proven effective in mice. Without a doubt, the contemporary couple has access to many more birth control options than their grandparents had.

4.7 What Kind of Future Are We Creating Now?

Because there is a lag between the time when a society reaches replacement birth rate and the end of population growth, we are deciding now what the world will look like in a hundred years. How many people will be in the world a century from now? Most demographers believe that world population will stabilize sometime during the twenty-first century. When we reach that equilibrium, the total number of humans is likely to be somewhere around 8 to 10 billion, depending on the success of family planning programs and the multitude of other factors affecting human populations. The United Nations Population Division projects four population scenarios (fig. 4.16). The optimistic (low) projection suggests that world population might stabilize by about 2030 and then drop back below current levels. This is not likely. The medium projections suggest that growth could continue until at

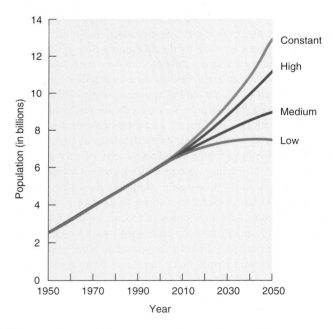

Figure 4.16 Population projections for different growth scenarios. Recent progress in family planning and economic development have led to significantly reduced estimates compared to a few years ago. The medium projection is 8.9 billion in 2050, compared to previous estimates of over 10 billion for that date.
Source: UN Population Division, 2004.

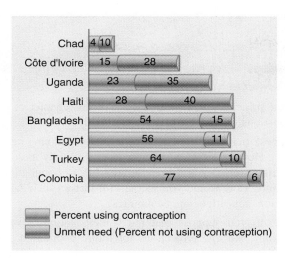

Figure 4.17 "Unmet need" is measured as percent of married women who want to control their fertility but do not use contraception. Globally, more than 100 million women in developing countries would prefer to avoid pregnancy but do not have access to family planning.
Source: Data from Population Reference Bureau, 2003.

least 2050, at which point the total world population could be 11 to 13 billion.

Which of these scenarios will we follow? As you have seen in this chapter, population growth is a complex subject. Stabilizing or reducing human populations will require substantial changes from business as usual. One impediment is that the United States refuses to release payments to the United Nations Family Planning Fund because some of the 135 countries that receive UN aid include abortion as part of population control programs. The UNFPA is the world's leading provider of funding for family planning and reproductive health programs. Since UNFPA began operations in 1969, fertility rates in developing countries have fallen from six children per woman to three.

An encouraging sign is that worldwide contraceptive use has increased sharply in recent years. About half of the world's married couples used some family planning techniques in 2000, compared with only 10 percent 30 years earlier, but another 100 million couples say they want, but do not have access to, family planning. If given a choice, people prefer smaller families.

The unmet need for family planning among married women in some countries is high (fig. 4.17). When people in developing countries are asked what they want most, men say they want better jobs, but the first choice for a vast majority of women is family planning assistance. In general, a 15 percent increase in contraceptive use equates to about 1 fewer births per woman per lifetime. In Chad, for example, where only 4 percent of all women use contraceptives, the average fertility is 6.6 children per woman. In Colombia, by contrast, where 77 percent of the women who would prefer not to be pregnant use contraceptives, the average fertility is 2.6 children.

Successful family planning programs often require significant societal changes. Among the most important of these are

(1) improved social, educational, and economic status for women (birth control and women's rights are often linked); (2) improved status for children (fewer children are born if they are not needed as a cheap labor source); (3) acceptance of calculated choice as a valid element in life in general and in fertility in particular (the belief that we have no control over our lives discourages a sense of responsibility); (4) social security and political stability that give people the means and the confidence to plan for the future; and (5) the knowledge, availability, and use of effective and acceptable means of birth control. Concerted efforts to bring about these types of societal changes can be effective. Twenty years of economic development and work by voluntary family planning groups in Zimbabwe, for example, have lowered total fertility rates from 8.0 to 5.5 children per woman on average. Surveys show that desired family sizes have fallen nearly by half (9.0 to 4.6) and that nearly all women and 80 percent of men in Zimbabwe use contraceptives.

The current world average fertility rate of 2.6 births per woman is less than half what it was 50 years ago. If similar progress could be sustained for the next half century, fertility rates could fall to the replacement rate of 2.1 children per woman. Whether this scenario comes true or not depends on choices that all of us make. In their approach to population issues, the people of Kerala may have indicated for the rest of the world an effective yet supremely humane direction to take.

Conclusion

A few decades ago, we were warned that a human population explosion was about to engulf the world. Exponential population growth was seen as a cause or corollary to nearly every important environmental problem. Some people still warn that the total number of humans might grow to 30 or 40 billion by the end of this century. Birth rates have fallen, however, almost everywhere, and most demographers now believe that we will reach an equilibrium around 9 billion people in about 2050. Some claim that if we promote equality, democracy, human development, and modern family planning techniques, population might even decline to below its current level of 6.7 billion in the next 50 years. How we should carry out family planning and birth control remains a controversial issue. Should we focus on political and economic reforms, and hope that a demographic transition will naturally follow; or should we take more direct action (or any action) to reduce births?

Whether our planet can support 9 billion—or even 6 billion—people on a long-term basis remains a vital question. If all those people try to live at a level of material comfort and affluence now enjoyed by residents of the wealthiest nations, using the old, polluting, inefficient technology that we now employ, the answer is almost certain that even 6 billion people is too many in the long run. If we find more sustainable ways to live, however, it may be that 9 billion people could live happy, comfortable, productive lives. If we don't find new ways to live, we probably face a crisis no matter what happens to our population size. We'll discuss pollution problems, energy sources, and sustainability in subsequent chapters of this book.

1. About how many years of human existence passed before the world population reached its first billion? What factors restricted population before that time, and what factors contributed to growth after that point?

2. Describe the pattern of human population growth over the past 200 years. What is the shape of the growth curve (recall chapter 3)?

3. Define *ecological footprint*. Why is it helpful, but why might it also be inaccurate?

4. Why do some economists consider human resources more important than natural resources in determining a country's future?

5. In which regions of the world will most population growth occur during the twenty-first century? What conditions contribute to rapid population growth in these locations?

6. Define *crude birth rate, total fertility rate, crude death rate,* and *zero population growth*.

7. What is the difference between life expectancy and life span? Why are they different?

8. What is the dependency ratio, and how might it affect the United States in the future?

9. What factors increase or decrease people's desire to have babies?

10. Describe the conditions that lead to a demographic transition.

Apply the principles you have learned in this chapter to discuss these questions with other students.

1. Suppose that you were head of a family planning agency in India. How would you design a scientific study to determine the effectiveness of different approaches to population stabilization? How would you account for factors such as culture, religion, education, and economics?

2. Why do you suppose that the United Nations gives high, medium, and low projections for future population growth? Why not give a single estimate? What factors would you consider in making these projections?

3. Some demographers claim that the total world population has already begun to slow, while others dispute this claim. How would you recognize a true demographic transition, as opposed to mere random fluctuations in birth and death rates?

4. Why do we usually express crude birth and death rates per thousand people? Why not give the numbers per person or for the entire population?

5. In northern Europe, the demographic transition began in the early 1800s, a century or more before the invention of modern antibiotics and other miracle drugs. What factors do you think contributed to this transition? How would you use historical records to test your hypothesis?

6. In chapter 3, we discussed carrying capacities. What do you think the maximum and optimum carrying capacities for humans are? Why is this a more complex question for humans than it might be for other species? Why is designing experiments in human demography difficult?

Data Analysis: Communicating with Graphs

Graphs are pictorial representations of data designed to communicate information. They are particularly useful in conveying patterns and relationships. Most of us, when confronted by a page of numbers, tend to roll our eyes and turn to something else. A graph can simplify trends and highlight important information. If you read a newspaper or news magazine these days, you will undoubtedly see numerous graphs portraying a wide variety of statistics. Why are they used so widely and why do they make such powerful impressions on us?

Graphs can be used to explain, persuade, or inform. Like photographs or text, however, they also can be used to manipulate or even mislead the viewer. When you create a graph, you

inevitably make decisions about what details to display, and which ones to omit. By selecting certain scales or methods of displaying data, you deliberately choose the message you intend to convey. It takes critical thinking skills to understand what a graph is (and isn't) showing.

Part A

Central questions addressed in this chapter are: How serious is the world population problem? Does it demand action? Many graphs present information about population growth, yet the impressions each conveys to the reader vary considerably, depending on the type of data used and the design of the graph. Reexamine figure

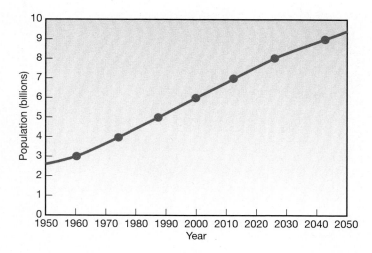

Figure 1 Growth of the world population, 1950–2050.
Source: U.S. Census Bureau, 2006.

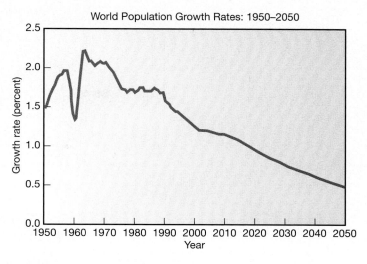

World Population Growth Rates: 1950–2050

Figure 2 World population growth rates, 1950–2050.
Source: U.S. Census Bureau, 2006.

4.3, which presents a historical perspective on human population growth. What impressions do you draw from it? What conclusions do you suppose the person who originally created this graph intended for you to take away from this presentation? There are a number of differences in graphs that control your impressions of the problem. Compare fig. 4.2 to figure 1, above. These two graphs look very different, but look again, carefully.

1. Do they show the same kind of data?

2. What aspects of these two graphs differ?

3. Which is more alarming, and why?

Part B

There are other ways to plot population change. Compare figure 2, above, to figure 1.

4. What do the two Y-axes show?

5. What is the approximate growth rate now? About how long would it be until the growth rate reached 0, if this trend is correct?

6. Which graph shows better news?

7. What would figure 2 look like if the Y axis values ranged from 0 to 25?

8. Are there policy purposes that would make you choose one of these graphs to show people? What might those purposes be?

Now look back at fig. 4.6, on p. 81.

9. Does fig. 4.6 show the same information as the others discussed here? Is it the same topic?

10. How does the time frame in fig. 4.6 compare to the other graphs discussed above?

11. What was the proportion of fertility in *more developed countries* to those in *less developed countries* in 1950? today? (Round to whole numbers.)

12. What is the proportion of fertility in 1950 compared to 2050 for *less developed countries*?

Wolves reintroduced to Yellowstone National Park may be helping to restore biodiversity.

5 Biomes and Biodiversity

Learning Outcomes

After studying this chapter, you should be able to answer the following questions:

- What are nine major terrestrial biomes, and what environmental conditions control their distribution?
- How does vertical stratification differentiate life zones in oceans?
- Why are coral reefs, mangroves, estuaries, and wetlands biologically important?
- What do we mean by *biodiversity*? List several regions of high biodiversity.
- What are four major benefits of biodiversity?
- What are the major human-caused threats to biodiversity?
- How can we reduce these threats to biodiversity?

In the end, we conserve only what we love.
We will love only what we understand.
We will understand only what we are taught.

—Baba Dioum

Predators Help Restore Biodiversity in Yellowstone

In the past dozen years, Yellowstone National Park has been the site of a grand experiment in managing biodiversity. The park is renowned for its wildlife and wilderness. But between about 1930 and 1990, park managers and ecologists observed that growing elk populations were over-browsing willow and aspen. Plant health and diversity were declining, and populations of smaller mammals, such as beaver, were gradually dwindling. Most ecologists blamed these changes on the eradication of one of the region's main predators, the gray wolf (*Canis lupus*). Now, a project to restore wolves has provided a rare opportunity to watch environmental change unfold.

Wolves were once abundant in the Greater Yellowstone ecosystem (fig. 5.1). In the 1890s, perhaps 100,000 wolves roamed the western United States. Farmers and ranchers, aided by government programs to eradicate predators, gradually poisoned, shot, and trapped nearly all the country's wolves. In the northern Rocky Mountains, elk and deer populations expanded rapidly without predators to restrain their numbers. Yellowstone's northern range elk herd grew to about 20,000 animals. Some ecologists warned that without a complete guild of predators, the elk, bison, and deer could damage one of our best-loved parks. After

much controversy, 31 wolves were captured in the Canadian Rockies and relocated to Yellowstone National Park in 1995 and 1996.

Wolves expanded quickly in their new home. With abundant elk for prey, as well as occasional moose and bison, the wolves tripled their population in just three years. The population now exceeds 100 animals in over 20 packs, well beyond the recovery goal of 10 packs.

Following the wolf reintroductions, elk numbers have fallen, and regenerating willow and aspen show signs of increased stature and abundance. Elk carcasses left by wolves provide a feast for scavengers, and wolves appear to have driven down numbers of coyotes, which prey on small mammals. Many ecologists see evidence of more songbirds, hawks, foxes, voles, and ground squirrels, which they attribute to wolf reintroductions. Other ecologists caution that Yellowstone is a complex system, and wolves are one of many factors that may influence regeneration. For example, recent mild winters may have allowed elk to find alternate food sources, reducing browsing pressure on willows and aspen, and drought may have contributed to declining elk numbers. Debate on the exact impact of wolves underscores the difficulty of interpreting change in complex ecosystems. Whether or not wolves are driving ecological change, most ecologists agree that wolves contribute to a diverse and healthy ecosystem.

New Technology

How do ecologists study these changes? For years, ecologists have skiied and snowshoed out to remote observation posts to record wolf kills and observe wolf behavior. But the difficulty of this field work has limited the amount of data that could be collected. One of the exciting new alternatives involves a remotely controlled video camera that allows researchers to monitor wolf ecology and behavior without intruding on the wolves.

Ecologist Dan MacNulty, of the University of Minnesota, together with Glenn Plumb of the National Park Service, has installed a pair of computer-controlled video cameras in an open meadow near hot springs where bison gather to seek exposed grass in the winter. Wolves, hoping to prey on the bison, frequent the area as well. Instead of skiing long, cold miles to the site, MacNulty can now turn on a camera from his office desktop (fig. 5.2). The camera scans an area of 17 km², and it can zoom in to identify individual animals up to 9.5 km away. In the past, his team was lucky to get a few weeks of winter wolf observation in a year, but now MacNulty and his 12 undergraduate research assistants can watch for wolves year-round, any time

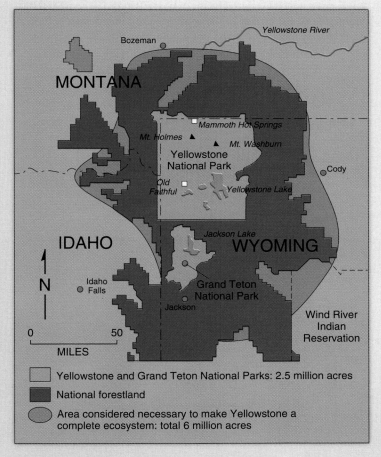

Figure 5.1 This map shows the Yellowstone ecosystem complex, or biogeographical region, which extends far beyond the park's boundaries. Park managers and ecologists believe that it is necessary to manage the entire region if the park itself is to remain biologically viable.

In the map:
- Yellowstone and Grand Teton National Parks: 2.5 million acres
- National forestland
- Area considered necessary to make Yellowstone a complete ecosystem: total 6 million acres

MONTANA
IDAHO
WYOMING
Bozeman
Yellowstone River
Mammoth Hot Springs
Mt. Holmes
Mt. Washburn
Yellowstone National Park
Cody
Old Faithful
Yellowstone Lake
Jackson Lake
Grand Teton National Park
Jackson
Idaho Falls
Wind River Indian Reservation
0   50
MILES
N

Figure 5.2 Dan MacNulty observes wolves from his office.

it's light enough to see them. Working in shifts, the group records sightings and behavior of wolves, bears, elk, bison, coyotes, and fox. (For discussion of additional research methods, see Exploring Science, p. 112.)

Because they can observe the animals at length without disturbing them, MacNulty's group has made some intriguing observations. Wolves are masterful at detecting when a bison is vulnerable. They almost never attack a large bull, unless it has wandered from the group or gotten into deep snow, where bison quickly get bogged down. Bison will also work together to fend off wolves—if they can avoid leaving the snow-free areas. Wolves can also be surprisingly persistent: in one case McNulty

says wolves harassed a vulnerable bison for 36 hours before finally killing it. All these observations help ecologists understand how wolves hunt, select their prey, and avoid injury in the chase.

Yellowstone's wolf restoration is exciting because it is almost unique as a large, biodiversity restoration project. As a natural laboratory, the park has allowed researchers to explore many questions about ecosystem function, stability, and resilience. In this chapter, we'll examine the major living systems that harbor the earth's biodiversity. We'll also consider the benefits of biodiversity, activities that threaten diversity, and opportunities to preserve it.

5.1 Terrestrial Biomes

As a top predator, Yellowstone's wolves appear to help maintain **biodiversity** (the number and variety of species) in the ecosystem they occupy. Biodiversity is important because it can help maintain stability in a system, or it can help a system recover from disturbance. Diversity of species also contributes to resources we humans need. Biodiversity depends on local conditions, such as the number of predators or recent drought; it also depends on larger regional patterns of temperature and precipitation. For example, Yellowstone's ecosystem experiences summers that are warm and often dry; its winters are cold, with deep snow. As a consequence, Yellowstone has coniferous trees that can withstand cold months, as well as elk, moose, and bison that can survive the winter by finding enough vegetation through the snow (especially near hot springs) to browse. These herbivores support wolves, whose cooperative hunting and family structure allow them to prey on large herbivores.

To begin understanding variations in biodiversity, it is important to recognize the characteristic types of environments that occur in different conditions of temperature and precipitation. We call these broad types of biological communities **biomes**. If we know the general temperature and precipitation ranges of an area, we can predict what kind of biological community is likely to occur there, in the absences of human disturbance (fig. 5.3). Identifying these general categories helps us see patterns in the infinite variety of local environments. Yellowstone, with warm summers, cold winters, and mostly pine forests, can be described as part of a temperate forest biome. You can find this biome and its range of temperature and precipitation in figure 5.3. What is the range of average annual temperatures for temperate forests? What is the range of precipitation? If conditions were drier, let's say less than 100 cm per year, what biomes might Yellowstone be in?

Understanding the global distribution of biomes, and knowing the differences in what grows where and why, is essential to the study of global environmental science. For example, biological productivity (the rate at which photosynthesizing plants produce biomass) varies greatly from warm to cold climates, and from wet to dry environments. The amount of resources we can extract, such as timber or crops, depends largely on a biome's biological productivity. Similarly, the ability of an ecosystem to recover from disturbance, or our ability to restore ecosystems (chapter 6) depends on the diversity and productivity of the biome. Clear-cut forests regrow relatively quickly in Georgia, a warm, moist, temperate forest biome in the southeastern United States. Forests regrow very slowly in Siberia, a boreal forest that is also home to one of the world's most rapidly expanding logging industries.

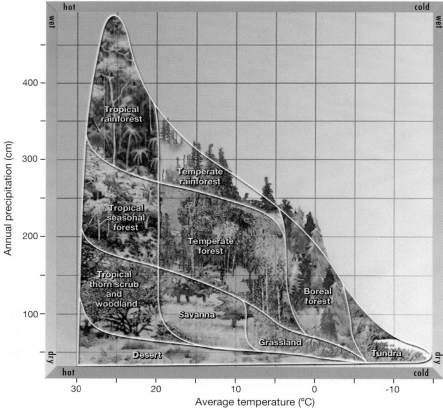

Figure 5.3 Biomes most likely to occur in the absence of human disturbance or other disruptions, according to average annual temperature and precipitation. *Note:* This diagram does not consider soil type, topography, wind speed, or other important environmental factors. Still, it is a useful general guideline for biome location.

Source: From *Communities & Ecosystems,* 2e by R. H. Whitaker, 1975. Reprinted by permission of Prentice Hall, Upper Saddle River, New Jersey.

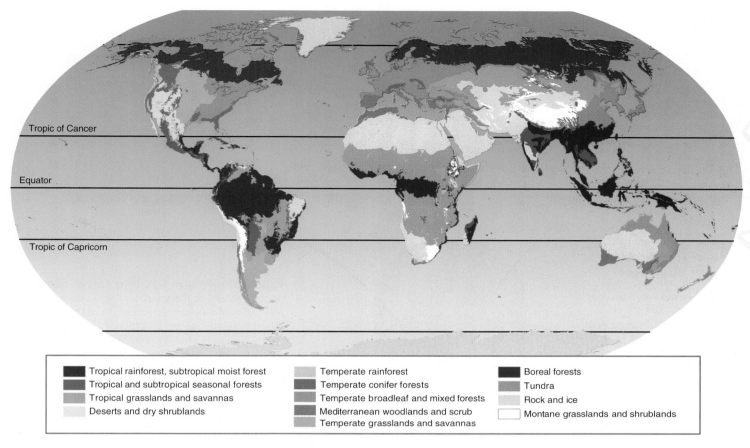

Tropic of Cancer

Equator

Tropic of Capricorn

■ Tropical rainforest, subtropical moist forest	■ Temperate rainforest	■ Boreal forests
■ Tropical and subtropical seasonal forests	■ Temperate conifer forests	■ Tundra
■ Tropical grasslands and savannas	■ Temperate broadleaf and mixed forests	■ Rock and ice
■ Deserts and dry shrublands	■ Mediterranean woodlands and scrub	□ Montane grasslands and shrublands
	■ Temperate grasslands and savannas	

Figure 5.4 Major world biomes. Compare this map with figure 5.3 for generalized temperature and moisture conditions that control biome distribution. Also compare it with the satellite image of biological productivity (fig. 5.15).
Source: WWF Ecoregions.

In the sections that follow, you will examine nine major biome types. These nine can be further divided into smaller classes: for example, some temperate forests have mainly cone-bearing trees (conifers), such as pines or fir, while other have mainly broad-leaf trees, such as maples or oaks (fig. 5.4). Many temperature-controlled biomes occur in latitudinal bands. A band of boreal (northern) forest crosses Canada and Siberia, tropical forests occur near the equator, and expansive grasslands lie near—or just beyond—the tropics. Many biomes are even named for their latitudes: tropical rainforests occur between the Tropic of Cancer (23° north) and the Tropic of Capricorn (23° south); arctic tundra lies near or above the Arctic Circle (66.6° north).

Temperature and precipitation change with elevation as well as with latitude. In mountainous regions, temperatures are cooler and precipitation is usually greater at high elevations. Mountains are cooler and often wetter, than low elevations. **Vertical zonation** is a term applied to vegetation zones defined by altitude. A 100 km transect from California's Central Valley up to Mount Whitney, for example, crosses as many vegetation zones as you would find on a journey from southern California to northern Canada (fig. 5.5).

As you consider the terrestrial biomes, compare the annual temperatures and precipitation levels that occur there. To begin,

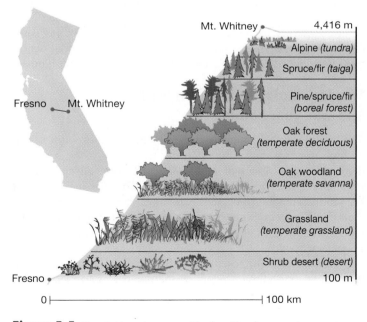

Figure 5.5 Vegetation changes with elevation because temperatures are lower and precipitation is greater high on a mountain side. A 100 km transect from Fresno, California, to Mount Whitney (California's highest point) crosses vegetation zones similar to about seven different biome types.

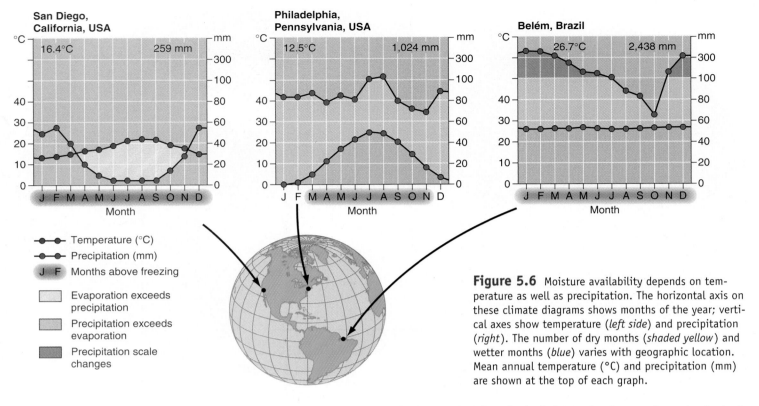

Figure 5.6 Moisture availability depends on temperature as well as precipitation. The horizontal axis on these climate diagrams shows months of the year; vertical axes show temperature (*left side*) and precipitation (*right*). The number of dry months (*shaded yellow*) and wetter months (*blue*) varies with geographic location. Mean annual temperature (°C) and precipitation (mm) are shown at the top of each graph.

examine the three climate graphs in figure 5.6. These graphs show annual trends in temperature and precipitation (rainfall and snowfall). They also indicate the relationship between potential evaporation, which depends on temperature, and precipitation. When evaporation exceeds precipitation, dry conditions result (marked yellow). Moist climates may vary in precipitation rates, but evaporation rarely exceeds precipitation. Months above freezing temperature (marked brown) have most evaporation. Comparing these climate graphs helps us understand the different seasonal conditions that control plant and animal growth in the different biomes.

Tropical moist forests are warm and wet year-round

The humid tropical regions support one of the most complex and biologically rich biome types in the world (fig. 5.7). Although there are several kinds of moist tropical forests, all have ample rainfall and uniform temperatures. Cool **cloud forests** are found high in the mountains where fog and mist keep vegetation wet all the time. **Tropical** rainforests occur where rainfall is abundant—more than 200 cm (80 in.) per year—and temperatures are warm to hot year-round.

The soil of both these tropical moist forest types tends to be thin, acidic, and nutrient-poor, yet the number of species present can be mind-boggling. For example, the number of insect species

in the canopy of tropical rainforests has been estimated to be in the millions! It is estimated that one-half to two-thirds of all species of terrestrial plants and insects live in tropical forests.

The nutrient cycles of these forests also are distinctive. Almost all (90 percent) of the nutrients in the system are contained in the bodies of the living organisms. This is a striking contrast to temperate forests, where nutrients are held within the soil and made

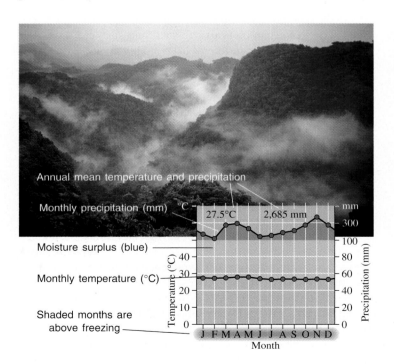

Figure 5.7 Tropical rainforests have luxuriant and diverse plant growth. Heavy rainfall in most months, shown in the climate graph, supports this growth.

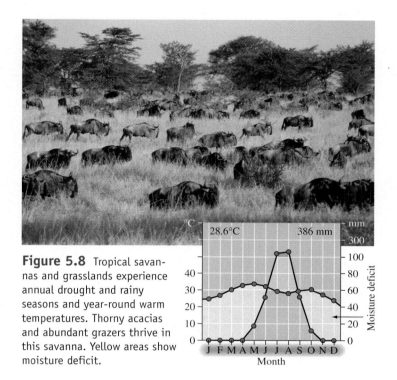

Figure 5.8 Tropical savannas and grasslands experience annual drought and rainy seasons and year-round warm temperatures. Thorny acacias and abundant grazers thrive in this savanna. Yellow areas show moisture deficit.

available for new plant growth. The luxuriant growth in tropical rainforests depends on rapid decomposition and recycling of dead organic material. Leaves and branches that fall to the forest floor decay and are incorporated almost immediately back into living biomass.

When the forest is removed for logging, agriculture, and mineral extraction, the thin soil cannot support continued cropping and cannot resist erosion from the abundant rains. And if the cleared area is too extensive, it cannot be repopulated by the rainforest community.

Tropical seasonal forests have annual dry seasons

Many tropical regions are characterized by distinct wet and dry seasons, although temperatures remain hot year-round. These areas support **tropical seasonal forests**: drought-tolerant forests that look brown and dormant in the dry season but burst into vivid green during rainy months. These forests are often called dry tropical forests because they are dry much of the year; however, there must be some periodic rain to support tree growth. Many of the trees and shrubs in a seasonal forest are drought-deciduous: they lose their leaves and cease growing when no water is available. Seasonal forests are often open woodlands that grade into savannas.

Tropical dry forests have typically been more attractive than wet forests for human habitation and have suffered greater degradation. Clearing a dry forest with fire is relatively easy during the dry season. Soils of dry forests often have higher nutrient levels and are more agriculturally productive than those of a rainforest. Finally, having fewer insects, parasites, and fungal diseases than a wet forest makes a dry or seasonal forest a healthier place for

humans to live. Consequently, these forests are highly endangered in many places. Less than 1 percent of the dry tropical forests of the Pacific coast of Central America or the Atlantic coast of South America, for instance, remain in an undisturbed state.

Tropical savannas and grasslands are dry most of the year

Where there is too little rainfall to support forests, we find open **grasslands** or grasslands with sparse tree cover, which we call **savannas** (fig. 5.8). Like tropical seasonal forests, most tropical savannas and grasslands have a rainy season, but generally the rains are less abundant or less dependable than in a forest. During dry seasons, fires can sweep across a grassland, killing off young trees and keeping the landscape open. Savanna and grassland plants have many adaptations to survive drought, heat, and fires. Many have deep, long-lived roots that seek groundwater and that persist when leaves and stems above the ground die back. After a fire, or after a drought, fresh green shoots grow quickly from the roots. Migratory grazers, such as wildebeest, antelope, or bison thrive on this new growth. Grazing pressure from domestic livestock is an important threat to both the plants and animals of tropical grasslands and savannas.

Deserts are hot or cold, but always dry

You may think of deserts as barren and biologically impoverished. Their vegetation is sparse, but it can be surprisingly diverse, and most desert plants and animals are highly adapted to survive long droughts, extreme heat, and often extreme cold. **Deserts** occur where precipitation is uncommon and slight, usually with less than

Figure 5.9 Deserts generally receive less than 300 mm (30 cm) of precipitation per year. Hot deserts, as in the American Southwest, endure year-round drought and extreme heat in summer.

22.5°C   86 mm

Figure 5.10 Grasslands occur at midlatitudes on all continents. Kept open by extreme temperature, dry conditions, and periodic fires, grasslands can have surprisingly high plant and animal diversity.

13.1°C   803 mm

30 cm of rain per year. Adaptations to these conditions include water-storing leaves and stems, thick epidermal layers to reduce water loss, and salt tolerance. As in other dry environments, many plants are drought-deciduous. Most desert plants also bloom and set seed quickly when a spring rain does fall.

Warm, dry, high-pressure climate conditions (chapter 9) create desert regions at about 30° north and south. Extensive deserts occur in continental interiors (far from oceans, which evaporate the moisture for most precipitation) of North America, Central Asia, Africa, and Australia (fig. 5.9). The rain shadow of the Andes produces the world's driest desert in coastal Chile. Deserts can also be cold. Antarctica is a desert; some inland valleys apparently get almost no precipitation at all.

Like plants, animals in deserts are specially adapted. Many are nocturnal, spending their days in burrows to avoid the sun's heat and desiccation. Pocket mice, kangaroo rats, and gerbils can get most of their moisture from seeds and plants. Desert rodents also have highly concentrated urine and nearly dry feces, which allow them to eliminate body waste without losing precious moisture.

Deserts are more vulnerable than you might imagine. Sparse, slow-growing vegetation is quickly damaged by off-road vehicles. Desert soils recover slowly. Tracks left by army tanks practicing in California deserts during World War II can still be seen today.

Deserts are also vulnerable to overgrazing. In Africa's vast Sahel (the southern edge of the Sahara Desert), livestock are destroying much of the plant cover. Bare, dry soil becomes drifting sand, and restabilization is extremely difficult. Without plant roots and organic matter, the soil loses its ability to retain what rain does fall, and the land becomes progressively drier and more bare. Similar depletion of dryland vegetation is happening in many desert areas, including Central Asia, India, and the American Southwest and Plains states.

Temperate grasslands have rich soils

As in tropical latitudes, temperate (midlatitude) grasslands occur where there is enough rain to support abundant grass but not enough for forests (fig. 5.10). Usually grasslands are a complex, diverse mix of grasses and flowering herbaceous plants, generally known as forbs. Myriad flowering forbs make a grassland colorful and lovely in summer. In dry grasslands, vegetation may be less than a meter tall. In more humid areas, grasses can exceed 2 m. Where scattered trees occur in a grassland, we call it a savanna.

Deep roots help plants in temperate grasslands and savannas survive drought, fire, and extreme heat and cold. These roots, together with an annual winter accumulation of dead leaves on the surface, produce thick, organic-rich soils in temperate grasslands. Because of this rich soil, many grasslands have been converted to farmland. The legendary tallgrass prairies of the central United States and Canada are almost completely replaced by corn, soybeans, wheat, and other crops. Most remaining grasslands in this region are too dry to support agriculture, and their greatest threat is overgrazing. Excessive grazing eventually kills even deep-rooted plants. As ground cover dies off, soil erosion results, and unpalatable weeds, such as cheatgrass or leafy spurge, spread.

Temperate shrublands have summer drought

Often, dry environments support drought-adapted shrubs and trees, as well as grass. These mixed environments can be highly variable. They can also be very rich biologically. Such conditions are often described as Mediterranean (where the hot season coincides with the dry season producing hot, dry summers and cool, moist winters). Evergreen shrubs with small, leathery, sclerophyllous (hard,

Figure 5.11 Temperate deciduous forests have year-round precipitation and winters near or below freezing.

produce brilliant colors in these forests in autumn (fig. 5.11). At lower latitudes, broad-leaf forests may be evergreen or drought-deciduous. Southern live oaks, for example, are broad-leaf evergreen trees.

Although these forests have a dense canopy in summer, they have a diverse understory that blooms in spring, before the trees leaf out. Spring ephemeral (short-lived) plants produce lovely flowers, and vernal (springtime) pools support amphibians and insects. These forests also shelter a great diversity of songbirds.

North American deciduous forests once covered most of what is now the eastern half of the United States and southern Canada. Most of western Europe was once deciduous forest but was cleared a thousand years ago. When European settlers first came to North America, they quickly settled and cut most of the eastern deciduous forests for firewood, lumber, and industrial uses, as well as to clear farmland. Many of those regions have now returned to deciduous forest, though the dominant species have changed.

Deciduous forests can regrow quickly because they occupy moist, moderate climates. But most of these forests have been occupied so long that human impacts are extensive, and most native species are at least somewhat threatened. The greatest threat to broad-leaf deciduous forests is in eastern Siberia, where deforestation is proceeding rapidly. Siberia may have the highest deforestation rate in the world. As forests disappear, so do Siberian tigers, bears, cranes, and a host of other endangered species.

Coniferous forests

Coniferous forests grow in a wide range of temperature and moisture conditions. Often they occur where moisture is limited: in cold climates, moisture is unavailable (frozen) in winter; hot climates may have seasonal drought; sandy soils hold little moisture, and they are often occupied by conifers. Thin, waxy leaves (needles) help these trees reduce moisture loss. Coniferous forests provide most wood products in North America. Dominant wood production regions include the southern Atlantic and Gulf coast states, the mountain West, and the Pacific Northwest (northern California to Alaska), but coniferous forests support forestry in many regions.

The coniferous forests of the Pacific coast grow in extremely wet conditions. The wettest coastal forests are known as **temperate rainforest**, a cool, rainy forest often enshrouded in fog (fig. 5.12). Condensation in the canopy (leaf drip) is a major form of precipitation in the understory. Mild year-round temperatures and abundant rainfall, up to 250 cm (100 in.) per year, result in luxuriant plant growth and giant trees such as the California redwoods, the largest trees in the world and the largest above-ground organism ever known to have existed. Redwoods once grew along the Pacific coast from California to Oregon, but logging has reduced them to a few small fragments.

Boreal forests lie north of the temperate zone

Because conifers can survive winter cold, they tend to dominate the **boreal forest**, or northern forests, that lie between about 50° and 60° north (fig. 5.13). Mountainous areas at lower latitudes may also have many characteristics and species of the boreal forest. Dominant trees are pines, hemlocks, spruce, cedar, and fir. Some

waxy) leaves form dense thickets. Scrub oaks, drought-resistant pines, or other small trees often cluster in sheltered valleys. Periodic fires burn fiercely in this fuel-rich plant assemblage and are a major factor in plant succession. Annual spring flowers often bloom profusely, especially after fires. In California, this landscape is called **chaparral**, Spanish for thicket. Resident animals are drought tolerant species, such as jackrabbits, kangaroo rats, mule deer, chipmunks, lizards, and many bird species. Very similar landscapes are found along the Mediterranean coast as well as southwestern Australia, central Chile, and South Africa. Although this biome doesn't cover a very large total area, it contains a high number of unique species and is often considered a "hot-spot" for biodiversity. It also is highly desired for human habitation, often leading to conflicts with rare and endangered plant and animal species.

Temperate forests can be evergreen or deciduous

Temperate, or midlatitudes, forests occupy a wide range of precipitation conditions but occur mainly between about 30° and 55° latitude (see fig. 5.4). In general we can group these forests by tree type, which can be broad-leaved **deciduous** (losing leaves seasonally) or evergreen **coniferous** (cone-bearing).

Deciduous forests

Broad-leaf forests occur throughout the world where rainfall is plentiful. In midlatitudes, these forests are deciduous and lose their leaves in winter. The loss of green chlorophyll pigments can deciduous trees are also present, such as maples, birch, aspen, and alder. These forests are slow-growing because of the cold temperatures and short frost-free growing season, but they are still an expansive resource. In Siberia, Canada, and the western United States, large regional economies depend on boreal forests.

The extreme, ragged edge of the boreal forest, where forest gradually gives way to open tundra, is known by its Russian name, **taiga**. Here extreme cold and short summer limits the growth rate of trees. A 10 cm diameter tree may be over 200 years old in the far north.

Tundra can freeze in any month

Where temperatures are below freezing most of the year, only small, hardy vegetation can survive. **Tundra**, a treeless landscape that occurs at high latitudes or on mountaintops, has a growing season of only two to three months, and it may have frost any month of the year. Some people consider tundra a variant of grass-lands because it has no trees; others consider it a very cold desert because water is unavailable (frozen) most of the year.

Arctic tundra is an expansive biome that has low productivity because it has a short growing season. During midsummer, however, 24-hour sunshine supports a burst of plant growth and an explosion of insect life. Tens of millions of waterfowl, shorebirds, terns, and songbirds migrate to the Arctic every year to feast on the abundant invertebrate and plant life and to raise their young on the brief bounty. These birds then migrate to wintering grounds, where they may be eaten by local predators—effectively they carry energy and protein from high latitudes to low latitudes. Arctic tundra is essential for global biodiversity, especially for birds.

Alpine tundra, occurring on or near mountaintops, has environmental conditions and vegetation similar to arctic tundra (fig. 5.14). These areas have a short, intense growing season. Often one sees a splendid profusion of flowers in alpine tundra; everything must flower at once in order to produce seeds in a few weeks before the snow comes again. Many alpine tundra plants also have deep pigmentation and leathery leaves to protect against the strong ultraviolet light in the thin mountain atmosphere.

Compared to other biomes, tundra has relatively low diversity. Dwarf shrubs, such as willows, sedges, grasses, mosses, and lichens tend to dominate the vegetation. Migratory musk-ox, caribou, or alpine mountain sheep and mountain goats can live on the vegetation because they move frequently to new pastures.

Because these environments are too cold for most human activities, they are not as badly threatened as other biomes. There are important problems, however. Global climate change may be altering the balance of some tundra ecosystems, and air pollution from distant cities tends to accumulate at high latitudes (chapter 9). In eastern Canada, coastal tundra is being badly

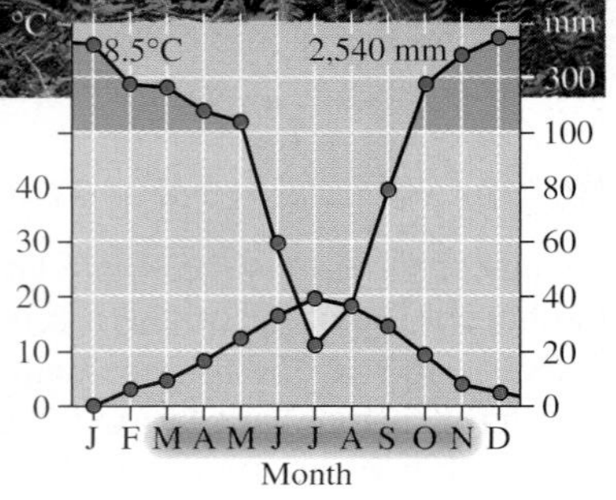

Figure 5.12 Temperate rainforests have abundant but often seasonal precipitation that supports magnificent trees and luxuriant understory vegetation. Often these forests experience dry summers.

°C   8.5°C   2,540 mm   mm

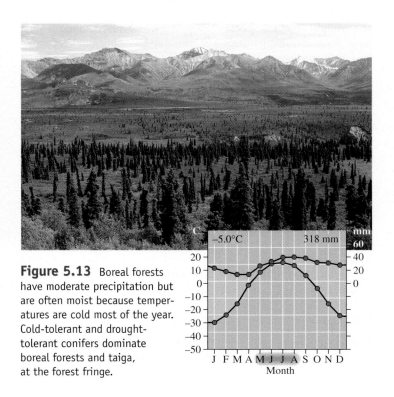

Figure 5.13 Boreal forests have moderate precipitation but are often moist because temperatures are cold most of the year. Cold-tolerant and drought-tolerant conifers dominate boreal forests and taiga, at the forest fringe.

°C   −5.0°C   318 mm   mm

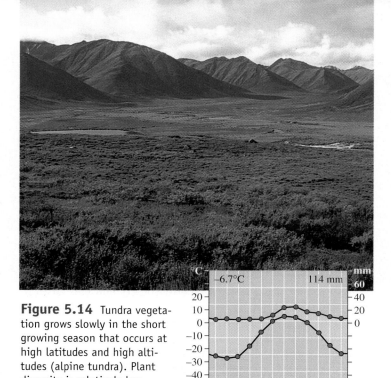

Figure 5.14 Tundra vegetation grows slowly in the short growing season that occurs at high latitudes and high altitudes (alpine tundra). Plant diversity is relatively low, and frost can occur even in summer.

depleted by overabundant populations of snow geese, whose numbers have exploded due to winter grazing on the rice fields of Arkansas and Louisiana. Oil and gas drilling—and associated truck traffic—threatens tundra in Alaska and Siberia. Clearly, this remote biome is not independent of human activities at lower latitudes.

5.2 Marine Ecosystems

The biological communities in oceans and seas are poorly understood, but they are probably as diverse and as complex as terrestrial biomes. In this section, we will explore a few facets of these fascinating environments. Oceans cover nearly three-fourths of the earth's surface, and they contribute in important, although often unrecognized, ways to terrestrial ecosystems. Like land-based systems, most marine

communities depend on photosynthetic organisms. Often it is algae or tiny, free-floating photosynthetic plants (**phytoplankton**) that support a marine food web, rather than the trees and grasses we see on land. In oceans, photosynthetic activity tends to be greatest near coastlines, where nitrogen, phosphorus, and other nutrients wash offshore and fertilize primary producers. Ocean currents also contribute to the distribution of biological productivity, as they transport nutrients and phytoplankton far from shore (fig. 5.15).

As plankton, algae, fish, and other organisms die, they sink toward the ocean floor. Deep-ocean ecosystems, consisting of crabs, filter-feeding organisms, strange phosphorescent fish, and

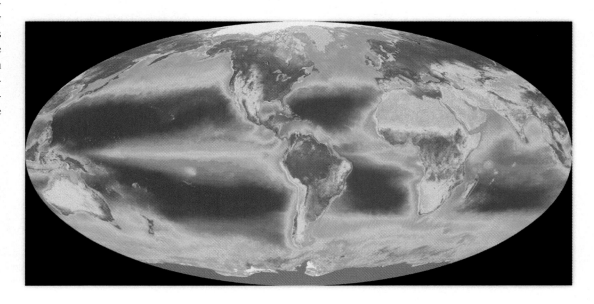

Figure 5.15 Satellite measurements of chlorophyll levels in the oceans and on land. Dark green to blue land areas have high biological productivity. Dark blue oceans have little chlorophyll and are biologically impoverished. Light green to yellow ocean zones are biologically rich.

many other life-forms, often rely on this "marine snow" as a primary nutrient source. Surface communities also depend on this material. Upwelling currents circulate nutrients from the ocean floor back to the surface. Along the coasts of South America, Africa, and Europe, these currents support rich fisheries.

Vertical stratification is a key feature of aquatic ecosystems. Light decreases rapidly with depth, and communities below the photic zone (light zone, often reaching about 20 m deep) must rely on energy sources other than photosynthesis to persist. Temperature also decreases with depth. Deep-ocean species often grow slowly in part because metabolism is reduced in cold conditions. In contrast, warm, bright, near-surface communities, such as coral reefs and estuaries, are among the world's most biologically productive environments. Temperature also affects the amount of oxygen and other elements that can be absorbed in water. Cold water holds abundant oxygen, so productivity is often high in cold oceans, as in the North Atlantic, North Pacific, and Antarctic.

Open ocean communities vary from surface to hadal zone

Ocean systems can be described by depth and proximity to shore (fig. 5.16). In general, **benthic** communities occur on the bottom, and **pelagic** (from "sea" in Greek) zones are the water column. The epipelagic zone (*epi* = on top) has photosynthetic organisms.

Below this are the mesopelagic (*meso* = medium) and bathypelagic (*bathos* = deep) zones. The deepest layers are the abyssal zone (to 4,000 m) and hadal zone (deeper than 6,000 m). Shorelines are known as littoral zones, and the area exposed by low tides is known as the intertidal zone. Often there is a broad, relatively shallow region along a continent's coast, which may reach a few kilometers or hundreds of kilometers from shore. This undersea area is the continental shelf.

We know relatively little about marine ecosystems and habitats, and much of what we know we have learned only recently. The open ocean has long been known as a biological desert, because it has relatively low productivity, or biomass production. Fish and plankton abound in many areas, however. Sea mounts, or undersea mountain chains and islands, support many commercial fisheries and much newly discovered biodiversity. In the equatorial Pacific and Antarctic oceans, currents carry nutrients far from shore, supporting biological productivity (fig. 5.17). The Sargasso Sea, a large region of the Atlantic near Bermuda, is known for its free-floating mats of brown algae. These algae mats support a phenomenal diversity of animals, including sea turtles, fish, and other species. Eels that hatch amid the algae eventually migrate up rivers along the Atlantic coasts of North America and Europe.

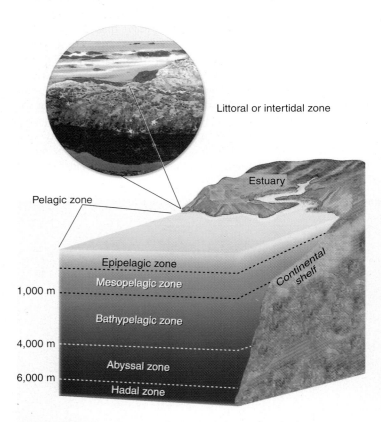

Figure 5.16 Light penetrates only the top 10–20 m of the ocean. Below this level, temperatures drop and pressure increases. Nearshore environments include the intertidal zone and estuaries.

Figure 5.17 Deep-ocean thermal vent communities were discovered only recently. They have great diversity and are unusual because they rely on chemosynthesis, not photosynthesis, for energy.

Deep-sea thermal vent communities are another remarkable type of marine system that was completely unknown until 1977, when the deep-sea submarine *Alvin* descended to the deep-ocean floor. These communities are based on microbes that capture chemical energy, mainly from sulfur compounds released from thermal vents—jets of hot water and minerals on the ocean floor. Magma below the ocean crust heats these vents. Tube worms, mussels, and microbes on the vents are adapted to survive both extreme temperatures, often above 350°C (700°F), and intense water pressure at depths of 7,000 m (20,000 ft) or more. Oceanographers have discovered thousands of different types of organisms, most of them microscopic, in these communities. Some estimate that the total mass of microbes on the seafloor represents one-third of all biomass on the planet.

Tidal shores support rich, diverse communities

As in the open ocean, shoreline communities vary with depth, light, and temperature. Some shoreline communities, such as estu-aries, have high biological productivity and diversity because they are enriched by nutrients washing from the land. Others, such as coral reefs, occur where there is little runoff from shore, but where shallow, clear warm water supports photosynthesis.

Coral reefs are among the best-known marine systems, because of their extraordinary biological productivity and their diverse and beautiful organisms (fig. 5.18a). Reefs are colonies of minute, colonial animals ("coral polyps") that live symbiotically with photosynthetic algae. Calcium-rich coral skeletons shelter the algae, and algae nourish the coral animals. The complex structure of a reef also shelters countless species of fish, worms, crustaceans, and other life-forms. Reefs occur where the water is shallow and clear enough for sunlight to reach the photosynthetic algae. They cannot tolerate abundant nutrients in the water, as nutrients support tiny floating plants and animals called plankton, which block sunlight.

Reefs are among the most endangered biological communities. Sediment from coastal development, farming, sewage, or other pollution can reduce water clarity and smother coral.

(a) Coral reefs

(b) Mangroves

(c) Estuary and salt marsh

(d) Tide pool

Figure 5.18 Coastal environments support incredible diversity and help stabilize shorelines. Coral reefs (a), mangroves (b), and estuaries (c) also provide critical nurseries for marine ecosystems. Tide pools (d) also shelter highly specialized organisms.

Destructive fishing practices, including dynamite and cyanide poison, have destroyed many Asian reefs. Reefs can also be damaged or killed by changes in temperature, by invasive fish, and by diseases. **Coral bleaching**, the whitening of reefs due to stress, often followed by coral death, is a growing and spreading problem that worries marine biologists (chapter 1).

Sea-grass beds, or eel-grass beds, occupy shallow, warm, sandy coastlines. Like reefs, these support rich communities of grazers, from snails to turtles to Florida's manatees.

Mangroves are a diverse group of salt-tolerant trees that grow along warm, calm marine coasts around the world (fig. 5.18b). Growing in shallow, tidal mudflats, mangroves help stabilize shorelines, blunt the force of storms, and build land by trapping sediment and organic material. After the devastating Indonesian tsunami of 2004, studies showed that mangroves, where they still stood, helped reduce the speed, height, and turbulence of the tsunami waves. Detritus, including fallen leaves, collects below mangroves and provides nutrients for a diverse community of animals and plants. Both marine species (such as crabs and fish) and terrestrial species (such as birds and bats) rely on mangroves for shelter and food.

Like coral reefs and sea-grass beds, mangrove forests provide sheltered nurseries for juvenile fish, crabs, shrimp, and other marine species on which human economies depend. However, like reefs and sea-grass beds, mangroves have been devastated by human activities. More than half of the world's mangroves that stood a century ago, perhaps 22 million ha, have been destroyed or degraded. They are clear-cut for timber or cleared to make room for fish and shrimp ponds. They are also poisoned by sewage and industrial waste near cities. Some parts of Southeast Asia and South America have lost 90 percent of their mangrove forests. Most have been cleared for fish and shrimp farming.

Estuaries are bays where rivers empty into the sea, mixing fresh water with salt water. **Salt marshes**, shallow wetlands flooded regularly or occasionally with seawater, occur on shallow coastlines, including estuaries (fig. 5.18c). Usually calm, warm, and nutrient-rich, estuaries and salt marshes are biologically diverse and productive. Rivers provide nutrients and sediments, and a muddy bottom supports emergent plants (whose leaves emerge above the water surface), as well as the young forms of crustaceans, such as crabs and shrimp, and mollusks, such as clams and oysters. Nearly two-thirds of all marine fish and shellfish rely on estuaries and saline wetlands for spawning and juvenile development.

Estuaries near major American cities once supported an enormous wealth of seafood. Oyster beds and clam banks in the waters adjacent to New York, Boston, and Baltimore provided free and easy food to early residents. Sewage and other contaminants long ago eliminated most of these resources, however. Recently, major efforts have been made to revive Chesapeake Bay, America's largest and most productive estuary. These efforts have shown some success, but many challenges remain (see related story "Restoring the Chesapeake" at www.mhhe.com/cunningham5e).

In contrast to the shallow, calm conditions of estuaries, coral reefs, and mangroves, there are violent, wave-blasted shorelines that support fascinating life-forms in **tide pools**. Tide pools are depressions in a rocky shoreline that are flooded at high tide but retain some water at low tide. These areas remain rocky where

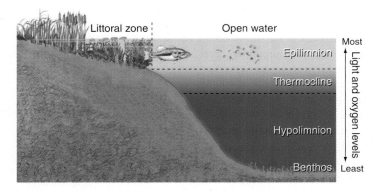

Figure 5.19 The layers of a deep lake are determined mainly by gradients of light, oxygen, and temperature. The epilimnion is affected by surface mixing from wind and thermal convections, while mixing between the hypolimnion and epilimnion is inhibited by a sharp temperature and density difference at the thermocline.

wave action prevents most plant growth or sediment (mud) accumulation. Extreme conditions, with frigid flooding at high tide and hot, dessicating sunshine at low tide, make life impossible for most species. But the specialized animals and plants that do occur in this rocky intertidal zone are astonishingly diverse and beautiful (fig. 5.18d).

5.3 Freshwater Ecosystems

Freshwater environments are far less extensive than marine environments, but they are centers of biodiversity. Most terrestrial communities rely, to some extent, on freshwater environments. In deserts, isolated pools, streams, and even underground water systems support astonishing biodiversity as well as provide water to land animals. In Arizona, for example, many birds are found in trees and bushes surrounding the few available rivers and streams.

Lakes have open water

Freshwater lakes, like marine environments, have distinct vertical zones (fig. 5.19). Near the surface a subcommunity of plankton, mainly microscopic plants, animals, and protists (single-celled organisms, such as amoebae), float freely in the water column. Insects such as water striders and mosquitoes also live at the air-water interface. Fish move through the water column, sometimes near the surface and sometimes at depth.

Finally, the bottom, or *benthos*, is occupied by a variety of snails, burrowing worms, fish, and other organisms. These make up the benthic community. Oxygen levels are lowest in the benthic environment, mainly because there is little mixing to introduce oxygen to this zone. Anaerobic (not using oxygen) bacteria may live in low-oxygen sediments. In the littoral zone, emergent plants, such as cattails and rushes, grow in the bottom sediment. These plants create important functional links between layers of an aquatic ecosystem, and they may provide the greatest primary productivity to the system.

Lakes, unless they are shallow, have a warmer upper layer that is mixed by wind and warmed by the sun. This layer is the *epilimnion*. Below the epilimnion is the hypolimnion (*hypo* = below), a colder, deeper layer that is not mixed. If you have gone swimming in a moderately deep lake, you may have discovered the sharp temperature boundary, known as the **thermocline**, between these layers. Below this boundary, the water is much colder. This boundary is also called the mesolimnion.

Local conditions that affect the characteristics of an aquatic community include (1) nutrient availability (or excess), such as nitrates and phosphates; (2) suspended matter, such as silt, that affects light penetration; (3) depth; (4) temperature; (5) currents; (6) bottom characteristics, such as muddy, sandy, or rocky floor; (7) internal currents; and (8) connections to, or isolation from, other aquatic and terrestrial systems.

Wetlands are shallow and productive

Wetlands are shallow ecosystems in which the land surface is saturated or submerged at least part of the year. Wetlands have vegetation that is adapted to grow under saturated conditions. These legal definitions are important because, although wetlands make up only a small part of most countries, they are disproportionately important in conservation debates and are the focus of continual legal disputes in North America and elsewhere around the world. Beyond these basic descriptions, defining wetlands is a matter of hot debate. How often must a wetland be saturated, and for how long? How large must it be to deserve legal protection? Answers can vary, depending on political, as well as ecological, concerns.

These relatively small systems support rich biodiversity, and they are essential for both breeding and migrating birds. Although wetlands occupy less than 5 percent of the land in the United States, the Fish and Wildlife Service estimates that one-third of all endangered species spend at least part of their lives in wetlands. Wetlands retain storm water and reduce flooding by slowing the rate at which rainfall reaches river systems. Floodwater storage is worth $3 billion to $4 billion per year in the United States. As water stands in wetlands, it also seeps into the ground, replenishing groundwater supplies. Wetlands filter, and even purify, urban and farm runoff, as bacteria and plants take up nutrients and contaminants in water. They are also in great demand for filling and development. They are often near cities or farms, where land is valuable, and, once drained, wetlands are easily converted to more lucrative uses. At least half of all the existing wetlands in the United States when Europeans first arrived have been drained, filled, or degraded. In some major farming states, losses have been even greater. Iowa, for example, has lost 99 percent of its original wetlands.

Wetlands are described by their vegetation. **Swamps**, also called forested wetlands, are wetlands with trees. **Marshes** are wetlands without trees (fig. 5.20). **Bogs** are areas of water-saturated ground, and usually the ground is composed of deep layers of accumulated, undecayed vegetation known as peat. **Fens** are similar to bogs except that they are mainly fed by groundwater, so that they have mineral-rich water and specially adapted plant species. Bogs are fed mainly by precipitation. Swamps and marshes have high biologi-

(a) Swamp, or wooded wetland

(b) Marsh

(c) Bog

Figure 5.20 Wetlands provide irreplaceable ecological services, including water filtration, water storage and flood reduction, and habitat. Forested wetlands (a) are often called swamps; marshes (b) have no trees; bogs (c) are acidic and accumulate peat.

cal productivity. Bogs and fens, which are often nutrient-poor, have low biological productivity. They may have unusual and interesting species, though, such as sundews and pitcher plants, which are adapted to capture nutrients from insects rather than from soil.

The water in marshes and swamps usually is shallow enough to allow full penetration of sunlight and seasonal warming. These mild conditions favor great photosynthetic activity, resulting in high productivity at all trophic levels. In short, life is abundant and varied. Wetlands are major breeding, nesting, and migration staging areas for waterfowl and shorebirds.

Streams and rivers are open systems

Streams form wherever precipitation exceeds evaporation and surplus water drains from the land. Within small streams, ecologists distinguish areas of riffles, where water runs rapidly over a rocky substrate, and pools, which are deeper stretches of slowly moving current. Water tends to be well mixed and oxygenated in riffles; pools tend to collect silt and organic matter. If deep enough, pools can have vertical zones similar to those of lakes. As streams collect water and merge, they form rivers, although there isn't a universal definition of when one turns into the other. Ecologists consider a river system to be a continuum of constantly changing environmental conditions and community inhabitants from the headwaters to the mouth of a drainage or watershed. The biggest distinction between stream and lake ecosystems is that, in a stream, materials, including plants, animals, and water, are continually moved downstream by flowing currents. This downstream drift is offset by active movement of animals upstream, productivity in the stream itself, and input of materials from adjacent wetlands or uplands.

5.4   Biodiversity

The biomes you've just learned about shelter an astounding variety of living organisms. From the driest desert to the dripping rainforests, from the highest mountain peaks to the deepest ocean trenches, life occurs in a marvelous spectrum of sizes, colors, shapes, life cycles, and interrelationships. The varieties of organisms and complex ecological relationships give the biosphere its unique, productive characteristics. **Biodiversity**, the variety of living things, also makes the world a more beautiful and exciting place to live. Three kinds of biodiversity are essential to preserve ecological systems and functions: (1) *genetic diversity* is a measure of the variety of versions of the same genes within individual species; (2) *species diversity* describes the number of different kinds of organisms within individual communities or ecosystems; and (3) *ecological diversity* means the richness and complexity of a biological community, including the number of niches, trophic levels, and ecological processes that capture energy, sustain food webs, and recycle materials within this system.

Increasingly, we identify species by genetic similarity

Species are a fundamental concept in understanding biodiversity, but what is a species? In general, species are distinct organisms that persist because they can produce fertile offspring. But many organisms reproduce asexually; others don't reproduce in nature just because they don't normally encounter one another.

Figure 5.21 Insects and other invertebrates make up more than half of all known species. Many, like this blue morpho butterfly, are beautiful as well as ecologically important.

Because of such ambiguities, evolutionary biologists favor the **phylogenetic species concept**, which identifies genetic similarity. Alternatively, the **evolutionary species concept** defines species according to evolutionary history and common ancestors. Both of these approaches rely on DNA analysis to define similarity among organisms.

How many organisms are there? Biologists have identified about 1.5 million species, but these probably represent only a small fraction of the actual number (table 5.1). Based on the rate of new discoveries by research expeditions—especially in the tropics—taxonomists estimate that somewhere between 3 million and 50 million different species may be alive today. About 70 percent of all known species are invertebrates (animals without backbones, such as insects, sponges, clams, and worms) (fig. 5.21).

Table 5.1	Current Estimates of Known Living Species by Taxonomic Group
Mammals	5,416
Birds	9,917
Reptiles	8,163
Amphibians	5,743
Fishes	28,500
Insects	950,000
Mollusks	70,000
Crustaceans	40,000
Other invertebrates	130,200
Ferns, mosses, and lichens	38,025
Gymnosperms	980
Grasses	59,300
Flowering plants	199,350
Total	1,545,594

Source: IUCN, 2005.

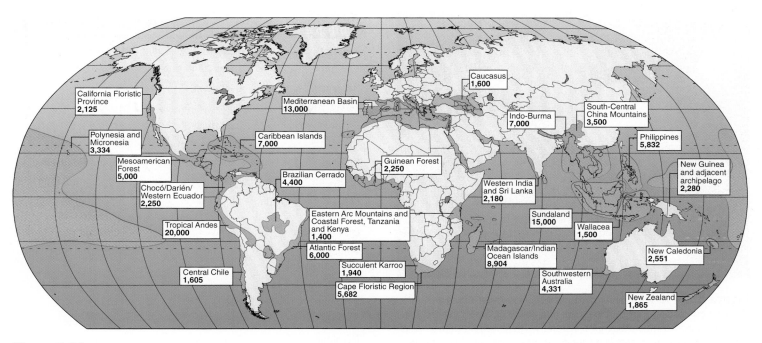

Figure 5.22 Biodiversity "hot spots" identified by Conservation International tend to be in tropical or Mediterranean climates and on islands, coastlines, or mountains where many habitats exist and physical barriers encourage speciation. Numbers represent estimated endemic (locally unique) species in each area.
Source: Data from Conservation International.

This group probably makes up the vast majority of organisms yet to be discovered and may constitute 90 percent of all species.

Biodiversity hot spots are rich and threatened

Most of the world's biodiversity concentrations are near the equator, especially tropical rainforests and coral reefs (fig. 5.22). Of all the world's species, only 10 to 15 percent live in North America and Europe. Many of the organisms in megadiversity countries have never been studied by scientists. The Malaysian Peninsula, for instance, has at least 8,000 species of flowering plants, while Britain, with an area twice as large, has only 1,400 species. There may be more botanists in Britain than there are species of higher plants. South America, on the other hand, has fewer than 100 botanists to study perhaps 200,000 species of plants.

Areas isolated by water, deserts, or mountains can also have high concentrations of unique species and biodiversity. Madagascar, New Zealand, South Africa, and California are all midlatitude areas isolated by barriers that prevent mixing with biological communities from other regions and produce rich, unusual collections of species.

5.5 Benefits of Biodiversity

We benefit from other organisms in many ways, some of which we don't appreciate until a particular species or community disappears. Even seemingly obscure and insignificant organisms can play irreplaceable roles in ecological systems or be the source of genes or drugs that someday may be indispensable.

All of our food comes from other organisms

Many wild plant species could make important contributions to human food supplies either as they are or as a source of genetic material to improve domestic crops. Noted tropical ecologist Norman Myers estimates that as many as 80,000 edible wild plant species could be utilized by humans. Villagers in Indonesia, for instance, are thought to use some 4,000 native plant and animal species for food, medicine, and other valuable products. Few of these species have been explored for possible domestication or more widespread cultivation. A 1975 study by the U.S. National Academy of Science found that Indonesia has 250 edible fruits, only 43 of which have been cultivated widely (fig. 5.23).

Rare species provide important medicines

Living organisms provide us with many useful drugs and medicines (table 5.2). More than half of all prescriptions contain some natural products. The United Nations Development Programme estimates the value of pharmaceutical products derived from developing world plants, animals, and microbes to be more than $30 billion per year.

Consider the success story of vinblastine and vincristine. These anticancer alkaloids are derived from the Madagascar periwinkle (*Catharanthus roseus*). They inhibit the growth of cancer cells and are very effective in treating certain kinds of cancer. Twenty years ago, before these drugs were introduced, childhood leukemias were invariably fatal. Now the remission rate for some childhood leukemias is 99 percent. Hodgkin's disease was 98 percent fatal a few years ago but is now only 40 percent fatal,

Figure 5.23 Mangosteens from Indonesia have been called the world's best-tasting fruit, but they are practically unknown beyond the tropical countries where they grow naturally. There may be thousands of other traditional crops and world food resources that could be equally valuable but are threatened by extinction.

Table 5.2	Some Natural Medicinal Products	
Product	**Source**	**Use**
Penicillin	Fungus	Antibiotic
Bacitracin	Bacterium	Antibiotic
Tetracycline	Bacterium	Antibiotic
Erythromycin	Bacterium	Antibiotic
Digitalis	Foxglove	Heart stimulant
Quinine	Chincona bark	Malaria treatment
Diosgenin	Mexican yam	Birth control drug
Cortisone	Mexican yam	Anti-inflammation treatment
Cytarabine	Sponge	Leukemia cure
Vinblastine, vincristine	Periwinkle plant	Anticancer drugs
Reserpine	Rauwolfia	Hypertension drug
Bee venom	Bee	Arthritis relief
Allantoin	Blowfly larva	Wound healer
Morphine	Poppy	Analgesic

thanks to these compounds. The total value of the periwinkle crop is roughly $15 million per year, although Madagascar gets little of those profits.

Biodiversity can support ecosystem stability

Human life is inextricably linked to ecological services provided by other organisms. Soil formation, waste disposal, air and water purification, nutrient cycling, solar energy absorption, and food production all depend on biodiversity (see chapters 2 and 7). Total value of these ecological services is at least $33 trillion per year, or more than double total world GNP. In many environments, high diversity may help biological communities withstand environmental stress better and recover more quickly than those with fewer species.

Because we don't fully understand the complex interrelationships between organisms, we often are surprised and dismayed at the effects of removing seemingly insignificant members of biological communities. For instance, it is estimated that 95 percent of the potential pests and disease-carrying organisms in the world are controlled by natural predators and competitors. Maintaining biodiversity is essential to preserving these ecological services.

Biodiversity has aesthetic and cultural benefits

Millions of people enjoy hunting, fishing, camping, hiking, wildlife watching, and other nature-based activities. These activities provide invigorating physical exercise, and contact with nature can be psychologically and emotionally restorative. In many cultures, nature carries spiritual connotations, and a particular species or landscape may be inextricably linked to a sense of identity and meaning. Observing and protecting nature has religious or moral significance for many people. Some religious organizations call for the protection of nature simply because it is God's creation.

Nature appreciation is economically important. The U.S. Fish and Wildlife Service estimates that Americans spend $104 billion every year on wildlife-related recreation (fig. 5.24). This compares to $81 billion spent each year on new automobiles. Forty percent of all adults enjoy wildlife, including 39 million who hunt or fish and 76 million who watch, feed, or photograph wildlife. Many communities are finding that local biodiversity can bring cash to remote areas through ecotourism (chapter 6).

For many people, the value of wildlife goes beyond the opportunity to photograph or shoot a particular species. They argue that

Figure 5.24 Birdwatching and other wildlife observation contribute more than $29 million each year to the U.S. economy.

existence value, the value of simply knowing that a species exists, is reason enough to protect and preserve it. Even if we never see a whooping crane or a redwood tree, many of us are happy to know that these species have not been destroyed.

5.6 What Threatens Biodiversity?

Extinction, the elimination of a species, is a normal process of the natural world. Species die out and are replaced by others, often their own descendants, as part of evolutionary change. In undisturbed ecosystems, the rate of extinction appears to be about one species lost every decade. Over the past century, however, human impacts on populations and ecosystems have accelerated that rate, possibly causing thousands of species, subspecies, and varieties to become extinct every year. Ecologist E. O. Wilson estimates that we are losing 10,000 species or subspecies a year—that makes more than 27 per day! If present trends continue, we may destroy millions of kinds of plants, animals, and microbes in the next few decades. In this section, we will look at some ways we threaten biodiversity.

In geologic history, extinctions are common. Studies of the fossil record suggest that more than 99 percent of all species that ever existed are now extinct. Most of those species were gone long before humans came on the scene. Periodically, mass extinctions have wiped out vast numbers of species and even whole families (table 5.3). The best studied of these events occurred at the end of the Cretaceous Period, when dinosaurs disappeared, along with at least 50 percent of existing species. An even greater disaster occurred at the end of the Permian Period, about 250 million years ago, when 90 percent of species and half of all families died out over a period of about 10,000 years—a mere moment in geologic time. Current theories suggest that these catastrophes were caused by climate changes, perhaps triggered when large asteroids struck the earth. Many ecologists worry that global climate change caused by our release of greenhouse gases in the atmosphere could have similarly catastrophic effects (chapter 9).

Human activities have sharply increased extinctions

The rate at which species are disappearing has increased dramatically over the past 150 years. Between A.D. 1600 and 1850, human activities appear to have eliminated two or three species per decade, about double the natural extinction rate. In the past 150 years, the extinction rate has increased to thousands per decade. If present trends continue, biologist Paul Ehrlich warns, somewhere between one-third to two-thirds of all current species could be extinct by the middle of the twenty-first century. Conservation biologists call this the sixth mass extinction but note that this time it's not asteroids or volcanoes but human impacts that are responsible. E. O. Wilson summarizes human threats to biodiversity with the acronym **HIPPO**, which stands for Habitat destruction, Invasive species, Pollution, Population (human), and Overharvesting. Let's look in more detail at each of these issues.

Table 5.3	Mass Extinctions	
Historic Period	**Time (Before Present)**	**Percent of Species Extinct**
Ordovician	444 million	85
Devonian	370 million	83
Permian	250 million	95
Triassic	210 million	80
Cretaceous	65 million	76
Quaternary	Present	33–66

Source: Data from W. W. Gibbs, 2001. "On the termination of species." *Scientific American* 285(5): 40–49.

Habitat destruction is the main threat for many species

The most important extinction threat for most species—especially terrestrial ones—is habitat loss. Perhaps the most obvious example of habitat destruction is conversion of forests and grasslands to farm land (fig. 5.25). Over the past 10,000 years, humans have transformed billions of hectares of former forests and grasslands to croplands, cities, roads, and other uses. These human-dominated spaces aren't devoid of wild organisms, but they generally favor weedy species adapted to coexist with us.

Today, forests cover less than half the area they once did, and only around one-fifth of the original forest retains its old-growth

Figure 5.25 Decrease in wooded area of Cadiz Township in southern Wisconsin during European settlement. Green areas represent the amount of land in forest each year.

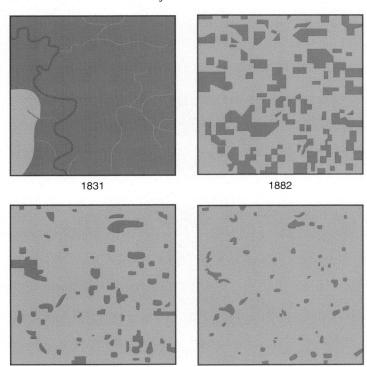

1831

1882

1902

1950

Exploring
SCIENCE:

Using Telemetry to Monitor Wildlife

Understanding wolf recovery in Yellowstone (opening case study) requires careful monitoring of these animals. Day-to-day movements provide some of the most important clues to an animal's habitat needs, behavior, and biology: How much territory do they need? How fast do they move? What kinds of habitat do they prefer? How readily will they risk crossing roads or human settlements? But wolves are fast-moving, wide-ranging animals, and biologists only spot them occasionally—outside of the camera's range. Beyond special locations such as a camera-monitored meadow, how do they monitor such elusive animals?

Many biologists rely on **telemetry**, or locating animals from a distance, to track the locations of wolves, deer, birds, or other mobile species. The most common method is radio telemetry, which involves placing a radio transmitter on an animal, usually on a collar (fig. 1). The collar radio sends out pulses of radio waves, and the researcher detects the collar's location using a radio receiver. The radio receiver can detect direction, but not distance, to the animal. To pinpoint the animal's location, the researcher must locate the signal from two directions, then triangulate the actual position. By locating an animal day after day, or week after week, biologists can study its behavior, habitat requirements, winter survival, and many other questions.

Radio collars were first developed for large animals, such as bears, deer, and wolves.

As radio technology has improved, transmitters have gotten small enough to attach to the backs of songbirds or small rodents (fig. 2). Tiny transmitters are even available that can be glued to the backs of beetles and cockroaches! Depending on the size of the animal, and the weight of batteries that it can easily carry, radios might transmit for a few weeks or more than a year. Depending on vegetation and terrain and the strength of the transmitter, signals may be detectable just within a few hundred meters or many kilometers away.

For animals that move too far or too fast to relocate with a radio receiver, such as migrating birds, polar bears, or sea turtles, biologists are increasingly using satellite telemetry. With satellite telemetry, the collar sends a radio signal that is picked up by a group of satellites that orbit the earth about 850 km above the ground surface. When two or more satellites detect the radio signal, they can triangulate the collar's location, essentially as a biologist would do on the ground. The satellites send data to a receiving station on the ground, which distributes data to researchers for mapping and analysis (fig. 3).

This technology has allowed researchers to map hundreds of species, including detailed

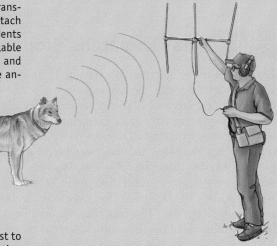

Figure 1 An ecologist carries a radio receiver and antenna to pick up the signal from a radio collar.

records of Yellowstone's wolf packs, and how their territories shift from year to year. Ornithologists have mapped migratory routes for cranes, mapped habitat use by songbirds, and located never-before-discovered arctic breeding grounds for waterfowl. Marine biologists have mapped the seasonal movements of sea turtles, sharks, and seals. The field of telemetry has made stunning advances in recent years and provided entirely new answers to old and difficult questions in wildlife studies.

Figure 2 Tiny radio transmitters let ecologists track small, mobile animals.

Figure 3 With satellite telemetry, a satellite sends data to a ground receiver, which sends data to the researcher for mapping.

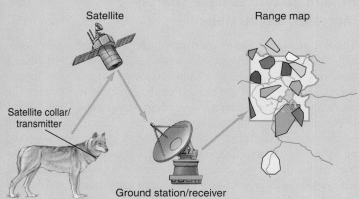

Satellite

Range map

Satellite collar/
transmitter

Ground station/receiver

characteristics. Species, such as the northern spotted owl (*Strix occidentalis caurina*), that depend on the varied structure and resources of old-growth forest vanish as their habitat disappears (chapter 6). Grasslands currently occupy about 4 billion ha (roughly equal to the area of closed-canopy forests). Much of the most highly productive and species-rich grasslands—for example, the tallgrass prairie that once covered the U.S. corn belt—has been converted to cropland. Much more may need to be used as farmland or pasture if human populations continue to expand.

Sometimes we destroy habitat as side effects of resource extraction, such as mining, dam-building, and indiscriminate fishing methods. Surface mining, for example, strips off the land covering along with everything growing on it. Waste from mining operations can bury valleys and poison streams with toxic material. Dam-building floods vital stream habitat under deep reservoirs and eliminates food sources and breeding habitat for some aquatic species. Our current fishing methods are highly unsustainable. One of the most destructive fishing techniques is bottom trawling, in which heavy nets are dragged across the ocean floor, scooping up every living thing and crushing the bottom structure to lifeless rubble. Marine biologist Jan Lubechenco says that trawling is "like collecting forest mushrooms with a bulldozer."

Fragmentation reduces habitat to small, isolated patches

In addition to the total area of lost habitat, another serious problem is habitat **fragmentation**—the reduction of habitat into small, isolated patches. Breaking up habitat reduces biodiversity because many species, such as bears and large cats, require large territories to subsist. Other species, such as forest interior birds, reproduce successfully only in deep forest far from edges and human settlement. Predators and invasive species often spread quickly into new regions following fragment edges.

Fragmentation also divides populations into isolated groups, making them much more vulnerable to catastrophic events, such as storms or diseases. A very small population may not have enough breeding adults to be viable even under normal circumstances. An important question in conservation biology is what is the **minimum viable population** size for various species.

Much of our understanding of fragmentation was elegantly outlined in the theory of **island biogeography**, developed by R. H. MacArthur and E. O. Wilson in the 1960s. Noticing that small islands far from a mainland have fewer terrestrial species than larger, nearer islands, MacArthur and Wilson proposed that species diversity is a balance between colonization and extinction rates. An island far from a population source naturally has a lower rate of colonization than a nearer island because it is harder for terrestrial organisms to reach. At the same time, fewer resources on small islands means that the population of any single species is likely to be small and more vulnerable to extinction. By contrast, a large island can support more individuals of a given species and is, therefore, less vulnerable to natural disasters, genetic problems, or demographic uncertainty (the chance that all the members of a single generation will be of the same sex).

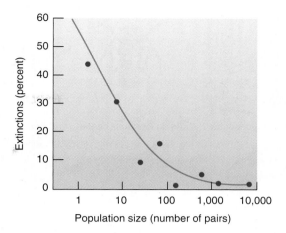

Figure 5.26 Extinction rates of bird species on the California Channel Islands as a function of population size over 80 years.
Source: Data from H. L. Jones and J. Diamond, "Short-term-base Studies of Turnover in Breeding Bird Populations on the California Coast Island," in *Condor*, vol. 78:526–549, 1976.

Island biogeographic effects have been observed in many places. Cuba, for instance is 100 times as large and has about 10 times as many amphibian species as its Caribbean neighbor, Montserrat. Similarly, in a study of bird species on the California Channel Islands, Jared Diamond observed that, on islands with fewer than 10 breeding pairs, 39 percent of the populations went extinct over an 80-year period, while only 10 percent of populations between 10 and 100 pairs went extinct in the same time (fig. 5.26). Only one species numbering between 100 and 1,000 pairs went extinct and no species with over 1,000 pairs disappeared over this time.

Many of the parks and wildlife refuges we establish are effectively islands of good habitat surrounded by oceans of inhospitable territory. Like small, remote oceanic islands, they are too isolated to be reached by new migrants, and they can't support large enough populations to survive catastrophic events or genetic problems. Large predators, such as tigers or wolves, need large expanses of contiguous range relatively free of human incursion to survive. Glacier National Park in Montana, for example, is excellent habitat for grizzly bears. It can support only about 100 bears, however, and if there isn't migration at least occasionally from other areas, this probably isn't a large enough population to survive in the long run.

The challenge of monitoring wildlife in fragmented habitat has led to important new techniques in wildlife research (see Exploring Science, p. 112).

Invasive species are a growing threat

A major threat to native biodiversity in many places is from accidentally or deliberately introduced species. Called a variety of names—alien, exotic, non-native, nonindigenous, unwanted, disruptive, or pests—**invasive species** are organisms that thrive in new territory where they are free of predators, diseases, or resource limitations that may have controlled their population in their native

Figure 5.27 A few of the approximately 50,000 invasive species in North America. Do you recognize any that occur where you live? What others can you think of?

(Labeled illustrations, left to right, top to bottom:) Purple loosestrife, Asian longhorn beetle, Round goby, Kudzu vine, Multiflora rose, Eurasian milfoil, Zebra mussel, Grass carp, Leafy spurge, Gypsy moth, Mongoose, Asian tiger mosquito, Scotch broom, Glossy buckthorn, Cheat grass, Sea lamprey, Water hyacinth, Boll weevil, Nutria, *Caulerpa taxifolia* alga, European green crab, Canadian thistle, Russian thistle.

habitat. Although humans have probably transported organisms into new habitats for thousands of years, the rate of movement has increased sharply in recent years with the huge increase in speed and volume of travel by air, water, and land. We move species around the world in a variety of ways. Some are deliberately released because people believe they will be aesthetically pleasing or economically beneficial. Others hitch a ride in ship ballast water, in the wood of packing crates, inside suitcases or shipping containers, in the soil of potted plants, even on people's shoes (fig. 5.27).

Over the past 300 years, approximately 50,000 non-native species have become established in the United States. Many of these introductions, such as corn, wheat, rice, soybeans, cattle, poultry, and honeybees, have proved to be both socially and economically beneficial. At least 4,500 of these species have established free-living populations, of which 15 percent cause environmental or economic damage (fig. 5.28). Invasive species are estimated to cost the United States $138 billion annually and are forever changing a variety of ecosystems (see What Can You Do? p. 115).

A few important examples of invasive species include the following:

- Eurasian milfoil (*Myriophyllum spicatum* L.) is an exotic aquatic plant native to Europe, Asia, and Africa. Scientists believe that milfoil arrived in North America during the late nineteenth century in shipping ballast. It grows rapidly

Figure 5.28 Invasive leafy spurge (*Euphorbia esula*) blankets a formerly diverse pasture. Introduced accidentally, and inedible for most herbivores, this plant costs hundreds of millions of dollars each year in lost grazing and weed control.

and tends to form a dense canopy on the water surface, which displaces native vegetation, inhibits water flow, and obstructs boating, swimming, and fishing. Humans spread the plant between water body systems from boats and boat trailers carrying the plant fragments. Herbicides and mechanical harvesting are effective in milfoil control but can be expensive (up to $5,000 per hectare per year). There is also concern that the methods may harm nontarget organisms. A native milfoil weevil, *Euhrychiopsis lecontei*, is being studied as an agent for milfoil biocontrol.

- Water hyacinth (*Eichhornia crassipes*) is a free-floating aquatic plant that has thick, waxy, dark green leaves with bulbous, spongy stalks. It grows a tall spike of lovely blue or purple flowers. This South American native was introduced into the United States in the 1880s. Its growth rate is among the highest of any plant known: hyacinth populations can double in as little as 12 days. Many lakes and ponds are covered from shore to shore with up to 500 tons of hyacinths per hectare. Besides blocking boat traffic and preventing swimming and fishing, water hyacinth infestations also prevent sunlight and oxygen from getting into the water. Thus, water hyacinth infestations reduce fisheries, shade-out submersed plants, crowd-out immersed plants, and diminish biological diversity. Water hyacinth is controlled with herbicides, machines, and biocontrol insects.

- Kudzu vine (*Pueraria lobata*) has blanketed large areas of the southeastern United States. Long cultivated in Japan for edible roots, medicines, and fibrous leaves and stems used for paper production, kudzu was introduced by the U.S. Soil Conservation Service in the 1930s to control erosion. Unfortunately, it succeeded too well. In the ideal conditions of its new home, kudzu can grow 18 to 30 m in a single season. Smothering everything in its path, it kills trees, pulls down utility lines, and causes millions of dollars in damage every year.

- Asian tiger mosquitoes (*Aedes albopictus*) are unusually aggressive species that now infest many coastal states in the United States. These species have apparently arrived on container ships carrying used tires, a notorious breeding habitat for mosquitoes. Asian tiger mosquitoes spread West Nile virus (another species introduced with the mosquitoes), which is deadly to many wild birds and occasionally to people and livestock.

- Purple loosestrife (*Lythrum salicaria*) grows in wet soil. Originally cultivated by gardeners for its bright purple flower spikes, this tall wetland plant escaped into New England marshes about a century ago. Spreading rapidly across the Great Lakes, it now fills wetlands across much of the northern United States and southern Canada. Because it crowds out indigenous vegetation and has few native predators or symbionts, it tends to reduce biodiversity wherever it takes hold.

- Zebra mussels (*Dreissena polymorpha*) probably made their way from their home in the Caspian Sea to the Great

What Can You Do?

You Can Help Preserve Biodiversity

Our individual actions are some of the most important obstacles—and most important opportunities—in conserving biodiversity.

Pets and Plants

- Help control invasive species. Never release fish or vegetation from fish tanks into waterways or sewers. Pet birds, cats, dogs, snakes, lizards, and other animals, released by well-meaning owners, are widespread invasive predators.
- Keep your cat indoors. House cats are major predators of woodland birds and other animals.
- Plant native species in your garden. Exotic nursery plants often spread from gardens, compete with native species, and introduce parasites, insects, or diseases that threaten ecosystems. Local-origin species are an excellent, and educational, alternative.
- Don't buy exotic birds, fish, turtles, reptiles, or other pets. These animals are often captured, unsustainably, in the wild. The exotic pet trade harms ecosystems and animals.
- Don't buy rare or exotic house plants. Rare orchids, cacti, and other plants are often collected and sold illegally and unsustainably.

Food and Products

- When buying seafood, inquire about the source. Try to buy species from stable populations. Farm-raised catfish, tilapia, trout, Pacific pollack, Pacific salmon, mahimahi, squid, crabs, and crayfish are some of the stable or managed species that are good to buy. Avoid slow-growing top predators, such as swordfish, marlin, bluefin tuna, and albacore tuna.
- Buy shade-grown coffee and chocolate. These are also organic and often "fair trade" varieties that support workers' families as well as biodiversity in growing regions.
- Buy sustainably harvested wood products. Your local stores will start carrying sustainable wood products if you and your friends ask for them. Persistent consumers are amazingly effective forces of change!

Lakes in ballast water of transatlantic cargo ships, arriving sometime around 1985. Attaching themselves to any solid surface, zebra mussels reach enormous densities—up to 70,000 animals per square meter—covering fish-spawning beds, smothering native mollusks, and clogging utility intake pipes. Found in all the Great Lakes, zebra mussels have moved into the Mississippi River and its tributaries. Public and private costs for zebra mussel removal now amount to some $400 million per year. On the good side, mussels have improved water clarity in Lake Erie at least four-fold by filtering out algae and particulates.

Disease organisms, or pathogens, may also be considered predators. To be successful over the long term, a pathogen must establish a balance in which it is vigorous enough to reproduce, but not so lethal that it completely destroys its host. When a disease is introduced into a new environment, however, this balance may be lacking and an epidemic may sweep through the area.

The American chestnut was once the heart of many eastern hardwood forests. In the Appalachian Mountains, at least one of every four trees was a chestnut. Often over 45 m (150 ft) tall, 3 m (10 ft) in diameter, fast growing, and able to sprout quickly from a cut stump, it was a forester's dream. Its nutritious nuts were important for birds (such as the passenger pigeon), forest mammals, and humans. The wood was straight-grained, light, and rot-resistant, and it was used for everything from fence posts to fine furniture. In 1904 a shipment of nursery stock from China brought a fungal blight to the United States, and within 40 years, the American chestnut had all but disappeared from its native range. Efforts are now underway to transfer blight-resistant genes into the few remaining American chestnuts that weren't reached by the fungus or to find biological controls for the fungus that causes the disease.

The flow of organisms isn't just into the United States; Americans also send exotic species to other places. The Leidy's comb jelly, for example, which is native to the western Atlantic coast, has devastated the Black Sea, now making up more than 90 percent of all biomass at certain times of the year. Similarly, the bristle worm from North America has invaded the coast of Poland and now is almost the only thing living on the bottom of some bays and lagoons. A tropical seaweed named *Caulerpa taxifolia*, originally grown for the aquarium trade, has escaped into the northern Mediterranean, where it covers the shallow seafloor with a dense, meter-deep shag carpet from Spain to Croatia. Producing more than 5,000 leafy fronds per square meter, this aggressive weed crowds out everything in its path. Rarely growing in more than scattered clumps less than 25 cm (10 in.) high in its native habitat, this alga was transformed by aquarium breeding into a supercompetitor that grows over everything and can withstand a wide temperature range. Getting rid of these alien species once they dominate an ecosystem is difficult if not impossible.

Pollution poses many different types of risk

We have known for a long time that toxic pollutants can have disastrous effects on local populations of organisms. Pesticide-linked declines of fish-eating birds and falcons were well documented in the 1970s (fig. 5.29). Marine mammals, alligators, fish, and other declining populations suggest complex interrelations between pollution and health (chapter 8). Mysterious, widespread deaths of thousands of seals on both sides of the Atlantic in recent years are thought to be linked to an accumulation of persistent chlorinated hydrocarbons, such as DDT, PCBs, and dioxins, in fat, causing weakened immune systems that make animals vulnerable to infections. Similarly, mortality of Pacific sea lions, beluga whales in the St. Lawrence estuary, and striped dolphins in the Mediterranean is thought to be caused by accumulation of toxic pollutants.

Figure 5.29 Bald eagles, and other bird species at the top of the food chain, were decimated by DDT in the 1960s. Many such species have recovered since DDT was banned in the United States, and because of protection under the Endangered Species Act.

Lead poisoning is another major cause of mortality for many species of wildlife. Bottom-feeding waterfowl, such as ducks, swans, and cranes, ingest spent shotgun pellets that fall into lakes and marshes. They store the pellets, instead of stones, in their gizzards and the lead slowly accumulates in their blood and other tissues. The U.S. Fish and Wildlife Service (USFWS) estimates that 3,000 metric tons of lead shot are deposited annually in wetlands and that between 2 and 3 million waterfowl die each year from lead poisoning.

Human population growth threatens biodiversity

If our consumption patterns remain constant, with more people we will need to harvest more timber, catch more fish, plow more land for agriculture, dig up more fossil fuels and minerals, build more houses, and use more water. All of these demands impact wild species. Unless we find ways to dramatically increase the crop yield per unit area, it will take much more land than is currently domesticated to feed everyone if our population grows to 8 to 10 billion, as current projections predict. This will be especially true if we abandon intensive (but highly productive) agriculture and introduce more sustainable practices. The human population growth

Figure 5.30 A pair of stuffed passenger pigeons (*Ectopistes migratorius*). The last member of this species died in the Cincinnati Zoo in 1914.
Courtesy of Bell Museum, University of Minnesota.

curve is leveling off (chapter 4), but it remains unclear whether we can reduce global inequality and provide a tolerable life for all humans while also preserving healthy natural ecosystems and a high level of biodiversity.

Overharvesting has depleted or eliminated many species

Overharvesting involves taking more individuals than reproduction can replace. A classic example is the extermination of the American passenger pigeon (*Ectopistes migratorius*). Even though it inhabited only eastern North America, 200 years ago this was the world's most abundant bird, with a population of between 3 and 5 billion animals (fig. 5.30). It once accounted for about one-quarter of all birds in North America. In 1830 John James Audubon saw a single flock of birds estimated to be ten miles wide, hundreds of miles long, and thought to contain perhaps a billion birds. In spite of this vast abundance, market hunting and habitat

destruction caused the entire population to crash in only about 20 years between 1870 and 1890. The last known wild bird was shot in 1900 and the last existing passenger pigeon, a female named Martha, died in 1914 in the Cincinnati Zoo.

At about the same time that passenger pigeons were being extirpated, the American bison, or buffalo (*Bison bison*), was being hunted to near extinction on the Great Plains. In 1850 some 60 million bison roamed the western plains. Many were killed only for their hides or tongues, leaving millions of carcasses to rot. Much of the bison's destruction was carried out by the U.S. Army to deprive native peoples who depended on bison for food, clothing, and shelter of these resources, thereby forcing them onto reservations. After 40 years, there were only about 150 wild bison left and another 250 in captivity.

Fish stocks have been seriously depleted by overharvesting in many parts of the world. A huge increase in fishing fleet size and efficiency in recent years has led to a crash of many oceanic populations. Worldwide, 13 of 17 principal fishing zones are now reported to be commercially exhausted or in steep decline. At least three-quarters of all commercial oceanic species are overharvested. Canadian fisheries biologists estimate that only 10 percent of the top predators, such as swordfish, marlin, tuna, and shark, remain in the Atlantic Ocean. Groundfish, such as cod, flounder, halibut, and hake, also are severely depleted. You can avoid adding to this overharvest by eating only abundant, sustainably harvested varieties.

Perhaps the most destructive example of harvesting terrestrial wild animal species today is the African bushmeat trade. Wildlife biologists estimate that 1 million tons of bushmeat, including antelope, elephants, primates, and other animals, are sold in African markets every year. For many poor Africans, this is the only source of animal protein in their diet. If we hope to protect the animals targeted by bushmeat hunters, we will need to help them find alternative livelihoods and replacement sources of high-quality protein. Bushmeat can endanger those who eat it, too. Outbreaks of ebola and other diseases are thought to result from capturing and eating wild monkeys and other primates. The emergence of SARS in 2003 (chapter 8) resulted from the wild food trade in China and Southeast Asia, where millions of civets, monkeys, snakes, turtles, and other animals are consumed each year as luxury foods.

Commercial collection serves medicinal and pet trades

In addition to harvesting wild species for food, we also obtain a variety of valuable commercial products from nature. Much of this represents sustainable harvest, but some forms of commercial exploitation are highly destructive, however, and represent a serious threat to certain rare species (fig. 5.31). Despite international bans on trade in products from endangered species, smuggling of furs, hides, horns, live specimens, and folk medicines amounts to millions of dollars each year.

Developing countries in Asia, Africa, and Latin America with the richest biodiversity in the world are the main sources of wild

Figure 5.31 Parts from rare and endangered species for sale on the street in China. Use of animal products in traditional medicine and prestige diets is a major threat to many species.

Figure 5.32 A diver uses cyanide to stun tropical fish being caught for the aquarium trade. Many fish are killed by the method itself, while others die later during shipment. Even worse is the fact that cyanide kills the coral reef itself.

animals and animal products, while Europe, North America, and some of the wealthy Asian countries are the principal importers. Japan, Taiwan, and Hong Kong buy three-quarters of all cat and snake skins, for instance, while European countries buy a similar percentage of live birds. The United States imports 99 percent of all live cacti and 75 percent of all orchids sold each year.

The profits to be made in wildlife smuggling are enormous. Tiger or leopard fur coats can bring $100,000 in Japan or Europe. The population of African black rhinos dropped from approximately 100,000 in the 1960s to about 3,000 in the 1980s because of a demand for their horns. In Asia, where it is prized for its supposed medicinal properties, powdered rhino horn fetches $28,000 per kilogram. In Yemen, a rhino horn dagger handle can sell for up to $1,000.

Plants also are threatened by overharvesting. Wild ginseng has been nearly eliminated in many areas because of the Asian demand for the roots, which are used as an aphrodisiac and folk medicine. Cactus "rustlers" steal cacti by the ton from the American Southwest and Mexico. With prices as high as $1,000 for rare specimens, it's not surprising that many are now endangered.

The trade in wild species for pets is an enormous business. Worldwide, some 5 million live birds are sold each year for pets, mostly in Europe and North America. Currently, pet traders import (often illegally) into the United States some 2 million reptiles, 1 million amphibians and mammals, 500,000 birds, and 128 million tropical fish each year. About 75 percent of all saltwater tropical aquarium fish sold come from coral reefs of the Philippines and Indonesia.

Many of these fish are caught by divers using plastic squeeze bottles of cyanide to stun their prey (fig. 5.32). Far more fish die with this technique than are caught. Worst of all, it kills the coral animals that create the reef. A single diver can destroy all of the life on 200 m² of reef in a day. Altogether, thousands of divers currently destroy about 50 km² of reefs each year. Net fishing would

prevent this destruction, and it could be enforced if pet owners would insist on net-caught fish. More than half the world's coral reefs are potentially threatened by human activities, with up to 80 percent at risk in the most populated areas.

Predator and pest control is expensive but widely practiced

Some animal populations have been greatly reduced, or even deliberately exterminated, because they are regarded as dangerous to humans or livestock or because they compete with our use of resources. Every year, U.S. government animal control agents trap, poison, or shoot thousands of coyotes, bobcats, prairie dogs, and other species considered threats to people, domestic livestock, or crops.

This animal control effort costs about $20 million in federal and state funds each year and kills some 700,000 birds and mammals, about 100,000 of which are coyotes. Defenders of wildlife regard this program as cruel, callous, and mostly ineffective in reducing livestock losses. Protecting flocks and herds with guard dogs or herders or keeping livestock out of areas that are home range of wild species would be a better solution, they believe. Ranchers and trappers, on the other hand, argue that, without predator control, western livestock operations would be uneconomical.

5.7 Endangered Species Management and Biodiversity Protection

Over the years, we have gradually become aware of the harm we have done—and continue to do—to wildlife and biological resources. Slowly, we are adopting national legislation and

Table 5.4	Endangered and Threatened Species Worldwide
Mammals	1,101
Birds	1,213
Reptiles	304
Amphibians	1,770
Fish	800
Insects and other invertebrates	1,992
Total fauna	7,180
Plants	8,321

Source: Data from *IUCN Red List*, 2005.

international treaties to protect these irreplaceable assets. Parks, wildlife refuges, nature preserves, zoos, and restoration programs have been established to protect nature and rebuild depleted populations. There has been encouraging progress in this area, but much remains to be done. While most people favor pollution control or protection of favored species, such as whales or gorillas, surveys show that few understand what biological diversity is or why it is important.

Hunting and fishing laws protect reproductive populations

In 1874 a bill was introduced in the United States Congress to protect the American bison, whose numbers were already falling to dangerous levels. This initiative failed, however, because most legislators believed that all wildlife—and nature in general—was so abundant and prolific that it could never be depleted by human activity. As we discussed earlier in this chapter, however, by the end of the nineteenth century, bison had plunged from some 60 million to only a few hundred animals.

By the 1890s most states had enacted some hunting and fishing restrictions. The general idea behind these laws was to conserve the resource for future human use rather than to preserve wildlife for its own sake. The wildlife regulations and refuges established since that time have been remarkably successful for many species. A hundred years ago, there were an estimated half a million white-tailed deer in the United States; now there are some 14 million—more in some places than the environment can support. Wild turkeys and wood ducks were nearly gone 50 years ago. By restoring habitat, planting food crops, transplanting breeding stock, building shelters or houses, protecting these birds during breeding season, and using other conservation measures, we have restored populations of these beautiful and interesting birds to several million each. Snowy egrets, which were almost wiped out by plume hunters 80 years ago, are now common again.

The Endangered Species Act protects habitat and species

Hunting and fishing laws stabilized populations of many game species, but nongame species are also at risk. Bald eagles, gray whales, sea otters, and gray wolves are among thousands of species threatened by habitat loss, pollution, hunting, or other factors. The Endangered Species Act (ESA) of 1973 represented a powerful new approach to protecting these species. This law protects declining species even if they are not directly useful to humans. Part of the reason for reintroducing wolves to Yellowstone National Park (opening case study) was that the species was officially listed by the ESA, so that its recovery was recognized as a priority. Recovery of gray wolf populations has little direct usefulness for people, but benefits to the ecosystem, and indirect benefits for tourists, hunters, birdwatchers, and others, may be great. Bald eagles have also received protection under the ESA. Even though they are of no practical value for hunters, their symbolic value makes their conservation important to many Americans.

The ESA provides rules for identifying species at risk, planning for their recovery, and legally enforcing steps needed for recovery. The act identifies three degrees of risk: **endangered species** are those considered in imminent danger of extinction; **threatened species** are likely to become endangered, at least locally, within the forseeable future. **Vulnerable species** are naturally rare or have been locally depleted by human activities to a level that puts them at risk. Vulnerable species are often candidates for future listing as endangered species. For vertebrates, a protected subspecies or a local race or ecotype can be listed, as well as an entire species.

To protect species, the Congress gave the ESA power to regulate a wide range of activities involving endangered species, including "taking" (harassing, harming, pursuing, hunting, shooting, killing, or collecting) a species, either accidentally or on purpose. It is also illegal to possess, sell, or transport live organisms, body parts, or products from endangered species. Violators of the ESA are subject to fines up to $100,000 and a year in prison. Vehicles and equipment used in violations may also be subject to forfeiture.

Listing a new species usually takes years. Concerned ecologists or biologists may petition to put a species on the list of endangered, threatened, or vulnerable species; they must then produce data and documents to prove that the species is at risk. The U.S. Fish and Wildlife Service (USFWS) must then review the petition, a process that has been slowed by limited funding, political pressure, moratoria on listing, and changeable administrative priorities. Hundreds of species are classified as "warranted but precluded," or deserving of protection but lacking funding or local support. At least 18 species have gone extinct since being nominated for protection.

Currently, the United States has 1,300 species on its endangered and threatened species lists and about 250 candidate species waiting to be considered. The number of listed species in different taxonomic groups reflects much more about the kinds of organisms that humans consider interesting and desirable than the actual number in each group. In the United States, invertebrates make up about three-quarters of all known species but only 9 percent of those deemed worthy of protection. Worldwide, the International Union for Conservation of Nature and Natural Resources (IUCN) lists a total of 16,496 endangered and threatened species, including nearly one-quarter of all known bird varieties (table 5.4).

Figure 5.33 Gray wolves have been protected by the Endangered Species Act. Protecting habitat for this species provides habitat for many other species and maintains ecosystem diversity.

Recovery plans aim to rebuild populations

Once a species is officially listed as endangered, the Fish and Wildlife Service is required to prepare a recovery plan detailing how populations will be rebuilt to sustainable levels. It usually takes years to reach agreement on specific recovery plans. Among the difficulties are costs, politics, interference on local economic interests, and the fact that, once a species is endangered, much of its habitat and ability to survive is likely compromised. The total cost of recovery plans for all currently listed species is estimated to be nearly $5 billion.

The United States currently spends about $150 million per year on endangered species protection and recovery. About half that amount is spent on a dozen charismatic species, such as the California condor and the Florida panther and grizzly bear, which receive around $13 million per year. By contrast, the 137 endangered invertebrates and 532 endangered plants get less than $5 million per year altogether. Our funding priorities often are based more on emotion and politics than biology. A variety of terms are used for rare or endangered species thought to merit special attention:

- *Keystone species* are those with major effects on ecological functions and whose elimination would affect many other members of the biological community; examples are prairie dogs (*Cynomys ludovicianus*) and bison (*Bison bison*).
- *Indicator species* are those tied to specific biotic communities or successional stages or environmental conditions. They can be reliably found under certain conditions but not others; an example is brook trout (*Salvelinus fontinalis*).
- *Umbrella species* require large blocks of relatively undisturbed habitat to maintain viable populations. Saving this habitat also benefits other species. Examples of umbrella species are the northern spotted owl (*Strix occidentalis cau-*

rina), tiger (*Panthera tigris*), and gray wolf (*Canis lupus*) (fig. 5.33).

- *Flagship species* are especially interesting or attractive organisms to which people react emotionally. These species can motivate the public to preserve biodiversity and contribute to conservation; an example is the giant panda (*Ailuropoda melanoleuca*).

Although the ESA is imperfect, it has slowed the loss of many species. A few have even been designated as "recovered" and removed from the list of threatened and endangered species. The gray wolf was delisted in 2007 in northern Minnesota, where its population has grown to a stable level. The bald eagle was also delisted in 2007. In 1967, before the ESA was passed, only about 800 bald eagles remained in the contiguous United States. DDT poisoning, which prevented the hatching of young eagles, was the main cause. By 1994, after the banning of DDT, the population rebounded to 8,000 birds; by 2007 the population was up to about 20,000, enough to ensure a stable breeding population. Similarly, peregrine falcons, which had been down to 39 breeding pairs in the 1970s, had rebounded to 1,650 pairs by 1999 and were taken off the list. The American alligator was listed as endangered in 1967 because hunting and habitat destruction had reduced populations to precarious levels. Protection has been so effective that the species is now plentiful throughout its entire southern range. Florida alone may have a population of 1 million or more.

Controversy persists in species protection

In 1995 the Supreme Court ruled that critical habitat—habitat essential for a species' survival—must be protected, whether on public or private land. Protection on private land is often necessary for a species' survival, but it is also a cause of controversy around the ESA. Especially in conservation of obscure or cryptic species, such as the Delhi Sands flower-loving fly, or the orange-footed pimple-back pearlymussel, landowners object to restrictions on the development of their lands.

Bitter legal disputes often hinge on which is more valuable, a landowner's profits or the existence of a species with no clear utility. Opponents of the ESA have repeatedly tried to require that the economic costs and benefits of species preservation be incorported into planning for the species. An important test of the ESA occurred in 1978 in Tennessee, when construction of the Tellico Dam threatened a tiny fish called the snail darter. The dam was deemed more important, and as a result of this case a new federal committee was given power to override the ESA for economic reasons. (This committee subsequently became known as the God Squad.) Another important debate over the economics of endangered species protection was that of the northern spotted owl (chapter 6). Preserving this owl requires the conservation of expansive, undisturbed areas of old-growth temperate rainforest in the Pacific Northwest, where old-growth timber is extremely valuable and increasingly scarce (fig. 5.34). Timber industry economists calculated the cost of conserving a population of 1,600 to 2,400

"DAMN SPOTTED OWL!"

TIMBER MINING CO.

©1990 HERBLOCK

Figure 5.34 Endangered species often serve as a barometer for the health of an entire ecosystem and as surrogate protector for a myriad of less well-known creatures.
"Damn Spotted Owl" A 1990 Herblock Cartoon, copyright by *The Herb Block Foundation*.

owls at $33 billion. Ecologists countered that this number was highly inflated and that forest conservation would preserve countless other species and ecosystem services, whose values are almost impossible to calculate.

Sometimes the value of conserving a species is easier to calculate. Salmon and steelhead in the Columbia River are endangered by hydropower dams and water storage reservoirs that block their migration to the sea. Opening the floodgates could allow young fish to run downriver and adults to return to spawning grounds, but at high costs to electricity consumers, barge traffic, and farmers who depend on cheap water and electricity. On the other hand, commercial and sport fishing for salmon is worth over $1 billion per year and employs about 60,000 people directly or indirectly.

Reauthorizing the ESA has been contentious

The ESA officially expired in 1992. Since then, Congress has debated many alternative proposals, from outright eliminating it to substantially strengthening it. Emotions often run high over endangered species protection. In some western states, where traditions of individual liberty are strong and the federal government is viewed with considerable suspicion and hostility, the ESA seems to many an outsiders' plot to take away private property and trample on individual rights. Farmers, loggers, miners, rangers, developers, and other opponents of the ESA have repeatedly tried to scuttle the law or greatly reduce its power. Proponents of species conservation, meanwhile, often see the ESA as too politically embattled and chronically underfunded to be effective. Proponents of species protection would like to see stiffer penalties and broader ecosystem protection, rather than a single-species approach. Opponents argue for exemptions to ESA regulation for some landowners and a higher priority for economic costs, to allow more development of habitat. What do you think? Under what conditions would you favor the protection of rare species? Under what conditions would you rather let a species go extinct because it was too costly to preserve it?

Many countries have laws for species protection

In the past 25 years or so, many countries have recognized the importance of legal protection for endangered species. Rules for listing and protecting endangered species are established by Canada's Committee on the Status of Endangered Wildlife in Canada (COSEWIC) of 1977, the European Union's Birds Directive (1979) and Habitat Directive (1991), and Australia's Endangered Species Protection Act (1992). International agreements have also been developed, including the Convention on Biological Diversity (1992).

The Convention on International Trade in Endangered Species (CITES) of 1975 provides a critical conservation strategy by blocking the international sale of wildlife and their parts. The Convention makes it illegal to export or import elephant ivory, rhino horns, tiger skins, or live endangered birds, lizards, fish, and orchids. CITES enforcement has been far from perfect: smugglers hide live animals in their clothing and luggage; the volume of international shipping makes it impossible to inspect transport containers and ships; documents may be falsified. The high price of these products in North America and Europe, and increasingly in wealthy cities of China, make the risk of smuggling worthwhile: a single rare parrot can be worth tens of thousands of dollars, even though its sale is illegal. Even so, CITES provides a legal structure for restricting this trade, and it also raises public awareness of the real costs of the trade in endangered species.

Habitat protection may be better than species protection

Over the past decade, growing numbers of scientists, land managers, policymakers, and developers have been making the case that it is time to focus on a rational, continentwide preservation of ecosystems that supports maximum biological diversity rather than a species-by-species battle for the rarest or most popular organisms. By focusing on populations already reduced to only a few individuals, we spend most of our conservation funds on species that may be genetically doomed no matter what we do. Furthermore, by concentrating on individual species, we spend millions of dollars to breed plants or animals in captivity that have no natural habitat where they can be released. While flagship species, such as mountain gorillas and Indian tigers, are reproducing well in zoos and wild animal parks, the ecosystems that they formerly inhabited have largely disappeared.

A leader of this new form of conservation is J. Michael Scott, who was project leader of the California condor recovery program in the mid-1980s and had previously spent ten years working on endangered species in Hawaii. In making maps of endangered species, Scott discovered that even Hawaii, where more than 50 percent of the land is federally owned, has many vegetation types completely outside of natural preserves (fig. 5.35). The gaps between protected areas may contain more endangered species than are preserved within them.

This observation has led to an approach called **gap analysis**, in which conservationists and wildlife managers look for unprotected landscapes, or gaps in the network of protected lands, that are rich in species. Gap analysis uses GIS to overlay protected conservation areas with high-biodiversity areas. This overlay makes it easy to identify priority spots for conservation efforts. Maps also help biologists and land-use planners communicate about threats to biodiversity. This broad-scale, holistic approach seems likely to save more species than a piecemeal approach.

Conservation biologist R. E. Grumbine suggests four remanagement principles for protecting biodiversity in a large-scale, long-range approach:

1. Protect enough habitat for viable populations of all native species in a given region.

2. Manage at regional scales large enough to accommodate natural disturbances (fire, wind, climate change, etc.).

3. Plan over a period of centuries, so that species and ecosystems can continue to evolve.

4. Allow for human use and occupancy at levels that do not result in significant ecological degradation.

Conclusion

The types and varieties of living things in an area can be predicted, in general terms, by the temperature and precipitation conditions. Nine general biomes, defined in terms of their climate conditions, can be described. Humid climates tend to have tropical, temperate, or boreal forests; dry climates may have savanna or desert conditions. Marine life zones are differentiated mainly according to light and depth. Littoral (shoreline) zones have great diversity that reflects rapid transitions of environmental conditions. Pelagic, abyssal, and hadal zones are deeper and farther from shore. Wetlands are described in terms of the presence of trees, emergent plants, or mainly open water.

Biodiversity is important to us because it can aid ecosystem stability and because we rely on many different organisms for foods, medicines, and other products. We do not know what kinds of undiscovered species may provide medicines and foods in the future. Biodiversity also has important cultural and aesthetic benefits.

Many factors threaten biodiversity, including habitat loss, invasive species, pollution, population growth, and overharvesting. The acronym HIPPO has been used to describe these threats together as a whole. You can reduce these threats by avoiding exotic pets and unsustainable fisheries or wood products that are harvested unsustainably. The Endangered Species Act and other laws that protect biodiversity have been controversial. They have also provided essential support for conserving species, including the bald eagle and gray wolf. Often such laws protect umbrella species, whose habitat also protects many other species.

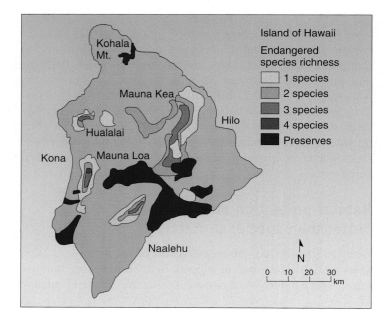

Figure 5.35 Protected lands (*green*) are often different from biologically diverse areas (*red shades*), as shown here on the island of Hawaii.

Practice Quiz

1. Why did ecologists want to reintroduce wolves to Yellowstone Park? What goals did they have, and have their goals been achieved?

2. Describe nine major types of terrestrial biomes.

3. Explain how climate graphs (as in fig. 5.6) should be read.

4. Describe conditions under which coral reefs, mangroves, estuaries, and tide pools occur.

5. Throughout the central portion of North America is a large biome once dominated by grasses. Describe how physical conditions and other factors control this biome.

6. Explain the difference between swamps, marshes, and bogs.

7. How do elevation (on mountains) and depth (in water) affect environmental conditions and life-forms?

8. Figure 5.15 shows chlorophyll (plant growth) in oceans and on land. Explain why green, photosynthesizing organisms occur in long bands at the equator and along the edges of continents. Explain the very dark green areas and yellow/orange areas on the continents.

9. Define *biodiversity* and give three types of biodiversity essential in preserving ecological systems and functions.

10. What is a biodiversity "hot spot"? List several of them (see fig. 5.22).

11. How do humans benefit from biodiversity?

12. What does the acronym HIPPO refer to?

13. Have extinctions occurred in the past? Is there anything unusual about current extinctions?

14. Why are exotic or invasive species a threat to biodiversity? Give several examples of exotic invasive species (see fig. 5.27).

15. What is the Endangered Species Act? Describe some of the main arguments of its proponents and opponents.

16. What is a flagship or umbrella species? Why are they often important, even though they are costly to maintain?

Critical Thinking and Discussion Questions

Apply the principles you have learned in this chapter to discuss these questions with other students.

1. Many poor tropical countries point out that a hectare of shrimp ponds can provide 1,000 times as much annual income as the same area in an intact mangrove forest. Debate this point with a friend or classmate. What are the arguments for and against saving mangroves?

2. Genetic diversity, or diversity of genetic types, is believed to enhance stability in a population. Most agricultural crops are genetically very uniform. Why might the usual importance of genetic diversity *not* apply to food crops? Why *might* it apply?

3. Scientists need to be cautious about their theories and assumptions. What arguments could you make for *and* against the statement that humans are causing extinctions unlike any in the history of the earth?

4. A conservation organization has hired you to lead efforts to reduce the loss of biodiversity in a tropical country. Which of the following problems would you focus on first and why: habitat destruction and fragmentation, hunting and fishing activity, harvesting of wild species for commercial sale, or introduction of exotic organisms?

5. Many ecologists and resource scientists work for government agencies to study resources and resource management. Do these scientists serve the public best if they try to do pure science, or if they try to support the political positions of democratically elected representatives, who, after all, represent the positions of their constituents?

6. You are a forest ecologist living and working in a logging community. An endangered salamander has recently been discovered in your area. What arguments would you make for and against adding the salamander to the official endangered species list?

 **Data Analysis:** Confidence Limits in the Breeding Bird Survey

A central principle of science is the recognition that all knowledge involves uncertainty. No study can observe every possible event in the universe, so there is always missing information. Scientists try to define the limits of their uncertainty, in order to allow a realistic assessment of their results. A corollary of this principle is that the more data we have, the less uncertainty we have. More data increase our confidence that our observations represent the range of possible observations.

One of the most detailed records of wildlife population trends in North America is the Breeding Bird Survey (BBS). Every June, volunteers drive more than 4,000 established 25-mile routes. They stop every half mile and count every bird they see or hear.

The accumulated data from thousands of routes, over more than 40 years of the study, indicate population *trends*, telling which populations are increasing, decreasing, or expanding into new territory.

Because many scientists use BBS data, it is essential to communicate how much confidence there is in the data. The online BBS database reports measures of data quality, including:

- **N**: the number of survey routes from which population trends are calculated.

- **Confidence limits**: because the reported trend is an average of a small *sample* of year-to-year changes on routes, confidence limits tell us how close the sample's average probably is to the average for the *entire population* of that species. Statistically, 95 percent of all samples should fall between the confidence limits. In effect, we can be 95 percent sure that the entire population's actual trend falls between the upper and lower confidence limits.

1. Examine the table at right, which shows 10 species taken from the online BBS database. How many species have a positive population trend (>0)? If a species had a trend of 0, how much would it change from year to year?

2. Which species has the greatest decline per year? For every 100 birds this year, how many fewer will there be next year?

3. If the distance between upper and lower confidence limits (the *confidence interval*) is narrow, then we can be reasonably sure the trend in our *sample* is close to the trend for the *total population* of that species. What is the reported trend for ring-necked pheasant? What is the number of routes (N) on which this trend is based? What is the range in which the pheasant's true population trend probably falls? Is there a reasonable chance that the pheasant population's average annual change is 0? That the population trend is actually –5?

   Now look at the ruffed grouse, on either the table or the graph. Does the trend show that the population is increasing or decreasing? Can you be certain that the actual trend is not 7, or 17? On how many routes is this trend based?

4. In general, confidence limits depend on the number of observations (N), and how much all the observed values (trends on routes, in this case) differ from the average value. If the values vary greatly, the confidence interval will be wide. Examine the table and graph. Does a *large* N tend to widen or narrow the confidence interval?

5. A trend of 0 would mean no change at all. When 0 falls within the confidence interval, we have little certainty that the trend is *not* 0. In this case, we say that the trend is not significant. How many species have trends that are not significant (at 95 percent certainty)?

   Can we be certain that the sharp-tailed grouse and greater prairie-chicken are changing at different rates? How about the sharp-tailed grouse and wild turkey?

Species	Trend 1966-2004	N	Lower limit	Upper limit
Ring-necked pheasant	−0.5	397	−1.2	0.2
Ruffed grouse	0.7	26	−19.7	21.1
Sage grouse	−2.4	23	−5.6	0.9
Sharp-tailed grouse	0	104	−3.4	3.5
Greater prairie-chicken	−6.8	41	−12.8	−0.8
Wild turkey	8.2	308	5.1	11.2
Northern bobwhite	−2.2	541	−2.7	−1.8
Clapper rail	−0.7	8	−7.3	5.8
King rail	−6.8	22	−11.7	−1.8
Virginia rail	4.7	21	1.4	7.9

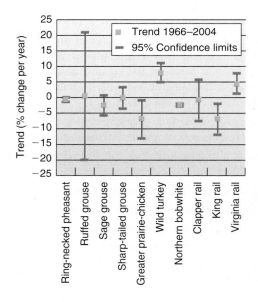

6. Does uncertainty in the data mean results are useless? Does reporting of confidence limits increase or decrease your confidence in the results?

For further information on the Breeding Bird Survey, see http://www.mbr-pwrc.usgs.gov/bbs/.

British Columbia's Great Bear Rainforest will preserve the home for rare, white-phase black (or spirit) bears along with salmon streams, misty fjords, rich tidal estuaries, and the largest remaining area of old-growth, coastal, temperate rainforest in the world.

6 Environmental Conservation:
Forests, Grasslands, Parks, and Nature Preserves

Learning Outcomes
After studying this chapter, you should be able to answer the following questions:

- What portion of the world's original forests remain?
- What activities threaten global forests? What steps can be taken to preserve them?
- Why is road construction a challenge to forest conservation?
- Where are the world's most extensive grasslands?
- How are the world's grasslands distributed, and what activities degrade grasslands?
- What are the original purposes of parks and nature preserves in North America?
- What is a *wilderness*? Why are wilderness areas both important and controversial?
- What are some steps to help restore natural areas?

What a country chooses to save is what a country chooses to say about itself.

–Mollie Beatty, former director, U.S. Fish and Wildlife Service

Saving the Great Bear Rainforest

The wild, rugged coast of British Columbia is home to one of the world's most productive natural communities: the temperate rainforest. Nurtured by abundant rainfall and mild year-round temperatures, forests in the deep, misty fjords shelter giant cedar, spruce, and fir trees. Since this cool, moist forest rarely burns, trees often live for 1,000 years or more, and can be 5 m (16 ft) in diameter and 70 m tall. In addition to huge, moss-draped trees, the forest is home to an abundance of wildlife. One animal, in particular, has come to symbolize this beautiful landscape: it's a rare, white or cream-colored black bear. Called a Kermode bear by scientists, these animals are more popularly known as "spirit bears," the name given to them by native Gitga'at people.

The wetlands and adjacent coastal areas also are biologically rich. Whales and dolphins feed in the sheltered fjords and interisland channels. Sea otters float on the rich offshore kelp forests. It's estimated that 20 percent of the world's remaining wild salmon migrate up the wild rivers of this coastline.

In 2006, officials from the provincial government, Native Canadian nations, logging companies, and environmental groups announced a historic agreement for managing the world's largest remaining intact temperate coastal rainforest. This Great Bear Rainforest encompasses about 6 million ha (15.5 million acres) or about the size of Switzerland (fig. 6.1). One-third of the area will be entirely protected from logging. In the rest of the land, only selective, sustainable logging will be allowed rather than the more destructive clear-cutting that has devastated surrounding forests. At least $120 million will be provided for conservation projects and ecologically sustainable business ventures, such as ecotourism lodges and an oyster farm.

A series of factors contributed to preserving this unique area. The largest environmental protest in Canadian history took place at Clayoquot Sound on nearby Vancouver Island in the 1980s, when logging companies attempted to clear-cut land claimed by First Nations people. This educated the public about the values of and threats to the coastal temperate rainforest. As a result of the lawsuits and publicity generated by this controversy, most of the largest logging companies have agreed to stop clear-cutting in the remaining virgin forest. The rarity of the spirit bears also caught the public imagination. Tens of thousands of schoolchildren from across Canada wrote to the provincial government begging them to set aside a sanctuary for this unique animal. And a growing recognition of the rights of native people also helped convince public officials that traditional lands and ways of living need to be preserved.

More than 60 percent of the world's temperate rainforest has already been logged or developed. The Great Bear Rainforest contains one-quarter of what's left. It also contains about half the estuaries, coastal wetlands, and healthy, salmon-bearing streams in British Columbia.

How did planners choose the areas to be within the protected area? One of the first steps was a biological survey. Where were the biggest and oldest trees? Which areas are especially valuable for wildlife? Protecting water quality in streams and coastal regions was also a high priority. Keeping logging and roads out of riparian habitats is particularly important. Interestingly, native knowledge of the area was also consulted in drawing boundaries. Which places are mentioned in oral histories? What are the traditional uses of the forest? While commercial logging is prohibited in the protected areas, First Nations people will be allowed to continue their customary harvest of selected logs for totem poles, longhouses, and canoes. They also will be allowed to harvest berries, catch fish, and hunt wildlife for their own consumption.

Because it's so remote, few people will ever visit the Great Bear Rainforest, yet many of us like knowing that some special places like this continue to exist. Although we depend on wildlands for many products and services, perhaps we don't need to exploit every place on the planet. Which areas we choose to set aside, and how we protect and manage those special places says a lot about who we are. In chapter 5, we looked at efforts to save individual endangered species. Many biologists believe that we should focus instead on saving habitat and representative biological communities. In this chapter, we'll look at how we use and preserve landscapes.

Figure 6.1 The Great Bear Rainforest stretches along the British Columbia coast (including the Queen Charlotte Islands) from Victoria Island to the Alaska border. The area will be managed as a unit with some pristine wilderness, some First Nations lands, and some commercial production.

6.1 World Forests

Forests and grasslands together occupy almost 60 percent of global land cover (fig. 6.2). These ecosystems provide many of our essential resources, such as lumber, paper pulp, and grazing lands for livestock. They also provide essential ecological services, including regulating climate, controlling water runoff, providing wildlife habitat, purifying air and water, and supporting rainfall. Forests and grasslands also have scenic, cultural, and historic values that deserve protection. Forests and grasslands are also among the most heavily disturbed ecosystems (chapter 5).

In many cases, these competing land uses and needs are incompatible. Most political debates over conservation have concerned protection or use of forests, prairies, and rangelands. This chapter examines the ways we use and abuse these biological communities, as well as some of the ways we can protect them and conserve their resources. We discuss forests first, followed by grasslands and then strategies for conservation, restoration, and preservation.

Boreal and tropical forests are most abundant

Forests are widely distributed, but most remaining forests are in the humid tropics and the cold boreal ("northern") or taiga regions (fig. 6.3). Assessing forest distribution is tricky, because forests vary in density and height, and many are inaccessible. The UN Food and Agriculture Organization (FAO) defines "forest" as any area where trees cover more than 10 percent of the land. This definition includes woodlands ranging from open **savannas**, whose

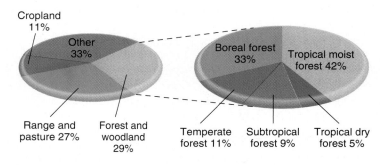

Figure 6.2 World land use and forest types. The "other" category includes tundra, desert, wetlands, and urban areas.
Source: UN Food and Agricultural Organization (FAO).

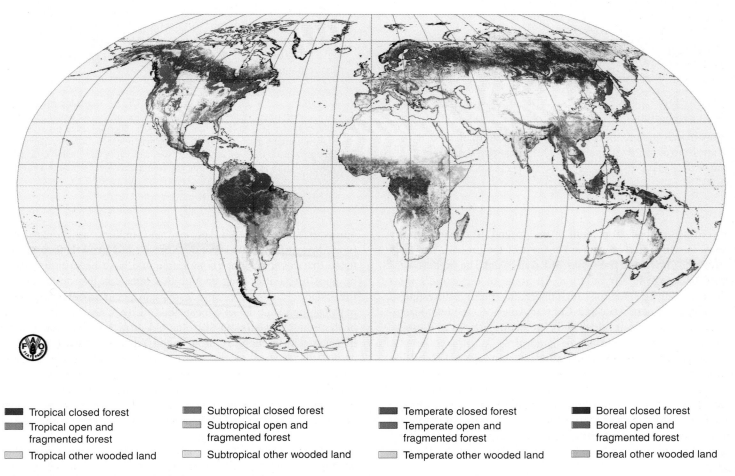

Tropical closed forest
Tropical open and fragmented forest
Tropical other wooded land

Subtropical closed forest
Subtropical open and fragmented forest
Subtropical other wooded land

Temperate closed forest
Temperate open and fragmented forest
Temperate other wooded land

Boreal closed forest
Boreal open and fragmented forest
Boreal other wooded land

Figure 6.3 Major forest types. Note that some of these forests are dense; others may have only 10 to 20 percent actual tree cover.
Source: Data from United Nations Food and Agriculture Organization, 2002.

Active Learning

Calculate Forest Area

Examine figures 6.3 and 6.4 to evaluate changes in global forest cover: Which region(s) originally had the greatest area of forest according to the FAO (fig. 6.4). What kinds of forest occupied those regions? Looking at figure 6.4, which region would you say had the greatest *percentage* forest cover? The lowest percent coverage?

Figures 6.3 and 6.4 show the extent of global forests in *map* form and *graph* form. You can also compare forest distribution *numerically*. You have just made a visual estimate of the percentage cover. Now try calculating percentage cover from the graph in figure 6.4. (Although we'll ask you to do some approximation of the data, the numbers you're comparing are so enormous that some generalization is acceptable for now.)

Starting with Asia, read the approximate size of the land area on the vertical axis (about 3,200 million ha). Then find the extent of original forest (about 1,900 million ha).

Now calculate the percentage of original forest, as follows:

$$\text{percentage} = \frac{\text{forest area}}{\text{land area}} = \frac{1,900 \text{ million ha}}{3,200 \text{ million ha}}$$

cancel out common terms and zeros:

$$\frac{1,900 \text{ million ha}}{3,200 \text{ million ha}} = \frac{19}{32}$$

Now divide the numbers by hand (notice that this is pretty close to 20/30, so you should expect a result not far from 2/3):

$$32\overline{)19} \;\rightarrow\; 32\overline{)19.0} \;\rightarrow\; \begin{array}{r} 0.5 \\ \hline 19.0 \\ 16.0 \end{array}$$

$$\begin{array}{r} 0.59 \\ 32\overline{)19.00} \\ \underline{160} \\ 300 \\ \underline{288} \\ 12 \end{array} \;\rightarrow\; 0.59$$

To look at this percentage as a whole number, multiply your result by 100:

$$0.59 \times 100 = 59\%$$

Repeat this process for the other regions in figure 6.4, and fill in this table:

Original percentage forest cover

Asia	Africa	Europe	N., C. America	S. America	Oceania
59%					

Working with round numbers makes comparison easy. How important are the details lost when you read approximate numbers from the graph? What kinds of generalization might have gone into producing the FAO's original data? What kinds of generalization might have been unavoidable?

Answers: Approximate percentages should be: Africa 70%; N., C. America 29%; S. America 86%; Oceania 44%. Usually approximate numbers provide a quick, useful comparison, and further detail isn't needed until further analysis is done. Defining forests and forest types, or determining extent and "original" extent all involve considerable, usually unavoidable, generalization.

---

trees cover less than 20 percent of the ground, to **closed-canopy forests**, in which tree crowns cover most of the ground. The largest tropical forest is in the Amazon River basin. The highest rates of forest loss are in Africa (fig. 6.4). Some of the world's most biologically diverse regions are undergoing rapid deforestation, including Southeast Asia and Central America (see related stories "Saving an African Eden" and "Protecting Forests to Preserve Rain" at www.mhhe.com/cunningham5e).

Forests are a huge carbon sink, storing some 422 billion metric tons of carbon in standing biomass. Clearing and burning of forests releases much of this carbon into the atmosphere (chapter 9) and may contribute substantially to global climate change. Moisture released from forests also contributes to rainfall. For example, recent climate studies suggest that deforestation of the Amazon could reduce precipitation in the American Midwest.

Among the forests of greatest ecological importance are the remnants of primeval forests that are home to much of the world's biodiversity, endangered species, and indigenous human cultures. Sometimes called frontier forests, **old-growth forests** are those that cover a large enough area and have been undisturbed by human activities long enough that trees can live out a natural life cycle and ecological processes can occur in relatively normal fashion. That doesn't mean that all trees need be enormous or thousands of years old. In some old-growth forests, most trees live less than a century before being killed by disease or some natural disturbance, such as a fire. Nor does it mean that humans have never been present. Where human occupation entails relatively little impact, an old-growth forest may have been inhabited by people for a very long time. Even forests that have been logged or converted to cropland often can revert to old-growth characteristics if left alone long enough.

While forests still cover about half the area they once did worldwide, only one-quarter of those forests retain old-growth features. The largest remaining areas of old-growth forest are in Russia, Canada, Brazil, Indonesia, and Papua New Guinea.

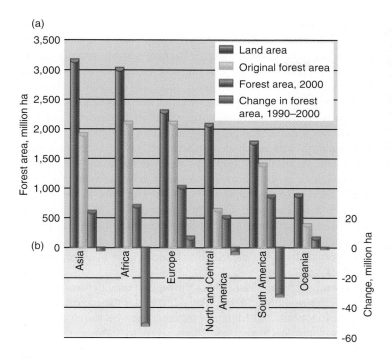

Figure 6.4 Regional distribution of original and current forests (a) and regional forest losses, 1990 to 2000 (b). Europe includes Russia's Siberian forests.
Source: Data from United Nations Food and Agriculture Organization, 2002.

Figure 6.5 Firewood accounts for almost half of all wood harvested worldwide and is the main energy source for nearly half of all humans.

Together, these five countries account for more than three-quarters of all relatively undisturbed forests in the world. In general, remoteness rather than laws protect those forests. Although official data describe only about one-fifth of Russian old-growth forest as threatened, rapid deforestation—both legal and illegal—especially in the Russian Far East, probably put a much greater area at risk.

Forests provide many valuable products

Wood plays a part in more activities of the modern economy than does any other commodity. There is hardly any industry that does not use wood or wood products somewhere in its manufacturing and marketing processes. Think about the amount of junk mail, newspapers, photocopies, and other paper products that each of us in developed countries handles, stores, and disposes of in a single day. Total annual world wood consumption is about 4 billion m³. This is more than steel and plastic consumption combined. International trade in wood and wood products amounts to more than $100 billion each year. Developed countries produce less than half of all industrial wood but account for about 80 percent of its consumption. Less-developed countries, mainly in the tropics, produce more than half of all industrial wood but use only 20 percent.

Paper pulp, the fastest growing type of forest product, accounts for nearly a fifth of all wood consumption. Most of the world's paper is used in the wealthier countries of North America, Europe, and Asia. Global demand for paper is increasing rapidly, however,

as other countries develop. The United States, Russia, and Canada are the largest producers of both paper pulp and industrial wood (lumber and panels). Much industrial logging in Europe and North America occurs on managed plantations, rather than in untouched old-growth forest. However, paper production is increasingly blamed for deforestation in Southeast Asia, West Africa, and other regions.

Fuelwood accounts for nearly half of global wood use. More than half of the people in the world depend on firewood or charcoal as their principal source of heating and cooking fuel (fig. 6.5). The average amount of fuelwood used in less-developed countries is about 1 m³ per person per year, roughly equal to the amount that each American consumes each year as paper products alone. Demand for fuelwood, which is increasing at slightly less than the global population growth rate, is causing severe fuelwood shortages and depleting forests in some developing areas, especially around growing cities. About 1.5 billion people have less fuelwood than they need, and many experts expect shortages to worsen as poor urban areas grow. Because fuelwood is rarely taken from closed-canopy forest, however, it does not appear to be a major cause of deforestation. Some analysts argue that fuelwood could be produced sustainably in most developing countries, with careful management.

Approximately one-quarter of the world's forests are managed for wood production. Ideally, forest management involves scientific planning for sustainable harvests, with particular attention paid to forest regeneration. In temperate regions, according to the UN Food and Agriculture Organization, more land is being replanted or allowed to regenerate naturally than is being permanently deforested. Much of this reforestation, however, is in large plantations of single-species, single-use, intensive cropping called **monoculture forestry**. Although this produces rapid growth and easier harvesting than a more diverse forest, a dense, single-species stand often supports little biodiversity and does poorly in providing the ecological services, such as soil erosion control and clean water production, that may be the greatest value of native forests.

Some of the countries with the most successful reforestation programs are in Asia. China, for instance, cut down most of its forests 1,000 years ago and has suffered centuries of erosion and terrible floods as a consequence. Recently, however, timber cutting in the headwaters of major rivers has been outlawed, and a massive reforestation project has begun. In the 1990s, China planted 42 billion trees, mainly in Xinjiang Province, to stop the spread of deserts. Korea and Japan also have had very successful forest restoration programs. After being almost totally denuded during World War II, both countries are now about 70 percent forested.

Tropical forests are being cleared rapidly

Tropical forests are among the richest and most diverse terrestrial systems. Although they now occupy less than 10 percent of the earth's land surface, these forests are thought to contain more than two-thirds of all higher plant biomass and at least half of all the plant, animal, and microbial species in the world.

A century ago, an estimated 12.5 million km² of tropical lands were covered with closed-canopy forest. This was an area larger than the entire United States. The FAO estimates that about 9.2 million ha, or about 0.6 percent, of the remaining tropical forest is cleared each year (fig. 6.6).

There is considerable debate about current rates of deforestation in the tropics. In 2003, satellite data showed more than 30,000 fires in a single month in Brazil. Remote sensing experts calculate that 3 million ha per year are now being cut and burned in the Amazon basin alone. However, there are different definitions of **deforestation**. Some scientists insist that it means a complete change from forest to agriculture, urban areas, or desert. Others include any area that has been logged, even if the cut was selective and regrowth will be rapid. Furthermore, savannas, open woodlands, and succession following natural disturbance are hard to distinguish from logged areas. Consequently, estimates for total tropical forest losses range from about 5 million to more than 20 million ha per year. The FAO estimates of 12.3 million ha deforested per year are generally the most widely accepted. To put that figure in perspective, it means that about 1 acre—or the area of a football field—is cleared every second, on average, around the clock.

By most accounts, Brazil has the highest total deforestation of any country in the world, but it also has by far the largest tropical forests. Indonesia and Malaysia together may be losing as much primary forest each year as Brazil, even though their original amount was far lower. In 1997 forest fires on Borneo and Sumatra, exacerbated by severe drought, burned 20,000 km² (8,000 mi²). The fires reportedly were set both to clear land for agriculture and to hide illegal logging.

In Africa, the coastal forests of Senegal, Sierra Leone, Ghana, Madagascar, Cameroon, and Liberia already have been mostly destroyed. Haiti was once 80 percent forested; today, essentially all that forest has been destroyed, and the land lies barren and eroded. India, Burma, Kampuchea (Cambodia), Thailand, and Vietnam all have little old-growth lowland forest left. In Central America, nearly two-thirds of the original moist tropical forest has been destroyed, mostly within the past 30 years and primarily due to conversion of forest to cattle range. (See related story on "Disappearing Butterfly Forests" at www.mhhe.com/cunningham5e.)

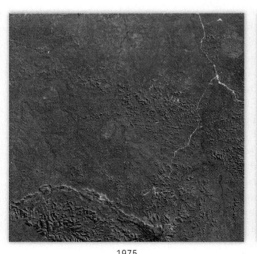

| 1975 | 1989 | 2001 |

Figure 6.6 Forest destruction in Rondonia, Brazil, between 1975 and 2001. Construction of logging roads creates a feather-like pattern that opens forests to settlement by farmers.

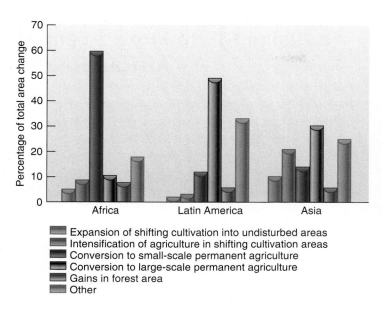

Expansion of shifting cultivation into undisturbed areas
Intensification of agriculture in shifting cultivation areas
Conversion to small-scale permanent agriculture
Conversion to large-scale permanent agriculture
Gains in forest area
Other

Figure 6.7 Tropical forest losses by different processes.
Source: Data from Food and Agriculture Organization of the United Nations, 2002.

Causes of deforestation

A variety of factors contribute to deforestation, and, as figure 6.7 shows, different forces predominate in various parts of the world. Logging for valuable tropical hardwoods, such as teak and mahogany, is generally the first step. Although loggers may take only one or two of the largest trees per hectare, the canopy of tropical forests is usually so strongly linked by vines and interlocking branches that felling one tree can bring down a dozen others. Building roads to remove logs kills more trees, but even more important, it allows entry to the forest by farmers, miners, hunters, and others who cause further damage.

In Africa, conversion of forest into small-scale agriculture accounts for nearly two-thirds of all tropical forest destruction. In Latin America, poor, landless farmers often start the deforestation but are bought out—or driven out—after a few years by large-scale farmers or ranchers (fig. 6.8). The thin, nutrient-poor tropical forest soils frequently are worn out after a few years of cropping, and much of this land becomes unproductive pasture or impoverished scrubland. Shifting cultivation (sometimes called "slash and burn" or milpa farming) is often blamed for forest destruction. This practice can be sustainable where population densities are low and individual plots are allowed to regenerate for a decade or two between cultivation periods. In some countries, however, growing populations and shrinking forests lead to short rotation cycles, and cropping intensification can lead to permanent deforestation.

As forests are cleared, rainfall patterns can change. A computer model created by Pennsylvania State University scientists suggests this phenomenon might create a kind of chain reaction. As forests are cut down, plant transpiration and rainfall decrease. Drought kills more vegetation, and fires become more numerous and extensive. In a worst-case scenario, an area as large as the entire Amazonian forest might be permanently damaged in just a few decades.

Figure 6.8 The number of termite mounds in this recently cleared section of the Brazilian Cerado (savanna) suggests how many trees once grew here. Expansion of soy farming in Brazil is pushing ranching into areas of once-intact forest.

Forest protection

What can be done to stop this destruction and encourage tropical forest protection? While much of the news is discouraging, there are some hopeful signs for forest conservation in the tropics. Many countries now recognize that forests are valuable resources. Intensive scientific investigations are under way to identify the best remaining natural areas (see Exploring Science, p. 132).

About 12 percent of all world forests are in some form of protected status, but the effectiveness of that protection varies greatly. Costa Rica has one of the best plans for forest guardianship in the world. Attempts are being made there not only to rehabilitate the land (make an area useful to humans) but also to restore the ecosystems to naturally occurring associations. One of the best-known of these projects is Dan Janzen's work in Guanacaste National Park. Like many dry, tropical forests, the northwestern part of Costa Rica had been almost completely converted to ranchland. By controlling fires, however, Janzen and his coworkers are bringing back the forest. One of the keys to this success is involving local people in the project. Janzen also advocates grazing in the park. The original forest evolved, he reasons, together with ancient grazing animals that are now extinct. Horses and cows can play a valuable role as seed dispersers.

People also are working on the grassroots level to protect and restore forests in other countries. India, for instance, has a long history of nonviolent, passive resistance movements—called *satyagrahas*—to protest unfair government policies. These protests go back to the beginning of Indian culture and often have been associated with forest preservation. Gandhi drew on this tradition in his protests of British colonial rule in the 1930s and 1940s. During the 1970s, commercial loggers began large-scale tree felling in the Garhwal region in the state of Uttar Pradesh in northern India. Landslides and floods resulted from stripping the forest cover from the hills. The firewood on which local people depended was destroyed, and the way of life of the traditional forest culture was threatened. In a remarkable display of courage and

Exploring

SCIENCE:

Using GIS to Protect Central African Forests

Protecting biodiversity-rich tropical forests often is difficult because information is so elusive. Deep, remote, swampy, tropical jungles are difficult to enter, map, and assess for their ecological value. Yet without information about their ecological importance, most people have little reason to care about these remote, trackless forests. How can you conserve ecosystems if you don't know what's there?

For most of history, understanding the extent and conditions of a remote area required an arduous trek to see the place in person. Even on publicly owned lands, only those who could afford the time, or who could afford to pay surveyors, might understand the resources. Over time, maps improved, but maps usually show only a few features, such as roads, rivers, and some boundaries.

In recent years, details about public lands and resources have suddenly burst into public view through the use of geographic information systems (**GIS**). A GIS consists of spatial data, such as boundaries or road networks, and software to display and analyze the data. Spatial data can include variables that are hard to see on the ground—watershed boundaries, annual rainfall, land ownership, or historical land use. Data can also represent phenomena much larger than we can readily see—land surface slopes and elevation, forested regions, river networks, and so on. By overlaying these layers, GIS analysts can investigate completely new questions about conservation, planning, and restoration.

You have probably used a GIS. Online mapping programs such as MapQuest or Google Earth organize and display spatial data. They let you turn layers on and off, or zoom in and out to display different scales. You can also use an online mapping program to calculate distances and driving directions between places. An ecologist, meanwhile, might use a GIS to calculate the extent of habitat areas, to monitor changes in area, to calculate the size of habitat fragments, or to calculate the length of waterways in a watershed.

Identifying Priority Areas

Recently, a joint effort of several conservation organizations has used GIS and spatial data to identify priority areas for conservation in Central Africa. The project was initiated because new data, including emerging GIS data, were showing dramatic increases in planned logging, in a region that contains the world's second greatest extent of tropical forest (fig. 1).

Researchers from the Wildlife Conservation Society, Worldwide Fund for Nature, World Resources Institute, USGS, and other agencies and

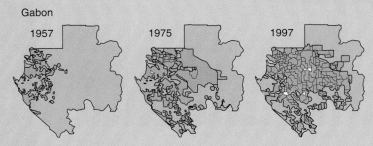

Figure 1 Gabon, Central Africa, has seen a steady increase in logging concessions.
Source: Wildlife Conservation Society.

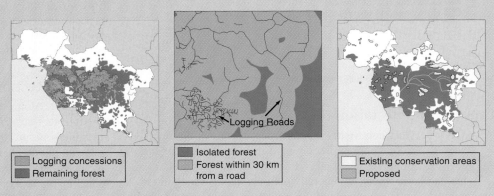

Figure 2 A few GIS layers used to identify priority conservation areas.
Source: Wildlife Conservation Society.

groups, began collecting GIS data on a variety of variables. They identified the range of great apes and other rare or threatened species. They identified areas of extreme plant diversity. They calculated the sizes of forest fragments to identify concentrations of intact, ancient forests. Using maps of logging roads, they calculated the area within a 30 km "buffer" around roads, since loggers, settlers, and hunters usually threaten biodiversity near roads. They also mapped existing and planned conservation areas (fig. 2).

By overlaying these and other layers, analysts identified priority conservation areas of extensive original forest, which have high biodiversity and rare species. Overlaying these priority areas with a map of protected lands and a map of timber concessions, they identified *threatened* priority areas (fig. 3).

Most of the unprotected priority areas may never be protected, but having this map provides two important guides for future conservation. First, it assesses the state of the problem. With this map, we know that most of the forest is unprotected but also that the region's primary forest is extensive. Second, this map provides priorities for conservation planning. In addition, maps are very effective tools for publicizing an issue. When a map like this is pub-

lished, more people become enthusiastic about joining the conservation effort.

GIS has become an essential tool for conserving forests, grasslands, ecosystems, and nature preserves. GIS has revolutionized the science of planning and conservation—examining problems using quantitative data—just as it may have revolutionized the way you plan a driving trip.

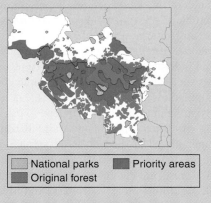

Figure 3 Priority areas outside of national parks.
Source: Wildlife Conservation Society.

determination, the village women wrapped their arms around the trees to protect them, sparking the Chipko Andolan movement (literally, movement to hug trees). They prevented logging on 12,000 km² of sensitive watersheds in the Alakanada basin. Today the Chipko Andolan movement has grown to more than 4,000 groups working to save India's forests.

Debt-for-nature swaps

Those of us in developed countries also can contribute toward saving tropical forests. Financing nature protection is often a problem in developing countries, where the need is greatest. One promising approach is called **debt-for-nature swaps**. Banks, governments, and lending institutions now hold nearly $1 trillion in loans to developing countries. There is little prospect of ever collecting much of this debt, and banks are often willing to sell bonds at a steep discount—perhaps as little as 10 cents on the dollar. Conservation organizations buy debt obligations on the secondary market at a discount and then offer to cancel the debt if the debtor country agrees to protect or restore an area of biological importance.

There have been many such swaps. Conservation International, for instance, bought $650,000 of Bolivia's debt for $100,000—an 85 percent discount. In exchange for canceling this debt, Bolivia agreed to protect nearly 1 million ha (2.47 million acres) around the Beni Biosphere Reserve in the Andean foothills. Ecuador and Costa Rica have had a different kind of debt-for-nature swap. They have exchanged debt for local currency bonds that fund activities of local private conservation organizations in the country. This has the dual advantage of building and supporting indigenous environmental groups while protecting the land. Critics, however, charge that these swaps compromise national sovereignty and do little to reduce the developing world's debt or to change the situations that led to environmental destruction in the first place.

Temperate forests also are at risk

Tropical countries aren't unique in harvesting forests at an unsustainable rate. Northern countries, such as the United States and Canada, also have allowed controversial forest management practices in many areas. For many years, the official policy of the U.S. Forest Service was "multiple use," which implied that the forests could be used for everything that we might want to do there simultaneously. Some uses are incompatible, however. Bird-watching, for example, isn't very enjoyable in a dirt bike racing course. And protecting species that need unbroken old-growth forest isn't feasible when you cut down the forest.

Old-growth forests

The most contentious forestry issues in the United States and Canada in recent years have centered on logging in old-growth forests in the Pacific Northwest. As you've learned in the opening case study for this chapter, these forests have incredibly high levels of biodiversity, and they accumulate more total biomass in standing vegetation per unit area than any other ecosystem on earth (fig. 6.9). Many endemic species, such as the northern spotted owl (fig. 6.10), Vaux's swift, and the marbled murrelet, are so highly adapted to the unique conditions of these ancient forests that they live nowhere else.

Figure 6.9 The temperate rainforests of the Pacific Northwest have the highest biomass per hectare of any landscape in the world.

Figure 6.10 Only about 2,000 pairs of northern spotted owls remain in the old-growth forests of the Pacific Northwest. Cutting old-growth forests threatens the endangered species, but reduced logging threatens the jobs of many timber workers.

Figure 6.11 Large clear-cuts, such as this, threaten species dependent on old-growth forest and expose steep slopes to soil erosion. Restoring something like the original forest will take hundreds of years.

Less than 10 percent of old-growth forest in the United States remains intact, and 80 percent of what remains is scheduled to be cut in the near future. Just a century ago, most of the coastal ranges of Washington, Oregon, northern California, British Columbia, and southeastern Alaska were clothed in a lush forest of huge trees, many a thousand years old or more. British Columbia has felled at least 60 percent of its richest and most productive ancient forests and is now cutting some 240 million ha (600 million acres) annually, about ten times the rate of old-growth harvest in the United States. At these rates, the only remaining ancient forests in North America in 50 years will be a fringe around the base of the mountains in a few national parks.

In 1989 wildlife conservationists sued the U.S. Forest Service over logging rates in Washington and Oregon, arguing that northern spotted owls are endangered and must be protected under the Endangered Species Act. Federal courts agreed and ordered some 1 million ha (2.5 million acres) of ancient forest set aside to preserve the last 2,000 pairs of owls. This is about half the remaining virgin forest in Washington and Oregon. The timber industry claimed that 40,000 jobs would be lost, although economic studies show that most logging jobs were lost as a result of log exports and mechanization. Still, outrage in the logging communities was loud and clear. Convoys of logging trucks converged on protest sites, while angry crowds burned conservationists in effigy. Bumper stickers urged, "Save a logger; eat an owl."

A compromise forest management plan was finally agreed upon that allows some continued logging but protects the most valuable forests, including selected old-growth preserves, riparian (streamside) buffer strips, and prime wildlife habitat. This plan may not be enough protection, however, to ensure the survival of endangered salmon and trout populations in northwestern rivers. An even bigger battle is shaping up as specific fish populations are considered for listing as endangered species. Several salmon and steelhead trout populations have been listed as endangered and more are under consideration. Forestry, farming and cheap hydropower are aligned on one side of this battle, while fishing, native rights, and wildlife protection join together on the other side.

Harvest methods

Most lumber and pulpwood in the United States and Canada currently are harvested by **clear-cutting**, in which every tree in a given area is cut, regardless of size (fig. 6.11). This method is effective for producing even-age stands of sun-loving species, such as aspen or pines, but often increases soil erosion and eliminates habitat for many forest species when carried out on large blocks. It was once thought that good forest management required immediate removal of all dead trees and logging residue. Research has shown, however, that standing snags and coarse woody debris play important ecological roles, including soil protection, habitat for a variety of organisms, and nutrient recycling.

Some alternatives to clear-cutting include **shelterwood harvesting**, in which mature trees are removed in a series of two or more cuts, and **strip-cutting**, in which all the trees in a narrow corridor are harvested. For many forest types, the least disruptive harvest method is **selective cutting**, in which only a small percentage of the mature trees are taken in each 10- or 20-year rotation. Ponderosa pine, for example, are usually selectively cut to thin stands and improve growth of the remaining trees. A forest managed by selective cutting can retain many of the characteristics of age distribution and groundcover of a mature old-growth forest. (See related story "Forestry for the Seventh Generation" at www.mhhe.com/cunningham5e.)

Should we subsidize logging on public lands?

An increasing number of people in the United States are calling for an end to all logging on federal lands. They argue that ecological services, from maintaining river levels for fish and irrigation to recreation, generate more revenue at lower costs. Many remote communities depend on logging jobs, but these jobs depend on subsidies. The federal government builds roads, manages forests, fights fires, and sells timber for less than the administrative costs of the sales. What is the best policy?

Some argue that logging should be restricted to privately owned lands. Just 4 percent of the nation's timber comes from national forests, and this harvest adds about $4 billion to the American economy per year. In contrast, recreation, fish and wildlife, clean water, and other ecological services provided by the forest, by their calculations, are worth at least $224 billion each year. Timber industry officials, on the other hand, dispute these claims, arguing that logging not only provides jobs and supports rural communities but also keeps forests healthy. What do you think? Could we make up for decreased timber production from public lands by more intensive management of private holdings and by substitution or recycling of wood products? Are there alternative ways you could suggest to support communities now dependent on timber harvesting?

Roads on public lands are another controversy. Over the past 40 years, the Forest Service has expanded its system of logging roads more than ten-fold, to a current total of nearly 550,000 km (343,000 mi), or more than ten times the length of the interstate highway system. Government economists regard road building as a benefit because it opens up the country to motorized recreation

What Can You Do?

Lowering Your Forest Impacts

For most urban residents, forests—especially tropical forests—seem far away and disconnected from everyday life. There are things that each of us can do, however, to protect forests.

- Reuse and recycle paper. Make double-sided copies. Save office paper, and use the back for scratch paper.

- Use email. Store information in digital form, rather than making hard copies of everything.

- If you build, conserve wood. Use wafer board, particle board, laminated beams, or other composites, rather than plywood and timbers made from old-growth trees.

- Buy products made from "good wood" or other certified sustainably harvested wood.

- Don't buy products made from tropical hardwoods, such as ebony, mahogany, rosewood, or teak, unless the manufacturer can guarantee that the hardwoods were harvested from agroforestry plantations or sustainable-harvest programs.

- Don't patronize fast-food restaurants that purchase beef from cattle grazing on deforested rainforest land. Don't buy coffee, bananas, pineapples, or other cash crops if their production contributes to forest destruction.

- Do buy Brazil nuts, cashews, mushrooms, rattan furniture, and other nontimber forest products harvested sustainably by local people from intact forests. Remember that tropical rainforest is not the only biome under attack. Contact the Taiga Rescue Network (www.taigarescue.org) for information about boreal forests.

- If you hike or camp in forested areas, practice minimum-impact camping. Stay on existing trails, and don't build more or bigger fires than you absolutely need. Use only downed wood for fires. Don't carve on trees or drive nails into them.

- Write to your congressional representatives, and ask them to support forest protection and environmentally responsible government policies. Contact the U.S. Forest Service, and voice your support for recreation and nontimber forest values.

Figure 6.12 By suppressing fires and allowing fuel to accumulate, we make major fires such as this more likely. The safest and most ecologically sound management policy for some forests may be to allow natural or prescribed fires, which don't threaten property or human life, to burn periodically.

Fire management

Following a series of disastrous fire years in the 1930s, in which hundreds of millions of hectares of forest were destroyed, whole towns burned to the ground, and hundreds of people died, the U.S. Forest Service adopted a policy of aggressive fire control in which every blaze on public land was to be out before 10 A.M. Smokey Bear was adopted as the forest mascot and warned us that "only you can prevent forest fires." Recent studies, however, of fire's ecological role suggest that our attempts to suppress all fires may have been misguided. Many biological communities are fire-adapted and require periodic burning for regeneration. Furthermore, eliminating fire from these forests has allowed woody debris to accumulate, greatly increasing the chances of a very big fire (fig. 6.12).

Forests that once were characterized by 50 to 100 mature, fire-resistant trees per hectare and an open understory now have a thick tangle of up to 2,000 small, spindly, mostly dead saplings in the same area. The U.S. Forest Service estimates that 33 million ha (73 million acres), or about 40 percent of all federal forestlands, are at risk of severe fires. To make matters worse, Americans increasingly live in remote areas where wildfires are highly likely. Because there haven't been fires in many of these places in living memory, many people assume there is no danger, but by some estimates, 40 million U.S. residents now live in areas with high wildfire risk.

A recent prolonged drought in the western United States has heightened fire danger. In 2007 nearly 80,000 wildfires burned 3.6 million ha (8.9 million acres) of forests and grasslands in the United States. Federal agencies spent almost $2 billion to fight these fires, nearly four times the previous ten-year average.

The dilemma is how to undo years of fire suppression and fuel buildup. Fire ecologists favor small, prescribed burns to clean out debris. Loggers decry this approach as a waste of valuable timber,

and industrial uses. Wilderness enthusiasts and wildlife supporters, however, see this as an expensive and disruptive program. In 2001 the Clinton administration announced a plan to protect 23.7 million ha (58.5 million acres) of de facto wilderness from roads. Timber, mining, and oil companies protested this rule. In 2002 a federal judge in Idaho ruled that blocking road construction would do "irreparable harm" to the forest. He also ruled that the 600 public meetings held over a 12-month period to discuss this plan, and the record 1.6 million written comments gathered (90 percent of which supported increased protection), did not represent adequate public input. The Bush administration overturned the "roadless rule" and ordered resource managers to expedite logging, mining, and motorized recreation. What do you think? How much of the remaining old growth should be protected as ecological reserves?

Forest Thinning and Salvage Logging

For more than 70 years, firefighting has been a high priority for forest managers. Unfortunately, our efforts have been so successful that dead wood and brush have now built up to dangerous levels in many forests. Lands that would once have been cleaned out by frequent low-temperature ground fires now have so much accumulated fuel that a catastrophic wildfire is all but inevitable. To make matters worse, increasing numbers of people are building cabins and homes in remote, fire-prone areas, where they expect to be protected from unavoidable risks.

People living in or near national forests demand protection from wildfires. In response, federal agencies are starting thinning programs to remove excess fuel. To make it profitable for loggers to remove fire-prone dead wood, small trees, and brush, the government is allowing them to harvest large, valuable, and fire-resistant trees located in the backcountry, often miles away from the nearest communities. And to avoid what many loggers claim is "red tape and litigation" that have tied up forest managers in "analysis paralysis," most thinning projects are exempt from public comment and administrative appeals, as well as from environmental reviews under the National Environmental Policy Act (NEPA).

Environmental groups denounce these projects as merely logging without laws. Thinning, they argue, looks very much like clear-cutting. Forest ecologists argue that thinning, unless it is repeated every few years, actually makes the forests more, rather than less, fire-prone. Removing big, old trees opens up the forest canopy and encourages growth of brush and new tree seedlings. Furthermore, logging compacts soil and introduces invasive species. Many forest ecologists maintain that small, prescribed burns can reduce fuel and produce healthier forests than commercial logging.

A 2005 study of an Oregon fire found that salvage logging actually increased fire susceptibility and decreased the ability of the forest to regenerate naturally. Post-fire logging harms the soil, introduces invasive species, and makes future fire much more likely (fig. 1). Dry branches and debris left on the ground provided much better fuel than did standing dead trees. In addition, seedling growth among standing dead trees was robust, with three times the density of seedlings in salvage areas.

Other research calls into question a main justification for thinning: protecting housing that has proliferated around the edges of federal forests. The only thinning needed to protect houses, according to fire experts, is within about 60 m around the building. Regardless of how intense the fire is, building damage can be minimized by installing a metal roof, clearing pine needles and brush around the house, and removing trees immediately around the building.

After a fire has burned a forest, local residents often call for emergency salvage logging to utilize dead trees before they fall and become fuel for future fires. Foes say that salvage sales allow timber companies to cut many trees that survive the flames and, like thinning, leave roads and scars from heavy equipment that last far longer than any effects of the fires themselves. This kind of logging costs far more than it yields. And, critics contend, the salvage contracts rarely require timber companies to remove the very fuel that stokes wildfires—the underbrush and downed timber for which there is little or no market. The result, some conservationists say, is the ecological equivalent of mugging a fire victim.

What do you think? How can we get out of the crisis we've created by preventing fires and letting dead wood accumulate? If you were a forest manager, would you authorize thinning and salvage logging, or can you think of other ways to remove forest fuels in an ecologically and economically sustainable manner? What research projects would you direct your staff to undertake in order to inform your decision in this dilemma?

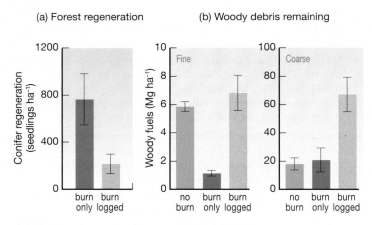

Figure 1 Research results from the 2005 Bisent fire in Oregon. (a) Post-fire logging reduces regeneration by about half. (b) Far more fine and coarse debris (fuelwood) are present in burned and logged forests than in burned only.
Data from Donato, et al., 2008.

and local residents of fire-prone areas fear that prescribed fires will escape and threaten them. Recently the Forest Service proposed a massive new program of forest thinning and emergency salvage operations (removing trees and flammable material from mature or recently burned forests) on 16 million ha (40 million acres) of national forest. Carried out over a 20-year period, this program could cost as much as $12 billion and would open up much roadless, de facto wilderness to large-scale logging. Critics complain that this program is ill advised and environmentally destructive. Proponents argue that the only way to save the forest is to log it (see What Do You Think?).

Ecosystem management

In the 1990s the U.S. Forest Service began to shift its policies from a timber production focus to **ecosystem management**, which attempts to integrate sustainable ecological, economic, and social goals in a unified, systems approach. Some of the principles of this new philosophy include:

- Managing across whole landscapes, watersheds, or regions over ecological time scales.

- Considering human needs and promoting sustainable economic development and communities.

Table 6.1	Draft Criteria for Sustainable Forestry
1.	Conservation of biological diversity
2.	Maintenance of productive capacity of forest ecosystems
3.	Maintenance of forest ecosystem health and vitality
4.	Maintenance of soil and water resources
5.	Maintenance of forest contribution to global carbon cycles
6.	Maintenance and enhancement of long-term socioeconomic benefits to meet the needs of legal, institutional, and economic framework for forest conservation and sustainable management

Source: Data from USFS, 2002.

- Maintaining biological diversity and essential ecosystem processes.
- Utilizing cooperative institutional arrangements.
- Generating meaningful stakeholder and public involvement and facilitating collective decision making.
- Adapting management over time, based on conscious experimentation and routine monitoring.

Some critics argue that we don't understand ecosystems well enough to make practical decisions in forest management on this basis. They argue we should simply set aside large blocks of untrammeled nature to allow for chaotic, catastrophic, and unpredictable events. Others see this new approach as a threat to industry and customary ways of doing things. Still, elements of ecosystem management appear in the *National Report on Sustainable Forests* prepared by the U.S. Forest Service. Based on the Montreal Working Group criteria and indicators for forest health, this report suggests goals for sustainable forest management (table 6.1).

6.2 Grasslands

After forests, grasslands are among the biomes most heavily used by humans. Prairies, savannas, steppes, open woodlands, and other grasslands occupy about one-quarter of the world's land surface. Much of the U.S. Great Plains and the Prairie Provinces of Canada fall in this category (fig. 6.13). The 3.8 billion ha (12 million mi²) of pastures and grazing lands in this biome make up about twice the area of all agricultural crops. When you add to this about 4 billion ha of other lands (forest, desert, tundra, marsh, and thorn scrub) used for raising livestock, more than half of all land is used at least occasionally for grazing. More than 3 billion cattle, sheep, goats, camels, buffalo, and other domestic animals on these lands make a valuable contribution to human nutrition. Sustainable pastoralism can increase productivity while maintaining biodiversity in a grassland ecosystem.

Because grasslands, chaparral, and open woodlands are attractive for human occupation, they frequently are converted to cropland, urban areas, or other human-dominated landscapes. Worldwide the rate of grassland disturbance each year is three times that of tropical forest. Although they may appear to be uniform and monotonous to the untrained eye, native prairies can be

Figure 6.13 This short-grass prairie in northern Montana is too dry for trees but, nevertheless, supports a diverse biological community.

highly productive and species-rich. According to the U.S. Department of Agriculture, more threatened plant species occur in rangelands than in any other major American biome.

Grazing can be sustainable or damaging

By carefully monitoring the numbers of animals and the condition of the range, ranchers and **pastoralists** (people who live by herding animals) can adjust to variations in rainfall, seasonal plant conditions, and the nutritional quality of forage to keep livestock healthy and avoid overusing any particular area. Conscientious management can actually improve the quality of the range.

When grazing lands are abused by overgrazing—especially in arid areas—rain runs off quickly before it can soak into the soil to nourish plants or replenish groundwater. Springs and wells dry up. Seeds can't germinate in the dry, overheated soil. The barren ground reflects more of the sun's heat, changing wind patterns, driving away moisture-laden clouds, and leading to further desiccation. This process of conversion of once fertile land to desert is called **desertification**.

This process is ancient, but in recent years it has been accelerated by expanding populations and the political conditions that force people to overuse fragile lands. According to the International Soil Reference and Information Centre in the Netherlands, nearly three-quarters of all rangelands in the world show signs of either degraded vegetation or soil erosion. Overgrazing is responsible for about one-third of that degradation (fig. 6.14). The highest percentage of moderate, severe, and extreme land degradation is in Mexico and Central America, while the largest total area is in Asia, where the world's most extensive grasslands occur. Can we reverse this process? In some places, people are reclaiming deserts and repairing the effects of neglect and misuse.

Overgrazing threatens many rangelands

As is the case in many countries, the health of most public grazing lands in the United States is not good. Political and economic

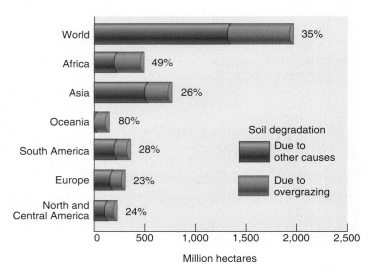

Figure 6.14 Rangeland soil degradation due to overgrazing and other causes. Notice that, in Europe, Asia, and the Americas, farming, logging, mining, urbanization, and so on are responsible for about three-quarters of all soil degradation. In Africa and Oceania, where more grazing occurs and desert or semiarid scrub make up much of the range, grazing damage is higher.
Source: Data from World Resource Institute.

Figure 6.15 More than half of all publicly owned grazing land in the United States is in poor or very poor condition. Overgrazing and invasive weeds are the biggest problems.

pressures encourage managers to increase grazing allotments beyond the carrying capacity of the range. Lack of enforcement of existing regulations and limited funds for range improvement have resulted in **overgrazing**, damage to vegetation and soil including loss of native forage species and erosion. The Natural Resources Defense Council claims that only 30 percent of public range-lands are in fair condition, and 55 percent are poor or very poor (fig. 6.15).

Overgrazing has allowed populations of unpalatable or inedible species, such as sage, mesquite, cheatgrass, and cactus, to build up on both public and private rangelands. Wildlife conservation groups regard cattle grazing as the most ubiquitous form of ecosystem degradation and the greatest threat to endangered species in the southwestern United States. They call for a ban on cattle and sheep grazing on all public lands, noting that it provides only 2 percent of the total forage consumed by beef cattle and supports only 2 percent of all livestock producers.

Like federal timber management policy, grazing fees charged for use of public lands often are far below market value and represent an enormous hidden subsidy to western ranchers. Holders of grazing permits generally pay the government less than 25 percent the amount of leasing comparable private land. The 31,000 permits on federal range bring in only $11 million in grazing fees but cost $47 million per year for administration and maintenance. The $36 million difference amounts to a massive "cow welfare" system of which few people are aware.

On the other hand, ranchers defend their way of life as an important part of western culture and history. Although few cattle go directly to market from their ranches, they produce almost all the beef calves subsequently shipped to feedlots. And without a viable ranch economy, they claim, even more of the western land-

scape would be subdivided into small ranchettes to the detriment of both wildlife and environmental quality. What do you think? How much should we subsidize extractive industries to preserve rural communities and traditional occupations?

Ranchers are experimenting with new methods

Where a small number of livestock are free to roam a large area, they generally eat the tender, best-tasting grasses and forbs first, leaving the tough, unpalatable species to flourish and gradually dominate the vegetation. In some places, farmers and ranchers find that short-term, intensive grazing helps maintain forage quality. As South African range specialist Allan Savory observed, wild ungulates (hoofed animals), such as gnus or zebras in Africa or bison (buffalo) in America, often tend to form dense herds that graze briefly but intensively in a particular location before moving on to the next area. Rest alone doesn't necessarily improve pastures and rangelands. Short-duration, **rotational grazing**—confining animals to a small area for a short time (often only a day or two) before shifting them to a new location—simulates the effects of wild herds (fig. 6.16). Forcing livestock to eat everything equally, to trample the ground thoroughly, and to fertilize heavily with manure before moving on helps keep weeds in check and encourages the growth of more desirable forage species. This approach doesn't work everywhere, however. Many plant communities in the U.S. desert Southwest, for example, apparently evolved in the absence of large, hoofed animals and can't withstand intensive grazing.

Restoring fire can be as beneficial to grasslands as it is to forests. In some cases, ranchers are cooperating with environmental groups in range management and preservation of a ranching economy (see Exploring Science, p. 140).

Another approach to ranching in some areas is to raise wild species, such as red deer, impala, wildebeest, or oryx (fig. 6.17). These animals forage more efficiently, resist harsh climates, often are more pest- and disease-resistant, and fend off predators better

Figure 6.16 Intensive, rotational grazing encloses livestock in a small area for a short time (often only one day) within a movable electric fence to force them to eat vegetation evenly and fertilize the area heavily.

Figure 6.17 Red deer (*Cervus elaphus*) are raised in New Zealand for antlers and venison.

than usual domestic livestock. Native species also may have different feeding preferences and needs for water and shelter than cows, goats, or sheep. The African Sahel, for instance, can provide only enough grass to raise about 20 to 30 kg (44 to 66 lbs) of beef per hectare. Ranchers can produce three times as much meat with wild native species in the same area because these animals browse on a wider variety of plant materials.

In the United States, ranchers find that elk, American bison, and a variety of African species take less care and supplemental feeding than cattle or sheep and result in a better financial return because their lean meat can bring a better market price than beef or mutton. Media mogul Ted Turner has become both the biggest private landholder in the United States and the owner of more American bison than anyone other than the government.

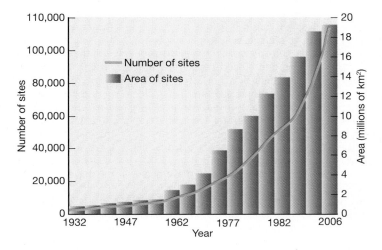

Figure 6.18 Growth of protected areas worldwide, 1932–2003.
Source: UN World Commission on Protected Areas.

6.3 Parks and Preserves

While most forests and grasslands serve useful, or utilitarian, purposes most societies also set aside some natural areas for aesthetic or recreational purposes. Natural preserves have existed for thousands of years. Ancient Greeks protected sacred groves for religious purposes. Royal hunting grounds have preserved forests in Europe for centuries. Although these areas were usually reserved for elite classes in society, they have maintained biodiversity and natural landscapes in regions where most lands are heavily used.

The first public parks open to ordinary citizens may have been the tree-sheltered agoras in planned Greek cities. But the idea of providing natural space for recreation, and to preserve natural environments, has really developed in the past 50 years (fig. 6.18). While the first parks were intended mainly for the recreation of growing urban populations, parks have taken on many additional purposes. Today we see our national parks as playgrounds for rest and recreation, as havens for wildlife, as places to experiment with ecological management, and as opportunities to restore ecosystems.

Currently, nearly 12 percent of the land area of the earth is protected in some sort of park, preserve, or wildlife management area. This represents about 19.6 million ha (7.6 million mi^2) in 107,000 different preserves. This is an encouraging environmental success story.

Many countries have created nature preserves

Different levels of protection are found in nature preserves. The World Conservation Union divides protected areas into five categories depending on the intended level of allowed human use (table 6.2). In the most stringent category (ecological reserves and wilderness areas) little or no human impacts are allowed. In some strict nature preserves, where particularly sensitive wildlife or natural features are located, human entry may be limited only to scientific research groups that visit on rare occasions. In some wildlife sanctuaries, for example, only a few people per year are

Exploring

SCIENCE:

Finding Common Ground on the Range

For decades, environmentalists have tried to limit grazing on public lands, where ranchers lease pastures from the government. Now some scientists and conservationists are saying that cattle ranches may be the last best hope for preserving habitat for many native species. Maintaining ranches may also be the only way to restore the periodic fires that keep brush and cactus from taking over western grasslands. In a number of places in the United States, ranchers are forming alliances with environmental organizations and government officials to find new ways to protect and manage rangelands.

One of the pressures driving this new model of collaboration is the growing popularity of western hobby ranches and rural homesteads for city folk. Falling commodity prices, drought, taxes, and other forces are causing many ranchers to consider selling their land. Why continue to struggle to make a living with ranching when you can make millions by cutting up your land into 40-acre ranchettes? But when the land is subdivided, invasive species move in along with people and their pets, and fewer native species can survive. Furthermore, it becomes much harder, if not impossible, to let fires burn across the land periodically, a process that is now thought to be essential in the southwestern landscape.

Some grazing practices clearly have been detrimental. Studies have found extensive damage from grazing in and around streams in the desert West, for instance. But few studies have compared the alternatives to ranching on these lands, which are home not only to ranchers but to many native animal and plant species. Recent research has found that ranches have at least as many species of birds, carnivores, and plants as similar areas protected as wildlife refuges. Ranches also have fewer invasive weeds.

An outstanding example of new cooperative relationships between ranchers, conservationists, and government agencies is the Malpai Borderlands Group (MBG). This community-based ecosystem management effort was created by landowners in a region called the "boot heel," where New Mexico, Arizona, and Mexico meet. Malpai is derived from the Spanish word for badlands. The craggy mountains, grassy plains, and scrub-covered desert hills of this region are home to more than 20 threatened species. Nearly 400,000 ha (about 1 million acres) are part of the collaboration, including private property, state trust lands, national forest, and Bureau of Land Management acreage.

This pioneering collaboration began in 1993 in an effort to address threats to ranching. Thirty-five neighbors got together to discuss common problems. They agreed that excluding wildfire from the range was contributing to increasing brush and declining grass cover, resulting in the loss of watershed stability, wildlife habitat, and livestock forage.

Early on, community leaders approached The Nature Conservancy (TNC) for assistance. TNC, in turn, brought in ecologists familiar with the borderlands ecosystems to aid in organizing a science program. This input from scientists was crucial in giving the MBG efforts a systems-based approach to range management. This approach emphasized conservation of natural processes rather than just a focus on the management of single species or particular resources as is typical of many conservation programs. Since 1993, the MBG and collaborators have established more than 200 monitoring plots to assess ecosystem health.

Using range science as a starting point, the MBG's goal has evolved to a comprehensive natural resource management and rural development agenda. Their stated goal is, "To preserve and maintain the natural processes that create and protect a healthy, unfragmented landscape to support a diverse, flourishing community of human, plant, and animal life in the borderlands region." One of the key features of this plan is to restore fire as a management tool.

With a large, contiguous area under common management, it's now possible to set prescribed fires that have a significant effect on vegetation. Before formation of the MBG coalition, the patchwork of ownership and management in the area made large-scale operations all but impossible. A key to MBG success was purchase of the 120,000-hectare Gray Ranch by TNC. This large landholding in the heart of the Malpai borderlands makes it possible to carry out an innovative program called "grass banking." If a neighbor rancher has a bad season (perhaps due to prolonged drought), he can move his cattle onto Gray Ranch until his own ranchland is able to recover. The rancher grants a conservation easement of equal value over to the MBG that prohibits future subdivision.

The MBG isn't the only innovative initiative in ranching country. The Quivira Coalition, also based in New Mexico, brings together ranchers, conservationists, and land managers from throughout the West to foster scientifically guided ranch management and riparian restoration. TNC-owned Matador Ranch in Montana also has a grass-banking program. Perhaps the success of these programs will inspire similar cooperation rather than confrontation and litigation, not only on rangeland problems, but also on other contentious environmental issues.

The Malpai borderlands are in the boot heel of New Mexico, where it meets Arizona and Mexico. Ranchers in this area have joined with government agents and conservation organizations to restore fire, protect wildlife habitat, and regenerate grasslands.

allowed to visit to avoid introducing invasive species or disrupting native species. In the least restrictive categories (national forests and other natural resource management areas), on the other hand, there may be a high level of human use.

Venezuela claims to have the highest proportion of its land area protected (70 percent) of any country in the world. About half this land is designated as preserves for indigenous people or for sustainable resource harvesting. With little formal management, protection from poaching by hunters, loggers, and illegal gold hunters is minimal. Unfortunately, it's not uncommon in the developing world to have "paper parks" that exist only as a line drawn on a map with no budget for staff, management, or infrastructure. The United States, by contrast, has only about 15.8 percent of its land area in protected status, and less than one-third of that amount is in IUCN categories I or II (nature reserves, wilderness areas, national parks). The rest is in national forests or wildlife management zones that are designated for sustainable use. With hundreds of thousands of state and federal employees, billions of dollars in

Table 6.2	IUCN Categories of Protected Areas
Category	**Allowed Human Impact or Intervention**
1. Ecological reserves and wilderness areas	Little or none
2. National parks	Low
3. Natural monuments and archaeological sites	Low to medium
4. Habitat and wildlife management areas	Medium
5. Cultural or scenic landscapes, recreation areas	Medium to high

Source: Data from World Conservation Union, 1990.

Table 6.3	Protected Areas by Region		
Region	**Total Area Protected (km²)**	**Protected Percent**	**Number of Areas**
North America	4,459,305	16.2%	13,447
South America	1,955,420	19.3%	1,456
North Eurasia	1,816,987	7.7%	17,724
East Asia	1,764,648	14.0%	3,265
Eastern and Southern Africa	1,696,304	14.1%	4,060
Brazil	1,638,867	18.7%	1,287
Australia/New Zealand	1,511,992	16.9%	9,549
North Africa and Middle East	1,320,411	9.8%	1,325
Western and Central Africa	1,131,153	8.7%	2,604
Southeast Asia	867,186	9.6%	2,689
Europe	785,012	12.4%	46,194
South Asia	320,635	6.5%	1,216
Central America	158,193	22.5%	781
Antarctic	70,323	0.5%	122
Pacific	67,502	1.9%	430
Caribbean	66,210	8.2%	958
Total	**19,630,149**	**11.6%**	**107,107**

Source: World Commission on Protected Areas, 2007.

public funding, and a high level of public interest and visibility, U.S. public lands are generally well managed.

Brazil, with more than one-quarter of all the world's tropical rainforest, is especially important in biodiversity protection. Currently, Brazil has the largest total area in protected status of any country. More than 1.6 million km² or 18.7 percent of the nation's land—mostly in the Amazon basin—is in some protected status. In 2006, the northern Brazilian state of Para, in collaboration with Conservation International (CI) and other nongovernmental organizations, announced the establishment of nine new protected areas along the border with Suriname and Guyana. These new areas, about half of which will be strictly protected nature preserves, will link together several existing indigenous areas and nature preserves to create the largest tropical forest reserve in the world. More than 90 percent of the new 15 million ha (58,000 mi², or about the size of Illinois) Guyana Shield Corridor is in pristine natural state. CI president Russ Mittermeir says, "If any tropical rainforest on earth remains intact a century from now, it will be this portion of northern Amazonia." In contrast to this dramatic success, the Pantanal, the world's largest wetland/savanna complex, which lies in southern Brazil and is richer in some biodiversity categories than the Amazon, is almost entirely privately owned. There are efforts to set aside some of this important wetland, but so far, little is in protected status.

Some other countries with very large reserved areas include Greenland (with a 972,000 km² national park that covers most of the northern part of the island), and Saudi Arabia (with a 640,000 km² wildlife management area in its Empty Quarter). These areas are relatively easy to set aside, however, being mostly ice covered (Greenland) or desert (Saudi Arabia). Canada's Quttinirpaaq National Park on Ellesmere Island is an example of a preserve with high wilderness values but little biodiversity. Only 800 km (500 miles) from the North Pole, this remote park gets fewer than 100 human visitors per year during its brief, three-week summer season (fig. 6.19). With little evidence of human occupation, it has abundant solitude and stark beauty, but very little wildlife and almost no vegetation. By contrast, the Great Bear Rainforest management area described in the opening case study for this chapter has a rich diversity of both marine and terrestrial life, but the valuable timber, mineral, and wildlife resources in the area make protecting it expensive and controversial.

Figure 6.19 Canada's Quttinirpaaq National Park at the north end of Ellesmere Island has plenty of solitude and pristine landscapes, but little biodiversity.

Collectively, according to the World Commission on Protected Areas, Central America has 22.5 percent of its land area in some protected status (table 6.3). The Pacific region, at 1.9 percent in nature reserves, has both the lowest percentage and the lowest total area. With land scarce on small islands, it's hard to find space to set aside for nature sanctuaries. Some biomes are well represented in nature preserves, while others are relatively under-protected. Figure 6.20 shows a comparison between the percent of each major biome in protected status. Not surprisingly, there's an inverse relationship between the percentage converted to human use (and where people live) and the percentage protected. Temperate grasslands and savannas (such as the American Midwest) and Mediterranean woodlands and scrub (such as the French Riviera or the coast of southern California) are highly domesticated, and, therefore, expensive to set aside in large areas. Temperate conifer forests (think of Siberia, or Canada's vast expanse of boreal forest) are relatively uninhabited, and therefore easy to put into some protected category.

Not all preserves are preserved

Even parks and preserves designated with a high level of protection aren't always safe from exploitation or changes in political priorities. Serious problems threaten natural resources and environmental quality in many countries. In Greece, the Pindus National Park is threatened by plans to build a hydroelectric dam in the center of the park. Furthermore, excessive stock grazing and forestry exploitation in the peripheral zone are causing erosion and loss of wildlife habitat. In Columbia, dam building also threatens the Paramillo National Park. Ecuador's largest nature preserve, Yasuni National Park, which contains one of the world's most megadiverse regions of lowland Amazonian forest, has been opened to oil drilling, while miners and loggers in Peru have invaded portions of Huascaran National Park. In Palau, coral reefs identified as a potential biosphere reserve are damaged by dynamite fishing, while on some beaches in Indonesia, almost every egg laid by endangered sea turtles is taken by egg hunters. These are just a few of the many problems faced by parks and preserves around the world. Often countries with the most important biomes lack funds, trained personnel, and experience to manage the areas under their control.

Even in rich countries, such as the United States, some of the "crown jewels" of the National Park System suffer from overuse and degradation. Yellowstone and Grand Canyon National Parks, for example, have large budgets and are highly regulated, but are being "loved to death" because they are so popular. When the U.S. National Park Service was established in 1916, Stephen Mather, the first director, reasoned that he needed to make the parks comfortable and entertaining for tourists as a way of building public support. He created an extensive network

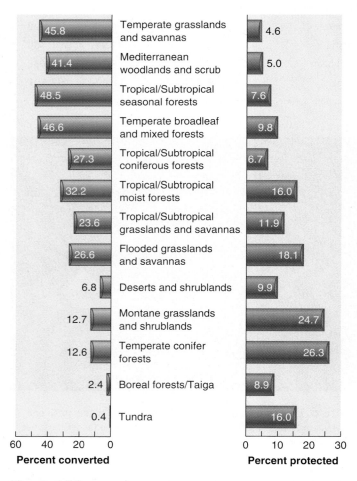

Figure 6.20 With few exceptions, the percent of each biome converted to human use is roughly inverse to the percent protected in parks and preserves.
Data from J. Hoekstra, et al., 2005.

Figure 6.21 Wild animals have always been one of the main attractions in national parks. Many people lose all common sense when interacting with big, dangerous animals. This is not a petting zoo.

of roads in the largest parks so that visitors could view famous sights from the windows of their automobiles, and he encouraged construction of grand lodges in which guests could stay in luxury.

His plan was successful; the National Park System is cherished and supported by many American citizens. But sometimes entertainment seems to have trumped over nature protection. Visitors were allowed—in some cases even encouraged—to feed wildlife. Bears lost their fear of humans and became dependent on an unhealthy diet of garbage and handouts (fig. 6.21). In Yellowstone and the Grand Teton National Parks, the elk herd was allowed to grow to 25,000 animals, or about twice the carrying capacity of the habitat. The excess population overgrazed the vegetation to the detriment of many smaller species and the biological community in general. As we discussed earlier in this chapter, 70 years of fire suppression resulted in changes of forest composition and fuel buildup that made huge fires all but inevitable. In Yosemite, you can stay in a world-class hotel, buy a pizza, play video games, do laundry, play golf or tennis, and shop for curios, but you may find it difficult to experience the solitude or enjoy the natural beauty extolled by John Muir as a prime reason for creating the park.

In many of the most famous parks, traffic congestion and crowds of people stress park resources and detract from the experience of unspoiled nature (fig. 6.22). Some parks, such as Yosemite, and Zion National Park, have banned private automobiles from the most congested areas. Visitors must park in remote lots and take clean, quiet electric or natural gas-burning buses to popular sites. Other parks are considering limits on the number of visitors admitted each day. How would you feel about a lottery system that might allow you to visit some famous parks only once in your lifetime, but to have an uncrowded, peaceful experience on your one allowed visit? Or would you prefer to be able to visit whenever you wish even if it means fighting crowds and congestion?

Originally, the great wilderness parks of Canada and the United States were distant from development and isolated from most human impacts. This has changed in many cases. Forests are clear-cut right up to some park boundaries. Mine drainage contaminates streams and groundwater. At least 13 U.S. National Monuments are open to oil and gas drilling, including Texas's Padre Island, the only breeding ground for endangered Kemps Ridley sea turtles. Even in the dry desert air of the Grand Canyon, where visibility was once up to 150 km, it's often too smoggy now to see across the canyon due to air pollution from power plants just outside the park. Snowmobiles and off-road vehicles (ORV) create pollution and noise and cause erosion while disrupting wildlife in many parks (fig. 6.23).

Chronically underfunded, the U.S. National Park System now has a maintenance backlog estimated to be at least $5 billion. Politicians from both major political parties vow to repair park facilities during election campaigns, but then find other uses for public funds once in office. Ironically, a recent study found that, on average, parks generate $4 in user fees for every $1 they receive in federal subsidies. In other words, they more than pay their own way, and should have a healthy surplus if they were allowed to retain all the money they generate.

In recent years, the U.S. National Park System has begun to emphasize nature protection and environmental education over entertainment. This new agenda is being adopted by other countries as well. The IUCN has developed a **world conservation strategy** for protecting natural resources that includes the following three objectives: (1) to maintain essential ecological processes and life-support systems (such as soil regeneration and protection, nutrient recycling, and water purification) on which human survival and development depend; (2) to preserve genetic diversity essential for breeding programs to improve cultivated plants and domestic animals; and (3) to ensure that any utilization of wild species and ecosystems is sustainable.

Figure 6.22 Thousands of people wait for an eruption of Old Faithful geyser in Yellowstone National Park. Can you find the ranger who's giving a geology lecture?

Figure 6.23 Off-road vehicles cause severe, long-lasting environmental damage when driven through wetlands.

Marine ecosystems need greater protection

As ocean fish stocks become increasingly depleted globally, biologists are calling for protected areas where marine organisms are sheltered from destructive harvest methods. As the opening case study for chapter 1 describes, limiting the amount and kind of fishing in marine reserves can quickly replenish fish stocks in surrounding areas. In a study of 100 marine refuges around the world, researchers found that, on average, the number of organisms inside no-take preserves was twice as high as surrounding areas where fishing was allowed. In addition, the biomass of organisms was three times as great and individuals animals were, on average, 30 percent larger inside the refuge compared to outside. Recent research has shown that closing reserves to fishing even for a few months can have beneficial results in restoring marine populations. The size necessary for a safe haven to protect flora and fauna depends on the species involved, but some marine biologists call on nations to protect at least 20 percent of their nearshore territory as marine refuges.

Coral reefs are among the most threatened marine ecosystems in the world. Remote sensing surveys show that worldwide living coral covers only about 285,000 km^2 (110,000 mi^2), or an area about the size of Nevada. This is less than half of previous estimates, and 90 percent of all reefs face threats from rising sea temperatures, destructive fishing methods, coral mining, sediment runoff, and other human disturbance. In many ways, coral reefs are the old-growth rainforests of the ocean (fig. 6.24).

Figure 6.24 Coral reefs are among both the most biologically rich and endangered ecosystems in the world. Marine reserves are being established in many places to preserve and protect these irreplaceable resources.

Biologically rich, these sensitive communities can take a century or more to recover from damage. If current trends continue, some researchers predict that in 50 years there will be no viable coral reefs anywhere in the world.

What can be done to reverse this trend? Some countries are establishing large marine reserves specifically to protect coral reefs. Australia has one of the largest marine reserves in the world in its 344,000 km^2 Great Barrier Reef (a large portion of which is open ocean). Kiribati's recently established Phoenix Island Protected Area is even larger. Altogether, however, aquatic reserves make up less than 10 percent of all the world's protected areas despite the fact that 70 percent of the earth's surface is water. A survey of marine biological resources identified the ten richest and most threatened "hot spots," including the Philippines, the Gulf of Guinea and Cape Verde Islands (off the west coast of Africa), Indonesia's Sunda Islands, the Mascarene Islands in the Indian Ocean, South Africa's coast, southern Japan and the east China Sea, the western Caribbean, and the Red Sea and Gulf of Aden. We urgently need more no-take preserves to protect marine resources.

Conservation and economic development can work together

Many of the most biologically rich communities in the world are in developing countries, especially in the tropics. These countries are the guardians of biological resources important to all of us. Unfortunately, where political and economic systems fail to provide residents with land, jobs, food, and other necessities of life, people do whatever necessary to meet their own needs. Immediate survival takes precedence over long-term environmental goals. Clearly the struggle to save species and ecosystems can't be divorced from the broader struggle to meet human needs.

People in some developing countries are beginning to realize that their biological resources may be their most valuable assets, and that their preservation is vital for sustainable development. **Ecotourism** (tourism that is ecologically and socially sustainable) can be more beneficial in many places over the long term than extractive industries, such as logging and mining. The What Can You Do? box (p. 145) suggests some ways to ensure that your vacations are ecologically responsible.

Native people can play important roles in nature protection

The American ideal of wilderness parks untouched by humans is unrealistic in many parts of the world. As we mentioned earlier, some biological communities are so fragile that human intrusions have to be strictly limited to protect delicate natural features or particularly sensitive wildlife. In many important biomes, however, indigenous people have been present for thousands of years and have a legitimate right to pursue traditional ways of life. Furthermore, many of the approximately 5,000 indigenous or native people that remain today possess ecological knowledge about their ancestral homelands that can be valuable in ecosystem management. According to author Alan Durning, "encoded in indigenous languages, customs, and practices may be as much understanding of nature as is stored in the libraries of modern science."

Some countries have adopted draconian policies to remove native people from parks (fig. 6.25). In South Africa's Kruger National Park, for example, heavily armed solders keep intruders out with orders to shoot to kill. This is very effective in protecting wildlife. In all fairness, before this policy was instituted there was a great deal of poaching by mercenaries armed with automatic weapons. But it also means that people who were forcibly displaced from the park could be killed on sight merely for returning to their former homes to collect firewood or to hunt for rabbits and other

Figure 6.25 Some parks take draconian measures to expel residents and prohibit trespassing. How can we reconcile the rights of local or indigenous people with the need to protect nature?

small game. Similarly, in 2006, thousands of peasant farmers on the edge of the vast Mau Forest in Kenya's Rift Valley were forced from their homes at gun point by police who claimed that the land needed to be cleared to protect the country's natural resources. Critics claimed that the forced removal amounted to "ethnic cleansing" and was based on tribal politics rather than nature protection.

Other countries recognize that finding ways to integrate local human needs with those of nature is essential for successful conservation. In 1986, UNESCO (United Nations Educational, Scientific, and Cultural Organization) initiated its **Man and Biosphere (MAB) program**, which encourages the designation of **biosphere reserves**, protected areas divided into zones with different purposes. Critical ecosystem functions and endangered wildlife are protected in a central core region, where limited scientific study is the only human access allowed. Ecotourism and research facilities are located in a relatively pristine buffer zone around the core, while sustainable resource harvesting and permanent habitation are allowed in multiple-use peripheral regions (fig. 6.26).

While not yet given a formal MAB designation, the Great Bear Rainforest described in the opening case study for this chapter is organized along this general plan. A well established example of a biosphere reserve is Mexico's 545,000 ha (2,100 mi²) Sian Ka'an Reserve on the Tulum Coast of the Yucatán. The core area includes 528,000 ha (1.3 million acres) of coral reefs, bays, wetlands, and lowland tropical forest. More than 335 bird species have been observed within the reserve, along with endangered manatees, five types of jungle cats, spider and howler monkeys, and four species of increasingly rare sea turtles. Approximately 25,000 people (about the same number who live in the Great Bear Rainforest)

Figure 6.26 A model biosphere reserve. Traditional parks and wildlife refuges have well-defined boundaries to keep wildlife in and people out. Biosphere reserves, by contrast, recognize the need for people to have access to resources. Critical ecosystem is preserved in the core. Research and tourism are allowed in the buffer zone, while sustainable resource harvesting and permanent habitations are situated in the multiple-use area around the perimeter.

What Can You Do?

Being a Responsible Ecotourist

1. *Pretrip preparation.* Learn about the history, geography, ecology, and culture of the area you will visit. Understand the do's and don'ts that will keep you from violating local customs and sensibilities.

2. *Environmental impact.* Stay on designated trails and camp in established sites, if available. Take only photographs and memories and leave only goodwill wherever you go.

3. *Resource impact.* Minimize your use of scarce fuels, food, and water resources. Do you know where your wastes and garbage go?

4. *Cultural impact.* Respect the privacy and dignity of those you meet and try to understand how you would feel in their place. Don't take photos without asking first. Be considerate of religious and cultural sites and practices. Be as aware of cultural pollution as you are of environmental pollution.

5. *Wildlife impact.* Don't harass wildlife or disturb plant life. Modern cameras make it possible to get good photos from a respectful, safe distance. Don't buy ivory, tortoise shell, animal skins, feathers, or other products taken from endangered species.

6. *Environmental benefit.* Is your trip strictly for pleasure, or will it contribute to protecting the local environment? Can you combine ecotourism with work on cleanup campaigns or delivery of educational materials or equipment to local schools or nature clubs?

7. *Advocacy and education.* Get involved in letter writing, lobbying, or educational campaigns to help protect the lands and cultures you have visited. Give talks at schools or to local clubs after you get home to inform your friends and neighbors about what you have learned.

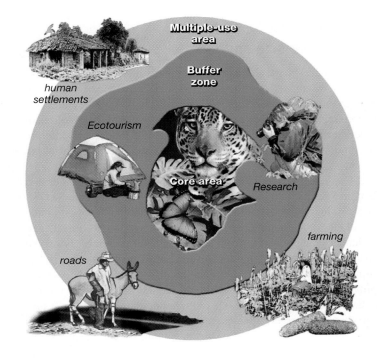

Figure 6.27 Corridors serve as routes of migration, linking isolated populations of plants and animals in scattered nature preserves. Although individual preserves may be too small to sustain viable populations, connecting them through river valleys and coastal corridors can facilitate interbreeding and provide an escape route if local conditions become unfavorable.

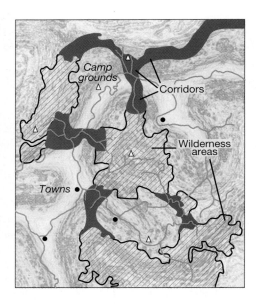

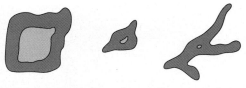

(a) Patch size and shape: these influence core/edge ratio

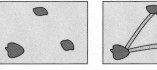

(b) Number of habitat patches, and patch proximity and connectivity

(c) Contrast between habitat patches and surrounding landscape

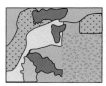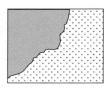

(d) Complexity of the landscape mosaic

Figure 6.28 Some spatial variables examined in landscape ecology. Size, arrangement, context, and other factors often vary simultaneously. Together, they can strongly influence the ability of a wildlife preserve to support rare species.

reside in communities and the countryside around Sian Ka'an. In addition to tourism, the economic base of the area includes lobster fishing, small-scale farming, and coconut cultivation.

The Amigos de Sian Ka'an, a local community organization, played a central role in establishing the reserve and is working to protect natural resources while also improving living standards for local people. New intensive farming techniques and sustainable harvesting of forest products enable residents to make a living without harming their ecological base. Better lobster harvesting techniques developed at the reserve have improved the catch without depleting native stocks. Local people now see the reserve as a benefit rather than an imposition from the outside. Similar success stories from many parts of the world show how we can support local people and recognize indigenous rights while still protecting important environmental features.

Species survival can depend on preserve size and shape

Many natural parks and preserves are increasingly isolated, remnant fragments of ecosystems that once extended over large areas. As park ecosystems are shrinking, however, they are also becoming more and more important for maintaining biological diversity. Principles of landscape design and landscape structure become important in managing and restoring these shrinking islands of habitat.

For years, conservation biologists have disputed whether it is better to have a single *large* or *several small* reserves (the SLOSS debate). Ideally, a reserve should be large enough to support viable populations of endangered species, keep ecosystems intact, and isolate critical core areas from damaging external forces. For some species with small territories, several small, isolated refuges can support viables populations, and having several small reserves provides insurance against a disease, habitat destruction, or other calamities that might wipe out a single population. But small preserves can't support species such as elephants or tigers, which need large amounts of space. Given human needs and pressures,

however, big preserves aren't always possible. One proposed solution has been to create **corridors** of natural habitat that can connect to smaller habitat areas (fig. 6.27). Corridors could effectively create a large preserve from several small ones. Corridors could also allow populations to maintain genetic diversity or expand into new breeding territory. The effectiveness of corridors probably depends on how long and wide they are, and on how readily a species will use them.

One of the reasons large preserves are considered better than small preserves is that they have more **core habitat**, areas deep in the interior of a habitat area, and that core habitat has better conditions for specialized species than do edges. **Edge effects** is a term generally used to describe habitat edges: for example, a forest edge is usually more open, bright, and windy than a forest interior, and temperatures and humidity are more varied. For a grassland, on the other hand, edges may be wooded, with more shade, and perhaps more predators, than in the core of the grassland area. As human disturbance fragments an ecosystem, habitat is broken into increasingly isolated islands, with less core and more edge. Small, isolated fragments of habitat often support fewer species, especially fewer rare species, than do extensive, uninterrupted ecosystems. The size and isolation of a wildlife preserve, then, may be critical to the survival of rare species.

Landscape ecology is a science that examines the relationship between these spatial patterns and ecological processes, such as species movement or survival. Landscape ecologists measure variables like habitat size, shape, and the relative amount of core habitat and edge (fig. 6.28). Landscape ecologists also examine the kind of land cover that surrounds habitat areas—is it a city, farm fields, or clear-cut forest? And they include utilitarian landscapes, such as farm fields, in their analysis, as well as pristine wilderness. By quantifying factors such as habitat shape and landscape complexity, and by monitoring the number of species or the size of popuations, landscape ecologists try to guide more effective design of nature preserves and parks.

A dramatic experiment in reserve size, shape, and isolation is being carried out in the Brazilian rainforest. In a project funded by the World Wildlife Fund and the Smithsonian Institution, loggers left 23 test sites when they clear-cut a forest. Test sites range from 1 ha (2.47 acres) to 10,000 ha. Clear-cuts surround some, and newly created pasture surrounds others (fig. 6.29); others remain connected to the surrounding forest. Selected species are regularly inventoried to monitor their survival after disturbance. As expected, some species disappear very quickly, especially from small areas. Sun-loving species flourish in the newly created forest edges, but deep-forest, shade-loving species disappear, particularly when the size or shape of a reserve reduces availability of core habitat. This experiment demonstrates the importance of maintaining core habitat in preserves.

Conclusion

Forests and grasslands cover nearly 60 percent of global land area. The vast majority of humans live in these biomes, and we obtain many valuable materials from them. And yet, these biomes also are the source of much of the world's biodiversity on which we depend for life-supporting ecological services. How we can live sustainably on our natural resources while also preserving enough nature so those resources can be replenished represents one of the most important questions in environmental science.

There is some good news in our search for a balance between exploitation and preservation. Although deforestation and land degradation are continuing at unacceptable rates—particularly

Figure 6.29 How small can a nature preserve be? In an ambitious research project, scientists in the Brazilian rainforest are carefully tracking wildlife in plots of various sizes, either connected to existing forests or surrounded by clear-cuts. As you might expect, the largest and most highly specialized species are the first to disappear.

in some developing countries—many countries are more thickly forested now than they were two centuries ago. Protection of the Great Bear Rainforest in Canada and Australia's great barrier reef shows that we can choose to protect some biodiverse areas in spite of forces that want to exploit them. Overall, nearly 12 percent of the earth's land area is now in some sort of protected status. While the level of protection in these preserves varies, the rapid recent increase in number and area in protected status exceeds the goals of the United Nations Millennium Project.

While we haven't settled the debate between focusing on individual endangered species versus setting aside representative samples of habitat, pursuing both strategies seems to be working. Protecting charismatic umbrella organisms, such as the "spirit bears" of the Great Bear Rainforest can result in preservation of innumerable unseen species. At the same time, protecting whole landscapes for aesthetic or recreational purposes can also achieve the same end.

Practice Quiz

1. What do we mean by *closed-canopy* forest and *old-growth* forest?

2. What land use is responsible for most forest losses in Africa? In Latin America? In Asia? (fig. 6.7).

3. What is a *debt-for-nature swap*?

4. Why is fire suppression a controversial strategy? Why are forest thinning and salvage logging controversial?

5. What portion of the United States' public rangelands are in poor or very poor condition due to overgrazing? Why do some groups say grazing fees amount to a "hidden subsidy"?

6. What is *rotational grazing*, and how does it mimic natural processes?

7. How do the size and design of nature preserves influence their effectiveness? What do landscape ecologists mean by *interior habitat* and *edge effects*?

8. What percentage of the earth's land area has some sort of protected status? How has the amount of protected areas changed globally (fig. 6.18)?

9. What is *ecotourism*, and why is it important?

10. What is a *biosphere reserve*, and how does it differ from a wilderness area or wildlife preserve?

Critical Thinking and Discussion Questions

Apply the principles you have learned in this chapter to discuss these questions with other students.

1. Conservationists argue that watershed protection and other ecological functions of forests are more economically valuable than timber. Timber companies argue that continued production supports stable jobs and local economies. If you were a judge attempting to decide which group was right, what evidence would you need on both sides? How would you gather this evidence?

2. Divide your class into a ranching group, a conservation group, and a suburban home-builders group, and debate the merits of subsidized grazing in the American West. What is the best use of the land? What landscapes are most desirable? Why? How do you propose to maintain these landscapes?

3. Calculating forest area and forest losses is complicated by the difficulty of defining exactly what constitutes a forest. Outline a definition for what counts as forest in your area, in terms of size, density, height, or other characteristics. Compare your definition to those of your colleagues. Is it easy to agree? Would your definition change if you lived in a different region?

4. There is considerable uncertainty about the extent of degradation on grazing lands. Suppose you were a range management scientist, and it was your job to evaluate degradation for the state of Montana. What data would you need? With an infinite budget, how would you gather the data you need? How would you proceed if you had a very small budget?

5. Why do you suppose dry tropical forest and tundra are well represented in protected areas, while grasslands and wetlands are protected relatively rarely? Consider social, cultural, geographic, and economic reasons in your answer.

6. Oil and gas companies want to drill in several parks, monuments, and wildlife refuges. Do you think this should be allowed? Why or why not? Under what conditions would drilling be allowable?

 **Data Analysis:** Detecting Edge Effects

Edge effects are a fundamental consideration in nature preserves. We usually expect to find dramatic edge effects in pristine habitat with many specialized species. But you may be able to find interior-edge differences on your own college campus, or in a park or other unbuilt area near you. Here are three testable questions you can examine using your own local patch of habitat: (1) Can an edge effect be detected or not? (2) Which species will indicate the difference between edge and interior conditions? (3) At what distance can you detect a difference between edge and interior conditions? To answer these questions, you can form a hypothesis and test it as follows:

1. Choose a study area. Find a distinct patch of habitat, such as woods, unmowed grass, or marshy but walkable wetland, about 50 m wide or larger. With other students, list some local, familiar plant species that you would expect to find in your study area. If possible, make this list on a visit to your site.

2. Form a hypothesis. Examine your list, and predict which species will occur most on edges, and (if you can) which you think will occur more in the interior. Form a hypothesis, or a testable statement, based on one of the three questions above. For example, "I will be able to detect an edge effect in my patch," or "I think an edge effect will be indicated by these species: _____," or "I think changes in species abundance will indicate an edge-interior change at _____ m from the edge of the patch."

3. Gather data. Get a meter tape and lay it along the ground from the edge of your habitat patch toward the interior. (You can also use a string and pace distances: treat one pace as a meter.) This line is your *transect*. At the edge end of the tape (or string), count the number of different species you can see within 1 m^2 on either side of your line. Repeat this count at each 5 m interval, up to 25 m. Thus you will create a list of species at 0, 5, 10, 15, 20, and 25 m in from the edge.

4. Examine your lists, and determine whether your hypothesis was correct. Can you see a change in species presence/absence from 0 to 25 m? Were you correct in your prediction of which species disappeared with distance from the edge? If you can identify an edge effect, at what distance did it occur?

5. Consider ways that your test could be improved. Should you take more frequent samples? Larger samples? Should you compare abundance rather than presence/absence? Might you have gotten different results if you had chosen a different study site? Or if your class had examined many sites and averaged the results? How else might you modify your test to improve the quality of your results?

For Additional Help in Studying This Chapter, please visit our website at www.mhhe.com/cunningham5e. You will find practice quizzes, key terms, answers to end of chapter questions, additional case studies, an extensive reading list, and Google Earth™ mapping quizzes.

Enormous farms have been carved out of Brazil's Cerrado (savanna), which once was the most biodiverse grassland and open tropical forest complex in the world.

Food and Agriculture

Learning Outcomes

After studying this chapter, you should be able to answer the following questions:

- How have global food production and population changed?
- How many people are chronically hungry, and why does hunger persist in a world of surpluses?
- What are some health risks of undernourishment, poor diet, and overeating?
- What are our primary food crops?
- Describe five components of soil.
- What was the green revolution?
- What are GMOs, and what traits are most commonly introduced with GMOs?
- Describe some environmental costs of farming, and ways we can minimize these costs.

We can't solve problems by using the same kind of thinking we used when we created them.

–Albert Einstein

Farming the Cerrado

A soybean boom is sweeping across South America. Inexpensive land, availability of new crop varieties, and government policies that favor agricultural expansion have made South America the fastest growing agricultural area in the world. The center of this rapid expansion is the Cerrado, a huge area of grassland and tropical forest stretching from Bolivia and Paraguay across the center of Brazil almost to the Atlantic Ocean (fig. 7.1). Biologically, this rolling expanse of grasslands and tropical woodland is the richest savanna in the world, with at least 130,000 different plant and animal species, many of which are threatened by agricultural expansion.

Until recently, the Cerrado, which is roughly equal in size to the American Midwest, was thought to be unsuitable for cultivation. Its red iron-rich soils are highly acidic and poor in essential plant nutrients. Furthermore, the warm, humid climate harbors many destructive pests and pathogens. For hundreds of years, the Cerrado was primarily cattle country with many poor-quality pastures producing low livestock yields.

In the past few decades, however, Brazilian farmers have learned that modest applications of lime and phosphorus can quadruple yields of soybeans, maize, cotton, and other valuable crops. Researchers have developed more than 40 varieties of soybeans—mostly through conventional breeding, but some created with molecular techniques—specially adapted for the soils and climate of the Cerrado. Until about 30 years ago, soybeans were a relatively minor crop in Brazil. Since 1975, however, the area planted with soy has doubled about every four years, reaching more than 25 million ha (60 million acres) in 2005. Although that's a large area, it represents only one-eighth of the Cerrado, more than half of which is still occupied by pasture.

Brazil is now the world's top soy exporter, shipping some 27 million metric tons per year, or about 10 percent more than the United States. With two crops per year, cheap land, low labor costs, favorable tax rates, and yields per hectare equal to those in the American Midwest, Brazilian farmers can produce soybeans for less than half the cost in America. Agricultural economists predict that, by 2020, the global soy crop will double from the current 160 million metric tons per year, and that South America could be responsible for most of that growth. In addition to soy, Brazil now leads the world in beef, corn (maize), oranges, and coffee exports. This dramatic increase in South American agriculture helps answer the question of how the world may feed a growing human population.

A major factor in Brazil's current soy expansion is rising income in China. With more money to spend, the Chinese are consuming more soy, both directly as tofu and other soy products, and indirectly as animal feed. Between 2002 and 2004, China's soy imports doubled to more than 21 million metric tons, or about one-third of total global soy shipments. The outbreak of mad cow disease (or BSE, see chapter 8) in Europe, Canada, and Japan also fueled increased worldwide demand for soybeans. Rather than feed livestock meat-processing wastes that may convey mad cow disease, producers are turning to protein- and lipid-rich soy meal. With 175 million free-range, grass-fed (and presumably BSE-free) cattle, Brazil has become the world's largest beef exporter.

Increasing demand for both soybeans and beef create land conflicts in Brazil. The pressure for more cropland and pasture is a leading cause of deforestation and habitat loss, most of which is occurring in the "arc of destruction" between the Cerrado and the Amazon. Small family farms are being gobbled up, and farm workers, displaced by mechanization, often migrate either to the big cities or to frontier forest areas. Increasing conflicts between poor farmers and big landowners have led to violent confrontations. The Landless Workers Movement claims that 1,237 rural workers died in Brazil between 1985 and 2000 as a result of assassinations and clashes over land rights. In 2005, a 74-year-old Catholic nun, Sister Dorothy Stang, was shot by gunmen hired by ranchers who resented her advocacy for native people, workers, and environmental protection. Over the past 20 years, Brazil claims to have resettled 600,000 families from the Cerrado. Still, tens of thousands of landless farm workers and displaced families live in unauthorized squatter camps and shantytowns across the country awaiting relocation.

As you can see, rapid growth of beef and soy production in Brazil have both positive and negative aspects. On one hand, more high-quality food is now available to feed the world. The 2 million km² of the Cerrado

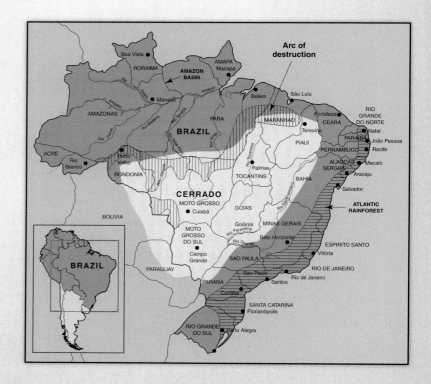

Figure 7.1 Brazil's Cerrado, 2 million ha of savanna (grassland) and open woodland is the site of the world's fastest growing soybean production. Cattle ranchers and agricultural workers, displaced by mechanized crop production, are moving northward into the "arc of destruction" at the edge of the Amazon rainforest, where the continent's highest rate of forest clearing is occurring.

represents one of the world's last opportunities to open a large area of new, highly productive cropland. On the other hand, the rapid expansion and mechanization of agriculture in Brazil is destroying biodiversity and creating social conflicts as people move into formerly pristine lands. The issues raised in this case study illustrate many of the major themes in this chapter. Will there be enough food for everyone in the world? What will be the environmental and social consequences of producing the nutrition we need? In this chapter, we'll look at world food supplies, agricultural inputs, and sustainable approaches that can help solve some of the difficult questions we face.

7.1 Global Trends in Food and Nutrition

The story of agricultural production in Brazil is one of the most dramatic cases of environmental change today. Brazil's Cerrado is a long way from Iowa and Illinois, in the heart of the U.S. corn belt, but if you live in an agricultural state you might recognize some of the changes in the soy fields of South America. Food production has been transformed from small-scale, diversified, family operations to expansive farms of thousands of hectares, growing one or two genetically modified crops, with abundant inputs of fuel and fertilizer, for a competitive global market. These changes have dramatically increased production, lowered food prices, and provided affordable meat protein in developing countries from Brazil to China. Food production has increased so dramatically that we now use edible corn and sugar to run our cars (chapter 12). According to the International Monetary Fund, 2005 global food costs (in inflation-adjusted dollars) were the lowest ever recorded, less than one-quarter of the cost in the mid-1970s. In the United States and Europe, overproduction has driven prices low enough that we pay farmers billions of dollars each year to take land out of production.

Revolutions in production have also profoundly altered our environment and our diets. In this chapter we'll examine these changes, as well as the ways farmers have managed to feed more and more of the world's growing population. We'll also consider some of the strategies needed for long-term sustainability in food production.

Food security is unevenly distributed

Four decades ago, hunger was one of the world's most prominent, persistent problems. In 1960, nearly 60 percent of people in developing countries were chronically undernourished, and the world's population was increasing by more than 2 percent every year. Today, some conditions have changed dramatically; others have changed very little. The world's population has risen from 3 billion to over 6.5 billion, but food production has increased even faster. While the average population growth in the past 45 years has been 1.7 percent per year, food production has increased by an average of 2.2 percent per year. Food availability has increased in most countries to well over 2,200 kilocalories, the amount generally considered necessary for a healthy and productive life (fig. 7.2a). Protein intake has also increased in most countries,

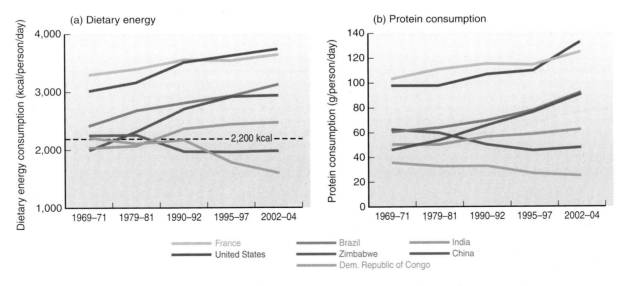

Figure 7.2 Changes in dietary energy (kcal) and protein consumption in selected countries.
Source: Data from Food and Agriculture Organization (FAO), 2008.

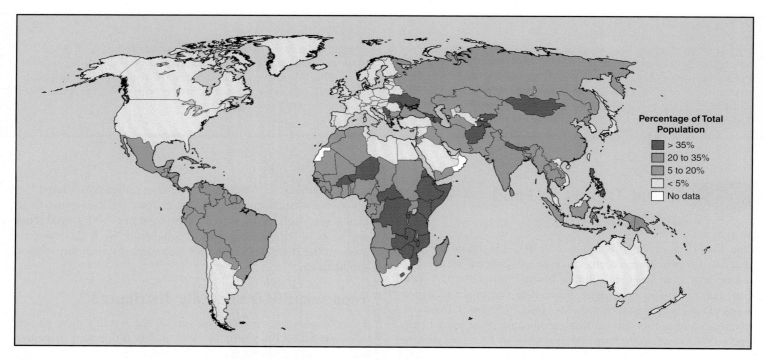

Figure 7.3 Hunger around the world. In 2007, the United Nations reported that 854 million people—815 million of them in developing countries—suffered from chronic hunger and malnutrition. Africa has the largest number of countries with food shortages.

Source: Hunger map from Food and Agriculture Organization of the United Nations website. Used by permission.

including China and India, the two most populous countries (fig. 7.2b). Less than 20 percent of people in developing countries now face chronic food shortages.

But hunger is still with us. An estimated 854 million people—almost one in every eight people on earth—suffer chronic hunger (fig. 7.3). This number is up slightly from a few years ago. However, because of population growth, the *percentage* of malnourished people is still falling (fig. 7.4).

About 95 percent of hungry people are in developing countries. Hunger is especially serious in sub-Saharan Africa, a region plagued by political instability (figs. 7.2, 7.3). Increasingly, we are

coming to understand that **food security**, or the ability to obtain sufficient, healthy food on a day-to-day basis, is a combined problem of economic, environmental, and social conditions. Even in wealthy countries such as the United States, millions lack a sufficient, healthy diet. Poverty, job losses, lack of social services, and other factors lead to persistent hunger—and even more to persistent poor diets—despite the fact that we have more, cheaper food (in terms of the work needed to acquire it) than almost any society in history.

Food security is important at multiple scales. In the poorest countries, entire national economies can suffer from a severe

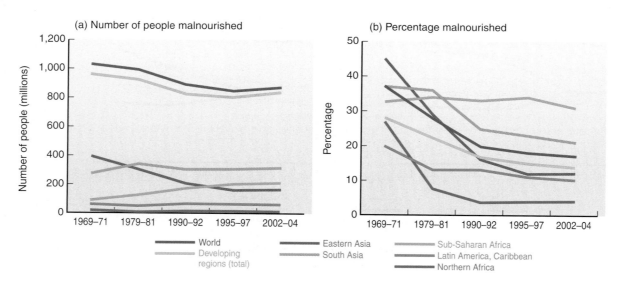

Figure 7.4 Changes in numbers and rates of malnourishment, by region.

Source: Data from UN Food and Agriculture Organization, 2008.

Figure 7.5 Children wait for their daily ration of porridge at a feeding station in Somalia. When people are driven from their homes by hunger or war, social systems collapse, diseases spread rapidly, and the situation quickly becomes desperate.

drought, flood, or insect outbreak. Individual villages also suffer from a lack of food security. In extremely poor areas, a single bad crop can devastate a family or a village, and local economies can collapse if farmers cannot produce a crop to eat and sell. Even within families there can be unequal food security. Males often get both the largest share and the most nutritious food, while women and children—who need food most—all too often get the poorest diet. At least 6 million children under 5 years old die every year of diseases exacerbated by hunger and malnutrition. Providing a healthy diet might eliminate as much as 60 percent of all premature deaths worldwide.

Hungry people can't work their way out of poverty. Nobel Prize-winning economist Robert Fogel estimates that in 1790, 20 percent of the population of England and France were effectively excluded from the labor force because they were too weak and hungry to work. Fogel calculates that improved nutrition can account for about half of all European economic growth during the nineteenth century. This analysis suggests that reducing hunger in today's poor countries could yield more than $120 billion (U.S.) in economic growth, by producing a healthier, longer-lived, and more productive work force.

Famines usually have political and social roots

Globally, widespread hunger arises when political instability, war, and conflict displace populations, removing villagers from their farms or making farming too dangerous to carry on. Economic disparities may drive peasants from the land, when landowners find it more profitable to raise commercial soybeans, than to collect rent from peasant farmers. Landlessness is a desperate problem for peasants in many countries, including Brazil (opening case study). Displaced farmers often have no choice but to migrate to the already overcrowded slums of major cities when they lose their land (chapter 14).

Famines are large-scale food shortages, with widespread starvation, social disruption, and economic chaos. Starving people are forced to eat their seed grain and slaughter breeding livestock in a desperate attempt to keep themselves and their families alive.

Even when better conditions return, they have often sacrificed their productive capacity and will take a long time to recover. Famines often trigger mass migrations to relief camps, where people survive but cannot maintain a healthy and productive life (fig. 7.5).

Nobel Prize-winning economist Amartya K. Sen, of Harvard, has shown that, while natural disasters often precipitate famines, farmers have almost always managed to survive these events if they aren't thwarted by inept or corrupt governments or greedy elites. Professor Sen points out that armed conflict and political oppression are almost always at the root of famine. No democratic country with a relatively free press, he says, has ever had a major famine.

China's recovery from the famines of the 1960s is a dramatic example of this relationship (fig. 7.2). Misguided policies from the central government destabilized farming economies all across China. Two years of bad crops in 1959–60 precipitated famines that may have killed 30 million people. In recent years, political and economic changes have transformed access to food, even while the population has doubled, from 650 million in 1960 to 1.3 billion in 2007. China now consumes almost twice as much meat (pork, chicken, and beef) as the United States—increasingly this meat comes from livestock fed on soybeans bought from Brazil.

7.2 Eating Right to Stay Healthy

A good diet is essential to keep you healthy. You need a balance of foods to provide the right nutrients, as well as enough calories for a productive and energetic lifestyle. The United Nations Food and Agriculture Organization (FAO) estimates that nearly 3 billion people (almost half the world's population) suffer from vitamin, mineral, or protein deficiencies. These shortages result in devastating illnesses and death, as well as reduced mental capacity, developmental abnormalities, and stunted growth.

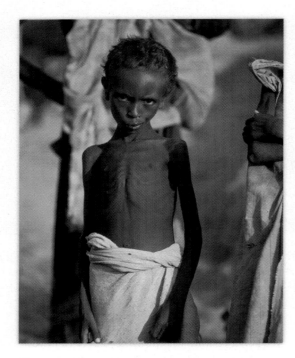

Figure 7.6 Marasmus is caused by combined energy (calorie) and protein deficiencies. Children with marasmus have the wizened look and dry, flaky skin of an old person.

A healthy diet includes the right nutrients

Malnourishment is a general term for nutritional imbalances caused by a lack of specific nutrients. In conditions of extreme food shortages, a lack of protein in young children can cause kwashiorkor, which is characterized by a bloated belly and discolored hair and skin. *Kwashiorkor* is a West African word meaning "displaced child." (A young child is displaced—and deprived of nutritious breast milk—when a new baby is born.) Marasmus (from Greek, "to waste away") is another severe condition in children who lack both protein and calories. A child suffering from severe marasmus is generally thin and shriveled, like a tiny, very old, starving person (fig. 7.6). Children with these deficiencies have low resistance to disease and infections, and

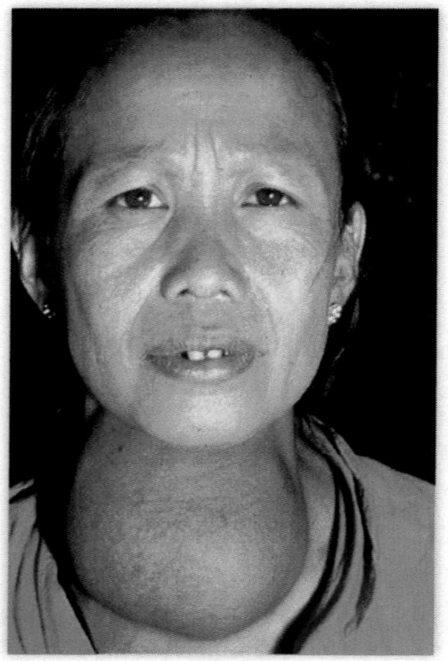

Figure 7.7 Goiter, a swelling of the thyroid gland at the base of the neck, is often caused by an iodine deficiency. It is a common problem in many parts of the world, particularly where soil iodine is low and seafood is unavailable.

they may suffer permanent debilities in mental, as well as physical, development.

Deficiencies in vitamin A, folic acid, and iodine are more widespread problems. Both are found in vegetables, especially dark green leafy vegetables. Deficiencies in folic acid have been linked to neurological problems in babies. Effects of vitamin A shortages cause an estimated 350,000 people to go blind every year. Dr. Alfred Sommer, an ophthalmologist from Johns Hopkins University, has shown that giving children just two cents' worth of vitamin A twice a year could prevent almost all cases of childhood blindness and premature death associated with shortages of vitamin A. Vitamin supplements also reduced maternal mortality by nearly 40 percent in one study in Nepal.

Iodine deficiencies can cause goiter (fig. 7.7), a swelling of the thyroid gland. Iodine is essential for synthesis of thyroxin, an endocrine hormone that regulates metabolism and brain development, among other things. The FAO estimates that 740 million people, mostly in Southeast Asia, suffer from iodine deficiency, including 177 million children whose development and growth have been stunted. Developed countries have largely eliminated this problem by adding a few pennies' worth of iodine to our salt.

Starchy foods, such as maize, polished rice, and manioc (tapioca), form the bulk of the diet for many poor people, but these foods are low in several essential vitamins and minerals. One celebrated effort to deliver crucial nutrients has been through genetic engineering of common foods, such as "golden rice," developed by Monsanto to include a gene for producing vitamin A. This strategy has shown promise, but it also has critics, who argue that genetically modified rice is too expensive for poor peasants. In addition, the herbicides needed to grow the golden rice kill the greens that villagers rely on to provide their essential nutrients. (See the related story "Golden Rice" at www.mhhe.com/cunningham5e.)

The best way to make sure you're healthy is to eat lots of vegetables and grains, moderate amounts of eggs and dairy products, and sparing amounts of meat, oils, and processed foods. Modest amounts of fats provide lipids (chapter 2) that are essential for healthy skin, cell function, and metabolism. But your body is not designed to process excessive amounts of fats (or sugars). Unsaturated plant-based oils, such as olive oil, are recommended by dietitians; trans fats (found in hydrogenated margarine) are not recommended. A food pyramid developed by Harvard dieticians provides a useful guide for healthy eating, together with a solid base of regular exercise (fig. 7.8).

Overeating is a growing world problem

While increasing world food supplies, and low prices, are good news for most of us, the downside is increasing overweight and obese populations. In the U.S., and increasingly in Europe, China, and developing countries, highly processed foods rich in sugars and fats have become a large part of our diet. Some 64 percent of adult Americans are overweight, up from 40 percent only a decade ago. About one-third of us are seriously overweight, or **obese**—generally considered to mean more than 20 percent over the ideal weight for a person's height and sex.

Being overweight substantially increases your risk of hypertension, diabetes, heart attacks, stroke, gallbladder disease, osteo-

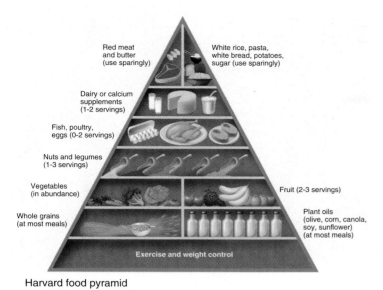

Harvard food pyramid

Figure 7.8 The Harvard food pyramid emphasizes fruits, vegetables, and whole grains as the basis of a healthy diet. Red meat, white rice, pasta, and potatoes should be consumed sparingly.
Source: Data from Willett and Stampfer, 2002.

Table 7.1	Some Important Food Sources
Crop	**2004 Yield (million metric tons)**
Wheat	620
Rice (paddy)	610
Maize (corn)	643
Potatoes	310
Coarse grains*	1,013
Soybeans	221
Cassava and sweet potatoes	449
Sugar (cane and beet)	144
Pulses (beans, peas)	55
Oil seeds	375
Vegetables and fruits	1,206
Meat and milk	870
Fish and seafood	140

*Barley, oats, sorghum, rye, millet.
Source: Data from Food and Agriculture Organization (FAO), 2005.

arthritis, respiratory problems, and some cancers. Every year, about 400,000 people in the United States die from illnesses related to obesity. This number is approaching the number related to smoking (435,000 annually). Paradoxically, food insecurity and poverty can contribute to obesity. In one study, more than half the women who reported not having enough to eat were overweight, compared with one-third of the food-secure women. Lack of good quality food may contribute to a craving for carbohydrates in people with a poor diet. A lack of time for cooking, limited access to healthy food choices, and ready availability of fast-food snacks and calorie-laden soft drinks, also lead to dangerous dietary imbalances for many people.

This trend isn't limited to richer countries. Obesity is spreading around the world (fig. 7.9). For the first time in history, there are probably more overweight people (more than 1 billion) than under-

weight. Diseases once thought to afflict only wealthy nations, such as heart attack, stroke, and diabetes, are now becoming the most prevalent causes of death and disability everywhere (chapter 8).

7.3 The Foods We Eat

Of the thousands of edible plants and animals in the world, only a few provide almost all our food. About a dozen types of grasses, three root crops, twenty or so fruits and vegetables, six mammals, two domestic fowl, and a few fish species make up almost all the food we eat (table 7.1). Two grasses, wheat and rice, are especially important because they are the staple foods for most of the 5 billion people in developing countries.

In the United States, corn (another grass, also known as maize) and soybeans have become our primary staples. We rarely eat either corn or soybeans directly, but corn provides the corn sweeteners, corn oil, and the livestock feed for producing our beef, chicken, and pork, as well as industrial starches and many synthetic vitamins. Soybeans are also fed to livestock, and soy provides protein and oils for processed foods. Because we have developed so many uses for corn, it now accounts for nearly two-thirds of our bulk commodity crops (corn, soy, wheat, rice). Together, corn and soy make up about 85 percent of these crops, with an annual production of 268 million tons of corn and 88 million tons of soy (fig. 7.10).

A boom in meat production brings costs and benefits

Because of dramatic increases in corn and soy production, meat consumption has grown in both developed and developing countries. In developing countries, meat consumption has risen from just 10 kg per person per year in the 1960s to over 26 kg today (fig. 7.11). In

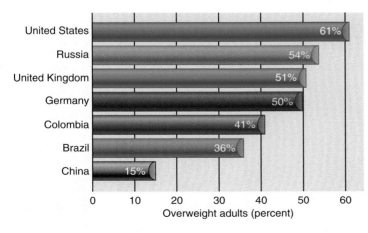

Figure 7.9 While nearly a billion people are chronically undernourished, people in wealthier countries are at risk from eating too much.
Source: Data from Worldwatch Institute, 2001.

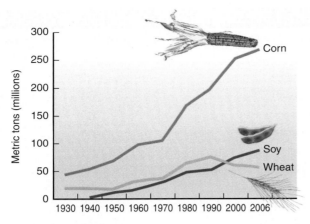

Figure 7.10 United States production of our three dominant crops, corn, soybeans, and wheat.
Source: Data from USDA and UN FAO, 2008.

Figure 7.11 Meat and dairy consumption has quadrupled in the past 40 years, and China represents about 40 percent of that increased demand.

the United States, our meat consumption has risen from 90 kg to 136 kg per person per year in the same interval. Meat is a concentrated, high-value source of protein, iron, fats, and other nutrients that give us the energy to lead productive lives. Dairy products are also a key protein source: globally we consume more than twice as much dairy as meat. But dairy production per capita has declined slightly while global meat production has doubled in the past 45 years.

Meat is a good indicator of wealth because it is expensive to produce, in terms of the resources needed to grow an animal (fig. 7.12). As discussed in chapter 2, herbivores use most of the energy they consume in growing muscle and bone, moving around, staying warm, and metabolizing (digesting) food. Only a little food energy is stored for consumption by carnivores, at the next level of the food pyramid. It takes over 8 kilos of grain fed to a beef cow to produce a single kilo of meat. (Actually, we raise mainly steers, or neutered males, for beef.) Pigs, being smaller, are more efficient. Just three pounds of pig feed are needed to produce a kilo of pork. Chickens and herbivorous fish (such as catfish) are still more efficient. Globally, some 660 million metric tons of cereals are used as livestock feed each year. This represents just over a third of the world cereal use. As figure 7.12 suggests, we could feed at least 8 times as many people by eating those cereals directly. What differences do you suppose it would make if we did so?

A number of technological and breeding innovations have made this increased production possible. One of the most important is the **confined animal feeding operation (CAFO)**, where animals are housed and fed—mainly soy and corn—for rapid growth (fig. 7.13). These operations dominate livestock raising in the U.S., Europe, and increasingly in China and other countries. Animals are housed in giant enclosures, with up to 10,000 hogs or a million chickens in an enormous barn complex, or 100,000 cattle in a feed lot (fig. 7.14). Operators feed the animals specially prepared mixes of corn, soy, and animal protein that maximizes their growth rate. New breeds of livestock have been developed that produce meat rapidly, rather than simply getting fat. The turn-around time is getting shorter, too. A U.S. chicken producer can turn baby chicks into chicken nuggets after just 8 weeks of growth.

Steers reach full size by just 18 months of age. Increased use of antibiotics, which are mixed in daily feed, make it possible to raise large numbers of animals in close quarters. Nearly 90 percent of U.S. hogs receive antibiotics in their feed.

Seafood is both wild and farmed

The 140 million metric tons of seafood we eat every year is an important part of our diet. Seafood provides about 15 percent of all animal protein eaten by humans, and it is the main animal protein source for about 1 billion people in developing countries.

Figure 7.12 Number of kilograms of grain needed to produce 1 kg of bread or 1 kg live weight gain.

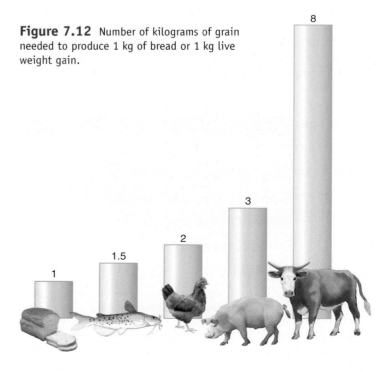

(a)

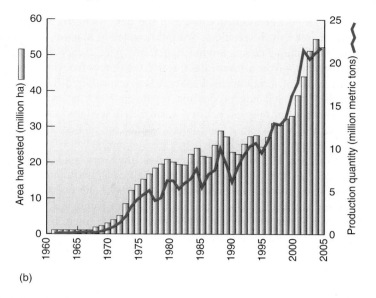

(b)

Figure 7.13 Confined animal feeding operations (a) have expanded the global market for corn and soybeans. Brazil's soy production (b) has grown from near zero to being the country's dominant agricultural product.
Source: Data from UN Food and Agriculture Organization.

Unfortunately, overharvesting and habitat destruction threaten most of the world's wild fisheries (see related stories "Saving the Reefs of Apo Island" in chapter 1 and "Shark Finning" at www.mhhe.com/cunningham5e). Annual catches of ocean fish rose by about 4 percent annually between 1950 and 1988. Since 1989, however, 13 of 17 major marine fisheries have declined dramatically or become commercially unsustainable. According to the United Nations, three-quarters of the world's edible ocean fish, crustaceans, and mollusks are declining and in urgent need of managed conservation.

The problem is too many boats using efficient but destructive technology to exploit a dwindling resource base. Boats as big as ocean liners travel thousands of kilometers and drag nets large enough to scoop up a dozen jumbo jets, sweeping a large patch of ocean clean of fish in a few hours. Long-line fishing boats set cables up to 10 km long with hooks every 2 meters that catch birds, turtles, and other unwanted "by-catch" along with targeted species. Trawlers drag heavy nets across the bottom, scooping up everything indiscriminately and reducing broad swaths of habitat to rubble. One marine biologist compared the technique to harvesting forest mushrooms with a bulldozer. In some operations, up to 15 kg of dead and dying by-catch are dumped back into the ocean for every kilogram of marketable food. The FAO estimates that operating costs for the 4 million boats now harvesting wild fish exceed sales by $50 billion (U.S.) per year. Countries subsidize fishing fleets to preserve jobs and to ensure access to this valuable resource.

Aquaculture is providing an increasing share of the world's seafood. Fish can be grown in farm ponds that take relatively little space but are highly productive. Cultivation of high-value carnivorous species, however, such as salmon, sea bass, and tuna, threaten wild stocks exploited to stock captive operations or to provide fish food. Building coastal fish-rearing ponds causes destruction of hundreds of thousands of hectares of mangrove forests and wetlands, which serve as irreplaceable nurseries for marine species. Net pens anchored in nearshore areas allow spread of diseases, escape of exotic species, and release of feces, uneaten food, antibiotics, and other pollutants into surrounding ecosystems (fig. 7.15).

Figure 7.14 Most livestock grown in the United States are raised in large-scale, concentrated animal-feeding operations. Up to a million animals can be held in a single facility. High population densities require heavy use of antibiotics and can cause severe local air and water pollution.

Increased production comes with increased risks

There are many environmental worries about this efficient production. Land conversion from pasture to soy and corn fields raises the rate of soil erosion (discussed in the next section). Bacteria in the manure in the feedlots, or liquid wastes in manure storage lagoons (holding tanks) around hog farms, can escape into the environment—from airborne dust around feedlots or from breaches in the walls of a manure tank. When Hurricane Floyd hit North Carolina's coastal hog production region in 1999, an estimated 10 million m³ of hog and poultry waste overflowed into local rivers, creating a dead zone in Pamlico Sound (see related story, "A Flood of Pigs," at www.mhhe.com/cunningham5e).

Constant use of antibiotics raises the very real risk of antibiotic-resistant diseases. More than half of all antibiotics used in the United States are administered to livestock. This massive and constant exposure produces antibiotic-resistant pathogens, strains that have adapted to survive antibiotics. This use, then, is slowly rendering our standard antibiotics useless for human health care. Next time you are prescribed an antibiotic by your doctor, you might ask whether she or he worries about antibiotic resistance, and you might think about how you would feel if your prescription was ineffective against your illness.

Although the public is increasingly aware of these environmental and health risks of concentrated meat production, we seem to be willing to accept these risks because this production system has made our favorite foods cheaper, bigger, and more available. A fast food hamburger today is more than twice the size it was in 1960, especially if you buy the kind with multiple patties and special sauce, and Americans love to eat them. At the same time, this larger burger costs less per pound, in constant dollars, than it did in 1960. As a consequence, for much of the world, consumption of protein and calories has climbed beyond what we really need to be healthy (figs. 7.2, 7.9).

As environmental scientists, we are faced with a conundrum, then. Improved efficiency has great environmental costs; it has also given us the abundant, inexpensive foods that we love. We have more protein, but also more obesity, heart disease, and diabetes than we ever had before. What do you think? Do the environmental risks balance a globally improved quality of life? Or should we consider reducing our consumption to reduce environmental costs? How might we go about making changes, if you think any are needed?

Figure 7.15 Pens for fish-rearing in Thailand.

We'll return to these questions at the end of the chapter. First we'll look more closely at how agricultural production works, to give you better grounds for deciding these issues.

7.4 Soil Is a Living Resource

Soil is a marvelous substance, a living resource of astonishing complexity and frailty. It is a complex mixture of mineral grains weathered from rocks, partially decomposed organic molecules, and a host of living organisms. Soil can be considered a living ecosystem by itself.

Building a few millimeters of soil can take anything from a few years (in a healthy grassland) to a few thousand years (in a desert or tundra). Under the best circumstances, topsoil accumulates at about 1 mm per year. With careful husbandry that prevents erosion and adds organic material, soil can be replenished and renewed indefinitely. But many farming techniques deplete soil. Crops consume the nutrients; plowing exposes the soil to erosion by wind or water. Severe erosion can carry away 25 mm or more of soil per year, far more than can accumulate under the best of conditions.

What is soil?

Soil is a complex mixture of six components:

1. *sand and gravel* (mineral particles from bedrock, either in place or moved from elsewhere, as in wind-blown sand)

2. *silts and clays* (extremely small mineral particles; many clays are sticky and hold water because of their flat surfaces and ionic charges; others give red color to soil)

3. *dead organic material* (decaying plant matter stores nutrients and gives soils a black or brown color)

4. *soil fauna and flora* (living organisms, including soil bacteria, worms, fungi, roots of plants, and insects, recycle organic compounds and nutrients)

5. *water* (moisture from rainfall or groundwater, essential for soil fauna and plants)

6. *air* (tiny pockets of air help soil bacteria and other organisms survive)

Figure 7.16 Soil ecosystems include numerous consumer organisms, as depicted here: (1) snail, (2) termite, (3) nematode and nematode-killing constricting fungus, (4) earthworm, (5) wood roach, (6) centipede, (7) carabid (ground) beetle, (8) slug, (9) soil fungus, (10) wireworm (click beetle larva), (11) soil protozoan, (12) sow bug, (13) ant, (14) mite, (15) springtail, (16) pseudoscorpion, and (17) cicada nymph.

Variations in these six components produce almost infinite variety in the world's soils. Abundant clays make soil sticky and wet. Abundant organic material and sand make the soil soft and easy to dig. Sandy soils drain quickly, often depriving plants of moisture. Silt particles are larger than clays and smaller than sand, so they aren't sticky and soggy, and they don't drain too quickly. Thus silty soils are ideal for growing crops, but they are also light and blow away easily when exposed to wind. Soils with abundant soil fauna quickly decay dead leaves and roots, making nutrients available for new plant growth. Compacted soils have few air spaces, making soil fauna and plants grow poorly. You can see some of these differences just looking at soil. Reddish soils are colored by iron-rich, rust-colored clays, the kind that store few nutrients for plants. Deep black soils are rich in decayed organic material, and thus are rich in nutrients.

Healthy soil fauna can determine soil fertility

Soil bacteria, algae, and fungi decompose and recycle leaf litter into plant-available nutrients, as well as helping to give soils structure and loose texture (fig. 7.16). Microscopic worms and nematodes process organic matter and create air spaces as they burrow through soil. These organisms mostly stay near the surface, often within the top few centimeters. The sweet aroma of freshly turned soil is caused by actinomycetes, bacteria that grow in fungus-like strands and give us the antibiotics streptomycin and tetracycline.

The health of the soil ecosystem depends on environmental conditions, including climate, topography, and parent material (the mineral grains or bedrock on which soil is built), and frequency of disturbance. Too much rain washes away nutrients and organic matter, but soil fauna cannot survive with too little rain. In extreme cold, soil fauna recycle nutrients extremely slowly; in extreme heat they may work so fast that leaf litter on the forest floor is taken up by plants in just weeks or months—so that the soil retains little organic matter. Frequent disturbance prevents the development of a healthy soil ecosystem, as does steep topography that allows rain to wash away soils. In the United States, the best farming soils tend to occur where the climate is not too wet or dry, on glacial silt deposits such as those in the upper Midwest, and on silt and clay-rich flood deposits, like those along the Mississippi River.

Most soil fauna occur in the uppermost layers of a soil, where they consume leaf litter accumulates. This layer is known as the "O" (organic) horizon. Just below the O horizon is a layer of mixed organic and mineral soil material, the A horizon (fig. 7.17), or **surface soil**.

SOIL HORIZONS

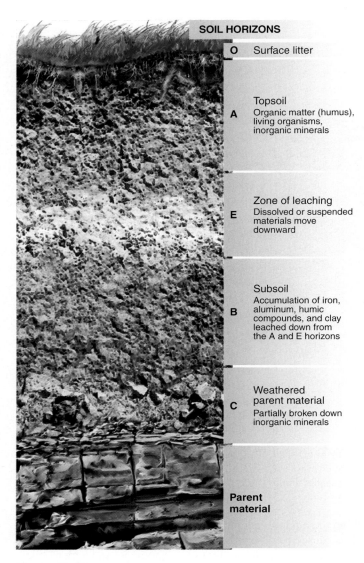

O — Surface litter

A — Topsoil
Organic matter (humus), living organisms, inorganic minerals

E — Zone of leaching
Dissolved or suspended materials move downward

B — Subsoil
Accumulation of iron, aluminum, humic compounds, and clay leached down from the A and E horizons

C — Weathered parent material
Partially broken down inorganic minerals

Parent material

Figure 7.17 Soil profile showing possible soil horizons. The actual number, composition, and thickness of these layers vary in different soil types.

The B horizon, or **subsoil**, tends to be richer in clays than the A; the B horizon is below most organic activity. The B layer accumulates clays that seep downward from the A horizon with rainwater that percolates through the soil. If you dig a hole, you may be able to tell where the B horizon begins because the soil tends to become slightly sticky. If you squeeze a handful of B soil, it should hold its shape better than a handful of A soil.

Sometimes an E (eluviated, or washed-out) layer lies between the A and B horizons. The E layer is loose and light-colored because most of its clays and organic material have been washed down to the B horizon. The C horizon, below the subsoil, is mainly decomposed rock fragments. Parent materials underlie the C layer. Parent material is the sand, wind-blown silt, bedrock, or other mineral material on which the soil is built. About 70 percent of the parent material in the United States was transported to its present site by glaciers, wind, and water, and is not related to the bedrock formations below it.

Figure 7.18 In many areas, soil or climate constraints limit soil development and agricultural production. These hungry goats in Sudan feed on a solitary Acacia shrub.

Your food comes mostly from the A horizon

Ideal farming soils have a thick, organic-rich A horizon. The soils that support the corn belt farm states of the U.S. Midwest have rich, black A horizon that can be over two meters thick, although a century of farming has washed much of this soil down the Mississippi River to the Gulf of Mexico. Most soils have less than half a meter of A horizon. Desert soils, with slow rates of organic activity, might have almost no O or A horizons (fig. 7.18).

Because topsoil is so important to our survival, we differentiate soils largely according to the thickness and composition of their upper layers. The U.S. Department of Agriculture classifies the thousands of different soil types into 11 soil orders (http://soils.usda.gov/technical/classification/orders/). Mollisols (*mollic* = soft, *sol* = soil), for example, have a thick, organic-rich A horizon that develops from the deep, dense roots of prairie grasses. Alfisols (*alfa* = first) have a slightly thinner A horizon, with slightly less organic matter. Alfisols develop in deciduous forests, where leaf litter is abundant. In contrast, the aridisols (*arid* = dry) of the desert Southwest have little organic matter, and they often have accumulations of mineral salts. Mollisols and alfisols dominate most of the farming regions of the United States (fig. 7.19).

7.5 Ways We Use and Abuse Soil

Only about 11 percent of the earth's land area (14.66 million km² out of a total of 132.4 million km²) is currently in agricultural production. Perhaps four times as much land could be converted to cropland, but much of this land serves as a refuge for cultural or biological diversity or suffers from constraints, such as steep slopes, shallow soils, poor drainage, tillage problems, low nutrient levels, metal toxicity, or excess soluble salts or acidity, that limit the types of crops that can be grown there.

Arable land is unevenly distributed

In parts of Canada and the United States, temperate climates, abundant water, and high soil fertility produce high crop yields that

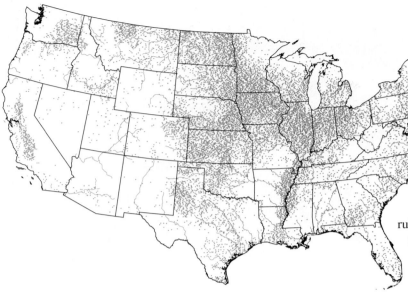

Each green dot represents 10,000 ha (25,000 ac) of cropland.
Total acres: 376,997,900.

Figure 7.19 Distribution of cropland in the United States. Glacial deposits and floodplain sediments make especially rich farmland.
Source: USDA Natural Resource Conservation Service.

contribute to high standards of living. Other countries, although rich in land area, lack suitable soil, topography, water, or climate to sustain our levels of productivity.

Arable land shrank from 0.38 ha (0.94 acre) per person in 1970 to 0.23 ha (0.56 acre) in 2000 (1 ha = 100 m $\times$ 100 m). If current population projections are correct, the amount of cropland per person will decline to 0.15 ha (0.37 acre) by 2050. In Asia, cropland will be even more scarce—0.09 ha (0.22 acre) per person—in 50 years. If you live on a typical quarter-acre suburban lot, look at your yard and imagine feeding yourself for a year on what you could produce there.

As the opening case study for this chapter shows, the largest increases in cropland over the past 30 years have occurred in South America, where forests and grazing lands are rapidly being converted to farms. Many developing countries are reaching the limit of lands that can be exploited for agriculture without unacceptable social and environmental costs, but others still have considerable potential for opening new agricultural lands. East Asia, for instance, already uses about three-quarters of its potentially arable land. Most of its remaining land has severe restrictions for agricultural use. Further increases in crop production will probably have to come from higher yields per hectare. Latin America, by contrast, uses only about one-fifth of its potential land, and Africa uses only about one-fourth of the land that theoretically could grow crops. However, there would be serious ecological trade-offs in putting much of this land into agricultural production.

While land surveys tell us that much more land in the world could be cultivated, not all of that land necessarily should be farmed. Much of it is more valuable in its natural state. The soils over much of tropical Asia, Africa, and South America are old,

weathered, and generally infertile. Most of the nutrients are in the standing plants, not in the soil. In many cases, clearing land for agriculture in the tropics has resulted in tragic losses of biodiversity and the valuable ecological services that it provides. Ultimately, much of this land is turned into useless scrub or semidesert.

Land degradation reduces crop yields

Agriculture both causes and suffers from environmental degradation. The International Soil Reference and Information Centre in the Netherlands estimates that, every year, 3 million ha (7.4 million acres) of cropland are ruined by erosion, 4 million ha are turned into deserts, and 8 million ha are converted to nonagricultural uses, such as homes, highways, shopping centers, factories, and reservoirs. Over the past 50 years, some 1.9 billion ha of agricultural land (an area greater than that now in production) have been degraded to some extent. About 300 million ha of this land are strongly degraded (meaning that the soil has deep gullies, severe nutrient depletion, or poor crop growth or that restoration is difficult and expensive). Some 910 million ha—about the size of China—are moderately degraded. Nearly 9 million ha of former croplands are so degraded that they no longer support any crops at all. The causes of this extreme degradation vary: in Ethiopia, it is water erosion; in Somalia, it is wind; and in Uzbekistan, salt and toxic chemicals are responsible. In Sweden and Finland, fallout from the Chernobyl nuclear reaction explosion has contaminated large amounts of grazing land and farmland.

Water and wind erosion provide the motive force for the vast majority of all soil degradation worldwide (fig. 7.20). Chemical degradation includes nutrient depletion, salt accumulation, acidification, and pollution. Physical degradation includes compaction by heavy machinery or trampling by cattle, water accumulation from excess irrigation and poor drainage, and laterization (solidification of iron and aluminum-rich tropical soil when exposed to sun and rain).

Farming accelerates erosion

Erosion is an important natural process, resulting in the redistribution of the products of geologic weathering, and it is part of both soil formation and soil loss. The world's landscapes have been sculpted by erosion. When the results are spectacular enough, we enshrine them in national parks, as we did with the Grand Canyon.

Where erosion has worn down mountains and spread soil over the plains or deposited rich alluvial silt in river bottoms, we farm it. Erosion is a disaster only when it occurs in the wrong place at the wrong time.

Some erosion occurs so rapidly that you can watch it happen. Deep gullies are created where water scours away the soil, leaving fenceposts and trees sitting on tall pedestals as the land erodes away around them. In most places, however, erosion is more subtle. It is a creeping disaster that occurs in small increments. A thin layer of topsoil washes off of fields year after year until,

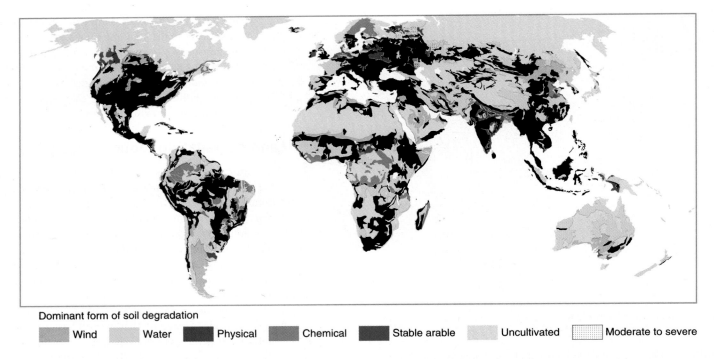

Dominant form of soil degradation

▨ Wind   ▨ Water   ■ Physical   ▨ Chemical   ■ Stable arable   ▨ Uncultivated   ▨ Moderate to severe

Figure 7.20 Global causes of soil erosion and degradation. Globally, 62 percent of eroded land is mainly affected by water; 20 percent is mainly affected by wind.
Source: ISRIC Global Assessment of Human-induced Soil Degradation, 2008.

eventually, nothing is left but poor-quality subsoil that requires more and more fertilizer and water to produce any crop at all.

An estimated 25 billion metric tons of soil are lost from croplands every year due to wind and water erosion. The net effect, worldwide, of this widespread topsoil erosion is a reduction in crop production equivalent to removing about 1 percent of the world's cropland *each year*. Many farmers are able to compensate for this loss by applying more fertilizer and by bringing new land into cultivation. Continuation of current erosion rates, however, could reduce agricultural production by 25 percent in Central America and Africa and by 20 percent in South America by 2020. The total annual soil loss from croplands is thought to be 25 billion metric tons. About twice that much soil is lost from rangelands, forests, and urban construction sites each year.

In addition to reduced land fertility, erosion results in sediment loading of rivers and lakes, siltation of reservoirs and harbors, smothering of wetlands and coral reefs, creation of "dead zones" in coastal regions, and clogging of water intakes and water-power turbines.

Wind and water move soil

Wind and water are the main agents that move soil around. Water flowing across a gently sloping, bare field removes a thin, uniform layer of soil in what is known as **sheet erosion**. When little rivulets of running water gather together and cut small channels in the soil, the process is called **rill erosion** (fig. 7.21a). If rills enlarge to form bigger channels or ravines that are too large to be removed by normal tillage operations, we call the process **gully erosion**

(fig. 7.21b). Streambank erosion is the washing away of soil from the banks of established streams, creeks, or rivers, often as a result of removing trees and brush along streambanks and of cattle damaging the banks.

Most soil erosion on agricultural land is rill erosion. Large amounts of soil can be transported a little bit at a time without being very noticeable. A farm field can lose 20 metric tons of soil per hectare during winter and spring runoff in rills so small that they are erased by the first spring cultivation. That represents a loss of only a few millimeters of soil over the whole surface of the field, hardly apparent to any but the most discerning eye. But it doesn't take much mathematical skill to see that, if you lose soil twice as fast as it is being replaced, it eventually will run out.

Wind can equal or exceed water in erosive force, especially in a dry climate and on relatively flat land. When plant cover and surface litter are removed from the land by agriculture or grazing, wind lifts loose soil particles and sweeps them away. In extreme conditions, windblown dunes encroach on useful land and cover roads and buildings (fig. 7.21c). Over the past 30 years, China has lost 93,000 km² (about the size of Indiana) to **desertification**, or conversion of productive land to desert. Advancing dunes from the Gobi desert are now only 160 km (100 mi) from Beijing. Every year more than 1 million tons of sand and dust blow from Chinese drylands, often traveling across the Pacific Ocean to the West Coast of North America. Since 1985, China has planted more than 40 billion trees to try to stabilize the soil and hold back deserts.

Many areas of the United States and Canada have very high erosion rates. The U.S. Department of Agriculture reports that

(a) Sheet and rill erosion

(b) Gullying

(c) Wind erosion and desertification

Figure 7.21 Land degradation affects more than 1 billion ha yearly, or about two-thirds of all global cropland. Water erosion (a and b) account for about half that total. Wind erosion affects a nearly equal area (c).

69 million ha (170 million acres) of U.S. farmland and range are eroding faster than 1 mm per year (10 tons/ha/year), the rate at which soil accumulates in the best conditions. These erosion rates steadily deplete long-term productivity.

Intensive farming practices are largely responsible for this situation. Row crops, such as corn and soybeans, leave soil exposed for much of the growing season. Deep plowing and heavy herbicide applications create weed-free fields that look neat but are subject to erosion. Because big machines cannot easily follow contours, they often go straight up and down the hills, creating ready-made gullies for water to follow. Farmers sometimes plow through grass-lined watercourses (low areas where water runs off after a rain) and pull out windbreaks and fencerows to accommodate the large machines and to get every last square meter into production. Consequently, wind and water carry away the topsoil.

7.6 Other Agricultural Resources

Soil is only part of the agricultural resource picture. Agriculture is also dependent upon water, nutrients, favorable climates to grow crops, productive crop varieties, and the mechanical energy to tend and harvest the crops.

Irrigation is necessary for high yields

Agriculture accounts for the largest single share of global water use. At least two-thirds of all fresh water withdrawn from rivers, lakes, and groundwater supplies is used for irrigation (chapter 10). Irrigation increases yields of most crops by 100 to 400 percent. Although estimates vary widely (as do definitions of irrigated land), about 15 percent of all cropland, worldwide, is irrigated.

Some countries are water rich and can readily afford to irrigate farmland, while other countries are water poor and must use water very carefully. The efficiency of irrigation water use varies greatly. High evaporative and seepage losses from unlined and

uncovered canals in some places can mean that as much as 80 percent of water withdrawn for irrigation never reaches its intended destination. Poor farmers may overirrigate because they lack the technology to meter water and distribute just the amount needed. In wealthier countries, farmers can afford sometimes-extravagant uses of water; they can also afford water-saving technologies such as drip irrigation or downward-facing sprinklers (fig. 7.22).

Excessive use not only wastes water but also often results in **waterlogging**. Waterlogged soil is saturated with water, and plant roots die from lack of oxygen. **Salinization**, in which mineral salts accumulate in the soil, is often a problem when irrigation water dissolves and mobilizes salts in the soil. As the water evaporates, it leaves behind a salty crust on the soil surface that is lethal to most plants. Flushing with excess water can wash away this salt accumulation, but the result is even more saline water for downstream users.

 Figure 7.22 Downward-facing sprinklers on this center-pivot irrigation system deliver water more efficiently than upward-facing sprinklers.

The FAO estimates that 20 percent of all irrigated land is damaged to some extent by waterlogging or salinity. Water conservation techniques can greatly reduce problems arising from excess water use. Conservation also makes more water available for other uses or for expanded crop production where water is in short supply (chapter 10).

Fertilizer boosts production

In addition to water, sunshine, and carbon dioxide, plants need small amounts of inorganic nutrients for growth. The major elements required by most plants are nitrogen, potassium, phosphorus, calcium, magnesium, and sulfur. Nitrogen is a key component of all living cells (chapter 2), and nitrogen is the most common limiting factor for plant growth. Thus nitrogen, potassium, and phosphorus are our primary fertilizers. Calcium and magnesium often are limited in areas of high rainfall and must be supplied in the form of lime. A good deal of the doubling in worldwide crop production since 1950 has come from increased inorganic fertilizer use. In 1950 the average amount of fertilizer used was 20 kg per hectare. In 2000 this had increased to an average of 90 kg per hectare worldwide.

There is considerable potential for increasing the world's food supply by increasing fertilizer use in low-production countries if ways can be found to apply fertilizer more effectively and to reduce pollution. Africa, for instance, uses an average of only 19 kg of fertilizer per hectare (17 lbs per acre), or about one-fourth of the world average. It has been estimated that the developing world could at least triple its crop production by raising fertilizer use to the world average.

Over-fertilizing is also a common problem. While European farmers use more than twice as much fertilizer per hectare as do North American farmers, their yields are not proportionally higher. Phosphates and nitrates from farm fields and cattle feedlots are a major cause of aquatic ecosystem pollution. Nitrate levels in groundwater have risen to dangerous levels in many areas where intensive farming is practiced. Young children are especially sensitive to the presence of nitrates. Using nitrate-contaminated water to mix infant formula can be fatal for newborns.

What are some alternative ways to fertilize crops? Manure and green manure (crops grown specifically to add nutrients to the soil) are important natural sources of soil nutrients. Nitrogen-fixing bacteria living symbiotically in root nodules of legumes are valuable for making nitrogen available as a plant nutrient (chapter 2). Interplanting and rotating beans or some other leguminous crop with such crops as corn and wheat are traditional ways of increasing nitrogen availability.

Modern agriculture runs on oil

Farming as it is generally practiced in the industrialized countries is highly energy-intensive. Fossil fuels supply almost all of this energy. Between 1920 and 1980, direct energy use on farms rose as gasoline and diesel fuels were consumed by increasing mechanization of agriculture. Indirect energy use, in the form of synthetic fertilizers, pesticides, and other agricultural chemicals, increased even more, especially after World War II. On intensively fertilized farms, indirect energy use may be five times that of direct energy. The energy price shocks of the 1970s encouraged energy conservation that has reduced farm energy use, even though total food production has continued to rise.

After crops leave the farm, additional energy is used in food processing, distribution, storage, and cooking. It has been estimated that the average food item in the American diet travels 2,000 km (1,250 mi) between the farm that grew it and the person who consumes it. The energy required for this complex processing and distribution system may be five times as much as is used directly in farming. As a whole, the food system in the United States consumes about 16 percent of the total energy we use. Most of our foods require more energy to produce, process, and get to market than they yield when we eat them.

Farmers are also beginning to produce energy, in the form of ethanol and biodiesel. There has been much debate about the environmental and economic costs and benefits of this, as discussed in chapter 12. State-of-the-art facilities do, in fact, appear to have a 150 percent net energy yield. Plant oils from sunflowers, soybeans, corn, and other oil seed crops can be burned directly in most diesel engines, and may be closer than ethanol to a net energy gain (fig. 7.23).

Brazil now makes about 16 billion liters of ethanol per year, mostly from sugarcane waste. Ethanol now accounts for 40 percent of the fuel sold in Brazil. The United States expects to produce over twice this amount by 2009.

Pest control saves crops

Biological pests reduce crop yields and spoil as much as half of the crops harvested every year in some areas. Modern agriculture largely depends on toxic chemicals to kill or drive away these pests, but there are many serious concerns about the types and amounts of pesticides now in use.

There are many traditional methods for protecting crops from pests, but our war against unwanted organisms entered a new phase in the twentieth century with the invention of synthetic organic chemicals, such as DDT (dichlorodiphenyltrichloroethane). These chemicals

Figure 7.23 Biofuel crops, such as these sunflowers, can provide a good cash crop for farmers and contribute to national energy supplies.

Figure 7.24 Spraying pesticides by air is quick and cheap, but toxins often drift to nearby fields.

have been an important part of our increased crop production and have helped control many disease-causing organisms. It's estimated that up to half our current crop yields might be lost if we had no pesticides to protect them. Indiscriminate and profligate pesticide use, however, also has caused many problems, such as killing nontarget species, creating new pests of organisms that previously were not a problem, and causing widespread pesticide resistance among pest species (fig. 7.24). Often highly persistent and mobile in the environment, many pesticides have moved through air, water, and soil and bioaccumulated or bioconcentrated in food chains, nearly exterminating several top predators, such as peregrine falcons.

Your exposure to pesticides is probably higher than you suspect. A study by the U.S. Department of Agriculture found that 73 percent of conventionally grown foods (out of 94,000 samples assayed) had residues from at least one pesticide. By contrast, only 23 percent of the organic samples of the same food groups had any residues. It is now clear that humans suffer adverse health effects from persistent organic pollutants (POPs), such as pesticides. In fact, most of the "dirty dozen" POPs recently banned globally are pesticides. See chapter 8 for more on health effects of POPs.

In 2002 Swiss researchers published results of a two-decades-long comparison of organic and conventional farming. Crops grown without synthetic fertilizers or pesticides produced, on average, about a 20 percent lower yield than similar crops grown by conventional methods, but they more than made up the difference with lower operating costs, premium crop prices, less ecological damage, and healthier farm families.

Some alternatives for reducing our dependence on dangerous chemical pesticides include management changes, such as using cover crops and mechanical cultivation, and planting mixed polycultures rather than vast monoculture fields. Consumers may have to learn to accept less than perfect fruits and vegetables if we want to avoid toxic chemicals in our diet. Biological controls, such as insect predators, pathogens, or natural poisons specific for a particular pest, can help reduce chemical use. Genetic breeding and biotechnology can produce pest-resistant crop and livestock strains as well. Integrated pest management (IPM) is a relatively new science that combines all of these alternative methods, together with judicious use of synthetic pesticides under precisely controlled conditions. There are ways that you can minimize pesticides in your diet (see What Can You Do?).

7.7 How We Have Managed to Feed Billions

In the developed countries, 95 percent of agricultural growth in the twentieth century came from improved crop varieties or increased fertilization, irrigation, and pesticide use, rather than from bringing new land into production. In fact, less land is being cultivated now than 100 years ago in North America, or 600 years ago in Europe. As more effective use of labor, fertilizer, and water and improved seed varieties have increased in the more-developed countries, productivity per unit of land has increased, and much marginal land has been retired, mostly to forests and grazing lands. In developing countries, at least two-thirds of recent production gains have come from new crop varieties and more intense cropping, rather than expansion into new lands.

Although at least 3,000 species of plants have been used for food at one time or another, most of the world's food now comes from only a few widely grown crops (see table 7.1). Many new or unconventional varieties might be valuable human food supplies,

Figure 7.25 Semidwarf wheat (*right*), bred by Norman Borlaug, has shorter, stiffer stems and is less likely to lodge (fall over) when wet than its conventional cousin (*left*). This "miracle" wheat responds better to water and fertilizer and has played a vital role in feeding a growing human population.

however, especially in areas where conventional crops are limited by climate, soil, pests, or other problems. Among the plants now being investigated as potential additions to our crop roster is the winged bean, a perennial plant that grows well in hot climates where other legumes will not grow. It is totally edible (pods, mature seeds, shoots, flowers, leaves, and tuberous roots), resistant to diseases, and enriches the soil. Another promising crop is triticale, a hybrid between wheat (*Triticum*) and rye (*Secale*) that grows in light, sandy, infertile soil. It is drought-resistant, has nutritious seeds, and is being tested for salt tolerance for growth in saline soils or irrigation with seawater. Some traditional crop varieties grown by Native Americans, such as tepary beans, amaranth, and Sonoran panic grass, are being collected by seed conservator Gary Nabhan, both as a form of cultural revival for native people and as possible food crops for harsh environments.

The green revolution has increased yields

So far, the major improvements in farm production have come from technological advances and modification of a few well-known species. Yield increases often have been spectacular. A century ago, when all maize (corn) in the United States was open pollinated, average yields were about 25 bushels per acre (bu/acre). In 2000 the average yield from rainfed fields in Iowa was 138 bu/acre, and irrigated maize in Arizona averaged 208 bu/acre. The highest yield ever recorded in field production was 370 bu/acre on an Illinois farm, but theoretical calculations suggest that 500 bu/acre (32 metric tons per hectare) could be possible. Most of this gain was accomplished by use of synthetic fertilizers along with conventional plant breeding: geneticists laboriously hand-pollinating plants and looking for desired characteristics in the progeny.

Starting about 50 years ago, agricultural research stations began to breed tropical wheat and rice varieties that would provide food for growing populations in developing countries. The first of the "miracle" varieties was a dwarf, high-yielding wheat developed by Norman Borlaug (who received a Nobel Peace Prize for his work) at a research center in Mexico (fig. 7.25). At about the same time, the International Rice Institute in the Philippines developed dwarf rice strains with three or four times the production of varieties in use at the time. The spread of these new high-yield varieties around the world has been called the **green revolution**. It is one of the main reasons that world food supplies have more than kept pace with the growing human population over the past few decades.

Most green revolution breeds really are "high responders," meaning that they yield well in response to optimal inputs of fertilizer, water, and chemical protection from pests and diseases (fig. 7.26). With fewer inputs, on the other hand, high responders may not produce as well as traditional varieties. Poor farmers who can't afford the expensive seed, fertilizer, and water required to become part of this movement usually are left out of the green revolution. In fact, they may be driven out of farming altogether as rising land values and falling commodity prices squeeze them from both sides.

Genetic engineering could have benefits and costs

Genetic engineering, splicing a gene from one organism into the chromosome of another, has the potential to greatly increase both the quantity and the quality of our food supply. It is now possible to build entirely new genes, and even new organisms, often called "transgenic" organisms or **genetically modified organisms (GMOs)**, by taking a bit of DNA from here, a bit from there, and even synthesizing artificial sequences to create desired characteristics in engineered organisms (fig. 7.27).

Proponents predict dramatic benefits from this new technology. Research is now underway to improve yields and create crops that resist drought, frost, or diseases. Other strains are being developed to tolerate salty, waterlogged, or low-nutrient soils, allowing degraded or marginal farmland to become productive. All of these could be important for reducing hunger in developing

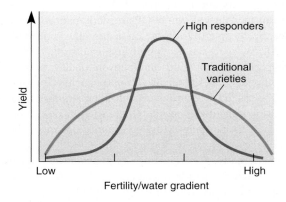

Figure 7.26 Green revolution miracle crops are really high responders, meaning that they have excellent yields under optimum conditions. For poor farmers who can't afford the fertilizer and water needed by high responders, traditional varieties may produce better yields.

countries. Plants that produce their own pesticides might reduce the need for toxic chemicals, while engineering for improved protein or vitamin content could make our food more nutritious. Attempts to remove specific toxins or allergens from crops also could make our food safer. Crops such as bananas and tomatoes have been altered to contain oral vaccines that can be grown in developing countries where refrigeration and sterile needles are unavailable. Plants have been engineered to make industrial oils and plastics. Animals, too, are being genetically modified to grow faster, gain weight on less food, and produce pharmaceuticals, such as insulin, in their milk. Pigs are now being engineered to produce omega-3 fatty acids for a "heart healthy" diet. It may even be possible to create animals with human cell-recognition factors that could serve as organ donors.

Opponents worry that moving genes willy-nilly could create a host of problems, some of which we can't even imagine. GMOs themselves might escape and become pests or they might interbreed wild relatives. In either case, we may create superweeds or reduce native biodiversity. Constant presence of pesticides in plants could accelerate pesticide resistance in insects or leave toxic residues in soil or our food. Genes for toxicity or allergies could be transferred along with desirable genes, or novel toxins could be created as genes are mixed together. This technology may be available only to the richest countries or the wealthiest corporations, making family farms uncompetitive and driving developing countries even further into poverty.

Both the number of GMO crops and the acreage devoted to growing them is increasing rapidly. Between 1987 and 2005, more than 10,500 field tests for some 750 different crop varieties were carried out in the United States (fig. 7.28). Worldwide, over 25 percent of all cropland (72 million ha) were planted with GMO crops in 2007. The United States accounted for 63 percent of that acreage, followed by Argentina with 21 percent. Canada,

Brazil, and China together make up 22 percent of all transgenic cropland.

Some transgenic crops have reached mainstream status. About 82 percent of all soybeans, one-quarter of all maize (corn), and 71 percent of all cotton grown in the United States are now GMOs. You've already probably eaten some genetically modified food. Estimates are that at least 60 percent of all processed food in America contains GMO ingredients. Since the United States, Argentina, and Brazil account for 90 percent of international trade in maize and soybeans (in which GMOs often are mixed with other grains), a large fraction of the world most likely has been exposed to some GMO products.

Most GMOs have been engineered for pest resistance or weed control

Biotechnologists have created plants with genes for endogenous insecticides. *Bacillus thuringiensis* (Bt), a bacterium, makes toxins lethal to Lepidoptera (butterfly family) and Coleoptera (beetle family). The genes for some of these toxins have been transferred into crops such as maize (to protect against European cutworms), potatoes (to fight potato beetles), and cotton (to protect against boll weevils). This allows farmers to reduce insecticide spraying. Arizona cotton farmers, for example, report reducing their use of chemical insecticides by 75 percent.

Most European nations have not approved Bt-containing varieties. Fear of transgenic products has been a sticking point in agricultural trade between the United States and the European Union.

In 2001 Starlink, a Bt-containing maize variety that had not yet been approved for human consumption, was found mixed into a wide variety of consumer products. By the time food was made from Starlink, no Bt could be detected, only the genes that code for it. There is no evidence that Bt is harmful to humans, but the form in Starlink may be allergenic. In fact, Bt is one of the few insecticides approved for organic farmers. Nevertheless, a public outcry ensued and seed companies, farmers, shippers, and food

Figure 7.27 A researcher measures growth of genetically engineered rice plants. Superior growth and yield, pest resistance, and other useful traits can be moved from one species to another by molecular biotechnology.

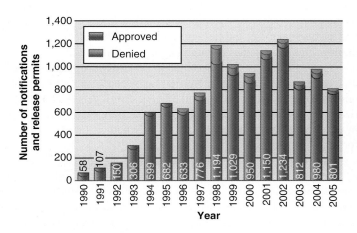

Figure 7.28 Transgenic crop field releases in the United States 1990 to 2005.
Source: Data from Information Systems for Biotechnology, Virginia Tech University.

processors spent more than $1 billion trying to remove it from the food chain. What do you think: was the fear of Starlink justified?

Entomologists worry that Bt plants churn out toxin throughout the growing season, regardless of the level of infestation, creating perfect conditions for selection of Bt resistance in pests. Already 500 species of insect, mite, or tick are resistant to one or more pesticides. Having Bt constantly available can only aggravate this dilemma. One solution is to plant at least a part of every field in non-Bt crops that will act as a refuge for nonresistant pests. The hope is that interbreeding between these "wild-type" bugs and those exposed to Bt will dilute out recessive resistance genes. Deliberately harboring pests and letting them munch freely on crops is something that many farmers find hard to do. In addition, devoting a significant part of their land to nonproductive crops lowers the total yield and counteracts the profitability of engineered seed.

The other major group of transgenic crops is engineered to tolerate high doses of herbicides. The two dominant products in this category are Monsanto's "Roundup Ready" crops—so-called because they can withstand treatment with Monsanto's best-selling herbicide, Roundup (glyphosate)—and AgrEvo's "Liberty Link" crops, which resist that company's Liberty (glufosinate) herbicide. Because crops with these genes can grow in spite of high herbicide doses, farmers can spray fields heavily to exterminate weeds. This allows for conservation tillage and leaving more crop residue on fields to protect topsoil from erosion—both good ideas—but it also can mean using more herbicide in higher doses than might otherwise be used.

Is genetic engineering safe?

The U.S. Food and Drug Administration has declined to require labeling of foods containing GMOs, saying that these new varieties are "substantially equivalent" to related varieties bred via traditional practices. After all, proponents say, we have been moving genes around for centuries through plant and animal breeding. All domesticated organisms should be classified as genetically modified, some people argue. Biotechnology may be a more precise way of creating novel organisms than normal breeding procedures, its supporters suggest. We're moving only a few genes—or even part of one gene—at a time with biotechnology, rather than hundreds of unknown genes through classical techniques.

There is already evidence that some GMO crops can spread their genes to nearby fields. Normal rape seed (canola oil) varieties in Canada were found to contain genes from genetically modified varieties in nearby fields. It isn't clear, however, if the genes involved will have any adverse affect or whether they will persist for many generations.

The first genetically modified animal designed to be eaten by humans is an Atlantic salmon (*Salmo salar*) containing extra growth hormone genes from an oceanic pout (*Macrozoarces americanus*). The greatest worry from this fish is not that it will introduce extra hormones into our diet—that's already being done by chickens and beef that get extra growth hormone via injections or their diet—but, rather, the ecological effects if the fish escape from captivity. The transgenic fish grow seven times faster and are more attractive to the opposite sex than a normal salmon. If they escape from captivity, they may outcompete already endangered wild relatives for food, mates, and habitat. Fish farmers say they will grow only sterile females and will keep them in secure net pens. Opponents point out that salmon frequently escape from aquaculture operations and that, if just a few fertile transgenic fish break out, it could be catastrophic for wild stocks.

Finally, there are social and economic implications of GMOs. Will they help feed the world, or will they lead to a greater consolidation of corporate power and economic disparity? Might higher yields and fewer losses to pests and diseases allow poor farmers in developing countries to stop using marginal land and avoid cutting down forests to create farmland? Is this simply a technological fix, or could it help promote agricultural sustainability? Critics suggest that there are simpler and cheaper ways other than high-tech crop varieties to provide vitamin A to children in developing countries or to increase the income of poor rural families. Adding a cow or a fishpond or training people in water harvesting or regenerative farming techniques (as we'll discuss in the next section) may have a longer-lasting impact than selling them expensive new seeds.

On the other hand, if we hope to reduce malnutrition and feed eight billion people in 50 years, maybe we need all the tools we can get. Where do you stand in this debate? What additional information would you need to reach a reasoned judgment about the risks and benefits of this new technology?

7.8 Alternatives in Food and Farming

How, then, shall we feed the world? Can we make agriculture compatible with sustainable ecological and social systems? Having discussed some of the problems that beset modern agriculture, we now consider some suggested ways to overcome problems and make farming and food production just and lasting enterprises. This goal is usually termed **sustainable agriculture**, or **regenerative farming**, both of which aim to produce food and fiber on a sustainable basis and repair the damage caused by destructive practices. Some alternative methods are developed through scientific research; others are discovered in traditional cultures and practices nearly forgotten in our mechanization and industrialization of agriculture.

Soil conservation is essential

With careful husbandry, soil is a renewable resource that can be replenished and renewed indefinitely. Since agriculture is the area in which soil is most essential and most often lost through erosion, agriculture offers the greatest potential for soil conservation and rebuilding. Some rice paddies in Southeast Asia, for instance, have been farmed continuously for a thousand years without any apparent loss of fertility. The rice-growing cultures that depend on these fields have developed management practices that return organic material to the paddies and carefully nurture the soil's ability to sustain life.

While American agriculture hasn't reached that level of sustainability, there is evidence that soil conservation programs are having a positive effect. In one Wisconsin study, erosion rates in one small watershed were 90 percent less in 1975–1993 than they were in the 1930s. Among the most important elements in soil conservation are terracing, ground cover, and reduced tillage.

Most erosion happens because water runs downhill. The faster it runs, the more soil it carries off the fields. Comparisons of erosion rates in Africa have shown that a 5 percent slope in a plowed field has three times the water runoff volume and eight times the soil erosion rate of a comparable field with a 1 percent slope. Water runoff can be reduced by grass strips in waterways and by **contour plowing**—that is, plowing across the hill rather than up and down. Contour plowing is often combined with **strip-farming**, the planting of different kinds of crops in alternating strips along the land contours (fig. 7.29). When one crop is harvested, the other is still present to protect the soil and keep water from running straight downhill. The ridges created by cultivation make little dams that trap water and allow it to seep into the soil, rather than running off. In areas where rainfall is very heavy, tied ridges are often useful. This method involves a series of ridges running at right angles to each other, so that water runoff is blocked in all directions and is encouraged to soak into the soil.

Terracing involves shaping the land to create level shelves of earth to hold water and soil (fig. 7.30). The edges of the terrace are planted with soil-anchoring plant species. This is an expensive procedure, requiring either much hand labor or expensive machinery, but it makes farming of very steep hillsides possible. The rice terraces in the Chico River Valley in the Philippines rise as much as 300 m (1,000 ft) above the valley floor. They are considered one of the wonders of the world. Some of these terraces have been farmed for 1,000 years without any apparent fertility loss. Will we be able to say the same thing a millenium from now?

Planting **perennial species** (plants that grow for more than two years) is the only suitable use for some lands and some soil types. Establishing forest, grassland, or crops such as tea, coffee, or other crops that do not have to be cultivated every year may be necessary to protect certain unstable soils on sloping sites or watercourses. The $180 billion U.S. Farm Bill includes $17 billion (over ten years) to encourage soil, water, and wildlife conservation.

Groundcover protects the soil

Annual row crops, such as corn or beans, generally cause the highest erosion rates because they leave soil bare for much of the year (table 7.2). Often the easiest way to provide cover that protects soil from erosion is to leave crop residues on the land after harvest. They not only cover the surface to break the erosive effects of wind and water but also reduce evaporation and soil temperature in hot climates and protect ground organisms that help aerate and rebuild soil. In some experiments, 1 ton of crop residue per acre (0.4 ha) increased water infiltration 99 percent, reduced runoff 99 percent, and reduced erosion 98 percent. Leaving crop residues on the field also can increase disease and pest problems, however, and may require increased use of pesticides and herbicides.

Figure 7.29 Contour plowing and strip cropping help prevent soil erosion on hilly terrain, as well as create a beautiful landscape.

Figure 7.30 Terracing, as in these Balinese rice paddies, can control erosion and make steep hillsides productive.

Where crop residues are not adequate to protect the soil or are inappropriate for subsequent crops or farming methods, such **cover crops** as rye, alfalfa, and clover can be planted immediately after harvest to hold and protect the soil. These cover crops can be plowed under at planting time to provide green manure.

Shade-Grown Coffee and Cocoa

Has it ever occurred to you that your purchases of coffee and chocolate may be contributing to the protection or destruction of tropical forests? Both coffee and cocoa are examples of food products grown exclusively in developing countries but consumed almost entirely in the wealthier, developed nations (vanilla and bananas are some other examples). Coffee grows in cool, mountain areas of the tropics, while cocoa is native to the warm, moist lowlands. Both are small trees of the forest understory, adapted to low light levels.

Until a few decades ago, most of the world's coffee and cocoa were **shade-grown** under a canopy of large forest trees. Recently, however, new varieties of both crops have been developed that can be grown in full sun. Because more coffee or cocoa trees can be crowded into these fields, and they get more solar energy than in a shaded plantation, yields for sun-grown crops are higher.

There are costs, however, in this new technology. Sun-grown trees die earlier from the stress and diseases common in these fields. Furthermore, ornithologists have found that the number of bird species can be cut in half in full-sun plantations, and the number of individual birds may be reduced by 90 percent. Shade-grown coffee and cocoa generally require fewer pesticides (or sometimes none) because the birds and insects residing in the forest canopy eat many of the pests. Shade-grown plantations also need less chemical fertilizer because many of the plants in these complex forests add nutrients to the soil. In addition, shade-grown crops rarely need to be irrigated because heavy leaf fall protects the soil, while forest cover reduces evaporation.

Currently, about 40 percent of the world's coffee and cocoa plantations have been converted to full-sun varieties and another 25 percent are in process. Traditional techniques for coffee and cocoa production are worth preserving. Thirteen of the world's 25 biodiversity hot spots occur in coffee or cocoa regions. If all the 20 million ha (49 million acres) of coffee and cocoa plantations in these areas are converted to monocultures, an incalculable number of species will be lost.

The Brazilian state of Bahia is a good example of both the ecological importance of these crops and how they might help preserve forest species. At one time, Brazil produced much of the world's cocoa, but in the early 1900s, the crop was introduced into West Africa. Now Côte d'Ivoire alone grows more than 40 percent of the world total, and the value of Brazil's harvest has dropped by 90 percent. Côte d'Ivoire is aided in this competition by a labor system that reportedly includes widespread child slavery. Even adult workers in Côte d'Ivoire get only about $165 (U.S.) per year (if they get paid at all), compared with a minimum wage of $850 (U.S.) per year in Brazil. As African cocoa production ratchets up, Brazilian landowners are converting their plantations to pastures or other crops.

Cocoa pods grow directly on the trunk and large branches of cocoa trees.

The area of Bahia where cocoa was once king is part of Brazil's Atlantic Forest, one of the most threatened forest biomes in the world. Only 8 percent of this forest remains undisturbed. Although cocoa plantations don't represent the full diversity of intact forests, they protect a surprisingly large sample of what once was there. And shade-grown cocoa can provide an economic rationale for preserving that biodiversity. Brazilian cocoa will probably never compete with that from other areas for lowest cost. There is room in the market, however, for specialty products. If consumers were willing to pay a small premium for organic, fair-trade, shade-grown chocolate and coffee, it might provide the incentive needed to preserve biodiversity. Wouldn't you like to know that your chocolate or coffee wasn't grown with child slavery and is helping protect plants and animal species that might otherwise go extinct?

Another method is to flatten cover crops with a roller and drill seeds through the residue to provide a continuous protective cover during early stages of crop growth.

In some cases, interplanting of two different crops in the same field not only protects the soil but also is a more efficient use of the land, providing double harvests. Native Americans and pioneer farmers, for instance, planted beans or pumpkins between the corn rows. The beans provided nitrogen needed by the corn, the

pumpkins crowded out weeds, and both crops provided foods that nutritionally balance corn. Traditional swidden (slash-and-burn) cultivators in Africa and South America often plant as many as 20 different crops together in small plots. The crops mature at different times, so that there is always something to eat, and the soil is never exposed to erosion for very long. Shade-grown coffee and cocoa can play important roles in conserving biodiversity (see What Do You Think? above).

Table 7.2	Soil Cover and Soil Erosion	
Cropping System	Average Annual Soil Loss (tons/ hectare)	Percent Rainfall Runoff
Bare soil (no crop)	41.0	30
Continuous corn	19.7	29
Continuous wheat	10.1	23
Rotation: corn, wheat, clover	2.7	14
Continuous bluegrass	0.3	12

Source: Based on 14 years of data from Missouri Experiment Station, Columbia, Missouri.

Reduced tillage can have many benefits

Farmers have traditionally used a moldboard plow to till the soil, digging a deep trench and turning the topsoil upside down. In the 1800s it was shown that tilling a field fully—until it was "clean"—increased crop production. It helped control weeds and pests, reducing competition; it brought fresh nutrients to the surface, providing a good seedbed; and it improved surface drainage and aerated the soil. This is still true for many crops and many soil types, but it is not always the best way to grow crops. Less plowing and cultivation often improves water management, preserves soil, saves energy, and increases crop yields.

There are three major **reduced tillage systems**. *Minimum till* involves reducing the number of times a farmer disturbs the soil by plowing, cultivating, and so on. This often involves a disc or chisel plow rather than a traditional moldboard plow. A chisel plow is a curved, chisel-like blade that doesn't turn the soil over but creates ridges on which seeds can be planted. It leaves up to 75 percent of plant debris on the surface between the rows, preventing erosion. *Conserv-till* farming uses a coulter (a sharp disc, like a pizza cutter), which slices through the soil, opening up a furrow, or slot, just wide enough to insert seeds. This disturbs the soil very little and leaves almost all plant debris on the surface. *No-till* planting is accomplished by drilling seeds into the ground directly through mulch and groundcover. This allows a cover crop to be interseeded with a subsequent crop (fig. 7.31).

Farmers who use these conservation tillage techniques often must depend on pesticides (insecticides, fungicides, and herbicides) to control insects and weeds. Increased use of toxic agricultural chemicals is a matter of great concern. Massive use of pesticides is not, however, a necessary corollary of soil conservation. It is possible to combat pests and diseases with integrated pest management that combines crop rotation, trap crops, natural repellents, and biological controls.

Low-input sustainable agriculture can benefit farmers, consumers, and the environment

In contrast to the trend toward industrialization and dependence on chemical fertilizers, pesticides, antibiotics, and artificial growth factors common in conventional agriculture, some farmers are going back to a more natural, agroecological farming style. Find-

Figure 7.31 No-till planting involves drilling seeds through debris from last year's crops. Here, soybeans grow through corn mulch. Debris keeps weeds down, reduces wind and water erosion, and keeps moisture in the soil.

ing that they can't—or don't want to—compete with factory farms, these folks are making money and staying in farming by returning to small-scale, low-input agriculture. The Minar family, for instance, operates a highly successful 150-cow dairy operation on 97 ha (230 acres) near New Prague, Minnesota. No synthetic chemicals are used on their farm. Cows are rotated every day between 45 pastures or paddocks to reduce erosion and maintain healthy grass. Even in the winter, livestock remain outdoors to avoid the spread of diseases common in confinement (fig. 7.32). Antibiotics

Figure 7.32 On the Minar family's 230-acre dairy farm near New Prague, Minnesota, cows and calves spend the winter outdoors in the snow, bedding down on hay. Dave Minar is part of a growing counterculture that is seeking to keep farmers on the land and bring prosperity to rural areas.
©2004 Star Tribune/Minneapolis-St. Paul.

are used only to fight diseases. Milk and meat from this operation are marketed through co-ops and a community-supported agriculture (CSA) program. Sand Creek, which flows across the Minar land, has been shown to be cleaner when leaving the farm than when it enters.

Similarly, the Franzen family, who raise livestock on their organic farm near Alta Vista, Iowa, allow their pigs to roam in lush pastures, where they can supplement their diet of corn and soybeans with grasses and legumes. Housing for these happy hogs is in spacious, open-ended hoop structures. As fresh layers of straw are added to the bedding, layers of manure beneath are composted, breaking down into odorless organic fertilizer.

Low-input farms such as these typically don't turn out the quantity of meat or milk that their intensive agriculture neighbors do, but their production costs are lower, and they get higher prices for their crops, so that the all-important net gain is often higher. The Franzens, for example, calculate that they pay 30 percent less for animal feed, 70 percent less for veterinary bills, and half as much for buildings and equipment as their neighboring confinement operations. And on the Minar's farm, erosion after an especially heavy rain was measured to be 400 times lower than on a conventional farm nearby.

Preserving small-scale family farms also helps preserve rural culture. As Marty Strange of the Center for Rural Affairs in Nebraska asks, "Which is better for the enrollment in rural schools, the membership of rural churches, and the fellowship of rural communities—two farms milking 1,000 cows each or twenty farms milking 100 cows each?" Family farms help keep rural towns alive by purchasing machinery at the local implement dealer, gasoline at the neighborhood filling station, and groceries at the mom-and-pop grocery store.

7.9 Consumer Choices Can Reshape Farming

Since the 1960s, U.S. farm policy and agricultural research has focused on developing large-scale production methods that use fertilizer, pesticides, breeding, and genetic engineering to provide abundant, inexpensive grain, meat, and milk. This strategy has had a complicated combination of very good and very problematic effects. As we discussed early in this chapter, we can now afford to eat more calories and more meat than ever before. Hunger still plagues many people and many regions, but increasing nutrition has

Figure 7.33 Your local farmers' market is a good source of locally grown organic produce.

reached most of the world's population, not just wealthy countries. For the first time, we now worry about weight-associated illnesses as a cause of mortality.

Cheap-food policies have raised production dramatically to feed growing populations both in the United States and abroad. Farm commodity prices have fallen so low, that we now spend billions of dollars every year to support prices or repay farmers whose production expenses are greater than the value of their crops. The United States buys millions of tons of surplus food every year, often using it as food aid to famine-stricken regions. This social good is also problematic, as cheap or free food donations undermine small farmers in other countries, who cannot compete with free or nearly free food on the local market.

These policies are deeply entrenched in our way of life, politics, and food systems. But there may be some steps that consumers can take to support the beneficial changes in farming methods and farm policies, while reducing their negative effects.

You can be a locavore

Every year the Oxford American Dictionary selects a word of the year, which represents an important trend. For 2007 that word was "locavore," or a person who consumes locally produced food. Supporting local farmers can have a variety of benefits, from keeping money in the local economy to ensuring a fresh and healthy diet. Maintaining a viable farm economy can also help slow the conversion of farmland into expanding suburban subdivisions. As many farmers point out, you don't need to eat 100 percent local. Converting just part of your shopping activity to locally produced goods can make a big difference to farmers.

Farmers' markets are usually the easiest way to eat locally (fig. 7.33). The produce is fresh, and profits go directly to the farmer who grows the crop. "Pick your own" farms also let you buy fresh fruit and other products—and they make a fun, social outing. Many conventional grocery stores also now offer locally produced, organic, and pesticide-free foods. Buying these products may (or may not) cost a little more than non-organic and non-local produce, but they can be better for you and they can help keep farming and fresh, local food in the community.

Many colleges and universities have adopted policies to buy as much locally grown food as possible. Because schools purchase a lot of vegetables, meat, eggs, and milk, this can mean a large amount of income for local and regional farm economies. Although this policy can take more effort and creativity than ordering from centralized, national distributors, many college food service administrators are happy to try to buy locally, if they see that students are interested. If your school doesn't have such a policy, perhaps you could talk to administrators about starting one.

Many areas also have "community supported agriculture" (CSA) projects, farms supported by local residents who pay ahead of time for shares of the farm's products, which can vary from vegetables to flowers to meat and eggs. While CSAs require a lump payment early in the season, the net cost of food by the end of the season is often less than you would have paid at the grocery store. You also get to meet interesting people and learn more about your local area by participating in a CSA. Usually you can find local CSA networks by searching online.

You can eat low on the food chain

Since there is less energy involved in producing food from plants than producing it from animals, one way you can reduce your impact on the world's soil and water is to eat a little more grains, vegetables, and dairy and a little less meat. This doesn't mean turning vegetarian—unless you choose to do so. Just returning to the level of protein and fat consumption of your grandparents could make a big difference for the environment and for your health.

You can eat organic, low-input foods

As you have read, agricultural production methods vary enormously in their effects on soil erosion, water contamination, and biodiversity. If you buy organic food, you are supporting farmers who use no pesticides or artificial fertilizers. Often these farmers use crop rotations to preserve soil nutrients and manage erosion carefully, to prevent loss of their topsoil.

Grass-fed beef and free-range poultry or pork can also be excellent low-input foods. Soil erosion is usually minimal on pastures, where vegetation keeps the soil covered year-round. At the same time, grass-fed livestock don't consume conventionally farmed, pesticide- or fertilizer-intensive corn or soy, or antibiotics.

Conclusion

Food production has grown faster than the human population in recent decades, and the percentage of people facing chronic hunger has declined, although the number has increased. Most of us consume more calories and protein than we need. But some 854 million people are still malnourished. Much, or perhaps most, hunger results from political instability, which displaces farmers, inhibits food distribution, and undermines local farming economies.

Increases in food production result from many innovations in agricultural production. The green revolution produced new varieties that yield more crops per hectare, although these crops often require extra inputs such as fertilizers, irrigation, or pesticides. Genetically modified organisms have also increased yields. Most GMOs are designed to tolerate herbicides. Many produce their own pesticides. Confined animal feeding operations, made possible by large-scale corn and soy production, have greatly increased the efficiency of producing meat. The rise of soy in Brazil, and of meat consumption in the United States, China, and elsewhere, result from these innovations.

These changes bring about important environmental effects, such as soil erosion and degradation, and water contamination from pesticide and fertilizer applications. Health effects are also a concern, including weight-related diseases, such as diabetes, exposure to agricultural chemicals, and antibiotic resistance.

Consumers can influence farm production by eating locally, eating low on the food chain, shopping at farmers' markets, and buying organic or grass-fed foods.

Practice Quiz

1. What is Brazil's Cerrado, and how is agriculture affecting it?

2. Explain how soybeans grown in Brazil are improving diets in China.

3. What does it mean to be chronically undernourished? How many people in the world currently suffer from this condition?

4. Why do nutritionists worry about food security? Who is most likely to suffer from food insecurity?

5. Describe the conditions that constitute a famine. Why does Amartya Sen say that famines are caused more by politics and economics than by natural disasters?

6. Define *malnutrition* and *obesity*. How many Americans are now considered obese?

7. What three crops provide most human caloric intake?

8. What are confined animal feeding operations, and why are they controversial?

9. What is *soil*? Why are soil organisms so important?

10. What are four dominant types of soil degradation? What is the primary cause of soil erosion?

11. What do we mean by the *green revolution*?

12. What is *genetic engineering*, or *biotechnology*, and how might it help or hurt agriculture?

13. What is *sustainable agriculture*?

14. How could your choices of coffee or cocoa help preserve forests, biodiversity, and local economies in tropical countries?

15. What are the economic advantages of low-input farming?

Critical Thinking and Discussion Questions

Apply the principles you have learned in this chapter to discuss these questions with other students.

1. Suppose that you were engaged in biotechnology, or genetic engineering; what environmental safeguards would you impose on your own research? Are there experiments that would be ethically off-limits for you?

2. Debate the claim that famines are caused more by human actions (or inactions) than by environmental forces. What scientific evidence would you need to have to settle this question? What hypotheses could you test to help resolve the debate?

3. Should farmers be forced to use ecologically sound techniques that serve farmers' best interests in the long run, regardless of short-term consequences? How could we mitigate hardships brought about by such policies?

4. How would you provide assurance to consumers, government officials, and other interested parties that new genetically engineered products were environmentally and socially safe? What experiments would you set up to test for unknown unknowns?

5. Former U.S. President Jimmy Carter said, "Responsible biotechnology is not the enemy; starvation is." What did he mean? Do you agree?

Data Analysis: Mapping Your Food Supply

Understanding where your food comes from helps you understand the environmental issues involved with producing the food you eat. Because this information is so important, culturally and economically, the United States Department of Agriculture (USDA) maintains a Natural Resources Inventory (NRI) that includes maps of our major crops. While these maps are updated only at long intervals, they describe large trends—the details change only a little from year to year, and the patterns you see in the maps are still current. Find this website and explore the maps to answer the following questions: http://www.nrcs.usda.gov/technical/NRI/maps/cropland.html

1. This page lists maps that show the use and erodibility of soils in the United States. Click on one or two map titles to see the maps that are available. List five topics that the USDA thinks are important to map.

2. Near the bottom of the list of maps, you can find the "Percent of Cultivated Cropland in Corn, 1997." This is the first of several crop maps. *Percentage* here means the fraction of watersheds planted in that crop. Click on that title, then click on the map image to see a larger version of the map. What do the two brownest shades mean? What are seven Midwestern states that have the most dark brown (greater than 50 percent of land in corn)?

3. Locate the Mississippi River. Also locate the Missouri and Ohio rivers, which enter the Mississippi as it passes between

Missouri and Illinois. How does corn production differ on the Missouri and the Ohio? What factors might explain this difference?

4. Return to the page listing map topics. Find the map of "Cultivated Cropland in Soybeans." Is the distribution of soybeans similar to corn, or very different?

5. Most Americans consume more potatoes than any other vegetable. Have you eaten a potato this week? Make a list of three states that you think may have produced the potatoes you eat.

6. Now find the map of land in potatoes. Note that the map shows *intensity* of potato production, in terms of percentage of cultivable land. List five of the states that have the most land in yellow and brown colors. Did your lists match?

7. Return to the list of maps again, and identify two crops that you think should grow in your state. Look at their maps. Were you correct in your guess? Does your state produce more of those two crops than other states do? Why or why not?

For Additional Help in Studying This Chapter, please visit our website at www.mhhe.com/cunningham5e. You will find practice quizzes, key terms, answers to end of chapter questions, additional case studies, an extensive reading list, and Google Earth™ mapping quizzes.

A public health worker explains to villagers that guinea worm is a parasitic disease, not the result of bad behavior or sorcery. Once the most guinea-worm-plagued region in the world, West Africa is now almost free of this terrible disease.

Environmental Health and Toxicology

Learning Outcomes

After studying this chapter, you should be able to answer the following questions:

- What is environmental health?
- What health risks should worry us most?
- Emergent diseases seem to be more frequent now. What human factors may be involved in this trend?
- Are there connections between ecology and our health?
- What are toxins, and how do they affect us?
- When Paracelsus said, "The dose makes the poison," what did he mean?
- What makes some chemicals dangerous and others harmless?
- How much risk is acceptable, and to whom?

To wish to become well is a part of becoming well.

–Seneca

Defeating the Fiery Serpent

Fighting back tears, an African child cradles her swollen foot as a thin, white worm emerges from an oozing sore. The unfortunate youngster has been invaded by a guinea worm (*Dracunculus mediensis*). The pain is intense. Her whole leg feels like it's on fire. It's difficult to walk or work when you're afflicted with one of these terrible invaders. The disease is known in Africa as "empty granary," because the worms usually erupt during harvest season, making it impossible to work in the fields and harvest crops on which the whole family depends.

The infection cycle starts when someone suffering from the pain of an emerging worm (fig. 8.1) bathes their wound in a local lake or pond. The worm, sensing water, emerges to release thousands of larvae, which are ingested by freshwater copepods ("water fleas"). Inside the water fleas, the larvae develop into the infective stage in about two weeks. When villagers drink the contaminated water, the copepods are digested, but the worms survive and penetrate the wall of the intestine and move to the abdominal cavity. Over the next year, the worm grows to about a meter long (3 feet) and as thick as a spaghetti noodle. When fully grown, the worm migrates to the site where it will erupt, usually in the legs or feet—or even eye sockets—of victims. It takes several weeks for the worm to emerge completely. If you pull too hard on it, the worm breaks, and the part left in your body festers and decays. If the suffering host soaks the lesion in water to soothe the pain, the cycle begins again.

This terrible parasite has plagued tropical countries for thousands of years. It has been found in Egyptian mummies and is thought to be the "fiery serpent" described in the Old Testament as torturing the Israelites in the desert. As recently as 1986, at least 3 million people in 16 countries suffered from this affliction, and more than 100 million people were at risk worldwide. But there's a happy ending to this story. In 1988, the world health community, led by former U.S. President Jimmy Carter, began a campaign to eliminate guinea worms.

Eradicating the parasite is fairly simple and relatively inexpensive. There is no medicine to cure the infection once the worm is inside your body, but wells can be drilled to prevent patients from contaminating drinking-water sources. And local ponds can be treated with pesticides that are safe for human consumption but kill the worm larvae and copepods. Furthermore, families can be taught to pour their drinking water through a fine cloth filter to remove any remaining water fleas. The problem is simply to get the proper materials and information to remote villages. Having the prestige of a former American president has been a powerful tool in convincing local officials that the world cares, and that a cure is possible.

Progress has been spectacular. Only a few countries still are afflicted by this horrendous disease. Nigeria is an example of this remarkable success. In 1986, the country was the most Guinea-worm-plagued country in the world, with more than 650,000 cases in 36 states. By 2006, more than 99.9 percent of the infections had been eliminated, and only 120 people still suffered from infections. Worldwide, there now are fewer than 12,000 cases, mostly in Sudan, where civil war has made public health intervention difficult. When guinea worms are finally eradicated, it will be only the second disease ever completely eliminated (smallpox, which was abolished in 1977 was first), and the only human parasite totally exterminated worldwide.

An encouraging outcome of this crusade is the demonstration that public health education and community organization can be effective, even in some of the poorest and most remote areas. Once people understand how the disease spreads and what they need to do to protect themselves and their families, they do change their behavior. And when the campaign is completed and guinea worms are completely vanquished, the health workers and volunteers will be available for further community development projects.

This case study reminds us of the importance of public health and how susceptible humans have always been to diseases and contaminants. In this chapter we'll look at the principles of environmental health to help you understand some of the risks we face and what we might do about them.

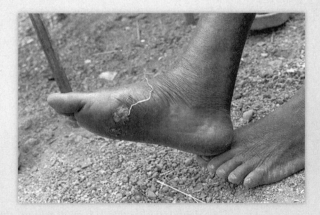

Figure 8.1 A guinea worm emerges from a patient's foot.

8.1 Environmental Health

What is health? The World Health Organization (WHO) defines **health** as a state of complete physical, mental, and social well-being, not merely the absence of disease or infirmity. By that definition, we all are ill to some extent. Likewise, we all can improve our health to live happier, longer, more productive, and more satisfying lives if we think about what we do.

What is disease? A **disease** is an abnormal change in the body's condition that impairs important physical or psychological functions. Diet and nutrition, infectious agents, toxic substances, genetics, trauma, and stress all play roles in **morbidity** (illness) and **mortality** (death). **Environmental health** focuses on factors that cause disease, including elements of the natural, social, cultural, and technological worlds in which we live. The WHO estimates that 24 percent of all global disease burden and 23 percent of premature mortality are due to environmental factors. Among children (0 to 14 years) deaths attributable to environmental factors may be as high as 36 percent. Figure 8.2 shows some of these environmental risk factors as well as the media through which we encounter them.

In ecological terms, your body is an ecosystem. Of the approximately 100 trillion cells that make up each of us, only 10 percent are actually human. The others are bacteria, fungi, protozoans, arthropods, and other microorganisms. Ideally, the various species in this complex system maintain a harmonious balance. Good organisms help regulate the dangerous ones. The health challenge shouldn't be to try to totally eradicate all these other species; we couldn't live without them. Rather, we need to find ways to live in equilibrium with our environment and our fellow travelers.

Ever since the publication of Rachel Carson's *Silent Spring* in 1962, the discharge, movement, fate, and effects of synthetic chemical toxins have been a special focus of environmental health, but, as the opening case study shows, infectious diseases remain a grave threat. In this chapter, we'll study these topics in detail. First, however, let's look at some of the major causes of illness worldwide.

Global disease burden is changing

In the past, health organizations have focused on the leading causes of death as the best summary of world health. Mortality data, however, fail to capture the impacts of nonfatal outcomes of disease and injury, such as dementia or blindness, on human well-being. When people are ill, work isn't done, crops aren't planted or harvested, meals aren't cooked, and children can't study and learn. Health agencies now calculate **disability-adjusted life years (DALYs)** as a measure of disease burden. DALYs combine premature deaths and loss of a healthy life resulting from illness or disability. This is an attempt to evaluate the total cost of disease, not simply how many people die. Clearly, many more years of expected life are lost when a child dies of neonatal tetanus than when an 80-year-old dies of pneumonia. Similarly, a teenager permanently paralyzed by a traffic accident will have many more years of suffering and lost potential than will a senior citizen who has a stroke. According to the WHO, chronic diseases now account for nearly 60 percent of the 56.5 million total deaths worldwide each year and about half of the global disease burden.

The world is now undergoing a dramatic epidemiological transition. Chronic conditions, such as cardiovascular disease and cancer, no longer afflict only wealthy people. Marvelous progress in eliminating communicable diseases, such as smallpox, polio, and malaria, is allowing people nearly everywhere to live longer. As chapter 6 points out, over the past century the average life expectancy worldwide has risen by about two-thirds. In some poorer countries, such as India, life expectancies nearly tripled in the twentieth century. Although the traditional killers in developing countries—infections, maternal and perinatal (birth) complications, and nutritional deficiencies—still take a terrible toll, diseases such as depression and heart attacks that once were thought to occur only in rich countries are rapidly becoming the leading causes of disability and premature death everywhere.

In 2020, the WHO predicts that heart disease, which was fifth in the list of causes of global disease burden a decade ago, will be the leading source of disability and deaths worldwide (table 8.1). Most of that increase will be in the poorer parts of the world where

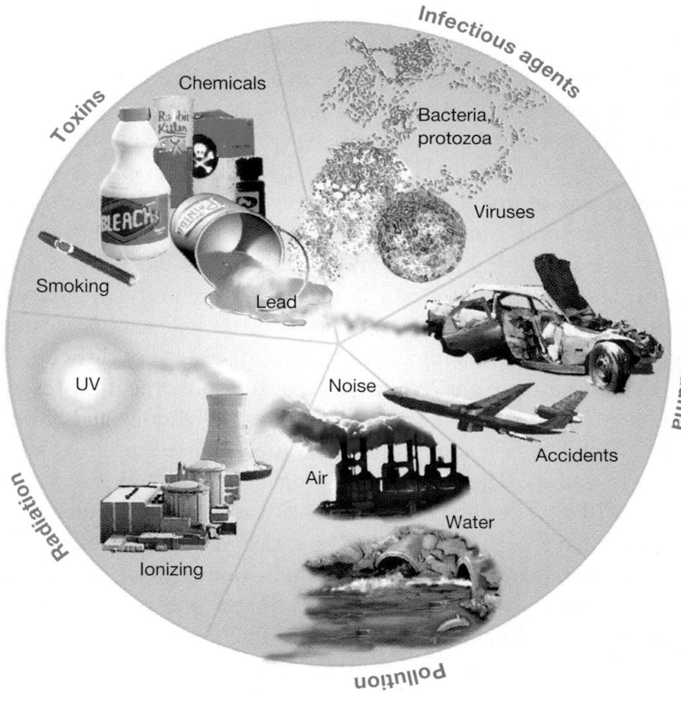

Figure 8.2 Major sources of environmental health risks.

Table 8.1	Leading Causes of Global Disease Burden		
Rank	1990	Rank	2020
1	Pneumonia	1	Heart disease
2	Diarrhea	2	Depression
3	Perinatal conditions	3	Traffic accidents
4	Depression	4	Stroke
5	Heart disease	5	Chronic lung disease
6	Stroke	6	Pneumonia
7	Tuberculosis	7	Tuberculosis
8	Measles	8	War
9	Traffic accidents	9	Diarrhea
10	Birth defects	10	HIV/AIDS
11	Chronic lung disease	11	Perinatal conditions
12	Malaria	12	Violence
13	Falls	13	Birth defects
14	Iron anemia	14	Self-inflicted injuries
15	Malnutrition	15	Respiratory cancer

Source: Data from World Health Organization, 2002.

people are rapidly adopting the lifestyles and diet of the richer countries. Similarly, global cancer rates will increase by 50 percent. By 2020 it's expected that 15 million people will have cancer and 9 million will die from it.

Taking disability as well as death into account in our assessment of disease burden reveals the increasing role of mental health as a worldwide problem. WHO projections suggest that psychiatric and neurological conditions could increase their share of the global burden from 10 percent currently to 15 percent of the total load by 2020. Again, this isn't just a problem of the developed world. Depression is expected to be the second largest cause of all years lived with disability worldwide, as well as the cause of 1.4 percent of all deaths. For women in both developing and developed regions, depression is the leading cause of disease burden, while suicide, which often is the result of untreated depression, is the fourth largest cause of female deaths.

Notice in table 8.1 that diarrhea, which was the second leading cause of disease burden in 1990, is expected to be ninth on the list in 2020, while measles and malaria are expected to drop out of the top 15 causes of disability. Tuberculosis, which is becoming resistant to antibiotics and is spreading rapidly in many areas (especially in Russia), is the only infectious disease whose ranking is not expected to change over the next 20 years. Traffic accidents are now soaring as more people drive. War, violence, and self-inflicted injuries similarly are becoming much more important health risks than ever before.

Chronic obstructive lung diseases (e.g., emphysema, asthma, and lung cancer) are expected to increase from eleventh to fifth in disease burden by 2020. A large part of the increase is due to rising use of tobacco in developing countries, sometimes called "the tobacco epidemic." Every day about 100,000 young people—most of them in poorer countries—become addicted to tobacco. At least 1.1 billion people now smoke, and this number is expected to increase at least 50 percent by 2020. If current patterns persist, about 500 million people alive today will eventually be killed by tobacco. This is expected to be the biggest single cause of death worldwide (because illnesses such as heart attack and depression are triggered by multiple factors). In 2003 the World Health Assembly adopted a historic tobacco-control convention that requires countries to impose restrictions on tobacco advertising, establish clean indoor air controls, and clamp down on tobacco smuggling. Dr. Gro Harlem Brundtland, former director-general of the WHO, predicted that the convention, if ratified by enough nations, could save billions of lives.

As chapter 7 points out, the world is now experiencing an epidemic of obesity. Poor diet and lack of exercise are now the second leading underlying cause of death in America, causing at least 400,000 deaths per year. Obesity is expected to overtake tobacco soon as the largest single health risk in many countries.

Emergent and infectious diseases still kill millions of people

Although the ills of modern life have become the leading killers almost everywhere in the world, communicable diseases still are responsible for about one-third of all disease-related mortality.

Figure 8.3 Millions of children die each year from easily prevented childhood diseases. This Guatemalan billboard urges that children be vaccinated against polio, diphtheria, TB, tetanus, pertussis (whooping cough), and scarlet fever (*left to right*).

Diarrhea, acute respiratory illnesses, malaria, measles, tetanus, and a few other infectious diseases kill about 11 million children under age five every year in the developing world. Better nutrition, clean water, improved sanitation, and inexpensive inoculations could eliminate most of those deaths (fig. 8.3).

A wide variety of **pathogens** (disease-causing organisms) afflict humans, including viruses, bacteria, protozoans (single-celled animals), parasitic worms, and flukes (fig. 8.4). The greatest loss of life from an individual disease in a single year was the great influenza pandemic of 1918. Epidemiologists now estimate that at least one-third of all humans living at the time were infected, and that between 50 to 100 million died. Businesses, schools, churches, and sports or entertainment events were shut down for months. If a similar pandemic were to occur now, it could sicken 2 billion people, kill up to 150 million, and bring the world economy to a standstill. Influenza is caused by a family of viruses (fig. 8.4a) that mutate rapidly and move from wild and domestic animals to humans, making control of this disease very difficult.

Every year there are 76 million cases of foodborne illnesses in the United States, resulting in 300,000 hospitalizations and 5,000 deaths. Both bacteria and intestinal protozoa cause these illnesses (fig. 8.4b and c). They are spread from feces through food and water. In 2006, a deadly strain of *E. coli* was spread in lettuce that appears to have been contaminated by a nearby cattle feedlot. At any given time, around 2 billion people—one-third of the world population—suffer from worms, flukes, and other internal parasites. Guinea worms (see opening case study) are one example, while people rarely die from parasites, they can be extremely debilitating, and can cause poverty that leads to other, more deadly, diseases.

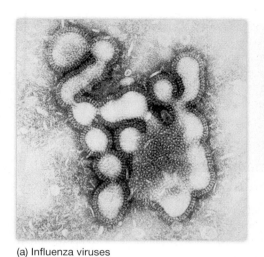

(a) Influenza viruses

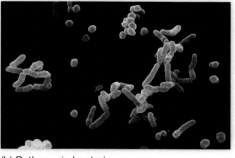

(b) Pathogenic bacteria

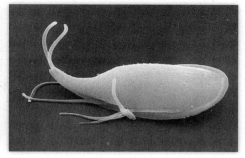

(c) *Giardia*

Figure 8.4 (a) A group of influenza viruses magnified about 300,000 times. (b) Pathogenic bacteria magnified about 50,000 times. (c) *Giardia*, a parasitic intestinal protozoan, magnified about 10,000 times.

Malaria is one of the most prevalent remaining infectious diseases. Every year about 500 million new cases of this disease occur, and about 2 million people die from it. The territory infected by this disease is expanding as global climate change allows mosquito vectors to move into new territory. Simply providing insecticide-treated bed nets and a few dollars worth of chloroquine pills could prevent tens of millions of cases of this debilitating disease every year. Tragically, some of the countries where malaria is most widespread tax both bed nets and medicine as luxuries, placing them out of reach for ordinary people.

Emergent diseases are those not previously known or that have been absent for at least 20 years. The new strain of bird flu now spreading around the world is a good example. There have been at least 39 outbreaks of emergent diseases over the past

two decades, including the extremely deadly Ebola and Marburg fevers, which have afflicted Central Africa in at least six different locations in the past decade. Similarly, cholera, which had been absent from South America for more than a century, reemerged in Peru in 1992 (fig. 8.5). Some other examples include a new drug-resistant form of tuberculosis, now spreading in Russia; dengue fever, which is spreading through Southeast Asia and the Caribbean; malaria, which is reappearing in places where it had been nearly eradicated; and a new human lymphotropic virus (HTLV), which is thought to have jumped from monkeys into people in Cameroon who handled or ate bushmeat. These HTLV strains are now thought to infect 25 million people.

Growing human populations push people into remote areas where they encounter diseases that may have existed for a long

Figure 8.5 Some recent outbreaks of highly lethal infectious diseases. Why are supercontagious organisms emerging in so many different places? **Source:** Data from U.S. Centers for Disease Control.

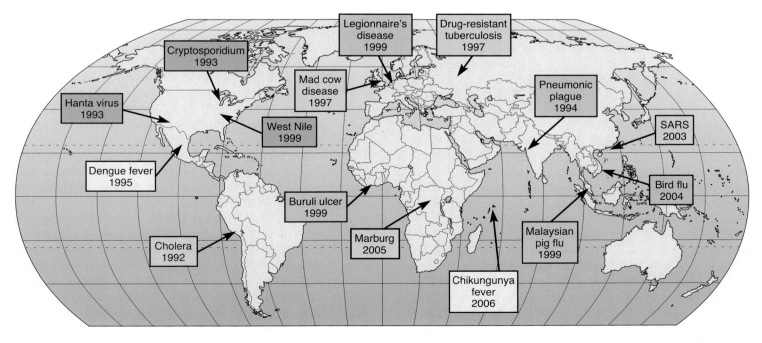

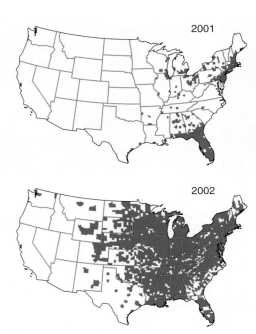

2001

2002

Figure 8.6 The spread of West Nile virus in birds, 2001–2002.

Source: Data from U.S. Centers for Disease Control and U.S. Geological Survey.

time, but only now are exposed to humans. Rapid international travel makes it possible for these new diseases to spread around the world at jet speed. Epidemiologists warn that the next deadly epidemic is only a plane ride away.

West Nile virus shows how fast new diseases can travel. West Nile belongs to a family of mosquito-transmitted viruses that cause encephalitis (brain inflammation). Although recognized in Africa in 1937, the West Nile virus was absent from North America until 1999, when it apparently was introduced by an imported bird or mosquito. The disease spread rapidly from New York, where it was first reported, throughout the eastern United States in only two years (fig. 8.6). Now, it's found everywhere in the lower 48 states. The virus infects at least 250 bird species and 18 mammalian species. In 2007, about 4,000 people contracted West Nile and about 100 died.

The largest recent human death toll from an emergent disease is due to HIV/AIDS. Although it was first recognized in the early 1980s, acquired immune deficiency syndrome has now become the fifth greatest cause of contagious deaths. The WHO estimates that about 33 million people are now infected with the human immune-deficiency virus and that 3 million die every year from AIDS complications. Although two-thirds of all current HIV infections are now in sub-Saharan Africa, the disease is spreading rapidly in South and East Asia. Over the next 20 years, there could be an additional 65 million AIDS deaths. In Swaziland, health officials estimate nearly 40 percent of all adults are HIV-positive and that two-thirds of all current 15-year-olds will die of AIDS before age 50. Without AIDS, the life expectancy in Swaziland would be expected to be 55.3 years. With AIDS, Swaziland's average life expectancy is now 35.7 years. Worldwide, more than 15 million children—the equivalent of every child under age five in America—have lost one or both parents to AIDS. The economic costs of treating patients and lost productivity from premature deaths resulting from this disease are estimated to be at least $35 billion (U.S.) per year, or about one-tenth of the total GDP of sub-Saharan Africa.

Conservation medicine combines ecology and health care

Humans aren't the only ones to suffer from new and devastating diseases. Domestic animals and wildlife also experience sudden and widespread epidemics, which are sometimes called **ecological diseases.** Ebola hemorrhagic fever is one of the most virulent viruses ever seen, killing up to 90 percent of its victims. In 2002, an outbreak of Ebola fever began killing humans along the Gabon-Congo border. A few months later, researchers found that 221 of the 235 western lowland gorillas they had been studying in this area disappeared in just a few months. Many chimpanzees also died. Although the study team could only find a few of the dead gorillas, 75 percent of those tested positive for Ebola. Altogether, researchers estimate that 5,000 gorillas died in this small area of the Congo. Extrapolating to all of central Africa, it's possible that Ebola has killed one-quarter of all the gorillas in the world. It's thought that the spread of this disease in humans resulted from the practice of hunting and eating primates.

A viral hemorrhagic septicemia much like Ebola is spreading through fish in North America's Great Lakes. It is highly infectious and also has an extremely high fatality rate. The virus infects about 40 species of fish. Shipments of game fish, such as salmon, trout, walleye, and perch, have been banned across much of America in an effort to slow the spread of this disease, which probably was introduced in ballast water of ships from Europe.

Botulism often causes disastrous die-offs among birds, especially when they're concentrated during migration. Over the past decade, large numbers of dead water birds have been discovered in the American Great Lakes (fig. 8.7). The spread of this epidemic from east to west through the lakes seems to be associated with the migration of the round goby, an invasive species from Europe. Algae and aquatic plants grow in lake areas fertilized by excessive nutrient runoff. When this plant matter falls to the lake bottom and decays, it creates anoxic conditions in which the bacteria that produce botulism toxin flourish. Invasive zebra and quagga mussels accumulate the toxin through filter feeding. The mussels are then eaten by round gobies, which are paralyzed by the toxin and easily caught by birds that also die.

Chronic wasting disease (CWD) is spreading through deer and elk populations in North America. Caused by a strange protein called a prion, CWD is one of a family of irreversible, degenerative neurological diseases known as transmissible spongiform encephalopathies (TSE) that include mad cow disease in cattle, scrapie in sheep, and Creutzfeldt-Jakob disease in humans. CWD probably started when elk ranchers fed contaminated animal by-products to their herds. Infected animals were sold to other ranches, and now the disease has spread to wild populations. First recognized in 1967 in Saskatchewan, CWD has been identified in wild deer populations and ranch operations in at least eleven American states.

No humans are known to have contracted TSE from deer or elk, but there is a concern that we might see something like the mad cow disaster that inflicted Europe in the 1990s. At least 100 people died, and nearly 5 million European cattle and sheep were slaughtered in an effort to contain that disease.

One thing that emergent diseases in humans and ecological diseases in natural communities have in common is environmental change that stresses biological systems and upsets normal ecological relationships. We cut down forests and drain wetlands, destroying habitat for native species. Invasive organisms and diseases are accidentally or intentionally introduced into new areas where they can grow explosively. Increasing incursion into former wilderness is spurred by human population growth and ecotourism. In 1950, only about 3 million people a year flew on commercial jets; by 2000, more than 300 million did. Diseases can spread around the globe in mere days as people pass through international travel hubs. In the case of severe acute respiratory syndrome (SARS), which emerged in southern China in 2003, one flight attendant is thought to have spread the virus to 160 people in seven countries before he was diagnosed with the disease (see related stories "The Cough Heard Round the World" and "The Next Pandemic" at www.mhhe.com/cunningham5e).

Climate change also facilitates—or forces—expansion of species into new territories. In 2001, a congressionally mandated assessment of climate change predicted greater incidence of malaria, yellow fever, and other tropical diseases in the United States, as mosquitoes, rodents, and other animals expand their range. This phenomenon is occurring worldwide. In the 1970s, only nine countries suffered from dengue fever. Today, dengue has spread to over 100 countries. Other diseases linked to environmental changes include lyme disease, schistosomiasis, tuberculosis, bubonic plague, and cholera.

We are coming to recognize that the delicate ecological balances that we value so highly—and disrupt so frequently—are important to our own health. **Conservation medicine** is an emerging discipline that attempts to understand how our environmental changes threaten our own health as well as that of the natural communities on which we depend for ecological services. While still small, this new field is gaining recognition from mainstream funding sources such as the World Bank, the World Health Organization, and the U.S. National Institutes of Health.

Figure 8.7 Hundreds of loons died in 2007 in the American Great Lakes. They were apparently killed by botulism toxin accumulated by zebra and quagga mussels and round gobies, invasive species from Europe currently spreading through the lakes.

Resistance to antibiotics and pesticides is increasing

In recent years, health workers have become increasingly alarmed about the rapid spread of methicillin-resistant *Staphylococcus aureus* (MRSA). Staphylococcus (or staph) is very common. Most people have at least some of these bacteria. They are a common cause of sore throats and skin infections, but are usually easily controlled. This new strain, however, is resistant to penicillin and related antibiotics, and can cause deadly infections, especially in people with weak immune systems. MRSA is most frequent in hospitals, nursing homes, correctional facilities, and other places where people are in close contact. It's generally spread through direct skin contact. School locker rooms, gymnasiums, and contact sports also are sources of infections. Several states have closed schools as a result of MRSA contamination. In 2006, the most recent date for which data are available, it's estimated that at least 100,000 MRSA infections in the United States resulted in about 19,000 deaths. A much worse situation is reported in China, where about half of the 5 million annual staph infections are thought to be methicillin-resistant.

Malaria, the most deadly of all insect-borne diseases, is another example of the return of a disease that once was thought to be nearly vanquished. Malaria now claims about 2 million lives every year—90 percent in Africa, and most of them children. With the advent of modern medicines and pesticides, malaria had nearly been wiped out in many places but recently has come roaring back. The protozoan parasite that causes the disease is now resistant to most antibiotics, while the mosquitoes that transmit it have developed resistance to many insecticides. Spraying of DDT in India and Sri Lanka, for instance, reduced malaria from millions of infections per year to only a few thousand in the 1950s and 1960s. Now South Asia is back near its pre-DDT levels with about half a million new cases of malaria every year. Other places that never had malaria cases now have them as a result of climate change and habitat alteration.

Why have vectors, such as mosquitoes, and pathogens, such as bacteria or the malaria parasite, become resistant to pesticides and antibiotics? Part of the answer is natural selection and the ability of many organisms to evolve rapidly. Another factor is the human tendency to use control measures carelessly. Many doctors prescribe penicillin and other antibiotics just in case they may do some good. Similarly, when we discovered that DDT and other insecticides could control mosquito populations, we spread them everywhere. This not only harmed wildlife and beneficial insects but also created perfect conditions for natural selection.

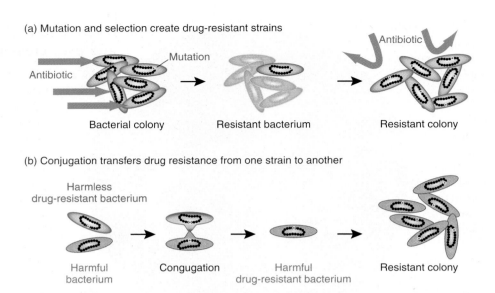

(a) Mutation and selection create drug-resistant strains

Antibiotic

Mutation

Bacterial colony   Resistant bacterium   Resistant colony

(b) Conjugation transfers drug resistance from one strain to another

Harmless
drug-resistant bacterium

Harmful   Congugation   Harmful   Resistant colony
bacterium   drug-resistant bacterium

Figure 8.8 How microbes acquire antibiotic resistance. (a) Random mutations make a few cells resistant. When challenged by antibiotics, only those cells survive to give rise to a resistant colony. (b) Sexual reproduction (conjugation) or plasmid transfer moves genes from one strain or species to another.

Many pests and pathogens were exposed only minimally to control measures, allowing those with natural resistance to survive and spread their genes through the population (fig. 8.8). After repeated cycles of exposure and selection, many microorganisms and their vectors have become insensitive to almost all our weapons against them.

Raising huge numbers of cattle, hogs, and poultry in densely packed barns and feedlots is another reason for widespread antibiotic resistance in pathogens. Confined animals are dosed constantly with antibiotics and steroid hormones to keep them disease-free and to make them gain weight faster. More than half of all antibiotics used in the United States each year are fed to livestock. A significant amount of these antibiotics and hormones are excreted in urine and feces, which are spread, untreated, on the land or discharged into surface water, where they contribute further to the evolution of supervirulent pathogens.

At least half of the 100 million antibiotic doses prescribed for humans every year in the United States are unnecessary or are the wrong ones. Furthermore, many people who start a course of antibiotic treatment fail to carry it out for the time prescribed. For your own health and that of the people around you, if you are taking an antibiotic, follow your doctor's orders. Finish your prescribed doses and don't stop taking the medicine as soon as you start feeling better.

Who should pay for health care?

The heaviest burden of illness is borne by the poorest people, who can afford neither a healthy environment nor adequate health care. Women in sub-Saharan Africa, for example, suffer six times the disease burden per 1,000 population of women in most European countries. The WHO estimates that 90 percent of all disease burden occurs in developing countries, where less than one-tenth of all health care dollars is spent. The group Medecins Sans Frontieres (MSF, or Doctors without Borders) calls this the 10/90 gap. While wealthy nations pursue drugs to treat baldness and obesity, depression in dogs, and erectile dysfunction, billions of people are sick or dying from treatable infections and parasitic diseases to which little attention is paid. Worldwide, only 2 percent of the people with AIDS have access to modern medicines. Every year, some 600,000 infants acquire HIV—almost all of them through mother-to-child transmission during birth or breast-feeding. Antiretroviral therapy costing only a few dollars can prevent most of this transmission. The Bill and Melinda Gates Foundation has pledged $200 million for medical aid to developing countries to help fight AIDS, TB, and malaria.

Dr. Jeffrey Sachs of the Columbia University Earth Institute says that disease is as much a cause as a consequence of poverty and political unrest, yet the world's richest countries now spend just $1 per person per year on global health. He predicts that raising our commitment to about $25 billion annually (about 0.1 percent of the annual GDP of the 20 richest countries) would not only save about 8 million lives each year but also would boost the world economy by billions of dollars. There also would be huge social benefits for the rich countries in not living in a world endangered by mass social instability; the spread of pathogens across borders; the spread of other ills, such as terrorism; and drug trafficking caused by social problems. Sachs also argues that reducing disease burden would help reduce population growth. When parents believe their offspring will survive, they have fewer children and invest more in food, health, and education for smaller families.

The United States is among the least generous of the world's rich countries, donating only about 12 cents per $100 of GDP on international development aid. Could this country do better? During this time of fear of terrorism and rising anti-American feel-

What Can You Do?

Tips for Staying Healthy

- Eat a balanced diet with plenty of fresh fruits, vegetables, legumes, and whole grains. Wash fruits and vegetables carefully; they may have come from a country where pesticide and sanitation laws are lax.

- Use unsaturated oils, such as olive or canola, rather than hydrogenated or semisolid fats, such as margarine.

- Cook meats and other foods at temperatures high enough to kill pathogens; clean utensils and cutting surfaces; store food properly.

- Wash your hands frequently. You transfer more germs from hand to mouth than any other means of transmission.

- When you have a cold or flu, don't demand antibiotics from your doctor—they aren't effective against viruses.

- If you're taking antibiotics, continue for the entire time prescribed— quitting as soon as you feel well is an ideal way to select for antibiotic-resistant germs.

- Practice safe sex.

- Don't smoke; avoid smoky places.

- If you drink, do so in moderation. Never drive when your reflexes or judgment are impaired.

- Exercise regularly: walk, swim, jog, dance, garden. Do something you enjoy that burns calories and maintains flexibility.

- Get enough sleep. Practice meditation, prayer, or some other form of stress reduction. Get a pet.

- Make a list of friends and family who make you feel more alive and happy. Spend time with one of them at least once a week.

Table 8.2	Top 20 Toxic and Hazardous Substances
Material	**Major Sources**
1. Arsenic	Treated lumber
2. Lead	Paint, gasoline
3. Mercury	Coal combustion
4. Vinyl chloride	Plastics, industrial uses
5. Polychlorinated biphenyls (PCBs)	Electric insulation
6. Benzene	Gasoline, industrial use
7. Cadmium	Batteries
8. Benzo(a)pyrene	Waste incineration
9. Polycyclic aromatic hydrocarbons	Combustion
10. Benzo(b)fluoranthene	Fuels
11. Chloroform	Water purification, industry
12. DDT	Pesticide use
13. Aroclor 1254	Plastics
14. Aroclor 1260	Plastics
15. Trichloroethylene	Solvents
16. Dibenz(a,h)anthracene	Incineration
17. Dieldrin	Pesticides
18. Chromium, hexavalent	Paints, coatings, welding, anticorrosion agents
19. Chlordane	Pesticides
20. Hexachlorobutadiene	Pesticides

Source: Data from U.S. Environmental Protection Agency.

ings around the globe, it's difficult to interest legislators in international aid, yet helping reduce disease might win the United States more friends and make the nation safer than buying more bombs and bullets. Improved health care in poorer countries may also help prevent the spread of emergent diseases, such as SARS, in a globally interconnected world.

Epidemiologists note that almost all of the 2.2 billion people expected to be added to the world population in the next 30 years will live in megacities of the developing world. The economic and environmental conditions in those cities will have a profound impact on global disease burden. Many world leaders urge us to address the "lethal disease of poverty." More discussion of urban areas and their problems is presented in chapter 14.

8.2 Toxicology

Toxicology is the study of **toxins** (poisons) and their effects, particularly on living systems. Because many substances are known to be poisonous to life (whether plant, animal, or microbial), toxicology is a broad field, drawing from biochemistry, histology, pharmacology, pathology, and many other disciplines. Toxins damage or kill living organisms because they react with cellular components to dis-

rupt metabolic functions. Because of this reactivity, toxins often are harmful even in extremely dilute concentrations. In some cases, billionths, or even trillionths, of a gram can cause irreversible damage.

All toxins are hazardous, but not all hazardous materials are toxic. Some substances, for example, are dangerous because they're flammable, explosive, acidic, caustic, irritants, or sensitizers. Many of these materials must be handled carefully in large doses or high concentrations, but they can be rendered relatively innocuous by dilution, neutralization, or other physical treatment. They don't react with cellular components in ways that make them poisonous at low concentrations.

Environmental toxicology, or ecotoxicology, specifically deals with the interactions, transformation, fate, and effects of natural and synthetic chemicals in the biosphere, including individual organisms, populations, and whole ecosystems. In aquatic systems the fate of the pollutants is primarily studied in relation to mechanisms and processes at interfaces of the ecosystem components. Special attention is devoted to the sediment/water, water/ organisms, and water/air interfaces. In terrestrial environments, the emphasis tends to be on effects of metals on soil fauna community and population characteristics.

Table 8.2 is a list of the top 20 toxins compiled by the U.S. Environmental Protection Agency from the 275 substances regulated by the Comprehensive Environmental Response, Compensation, and Liability Act (CERCLA), commonly known as the Superfund Act. These materials are listed in order of assessed importance in terms of human and environmental health.

How do toxins affect us?

Allergens are substances that activate the immune system. Some allergens act directly as **antigens**; that is, they are recognized as foreign by white blood cells and stimulate the production of specific antibodies (proteins that recognize and bind to foreign cells or chemicals). Other allergens act indirectly by binding to and changing the chemistry of foreign materials so they become antigenic and cause an immune response.

Formaldehyde is a good example of a widely used chemical that is a powerful sensitizer of the immune system. It is directly allergenic and can trigger reactions to other substances. Widely used in plastics, wood products, insulation, glue, and fabrics, formaldehyde concentrations in indoor air can be thousands of times higher than in normal outdoor air. Some people suffer from what is called **sick building syndrome**: headaches, allergies, and chronic fatigue caused by poorly vented indoor air contaminated by molds, carbon monoxide, nitrogen oxides, formaldehyde, and other toxic chemicals released by carpets, insulation, plastics, building materials, and other sources (fig. 8.9). The Environmental Protection Agency estimates that poor indoor air quality may cost the nation $60 billion a year in absenteeism and reduced productivity.

Neurotoxins are a special class of metabolic poisons that specifically attack nerve cells (neurons). The nervous system is so important in regulating body activities that disruption of its activities is especially fast-acting and devastating. Different types of neurotoxins act in different ways. Heavy metals, such as lead and mercury, kill nerve cells and cause permanent neurological damage. Anesthetics (ether, chloroform, halothane, etc.) and chlorinated hydrocarbons (DDT, Dieldrin, Aldrin) disrupt nerve cell membranes necessary for nerve action. Organophosphates (Malathion, Parathion) and carbamates (carbaryl, zeneb, maneb) inhibit acetylcholinesterase, an enzyme that regulates signal

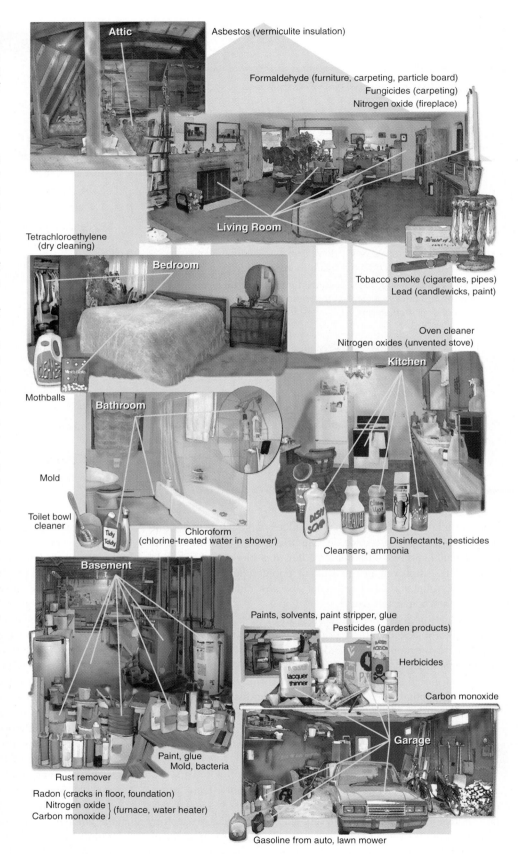

Figure 8.9 Some sources of toxic and hazardous substances in a typical home.

transmission between nerve cells and the tissues or organs they innervate (for example, muscle). Most neurotoxins are both acute and extremely toxic. More than 850 compounds are now recognized as neurotoxins.

Mutagens are agents, such as chemicals and radiation, that damage or alter genetic material (DNA) in cells. This can lead to birth defects if the damage occurs during embryonic or fetal growth. Later in life, genetic damage may trigger neoplastic (tumor) growth. When damage occurs in reproductive cells, the results can be passed on to future generations. Cells have repair mechanisms to detect and restore damaged genetic material, but some changes may be hidden, and the repair process itself can be flawed. It is generally accepted that there is no "safe" threshold for exposure to mutagens. Any exposure has some possibility of causing damage.

Teratogens are chemicals or other factors that specifically cause abnormalities during embryonic growth and development. Some compounds that are not otherwise harmful can cause tragic problems in these sensitive stages of life. Perhaps the most prevalent teratogen in the world is alcohol. Drinking during pregnancy can lead to **fetal alcohol syndrome**—a cluster of symptoms including craniofacial abnormalities, developmental delays, behavioral problems, and mental defects, that last throughout a child's life. Even one alcoholic drink a day during pregnancy has been associated with decreased birth weight.

Carcinogens are substances that cause **cancer**—invasive, out-of-control cell growth that results in malignant tumors. Cancer rates rose in most industrialized countries during the twentieth century, and cancer is now the second leading cause of death in the United States, killing more than half a million people in 2000. Twenty-three of the 28 compounds listed by the U.S. EPA as greatest risk to human health are probable or possible human carcinogens. More than 200 million people live in areas where the combined upper limit lifetime cancer risk from these carcinogens exceeds 10 in 1 million, or 10 times the risk normally considered acceptable.

The American Cancer Society calculates that one in two males and one in three females in the United States will have some form of cancer in their lifetime. Some authors blame this cancer increase on toxic synthetic chemicals in our environment and diet. Others argue that it is attributable mainly to lifestyle (smoking, sunbathing, drinking alcohol) or simply living longer.

Endocrine hormone disrupters are of special concern

One of the most recently recognized environmental health threats are **endocrine hormone disrupters**, chemicals that interrupt the normal endocrine hormone functions. Hormones are chemicals released into the bloodstream by glands in one part of the body to regulate the development and function of tissues and organs elsewhere in the body (fig. 8.10). You undoubtedly have heard about sex hormones and their powerful effects on how we look and behave, but these are only one example of the many regulatory hormones that rule our lives.

We now know that some of the most insidious effects of persistent chemicals, such as DDT and PCBs, are that they interfere

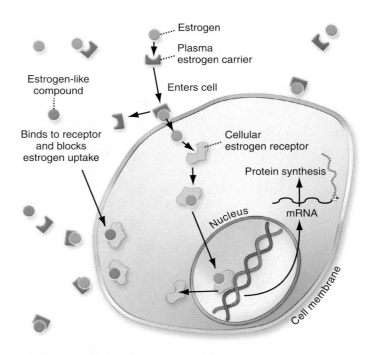

Figure 8.10 Steroid hormone action. Plasma hormone carriers deliver regulatory molecules to the cell surface, where they cross the cell membrane. Intracellular carriers deliver hormones to the nucleus, where they bind to and regulate expression of DNA.

with normal growth, development, and physiology of a variety of animals—presumably including humans—at very low doses. In some cases, picogram concentrations (trillionths of a gram per liter) may be enough to cause developmental abnormalities in sensitive organisms. These chemicals are sometimes called environmental estrogens or androgens, because they often cause sexual dysfunction (reproductive health problems in females or feminization of males, for example). They are just as likely, however, to disrupt thyroxin functions or those of other important regulatory molecules as they are to obstruct sex hormones.

8.3 Movement, Distribution, and Fate of Toxins

There are many sources of toxic and hazardous chemicals in the environment and many factors related to each chemical itself, its route or method of exposure, and its persistence in the environment, as well as characteristics of the target organism (table 8.3), that determine the danger of the chemical. We can think of both individuals and an ecosystem as sets of interacting compartments between which chemicals move, based on molecular size, solubility, stability, and reactivity (fig. 8.11). The dose (amount), route of entry, timing of exposure, and sensitivity of the organism all play important roles in determining toxicity. In this section, we will consider some of these characteristics and how they affect environmental health.

Air
Photolysis
Oxidation
Precipitation

Source
Industry
Agriculture
Domestic, etc.

Biota
Metabolism
Storage
Excretion

Soil and sediment
Photolysis and metabolism
Evaporation

Water
Hydrolysis
Oxidation
Microbial degradation
Evaporation
Sedimentation

Figure 8.11 Movement and fate of chemicals in the environment. Processes that modify, remove, or sequester compounds are shown below each compartment. Toxins also move directly from a source to soil and sediment.

Table 8.3	Factors in Environmental Toxicity
Factors Related to the Toxic Agent	
1.	Chemical composition and reactivity
2.	Physical characteristics (such as solubility, state)
3.	Presence of impurities or contaminants
4.	Stability and storage characteristics of toxic agent
5.	Availability of vehicle (such as solvent) to carry agent
6.	Movement of agent through environment and into cells
Factors Related to Exposure	
1.	Dose (concentration and volume of exposure)
2.	Route, rate, and site of exposure
3.	Duration and frequency of exposure
4.	Time of exposure (time of day, season, year)
Factors Related to the Organism	
1.	Resistance to uptake, storage, or cell permeability of agent
2.	Ability to metabolize, inactivate, sequester, or eliminate agent
3.	Tendency to activate or alter nontoxic substances so they become toxic
4.	Concurrent infections or physical or chemical stress
5.	Species and genetic characteristics of organism
6.	Nutritional status of subject
7.	Age, sex, body weight, immunological status, and maturity

Solubility and mobility determine when and where chemicals move

Solubility is one of the most important characteristics in determining how, where, and when a toxic material will move through the environment or through the body to its site of action. Chemicals can be divided into two major groups: those that dissolve more readily in water and those that dissolve more readily in oil. Water-soluble compounds move rapidly and widely through the environment because water is ubiquitous. They also tend to have ready access to most cells in the body because aqueous solutions bathe all our cells. Molecules that are oil- or fat-soluble (usually organic molecules) generally need a carrier to move through the environment and into or within the body. Once inside the body, however, oil-soluble toxins penetrate readily into tissues and cells because the membranes that enclose cells are themselves made of similar oil-soluble chemicals. Once inside cells, oil-soluble materials are likely to accumulate and to be stored in lipid deposits, where they may be protected from metabolic breakdown and persist for many years.

Exposure and susceptibility determine how we respond

Just as there are many sources of toxins in our environment, there are many routes for entry of dangerous substances into our bodies (fig. 8.12). Airborne toxins generally cause more ill health than any other exposure source. We breathe far more air every day than the volume of food we eat or water we drink. Furthermore, the lining of our lungs, which is designed to exchange gases very efficiently, also absorbs toxins very well. Epidemiologists estimate that 3 million people—two-thirds of them children—die each year from diseases caused or exacerbated by air pollution.

Still, food, water, and skin contact also can expose us to a wide variety of toxins. The largest exposures for many toxins are found in industrial settings, where workers may encounter doses thousands of times higher than would be found anywhere else. The European Agency for Safety and Health at Work warns that 32 million people (20 percent of all employees) in the European Union are exposed to unacceptable levels of carcinogens and other toxins in their workplace.

Condition of the organism and timing of exposure also have strong influences on toxicity. Healthy adults, for example, may be

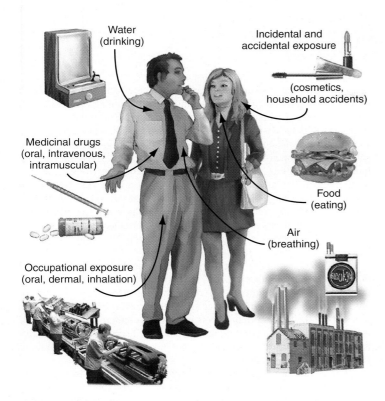

Figure 8.12 Routes of exposure to toxic and hazardous environmental factors.

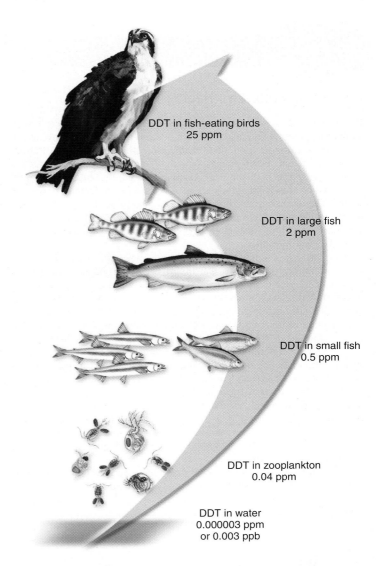

Figure 8.13 Bioaccumulation and biomagnification. Organisms lower on the food chain take up and store toxins from the environment. They are eaten by larger predators, who are eaten, in turn, by even larger predators. The highest members of the food chain can accumulate very high levels of the toxin.

relatively insensitive to doses that would be very dangerous for young children or for someone already weakened by disease (What Do You Think? p. 188). Similarly, exposure to a toxin may be very dangerous at certain stages of developmental or metabolic cycles but may be innocuous at other times. A single dose of the notorious teratogen thalidomide, for example, taken in the third week of pregnancy (a time when many women aren't aware they're pregnant) can cause severe abnormalities in fetal limb development. A complication in measuring toxicity is that great differences in sensitivity exist among species. Thalidomide was tested on a number of laboratory animals without any deleterious effects. Unfortunately, however, it is a powerful teratogen in humans.

Bioaccumulation and biomagnification increase chemical concentrations

Cells have mechanisms for **bioaccumulation**, the selective absorption and storage of a great variety of molecules. This allows them to accumulate nutrients and essential minerals, but at the same time, they also may absorb and store harmful substances through the same mechanisms. Toxins that are rather dilute in the environment can reach dangerous levels inside cells and tissues through this process of bioaccumulation.

The effects of toxins also are magnified in the environment through food webs. **Biomagnification** occurs when the toxic burden of a large number of organisms at a lower trophic level is accumulated and concentrated by a predator in a higher trophic level.

Phytoplankton and bacteria in aquatic ecosystems, for instance, take up heavy metals or toxic organic molecules from water or sediments (fig. 8.13). Their predators—zooplankton and small fish—collect and retain the toxins from many prey organisms, building up higher concentrations of toxins. The top carnivores in the food chain—game fish, fish-eating birds, and humans—can accumulate such high toxin levels that they suffer adverse health effects.

One of the first known examples of bioaccumulation and biomagnification was DDT, which accumulated through food chains, so that by the 1960s it was shown to be interfering with reproduction of peregrine falcons, brown pelicans, and other predatory birds at the top of their food chains.

Protecting Children's Health

Increasing evidence shows that children are much more vulnerable than adults to environmental health hazards. Pound for pound, children drink more water, eat more food, and breathe more air than do adults. Putting fingers, toys, and other objects into their mouths increases children's exposure to toxins in dust or soil. Compared to adults, children generally have less-developed immune systems or processes to degrade or excrete toxins. The brain is especially sensitive to disruption. From before birth to adolescence, the brain undergoes a highly complex series of changes. Obviously, anything that interferes with brain development can have tragic long-term results.

In 2006, an international team of epidemiologists, led by Drs. Phillippe Grandjean, of Harvard, and Philip Landrigan, of the Mount Sinai School of Medicine, both of whom are pioneers in studying long-term effects of toxins on children, surveyed toxicity data on common industrial chemicals. They identified 202 toxic substances—about half of them used in very large volumes—known to damage human brains. They found that effects of these chemicals on children have generally been neglected.

"The human brain is a precious and vulnerable organ. And because optimal brain function depends on the integrity of the organ, even limited damage may have serious consequences," says Dr. Grandjean. He estimates that one out of every six children in America has a developmental disability, usually involving the nervous system. Among the neurodevelopmental disorders that may be associated with toxic chemical exposure are autism, mental retardation, attention deficit disorder, shortened attention spans, slowed motor coordination, increased aggressiveness, and diminished social skills.

Despite years of evidence that many industrial substances are dangerous, few have been studied thoroughly for their effects on children.

Lead was the first toxicant identified as having effects at low levels on developing brains of children. Banning leaded gasoline and paint was one of the most successful steps ever taken to protect children's health. Between 1920 and 1970, virtually all children in industrialized countries were exposed to lead from house paint and leaded gasoline. Before these sources were banned in the 1970s, at least 4.4 million American children had more than 10 μg of lead per dl of blood, (the level considered safe by the U.S. EPA), and hundreds of children died each year from acute lead poisoning. Epidemiologists now estimate that for every 10 μg of lead per dl of blood, IQ scores drop 4.6 points. Grandjean and Landrigan suggest that high lead exposures in the United States reduced IQ scores above 130 (considered superior intelligence) by half, while the number of scores less than 70 increased comparably.

By 2006, the number of children with elevated blood lead levels had fallen more than 90 percent (fig. 1). This is a remarkable success story, but about 430,000 children—especially in poor urban areas—still have unsafe lead levels. Recent evidence suggests that even 10 μg/dl represents a risk of subtle but important effects in growing children's brains, including diminished learning ability, attention deficit disorder, and hyperactivity. Many public health experts argue that blood lead levels in all children should be below 2.5 μg/dl (fig. 2).

Grandjean and Landrigan point out that, in addition to lead, only four other toxins—methylmercury, arsenic, PDBs and toluene—are specifically regulated to protect children. Because industrialization has now become global, they warn that this represents a "silent pandemic" in which millions of children may be suffering reduced intelligence. They suggest that to protect children from industrial chemicals that cause mental retardation and developmental disabilities, we should adopt a precautionary approach to chemical testing and control. This philosophy is

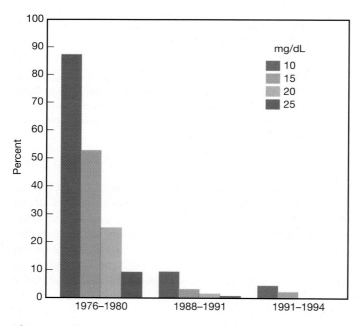

Figure 1 Blood lead levels in U.S. preschoolers fell 90 percent between the 1970s and 1990s. This is one of our greatest environmental health successes ever.
Source: J. Pickle et al., 1998.

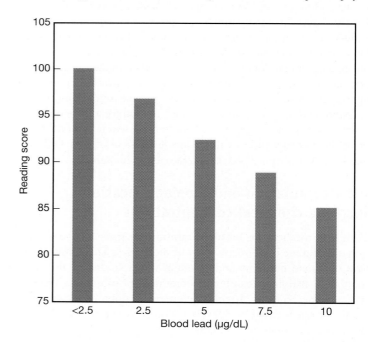

Figure 2 Reading scores of U.S. children are inversely related to blood lead levels even below 10 μg/dL, the amount considered safe by the U.S. EPA.
Source: J. Pickle et al., 1998.

already being applied in Europe, where industry is required to show that a chemical is safe before it is introduced, rather than requiring regulatory agencies to prove that it is unsafe in order to restrict its use.

What do you think? How important is it to protect children's mental development? What does it cost society to have large numbers of slow learners and behavioral problems? Industry claims that more rigorous testing and stricter regulations on industrial chemicals will drive up consumer prices and limit consumer choices. Would you support stronger regulations now, which could be relaxed later if hazards are less than anticipated, or the current regulations that require a high level of proof before restricting use of toxic products?

Persistence makes some materials a greater threat

Many substances degrade when exposed to sun, air, and water. This can destroy toxins or convert them to inactive forms, but some materials are persistent and can last for years or even centuries as they cycle through ecosytems. Even if released in minute concentrations, they can bioaccumulate in food webs to reach dangerous levels. Heavy metals, such as lead (see What Do You Think? p. 188) and mercury, are classic examples. Mercury, like lead, can destroy nerve cells and is particularly dangerous to children. The largest source of mercury in the United States is from burning coal. Every year, power plants in the United States release 48 tons of this toxic metal into the air (see "Controlling Mercury Pollution" at www.mhhe.com/cunningham5e). It works its way through food chains and is concentrated to dangerous levels in fish. Mercury contamination is the most common cause of lakes and rivers failing to meet pollution regulation standards. Forty-four states have issued warnings against eating local fish by children and pregnant women. In a nationwide survey of lakes and rivers in 2007, the Environmental Protection Agency found that 55 percent of the 2,500 fish sampled had mercury levels that exceeded dietary recommendations.

Many organic compounds, such as PVC plastics and chlorinated hydrocarbon pesticides, also are highly resistant to degradation. This makes them very useful, but it also allows them to accumulate in the environment and have unexpected effects far from the site of their original use. Some of these **persistent organic pollutants (POPs)** have become extremely widespread, being found now from the tropics to the Arctic. They often accumulate in food webs and reach toxic concentrations in long-living top predators, such as humans, sharks, raptors, swordfish, and bears. These are some of the greatest current concerns:

- Polybrominated diphenyl ethers (PBDE) are widely used as flame-retardants in textiles, foam in upholstery, and plastic in appliances and computers. This compound was first reported accumulating in women's breast milk in Sweden in the 1990s. It was subsequently found in humans and other species everywhere from Canada to Israel. Nearly 150 million metric tons (330 million lbs) of PBDEs are used every year worldwide. The toxicity and environmental persistence of PBDE are much like those of PCBs, to which it is closely related chemically. The dust at ground zero in New York City after September 11 was heavily laden with PBDE. The European Union has already banned this compound.

- Perfluorooctane sulfonate (PFOS) and perfluorooctanoic acid (PFOA, also known as C8) are members of a chemical family used to make nonstick, waterproof, and stain-resistant products such as Teflon, Gortex, Scotchguard, and Stainmaster. Industry makes use of their slippery, heat-stable properties to manufacture everything from airplanes and computers to cosmetics and household cleaners. Now these chemicals—which are reported to be infinitely persistent in the environment—are found throughout the world, even the most remote and seemingly pristine sites. Almost all Americans have one or more perfluorinated compounds in their blood. In one long-term study, workers exposed to high levels of PFOA were twice as likely to die of prostate cancer or stroke than colleagues with little or no exposure to the chemical. Heating some nonstick cooking pans above 260°C (500°F) can release enough PFOA to kill pet birds. This chemical family has been shown to cause liver damage as well as various cancers and reproductive and developmental problems in rats. Exposure may be especially dangerous to women and girls, who may be 100 times more sensitive than men to these chemicals.

- Perchlorate is a waterborne contaminant left over from propellants and rocket fuels. About 12,000 sites in the United States were used by the military for live munition testing and are contaminated with perchlorate. Polluted water used to irrigate crops such as alfalfa and lettuce has introduced the chemical into the human food chain. Tests of cow's milk and human breast milk detected perchlorate in nearly every sample from throughout the United States. Perchlorate can interfere with iodine uptake in the thyroid gland, disrupting adult metabolism and childhood development.

- Phthalates (pronounced *thalates*) are found in cosmetics, deodorants, and many plastics (such as soft polyvinyl chloride, or PVC) used for food packaging, children's toys, and medical devices. Some members of this chemical family are known to be toxic to laboratory animals, causing kidney and liver damage and possibly some cancers. In addition, many phthalates act as endocrine hormone disrupters and have been linked to reproductive abnormalities and decreased fertility. A correlation has been found between phthalate levels in urine and low sperm numbers and decreased sperm motility in men. Nearly everyone in the United States has phthalates in his or her body at levels reported to cause these problems. While not yet conclusive, these results could help explain a 50-year decline in semen quality in most industrialized countries. In 2007, California banned phthalates in products designed for children.

- Bisphenol A (BPA), a prime ingredient in polycarbonate plastic (commonly used for products ranging from water bottles to tooth-protecting sealants), has been widely found in humans. One possible source is canned food, which commonly has BPA in interior linings. So far, there is little direct evidence linking BPA exposure to human health risks, but studies in animals have found that the chemical can cause abnormal chromosome numbers, a condition called aneuploidy, which is the leading cause of miscarriages and several forms of mental retardation. It also is an environmental estrogen and may alter sexual development in both males and females.

- Atrazine is the most widely used herbicide in America. More than 60 million pounds of this compound are applied per year, mainly on corn and cereal grains, but also on golf courses, sugar cane, and Christmas trees. It has long been known to disrupt endocrine hormone functions in mammals, resulting in spontaneous abortions, low birth weights, and neurological disorders. Studies of families in corn-producing areas in the American Midwest have found higher rates of developmental defects among infants, and certain cancers in families with elevated atrazine levels in their drinking water. University of California professor Tyrone Hayes has shown that atrazine levels as low as 0.1 ppb (30 times less than the EPA maximum contaminant level) caused severe reproductive effects in amphibians, including abnormal gonadal development and hermaphroditism. Atrazine now is found in rain and surface waters nearly everywhere in the United States at levels that could cause abnormal development in frogs. In 2003, the European Union withdrew regulatory approval for this herbicide, and several countries banned its use altogether. Some toxicologists have suggested a similar rule in the United States.

Chemical interactions can increase toxicity

Some materials produce *antagonistic* reactions. That is, they interfere with the effects or stimulate the breakdown of other chemicals. For instance, vitamins E and A can reduce the response to some carcinogens. Other materials are *additive* when they occur together in exposures. Rats exposed to both lead and arsenic show twice the toxicity of only one of these elements. Perhaps the greatest concern is synergistic effects. **Synergism** is an interaction in which one substance exacerbates the effects of another. For example, occupational asbestos exposure increases lung cancer rates 20-fold. Smoking increases lung cancer rates by the same amount. Asbestos workers who also smoke, however, have a 400-fold increase in cancer rates. How many other toxic chemicals are we exposed to that are below threshold limits individually but combine to give toxic results?

8.4 Mechanisms for Minimizing Toxic Effects

A fundamental concept in toxicology is that every material can be poisonous under some conditions, but most chemicals have a safe level or threshold below which their effects are undetectable or insignificant. Each of us consumes lethal doses of many chemicals over the course of a lifetime. One hundred cups of strong coffee, for instance, contain a lethal dose of caffeine. Similarly, 100 aspirin tablets, 10 kg (22 lbs) of spinach or rhubarb, or a liter of alcohol would be deadly if consumed all at once. Taken in small doses, however, most toxins can be broken down or excreted before they do much harm. Furthermore, the damage they cause can be repaired. Sometimes, however, mechanisms that protect us from one type of toxin or at one stage in the life cycle become deleterious with another substance or in another stage of development. Let's look at how these processes help protect us from harmful substances, as well as how they can go awry.

Metabolic degradation and excretion eliminate toxins

Most organisms have enzymes that process waste products and environmental poisons to reduce their toxicity. In mammals, most of these enzymes are located in the liver, the primary site of detoxification of both natural wastes and introduced poisons. Sometimes, however, these reactions work to our disadvantage. Compounds such as benzepyrene, for example, that are not toxic in their original form are processed by the same liver enzymes into cancer-causing carcinogens. Why would we have a system that makes a chemical more dangerous? Evolution and natural selection are expressed through reproductive success or failure. Defense mechanisms that protect us from toxins and hazards early in life are "selected for" by evolution. Factors or conditions that affect postreproductive ages (such as cancer or premature senility) usually don't affect reproductive success or exert "selective pressure."

We also reduce the effects of waste products and environmental toxins by eliminating them from the body through excretion. Volatile molecules, such as carbon dioxide, hydrogen cyanide, and ketones, are excreted via breathing. Some excess salts and other substances are excreted in sweat. Primarily, however, excretion is a function of the kidneys, which can eliminate significant amounts of soluble materials through urine formation. Toxin accumulation in the urine can damage this vital system, however, and the kidneys and bladder often are subjected to harmful levels of toxic compounds. In the same way, the stomach, intestine, and colon often suffer damage from materials concentrated in the digestive system and may be afflicted by diseases and tumors.

Repair mechanisms mend damage

In the same way that individual cells have enzymes to repair damage to DNA and protein at the molecular level, tissues and organs that are exposed regularly to physical wear-and-tear or to toxic or hazardous materials often have mechanisms for damage repair. Our skin and the epithelial linings of the gastrointestinal tract, blood vessels, lungs, and urogenital system have high cellular reproduction rates to replace injured cells. With each reproduction cycle, however, there is a chance that some cells will lose normal growth controls and run amok, creating a tumor. Thus, any agent, such as smoking or drinking, that irritates tissues is likely to be carcinogenic. And tissues with high cell-replacement rates are among the most likely to develop cancers.

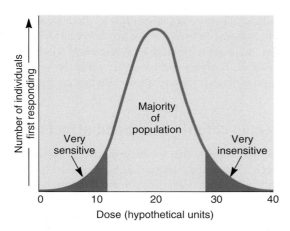

Figure 8.14 Probable variations in sensitivity to a toxin within a population. Some members of a population may be very sensitive to a given toxin, while others are much less sensitive. The majority of the population falls somewhere between the two extremes.

8.5 Measuring Toxicity

In 1540 the Swiss scientist Paracelsus said, "The dose makes the poison," by which he meant that almost everything is toxic at some level. This remains the most basic principle of toxicology. Sodium chloride (table salt), for instance, is essential for human life in small doses. If you were forced to eat a kilogram of salt all at once, however, it would make you very sick. A similar amount injected into your bloodstream would be lethal. How a material is delivered—at what rate, through which route of entry, and in what medium—plays a vital role in determining toxicity.

This does not mean that all toxins are identical, however. Some are so poisonous that a single drop on your skin can kill you. Others require massive amounts injected directly into the blood to be lethal. Measuring and comparing the toxicity of various materials are difficult because species differ in sensitivity, and individuals within a species respond differently to a given exposure. In this section, we will look at methods of toxicity testing and at how results are analyzed and reported.

We usually test toxins on lab animals

The most commonly used and widely accepted toxicity test is to expose a population of laboratory animals to measured doses of a specific substance under controlled conditions. This procedure is expensive, time-consuming, and often painful and debilitating to the animals being tested. It commonly takes hundreds—or even thousands—of animals, several years of hard work, and hundreds of thousands of dollars to thoroughly test the effects of a toxin at very low doses. More humane toxicity tests using computer simulations of model reactions, cell cultures, and other substitutes for whole living animals are being developed. However, conventional large-scale animal testing is the method in which scientists have the most confidence and on which most public policies about pollution and environmental or occupational health hazards are based.

In addition to humanitarian concerns, several other problems in laboratory animal testing trouble both toxicologists and policymakers. One problem is differences in toxin sensitivity among the members of a specific population. Figure 8.14 shows a typical dose/response curve for exposure to a hypothetical toxin. Some individuals are very sensitive to the toxin, while others are insensitive. Most, however, fall in a middle category, forming a bell-shaped curve. The question for regulators and politicians is whether we should set pollution levels that will protect everyone, including the most sensitive people, or only aim to protect the average person. It might cost billions of extra dollars to protect a very small number of individuals at the extreme end of the curve. Is that a good use of resources? Why or why not?

Dose/response curves are not always symmetrical, making it difficult to compare toxicity of unlike chemicals or different species of organisms. A convenient way to describe toxicity of a chemical is to determine the dose to which 50 percent of the test population is sensitive. In the case of a lethal dose (LD), this is called the **LD50** (fig. 8.15).

Unrelated species can react very differently to the same toxin, not only because body sizes vary but also because physiology and metabolism differ. Even closely related species can have very dissimilar reactions to a particular toxin. Hamsters, for instance, are nearly 5,000 times less sensitive to some dioxins than are guinea pigs. Of 226 chemicals found to be carcinogenic in either rats or mice, 95 cause cancer in one species but not the other. These variations make it difficult to estimate the risks for humans, since we don't consider it ethical to perform controlled experiments in which we deliberately expose people to toxins.

There is a wide range of toxicity

It is useful to group materials according to their relative toxicity. A moderate toxin takes about 1 g per kg of body weight (about 2 oz for an average human) to make a lethal dose. Very toxic materials take about one-tenth that amount, while extremely toxic substances take one-hundredth as much (only a few drops) to kill most people. Supertoxic chemicals are extremely potent; for some, a few micrograms (millionths of a gram—an amount invisible to the naked eye) make a lethal dose. These materials are not all synthetic. One of the most toxic chemicals known, for instance, is ricin, a protein found in castor bean seeds. It is so toxic that 0.3 billionths of a gram given intravenously will kill a mouse. If aspirin were this toxic for humans, a single tablet, divided evenly, could kill 1 million people.

Many carcinogens, mutagens, and teratogens are dangerous at levels far below their direct toxic effect because abnormal cell growth exerts a kind of biological amplification. A single cell, perhaps altered by a single molecular event, can multiply into millions

of tumor cells or an entire organism. Just as there are different levels of direct toxicity, however, there are different degrees of carcinogenicity, mutagenicity, and teratogenicity. Methanesulfonic acid, for instance, is highly carcinogenic, while the sweetener saccharin is a possible carcinogen whose effects may be vanishingly small.

Acute versus chronic doses and effects

Most of the toxic effects that we have discussed so far have been **acute effects**. That is, they are caused by a single exposure to the toxin and result in an immediate health crisis. Often, if the individual experiencing an acute reaction survives this immediate crisis, the effects are reversible. **Chronic effects**, on the other hand, are long-lasting, perhaps even permanent. A chronic effect can result from a single dose of a very toxic substance, or it can be the result of a continuous or repeated sublethal exposure.

We also describe long-lasting exposures as chronic, although their effects may or may not persist after the toxin is removed. It usually is difficult to assess the specific health risks of chronic exposures because other factors, such as aging or normal diseases, act simultaneously with the factor under study. It often requires very large populations of experimental animals to obtain statistically significant results for low-level chronic exposures. Toxicologists talk about "megarat" experiments in which it might take a million rats to determine the health risks of some supertoxic chemicals at very low doses. Such an experiment would be terribly expensive for even a single chemical, let alone for the thousands of chemicals and factors suspected of being dangerous.

An alternative to enormous studies involving millions of animals is to give massive amounts—usually the maximum tolerable dose—of a toxin being studied to a smaller number of individuals and then to extrapolate what the effects of lower doses might have been. This is a controversial approach because it is not clear that responses to toxins are linear or uniform across a wide range of doses.

Figure 8.16 shows three possible results from low doses of a toxin. Curve (a) shows a baseline level of response in the population, even at zero dose of the toxin. This suggests that some other factor in the environment also causes this response. Curve (b) shows a straight-line relationship from the highest doses to zero exposure. Many carcinogens and mutagens show this kind of response. Any exposure to such agents, no matter how small, carries some risks. Curve (c) shows a threshold for the response where some minimal dose is necessary before any effect can be observed. This generally suggests the presence of a defense mechanism that prevents the toxin from reaching its target in an active form or repairs the damage that the toxin causes. Low levels of exposure to the toxin in question may have no deleterious effects, and it might not be necessary to try to keep exposures to zero.

Which, if any, environmental health hazards have thresholds is an important but difficult question. The 1958 Delaney Clause to the U.S. Food and Drug Act forbids the addition of any amount of known carcinogens to food and drugs, based on the assumption that any exposure to these substances represents unacceptable risks. This standard was replaced in 1996 by a "no reasonable

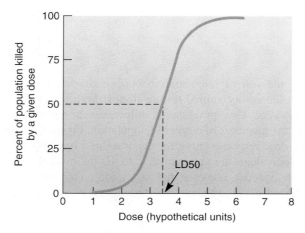

Figure 8.15 Cumulative population response to increasing doses of a toxin. The LD50 is the dose that is lethal to half the population.

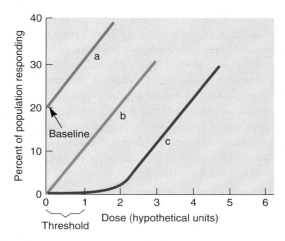

Figure 8.16 Three possible dose-response curves at low doses. (a) Some individuals respond, even at zero dose, indicating that some other factor must be involved. (b) Response is linear down to the lowest possible dose. (c) Threshold must be passed before any response is seen.

harm" requirement, defined as less than one cancer for every million people exposed over a lifetime. This change was supported by a report from the National Academy of Sciences concluding that synthetic chemicals in our diet are unlikely to represent an appreciable cancer risk. We will discuss risk analysis in the section "Risk Assessment and Acceptance."

Detectable levels aren't always dangerous

You may have seen or heard dire warnings about toxic materials detected in samples of air, water, or food. A typical headline announced recently that 23 pesticides were found in 16 food samples. What does that mean? The implication seems to be that any amount of dangerous materials is unacceptable and that counting the numbers of compounds detected is a reliable way to establish danger. We have seen, however, that the dose makes the poison. It matters not only what is there but also how much, where it is located, how accessible it is, and who is exposed. At some level, the mere presence of a substance is insignificant.

Toxins and pollutants may seem to be more widespread now than in the past, and this is surely a valid perception for many substances (fig. 8.17). The daily reports we hear of new materials found in new places, however, are also due, in part, to our more sensitive measuring techniques. Twenty years ago, parts per million were generally the limits of detection for most chemicals. Anything below that amount was often reported as "zero" or "absent," rather than more accurately as "undetected." A decade ago, new machines and techniques were developed to measure parts per billion. Suddenly, chemicals were found where none had been suspected. Now we can detect parts per trillion or even parts per quadrillion in some cases. Increasingly sophisticated measuring capabilities may lead us to believe that toxic materials have become more prevalent. In fact, our environment may be no more dangerous; we are just better at finding trace amounts.

Low doses can have variable effects

A complication in assessing risk is that the effects of low doses of some toxins and health hazards can be nonlinear. They may be either more or less dangerous than would be predicted from exposure to higher doses. For example, low doses of DHEP suppress activity of an enzyme essential for rat brain development. This is surprising because higher doses stimulated this enzyme. Thus, low doses are more damaging to brain development than expected.

On the other hand, very low amounts of radiation seem to be protective against certain cancers. This is perplexing, because ionizing radiation has long been recognized as a human carcinogen. It's thought now, however, that very low radiation exposure may stimulate DNA repair along with enzymes that destroy free radicals (atoms with unpaired, reactive electrons in their outer shells). Activating these repair mechanisms may defend us from other, unrelated hazards. These nonlinear effects are called **hormesis**.

Another complication is that some substances can have long-lasting effects on genetic expression. For example, researchers found that exposure of pregnant rats to the fungicide vinclozolin can have behavioral effects not only on the exposed rats, but on their daughters and granddaughters. A single dose given on a specific day in pregnancy can be expressed three generations later, even if those offspring have never been exposed to the toxin. This effect is called **epigenetics**. It doesn't require a permanent mutation in genes, but it can result in changes, both positive and negative, in expression of whole groups of critical genes over multiple generations. These epigenetic effects also can have different outcomes in variants of the same gene. Thus, exposure to a particular toxin could be very harmful to you but have no detectable effects in someone who has slightly different forms of the same genes. This may explain why, in a group of people exposed to the same carcinogen, some will get cancer while others don't. Or it could explain why a particular diet protects the health of some people but not others.

Figure 8.17 "Do you want to stop reading those ingredients while we're trying to eat?"
Source: Reprinted with permission of the *Star-Tribune*, Minneapolis-St. Paul.

8.6 Risk Assessment and Acceptance

Even if we know with some certainty how toxic a specific chemical is in laboratory tests, it still is difficult to determine **risk** (the probability of harm times the probability of exposure) if that chemical is released into the environment. As we have seen, many factors complicate the movement and fate of chemicals both around us and within our bodies. Furthermore, public perception of relative dangers from environmental hazards can be skewed so that some risks seem much more important than others.

Our perception of risks isn't always rational

A number of factors influence how we perceive relative risks associated with different situations.

- People with social, political, or economic interests—including environmentalists—tend to downplay certain risks and emphasize others that suit their own agendas. We do this individually as well, building up the dangers of things that don't benefit us, while diminishing or ignoring the negative aspects of activities we enjoy or profit from.

- Most people have difficulty understanding and believing probabilities. We feel that there must be patterns and connections in events, even though statistical theory says otherwise. If the coin turned up heads last time, we feel certain that it will turn up tails next time. In the same way, it is difficult to understand the meaning of a 1-in-10,000 risk of being poisoned by a chemical.

- Our personal experiences often are misleading. When we have not personally experienced a bad outcome, we feel it is more rare and unlikely to occur than it actually may be. Furthermore, the anxieties generated by life's gambles make us want to deny uncertainty and to misjudge many risks (fig. 8.18).

- We have an exaggerated view of our own abilities to control our fate. We generally consider ourselves above-average drivers, safer than most when using appliances or power tools, and less likely than others to suffer medical problems, such as heart attacks. People often feel they can avoid hazards because they are wiser or luckier than others.

- News media give us a biased perspective on the frequency of certain kinds of health hazards, overreporting some accidents or diseases, while downplaying or underreporting others. Sensational, gory, or especially frightful causes of death, such as murders, plane crashes, fires, or terrible accidents, receive a disproportionate amount of attention in the public media. Heart disease, cancer, and stroke kill nearly 15 times as many people in the United States as do accidents and 75 times as many people as do homicides, but the emphasis placed by the media on accidents and homicides is nearly inversely proportional to their relative frequency, compared with either cardiovascular disease or cancer. This gives us an inaccurate picture of the real risks to which we are exposed.

- We tend to have an irrational fear or distrust of certain technologies or activities that leads us to overestimate their dangers. Nuclear power, for instance, is viewed as very risky, while coal-burning power plants seem to be familiar and relatively benign; in fact, coal mining, shipping, and combustion cause an estimated 10,000 deaths each year in the United States, compared with none known so far for nuclear power generation. An old, familiar technology seems safer and more acceptable than does a new, unknown one.

- Alarmist myths and fallacies spread through society, often fueled by xenophobia, politics, or religion. For example, the World Health Organization campaign to eradicate polio worldwide has been thwarted by religious leaders in northern Nigeria—the last country where the disease remains widespread—who claim that oral vaccination is a U.S. plot to spread AIDS or infertility among Muslims.

How much risk is acceptable?

How much is it worth to minimize and avoid exposure to certain risks? Most people will tolerate a higher probability of occurrence of an event if the harm caused by that event is low. Conversely, harm of greater severity is acceptable only at low levels of frequency. A 1-in-10,000 chance of being killed might be of more concern to you than a 1-in-100 chance of being injured. For most people, a 1-in-100,000 chance of dying from some event or some factor is a threshold for changing what they do. That is, if the chance of death is less than 1 in 100,000, we are not likely to be worried enough to change our ways. If the risk is greater, we will probably do something about it. The Environmental Protection Agency generally assumes that a risk of 1 in 1 million is acceptable for most environmental hazards. Critics of this policy ask, acceptable to whom?

For activities that we enjoy or find profitable, we are often willing to accept far greater risks than this general threshold. Conversely, for risks that benefit someone else, we demand far higher protection. For

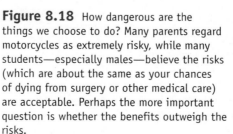

Figure 8.18 How dangerous are the things we choose to do? Many parents regard motorcycles as extremely risky, while many students—especially males—believe the risks (which are about the same as your chances of dying from surgery or other medical care) are acceptable. Perhaps the more important question is whether the benefits outweigh the risks.

Table 8.4	Lifetime Chances of Dying in the United States	
Source	**Odds (1 in x)**	
Heart disease	2	
Cancer	3	
Smoking	4	
Lung disease	15	
Pneumonia	30	
Automobile accident	100	
Suicide	100	
Falls	200	
Firearms	200	
Fires	1,000	
Airplane accident	5,000	
Jumping from high places	6,000	
Drowning	10,000	
Lightning	56,000	
Hornets, wasps, bees	76,000	
Dog bite	230,000	
Poisonous snakes, spiders	700,000	
Botulism	1 million	
Falling space debris	5 million	
Drinking water with EPA limit of trichloroethylene	10 million	

Source: Data from U.S. National Safety Council, 2003.

instance, your chance of dying in a motor vehicle accident in any given year are about 1 in 5,000, but that doesn't deter many people from riding in automobiles. Your chances of dying from lung cancer if you smoke one pack of cigarettes per day are about 1 in 1,000. By comparison, the risk from drinking water with the EPA limit of trichloroethylene is about 2 in 1 billion. Strangely, many people demand water with zero levels of trichloroethylene while continuing to smoke cigarettes.

More than 1 million Americans are diagnosed with skin cancer each year. Some of these cancers are lethal, and most are disfiguring, yet only one-third of teenagers routinely use sunscreen. Tanning beds more than double your chances of cancer, especially if you're young, but about 10 percent of all teenagers admit regularly using these devices.

Table 8.4 lists lifetime odds of dying from some leading causes. These are statistical averages, of course, and there clearly are differences in where one lives and how one behaves that affect the danger level of these activities. Although the average lifetime chance of dying in an automobile accident is 1 in 100, there are clearly things you can do—such as wearing a seat belt, driving defensively, and avoiding risky situations—to improve your odds. Still, it is interesting how we readily accept some risks while shunning others.

Our perception of relative risks is strongly affected by whether risks are known or unknown, whether we feel in control of the outcome, and how dreadful the results are. Risks that are unknown or unpredictable and results that are particularly gruesome or disgusting seem far worse than those that are familiar and socially acceptable.

Studies of public risk perception show that most people react more to emotion than to statistics. We go to great lengths to avoid some dangers while gladly accepting others. Factors that are involuntary, unfamiliar, undetectable to those exposed or catastrophic; those that have delayed effects; and those that are a threat to future generations are especially feared. Factors that are voluntary, familiar, detectable, or immediate cause less anxiety. Even though the actual number of deaths from automobile accidents, smoking, or alcohol, for instance, is thousands of times greater than those from pesticides, nuclear energy, or genetic engineering, the latter group preoccupies us far more than the former.

8.7 Establishing Public Policy

Risk management combines principles of environmental health and toxicology with regulatory decisions based on socioeconomic, technical, and political considerations (fig. 8.19). The biggest problem in making regulatory decisions is that we are usually exposed to many sources of harm, often unknowingly. It is difficult to separate the effects of all these different hazards and to evaluate their risks accurately, especially when the exposures are near the threshold of measurement and response. In spite of often

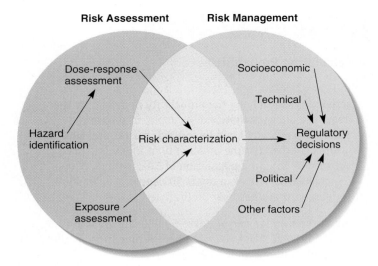

Risk Assessment   **Risk Management**

Figure 8.19 Risk assessment organizes and analyzes data to determine relative risk. Risk management sets priorities and evaluates relevant factors to make regulatory decisions.

Source: Data from D. E. Patton, "USEPA's Framework for Ecological Risk Assessment" in *Human Ecological Risk Assessment*, Vol. 1, No. 4.

Table 8.5	Relative Risks to Human Welfare
Relatively High-Risk Problems	
Habitat alteration and destruction	
Species extinction and loss of biological diversity	
Stratospheric ozone depletion	
Global climate change	
Relatively Medium-Risk Problems	
Herbicides/pesticides	
Toxins and pollutants in surface waters	
Acid deposition	
Airborne toxins	
Relatively Low-Risk Problems	
Oil spills	
Groundwater pollution	
Radionuclides	
Thermal pollution	

Source: Data from U.S. Environmental Protection Agency.

vague and contradictory data, public policymakers must make decisions.

The case of the sweetener saccharin is a good example of the complexities and uncertainties of risk assessment in public health. Studies in the 1970s at the University of Wisconsin and the Canadian Health Protection Branch suggested a link between saccharin and bladder cancer in male rats. Critics of these studies pointed out that humans would have to drink 800 cans of diet soda *per day* to get a saccharin dose equivalent to that given to the rats. Furthermore, they argued this response may be unique to male rats. In 2000 the U.S. Department of Health concluded a study that found no association between saccharin and cancer in humans. Congress ordered that all warnings be removed from saccharin-containing products. Still, some groups, such as the Center for Science in the Public Interest, consider this sweetener dangerous and urge us to avoid it if possible.

In setting standards for environmental toxins, we need to consider (1) combined effects of exposure to many different sources of damage, (2) different sensitivities of members of the population, and (3) effects of chronic as well as acute exposures. Some people argue that pollution levels should be set at the highest amount that does *not* cause measurable effects. Others demand that pollution be reduced to zero if possible, or as low as is technologically feasible. It may not be reasonable to demand that we be protected from every potentially harmful contaminant in our environment, no matter how small the risk. As we have seen, our bodies have mechanisms that enable us to avoid or repair many kinds of damage, so that most of us can withstand a minimal level of exposure without harm.

On the other hand, each challenge to our cells by toxic substances represents stress on our bodies. Although each individual stress may not be life-threatening, the cumulative effects of all the environmental stresses, both natural and human-caused, to which we are exposed may seriously shorten or restrict our lives. Furthermore, some individuals in any population are more susceptible to those stresses than others. Should we set pollution standards so that no one is adversely affected, even the most sensitive individuals, or should the acceptable level of risk be based on the average member of the population?

Finally, policy decisions about hazardous and toxic materials also need to be based on information about how such materials affect the plants, animals, and other organisms that define and maintain our environment. In some cases, pollution can harm or destroy whole ecosystems with devastating effects on the life-supporting cycles on which we depend. In other cases, only the most sensitive species are threatened. Table 8.5 shows the Environmental Protection Agency's assessment of relative risks to human welfare. This ranking reflects a concern that our exclusive focus on reducing pollution to protect human health has neglected risks to natural ecological systems. While there have been many benefits from a case-by-case approach in which we evaluate the health risks of individual chemicals, we have often missed broader ecological problems that may be of greater ultimate importance.

Conclusion

We have made marvelous progress in reducing some of the worst diseases that have long plagued humans. Smallpox is the first major disease to be completely eliminated. Guinea worms and polio are nearly eradicated worldwide; typhoid fever, cholera, yellow fever, tuberculosis, mumps, and other highly communicable diseases are rarely encountered in advanced countries. Childhood mortality has decreased 90 percent globally, and people almost everywhere are living twice as long, on average, as they did a century ago.

But the technological innovations and affluence that have diminished many terrible diseases, have also introduced new risks. Chronic conditions, such as cardiovascular disease, cancer, depression, dementia, diabetes, and traffic accidents, that once were confined to richer countries, now have become leading health problems nearly everywhere. Part of this change is that we no longer die at an early age of infectious disease, so we live long enough to develop the infirmities of old age. Another factor is that affluent lifestyles, lack of exercise, and unhealthy diets aggravate these chronic conditions.

New, emergent diseases are appearing at an increasing rate. With increased international travel, diseases can spread around the globe in a few days. Epidemiologists warn that the next deadly epidemic may be only a plane ride away. In addition, modern industry is introducing thousands of new chemical substances every year, most of which aren't studied thoroughly for health effects. Endocrine disrupters, neurotoxins, carcinogens, mutagens, teratogens, and other toxins can have tragic outcomes. The effects of lead on children's mental development is an example of both how we have introduced materials with unintended consequences, and a success story of controlling a serious health risk. Many other industrial chemicals could be having similar harmful effects.

Practice Quiz

1. Define the terms *health* and *disease*.
2. Name the five leading causes of global disease burden expected by 2020.
3. Define *emergent diseases* and give some recent examples.
4. What is *conservation medicine*?
5. What is the difference between toxic and hazardous? Give some examples of materials in each category.
6. What are *endocrine disrupters*, and why are they of concern?
7. What are *bioaccumulation* and *biomagnification*?
8. Why is atrazine a concern?
9. What is an *LD50*?
10. Distinguish between acute and chronic toxicity.

Critical Thinking and Discussion Questions

Apply the principles you have learned in this chapter to discuss these questions with other students.

1. Is it ever possible to be completely healthy?
2. How much would be appropriate for wealthy countries to contribute to global health? Why should we do more than we do now? What's in it for us?
3. Why do we spend more money on heart diseases or cancer than childhood diseases?
4. Why do we tend to assume that natural chemicals are safe while industrial chemicals are evil? Is this correct?
5. In the list of reasons why people have a flawed perception of certain risks, do you see things that you or someone you know sometimes do?
6. Do you agree that 1 in 1 million risk of death is an acceptable risk? Notice that almost everything in table 8.4 carries a greater risk than this. Does this make you want to change your habits?

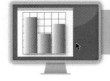

Data Analysis: Graphing Multiple Variables

Is it possible to show relationships between two dependent variables on the same graph? Sometimes that's desirable when you want to make comparisons between them. The graph on page 198 does just that. It's a description of how people perceive different risks. We judge the severity of risks based on how familiar they are and how much control we have over our exposure.

- Take a look at the different risks in Figure 1. Make a list of about 5 that you consider most dangerous. Then list about 5 that you think are least dangerous.
- Do most of your most dangerous activities/items fall into one quadrant? What are the axes of the graph? Do the ideas on the axes help explain why you consider some activities more dangerous than others?

- Are there other factors that help explain why you consider certain activities dangerous and others relatively safe? What are those other factors? If you compare your list to other peoples' lists, are they similar or different?

On this graph, which represents attitudes of many people, the Y-axis represents how mysterious, unknown, or delayed the risk seems to be. Things that are unobservable, unknown to those exposed, delayed in their effects, and unfamiliar or unknown to science tend to be more greatly feared than those that are observable, known, immediate, familiar, and known to science. The X-axis represents a measure of dread, which combines how much control we feel we have over the risk, how terrible the results could potentially be, and how equitably the risks are distributed. The size of the symbol for each risk indicates the combined effect of these two variables.

Notice that things such as DNA technology or nuclear waste, which have high levels of both mystery and dread, tend to be regarded with the greatest fear, while familiar, voluntary, personally rewarding behaviors such as riding in automobiles or on bicycles, or drinking alcohol are thought to be relatively minor risks. Actuarial experts (statisticians who gather mortality data) would tell you that automobiles, bicycles, and alcohol have killed far more people (so far) than DNA technology or radioactive waste. But this isn't just a question of data. It's a reflection of how much we fear various risks. Notice that this is a kind of scatter plot mapping categories of data that have no temporal sequence. Still, you can draw some useful inferences from this sort of graphic presentation.

For Additional Help in Studying This Chapter, please visit our website at www.mhhe.com/cunningham5e. You will find practice quizzes, key terms, answers to end of chapter questions, additional case studies, an extensive reading list, and Google Earth™ mapping quizzes.

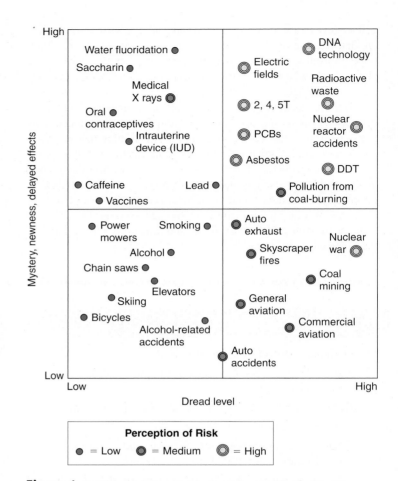

Figure 1 Public perception of risk depending on the familiarity, apparent potential for harm, and personal control over the risk.
Source: Data from Slovic, Paul. 1987. Perception of Risk, *Science*, 236 (4799): 286–290.

Planting trees is a good way to combat global warming while also beautifying your city or neighborhood.

9 Air:
Climate and Pollution

Learning Outcomes

After studying this chapter, you should be able to answer the following questions:

- What or where is the stratosphere, and why do we care?
- Explain the greenhouse effect and how it is changing our climate.
- What is the ENSO cycle, and how does it affect weather patterns?
- Is it too late to do anything about global climate change?
- Why has the United States refused to ratify the Kyoto Protocol?
- Why has stratospheric ozone been disappearing? Should we worry about it?
- What are the main sources and effects of air pollution?
- Has world air quality been getting better or worse?
- What is the "new source review"?

Climate is an angry beast, and we are poking it with sticks.

–Wallace Broecker

Ocean Fertilization

"Give me a half a tanker of iron and I'll give you an Ice Age." When oceanographer, John Martin, made that remark 25 years ago, most of his audience treated it as a joke, but now some people are taking his idea more seriously. Martin had carried out experiments showing that primary productivity in much of the ocean is limited by nutrient deficiencies. When he added iron to water collected a few hundred kilometers off the Antarctic coast, he found that chlorophyll levels increased by a factor of 10,000. Some subsequent experiments have produced even greater biomass growth (fig. 9.1).

The reason people are so interested in ocean fertilization is that it might offer a way to remove large amounts of carbon dioxide (CO_2) from the air. As we'll discuss in this chapter, a vast majority of climate scientists believe that our emissions of CO_2 and other heat-absorbing gases are now warming the atmosphere and changing our global climate. If these pollutants continue to accumulate at current rates, there could be truly disastrous consequences. Global climate change may well be the most serious threat—outside of nuclear war—that we currently face. We need urgently to reduce our discharges and to find ways to remove these gases from the air.

You'll learn in this chapter that numerous approaches for emissions reductions and carbon sequestration have been suggested. Many steps can be taken using existing technology and would actually save money. Some, like ocean enrichment, have uncertain benefits and potential complications. Critics worry that we might cause new problems in our attempt to solve old ones.

Will algal cells, stimulated by adding extra nutrients to the ocean, eventually die and sink to the bottom, or will they simply decay and release the CO_2 they've absorbed back into the atmosphere? It's possible that growing more algae in the ocean could help restore fish populations depleted by overharvesting. On the other hand, large-scale eutrophication often creates oxygen-depleted "dead zones" in which almost nothing can survive. Sometimes excess nutrients result in blooms of highly toxic algae and dinoflagellates. Many scientists worry that sensitive ecosystems, such as coral reefs, which already are being stressed by climate change and human intrusions, might be fatally damaged by efforts to change ocean chemistry.

However, at least two commercial companies have already announced plans for large-scale ocean fertilization. The Planktos Corporation of Foster City, California, expects to spread 50 to 100 tons of pulverized iron ore in a 50 to 100 km diameter area of the Pacific about 300 km west of the Galápagos Islands. And the Ocean Nourishment Corporation from New South Wales, Australia, already has approval to dump 500 tons of urea (as a nitrogen source) in the Sula Sea between the Philippines and Borneo. Ocean Nourishment also has plans for similar fertilization projects in Malaysia, Chile, and the United Arab Emirates. Both these companies hope to sell lucrative carbon offset credits on the global climate exchange as a result of their ocean nutrient enhancement.

In 2007, the International Maritime Organization, the international body that administers the Law of the Sea, declared that all ocean fertilization projects fall under their jurisdiction. They didn't ban these activities outright, but they expressed concern about unintended consequences and unknown ecological effects. Many environmental groups cheered this intervention, but others warn that we need to do everything in our power to combat global warming. What do you think? Is the risk of geoengineering worth its possible benefits? If you were a delegate to the Law of the Sea Convention, what monitoring steps and ecological safeguards would you impose on companies wanting to exploit this technology?

A major part of this chapter will be devoted to air pollutants and global climate change and what we might do about them. First, however, in order to understand our climate system better, we'll look at the factors that normally shape our weather and climate.

Figure 9.1 A massive plankton bloom fills the ocean off the coast of Norway.

9.1 The Atmosphere Is a Complex System

We live at the bottom of a virtual ocean of air that extends upward about 500 km (300 mi). In the layer closest to the earth's surface, known as the troposphere, the air moves ceaselessly, flowing, swirling, and continually redistributing heat and moisture from one part of the globe to another. The composition and behavior of the troposphere and other layers control our **weather** (daily temperature and moisture conditions in a place) and our **climate** (long-term weather patterns).

The earth's earliest atmosphere probably consisted mainly of hydrogen and helium. Over billions of years, most of that hydrogen and helium diffused into space. Volcanic emissions added carbon, nitrogen, oxygen, sulfur, and other elements to the atmosphere.

Virtually all of the molecular oxygen (O_2) we breathe was probably produced by photosynthesis in blue-green bacteria, algae, and green plants.

Clean, dry air is 78 percent nitrogen and almost 21 percent oxygen, with the remaining 1 percent composed of argon, carbon dioxide (CO_2), and a variety of trace gases. Water vapor concentrations vary from near 0 to 4 percent, depending on air temperature and available moisture. Minute particles and liquid droplets—collectively called **aerosols**—also are suspended in the air. Atmospheric aerosols play important roles in the earth's energy budget and in rain production.

The atmosphere has four distinct zones of contrasting temperature, due to differences in absorption of solar energy (fig. 9.2). The layer of air immediately adjacent to the earth's surface is called the **troposphere** (*tropein* means to turn or change, in Greek). Within the troposphere, air circulates in great vertical and horizontal **convection currents**, constantly redistributing heat and moisture around the globe. The troposphere ranges in depth from about 18 km (11 mi) over the equator to about 8 km (5 mi) over the poles, where air is cold and dense. Because gravity holds most air molecules close to the earth's surface, the troposphere is much denser than the other layers: it contains about 75 percent of the total mass of the atmosphere. Air temperature drops rapidly with increasing altitude in this layer, reaching about –60°C (–76°F) at the top of the troposphere. A sudden reversal of this temperature gradient creates a sharp boundary called the tropopause, which limits mixing between the troposphere and upper zones.

The **stratosphere** extends from the tropopause up to about 50 km (31 mi). It is vastly more dilute than the troposphere, but it has a similar composition—except that it has almost no water vapor and nearly 1,000 times more **ozone** (O_3). This ozone absorbs some wavelengths of ultraviolet solar radiation, known as UV-B (290–330 nm, see fig. 2.13). This absorbed energy makes the atmosphere warmer toward the top of the stratosphere. Since UV radiation damages living tissues, this UV absorption in the stratosphere also protects life on the surface. Recently discovered depletion of stratospheric ozone, especially over Antarctica, is allowing increased amounts of UV radiation to reach the earth's surface. If observed trends continue, this radiation could cause higher rates of skin cancer, genetic mutations, crop failures, and disruption of important biological communities, as you will see later in this chapter.

Unlike the troposphere, the stratosphere is relatively calm. There is so little mixing in the stratosphere that volcanic ash and human-caused contaminants can remain in suspension there for many years.

Above the stratosphere, the temperature diminishes again, creating the mesosphere, or middle layer. The thermosphere (heated layer) begins at about 50 km. This is a region of highly ionized (electrically charged) gases, heated by a steady flow of high-energy solar and cosmic radiation. In the lower part of the thermosphere, intense pulses of high-energy radiation cause electrically charged particles (ions) to glow. This phenomenon is what we know as the *aurora borealis* and *aurora australis*, or northern and southern lights.

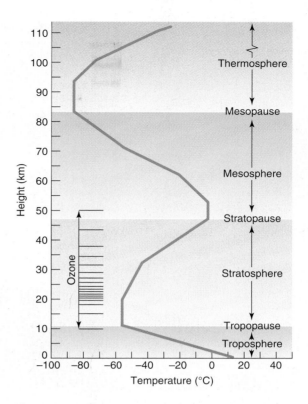

Figure 9.2 Temperatures change drastically in the four layers of the atmosphere. Bars in the ozone graph represent relative concentrations of stratospheric ozone with altitude.
Source: Courtesy of Dr. William Culver, St. Petersburg Junior College.

No sharp boundary marks the end of the atmosphere. Pressure and density decrease with distance from the earth until they become indistinguishable from the near vacuum of interstellar space.

The sun warms our world

The sun supplies the earth with an enormous amount of energy, but that energy is not evenly distributed over the globe. Incoming solar radiation (insolation) is much stronger near the equator than at high latitudes. Of the solar energy that reaches the outer atmosphere, about one-quarter is reflected by clouds and atmospheric gases, and another quarter is absorbed by carbon dioxide, water vapor, ozone, methane, and a few other gases (fig. 9.3). This energy absorption warms the atmosphere slightly. About half of incoming solar radiation (insolation) reaches the earth's surface. Most of this energy is in the form of light or infrared (heat) energy. Some of this energy is reflected by bright surfaces, such as snow, ice, and sand. The rest is absorbed by the earth's surface and by water. Surfaces that *reflect* energy have a high **albedo** (reflectivity). Fresh snow and dense clouds, for instance, can reflect as much as 85 to 90 percent of the light falling on them. Surfaces that absorb energy have a low albedo

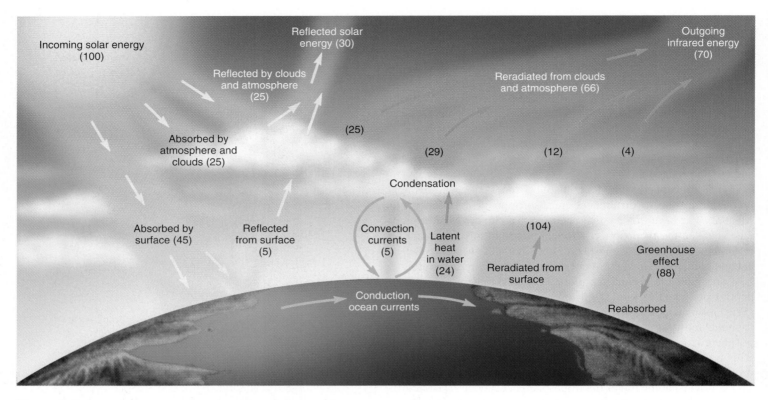

Figure 9.3 Energy balance between incoming and outgoing radiation. The atmosphere absorbs or reflects about half of the solar energy reaching the earth. Most of the energy reemitted from the earth's surface is long-wave, infrared energy. Most of this infrared energy is absorbed by aerosols and gases in the atmosphere and is re-radiated toward the planet, keeping the surface much warmer than it would otherwise be. This is known as the greenhouse effect.

and generally appear dark. Black soil, asphalt pavement, and water, for example, have low albedos (reflectivity as low as 3 to 5 percent).

Absorbed energy heats materials (such as an asphalt parking lot in summer), evaporates water, and provides the energy for photosynthesis in plants. Following the second law of thermodynamics, absorbed energy is gradually reemitted as lower-quality heat energy. A brick building, for example, absorbs energy in the form of light and reemits that energy in the form of heat.

The change in energy quality is very important because the atmosphere selectively absorbs longer wavelengths. Most solar energy comes in the form of intense, high-energy light or near-infrared wavelengths. This short-wavelength energy passes relatively easily through the atmosphere to reach the earth's surface. Energy re-released from the earth's warmed surface is lower-intensity, longer-wavelength energy in the far-infrared part of the spectrum. Atmospheric gases, especially carbon dioxide and water vapor, absorb much of this long-wavelength energy, re-releasing it in the lower atmosphere and letting it leak out to space only slowly. This re-irradiated energy provides most of the heat in the lower atmosphere. If the atmosphere were as transparent to infrared radiation as it is to visible light, the earth's average surface temperature would be about 20°C (36°F) colder than it is now.

This phenomenon is called the **greenhouse effect** because the atmosphere, loosely comparable to the glass of a greenhouse, transmits sunlight while trapping heat inside. The greenhouse effect is a natural atmospheric process that is necessary for life as we know it. However, too much greenhouse effect, caused by burning of fossil fuels and deforestation, may cause harmful environmental change (for a more detailed discussion of this phenomenon, see our web page www.mhhe.com/cunningham5e).

Water stores heat, and winds redistribute it

Much of the incoming solar energy is used to evaporate water. Every gram of evaporating water absorbs 580 calories of energy as it transforms from liquid to gas. Globally, water vapor contains a huge amount of stored energy, known as **latent heat**. When water vapor condenses, returning from a gas to a liquid form, the 580 calories of heat energy are released. Imagine the sun shining on the Gulf of Mexico in the winter. Warm sunshine and plenty of water allow continuous evaporation that converts an immense amount of solar (light) energy into latent heat stored in evaporated water. Now imagine a wind blowing the humid air north from the Gulf toward Canada. The air cools as it moves north (especially if it encounters cold air moving south). Cooling causes the water vapor to condense. Rain (or snow) falls as a consequence. Note

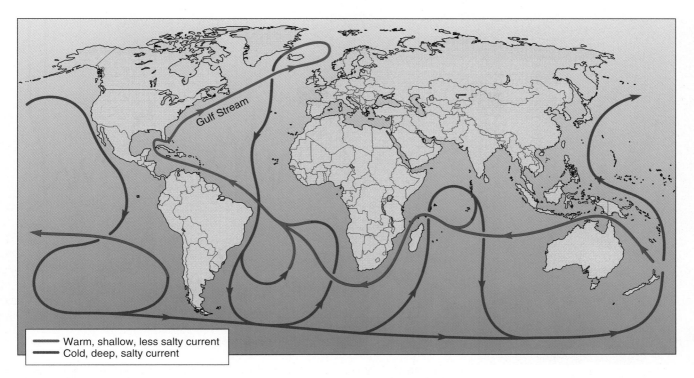

Warm, shallow, less salty current
Cold, deep, salty current

Figure 9.4 Ocean currents act as a global conveyor system, redistributing warm and cold water around the globe. These currents moderate our climate. For example, the Gulf Stream keeps northern Europe much warmer than northern Canada.

that it is not only water that has moved from the Gulf to the Midwest: 580 calories of heat have also moved with every gram of moisture. The heat and water have moved from a place with strong incoming solar energy to a place with much less solar energy and much less water. This redistribution of heat and water around the globe is essential to life on earth.

Ocean currents also modify our climate

Warm and cold ocean currents strongly influence climate conditions on land. Surface ocean currents result from wind pushing on the ocean surface. As surface water moves, deep water wells up to replace it, creating deeper ocean currents. Differences in water density—depending on the temperature and saltiness of the water—also drive ocean circulation. Huge cycling currents called gyres carry water north and south, redistributing heat from low latitudes to high latitudes (see appendix 3, p. 378, global climate map). The Alaska current, flowing from Alaska southward to California, keeps San Francisco cool and foggy during the summer.

The Gulf Stream, one of the best known currents, carries warm Caribbean water north past Canada's maritime provinces to northern Europe (fig. 9.4). This current is immense, some 800 times the volume of the Amazon, the world's largest river. The heat transported from the Gulf keeps Europe much warmer than it should be for its latitude. Stockholm, Sweden, for example, where temperatures rarely fall much below freezing, is at the same lati-

tude as Churchill, Manitoba, which is famous as one of the best places in the world to see polar bears. As the warm Gulf Stream passes Scandinavia and swirls around Iceland, the water cools and evaporates, becomes dense and salty, and plunges downward, creating a strong, deep, southward current.

Together, this surface- and deep-water circulation system is called the **thermohaline ocean conveyor**. Dr. Wallace Broecker of the Lamont Doherty Earth Observatory, who first described this great conveyor system, also found it can shut down suddenly. About 11,000 years ago, as the earth was gradually warming at the end of the Pleistocene ice age, a huge body of meltwater, called Lake Agassiz, collected along the south margin of the North American ice sheet. At its peak, Lake Agassiz contained more water than all the current freshwater lakes in the world. Drainage of the lake was blocked by ice lying over what is now the Great Lakes. When that ice dam suddenly gave way, it's estimated that 163,000 km^3 of fresh water roared down the St. Lawrence Seaway and out into the North Atlantic, where it layered on top of the ocean and prevented sinking of deep, cold, dense seawater. This stopped the oceanic conveyor and plunged the whole planet back into an ice age (called the Younger Dryas after a small tundra flower that became more common in colder conditions) that lasted for another 1,300 years. Cessation of the ocean conveyor may have occurred in years, maybe even in a few months.

Could this happen again? Certainly not as suddenly as 11,000 years ago. There's no meltwater buildup today similar to

Figure 9.5 Dr. Mark Twickler, of the University of New Hampshire, holds a section of the 3,000 m Greenland ice sheet core, which records 250,000 years of climate history.

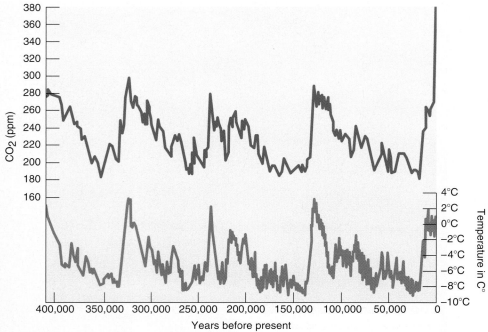

Figure 9.6 Atmospheric CO_2 concentrations and global mean temperatures estimated from the antarctic Vostok ice core. Note the relatively rapid changes and close correlation in both climate and atmospheric chemistry.

Source: Data from United Nations Environment Programme.

that of Lake Agassiz. But increased river flow, rain, and melting ice thought to be linked to global warming are now causing fresh water to accumulate on top of the Arctic Ocean. Changing ocean salinity seems to be slowing the ocean conveyor. British oceanographers recently reported that the deep-ocean return flow has slowed about 30 percent in the past 50 years. Much of the water that once traveled to the North Atlantic now seems trapped in a subtropical loop. If this situation persists, it could give Northern Europe a climate more like that of Siberia, not a pleasant prospect for many Europeans.

9.2 Climate Can Be an Angry Beast

When climatologist Wallace Broeker said that "climate is an angry beast, and we are poking it with sticks," he meant that we assume our climate is stable, but our thoughtless actions may be stirring it to sudden and dramatic changes. How stable is climate? That depends upon the time frame you consider. Over centuries and millennia, we know that climate shifts somewhat, but usually we expect little change on the scale of a human lifetime. The question now is whether that is a reasonable expectation. If climate does shift, how fast might it change, and what will those changes mean for the environmental systems we depend on?

Climates have changed dramatically throughout history

Ice cores from glaciers have revolutionized our understanding of climate history. In this research, a hollow tube is drilled down through the ice. Every 10 m, or so, the tube is pulled up and an ice cylinder is pushed out of the center (fig. 9.5). Bands of lighter and darker annual snow accumulation on top of the glacier. These annual bands allow us to tell the age of ice layers. Air bubbles trapped in the ice give us data on atmospheric composition when the snow was deposited. Ash layers and sulfate concentrations in the ice can be correlated with volcanic eruptions. The longest ice record ever collected is the Vostok ice core, which is 3,100 m long, and gives us a record of both global temperatures and atmospheric CO_2 over the past 420,000 years (fig. 9.6). A team of Russian scientists worked for 37 years at a site about 1,000 km from the South Pole to extract this ice core. A similar core, nearly as long as the Vostok, has been drilled from the Greenland ice sheet. Other glaciers throughout the world have also now been cored. All these ice samples show that climate has varied dramatically over time, but that there is a close correlation between atmospheric temperatures and CO_2 concentrations.

Major climatic changes, such as those of the Ice Ages, can have catastrophic effects on living organisms. If climatic change is gradual, species may have time to adapt or migrate to more

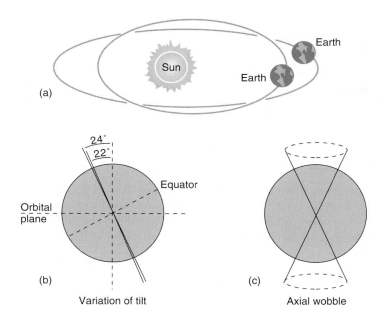

Figure 9.7 Milankovitch cycles that may affect long-term climate conditions: (a) changes in the eccentricity of the earth's orbit, (b) shifting tilt of the axis, and (c) wobble of the earth.

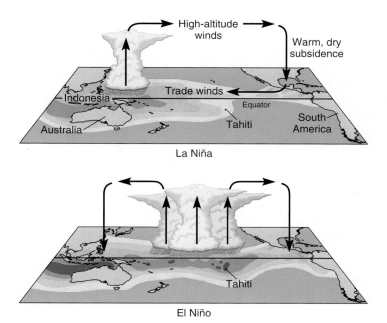

Figure 9.8 The El Niño/La Niña/Southern Oscillation cycle. Every three to five years, surface trade winds that normally push warm water westward toward Indonesia weaken and allow this pool of water to flow eastward toward South America.

suitable locations. Where climatic change is relatively abrupt, many organisms are unable to respond before conditions exceed their tolerance limits. Whole communities may be destroyed, and if the climatic change is widespread, many species may become extinct.

Geologic evidence suggests that there have been several great climatic changes, perhaps as many as a dozen, in which large numbers of species were exterminated (see table 5.3, p. 111).

What causes catastrophic climatic swings?

There are many explanations for climatic catastrophes. Asteroid impacts and massive volcanic eruptions have apparently caused some sudden die-offs. Changes in solar energy associated with 11-year sunspot cycles or 22-year solar magnetic cycles also appear to play a role. Furthermore, a regular 18.6-year cycle of shifts in the angle at which our moon orbits the earth alters tides and atmospheric circulation in a way that affects climate.

Milankovitch cycles, named after Serbian scientist Milutin Milankovitch, who first described them in the 1920s, are periodic shifts in the earth's orbit and tilt (fig. 9.7). The earth's elliptical orbit stretches and shortens in a 100,000-year cycle, while the axis of rotation changes its angle of tilt in a 40,000-year cycle. Furthermore, over a 26,000-year period, the axis wobbles like an out-of-balance spinning top. These variations change the distribution and intensity of sunlight reaching the earth's surface and, consequently, global climate. Bands of sedimentary rock laid in the oceans seem to match both these Milankovitch cycles and the periodic cold spells associated with worldwide expansion of glaciers every 100,000 years or so.

Volcanic eruptions can cause sudden climate flips. When Mount Toba exploded in western Sumatra about 73,000 years ago, for example, it was the largest volcanic cataclysm in the past 28 million years. It's estimated that Mount Toba ejected at least 2,800 km³ of material, compared to only 1 km³ emitted by Mount St. Helens in Washington State in 1980. Mount Toba produced enough ash to cover the entire earth an average of 10 cm deep. The signal from this eruption can be detected in the Greenland ice core. The sulfuric acid and particulate material ejected into the atmosphere from Mount Toba are estimated to have dimmed incoming sunlight by 75 percent and to have cooled the whole planet by as much as 16°C for more than 160 years. The ensuing volcanic winter is calculated to have reduced terrestrial biomass by 90 percent, and to have created a genetic bottleneck still evident in many species, including humans.

The El Niño/Southern Oscillation can have far-reaching effects

El Niño, **La Niña**, and the **Southern Oscillation** are all terms referring to a major ocean-current/climate connection that affects weather throughout the Pacific—and possibly throughout the world. The core of this system is a huge pool of warm surface water in the Pacific Ocean that sloshes slowly back and forth between Indonesia and South America like water in a giant bathtub. Most years, this pool is held in the western Pacific by steady equatorial trade winds that push ocean surface currents westward (fig. 9.8). From Southeast Asia to Australia, this concentration of warm equatorial water provides latent heat (water vapor) that drives strong upward convection (low pressure) in the atmosphere. Heavy rain

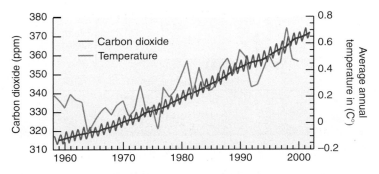

Figure 9.9 Carbon dioxide concentrations on Mauna Loa (monthly and annual averages) and global annual average temperature variations from 1960 to 2000.
Data from NASA Earth Observatory, 2002.

results, supporting dense tropical forests. On the American side of the Pacific, the westward-moving surface waters are replaced by cold water welling up along the South American coast. Cold, nutrient-rich waters support dense schools of anchovies and other fish. While the trade winds blow westward on the ocean's surface, returning winds high in the troposphere flow back from Indonesia to Chile and to Mexico and southern California. There the returning air sinks, creating dry, desert conditions.

Every three to five years, for reasons that we don't fully understand, Indonesian convection (rising air currents) weaken, and westward wind and ocean currents fail. Warm surface water surges back east across the Pacific. One theory is that the high cirrus clouds block enough sunshine to cool the ocean surface in Asia. Convection would then weaken, and trade winds—and ocean currents—would reverse, flowing eastward instead of westward. Another theory is that eastward-flowing deep currents periodically interfere with coastal upwelling, warming the sea surface off South America and eliminating the temperature gradient across the Pacific.

Fishermen in Peru were the first to notice irregular cycles of rising ocean temperatures because the fish disappeared when the water warmed. They named this event El Niño (Spanish for the Christ child) because they often occur around Christmastime. The counterpart to El Niño, when the eastern tropical Pacific cools, has come to be called La Niña (little girl). Together, these cycles are called the El Niño Southern Oscillation (ENSO).

ENSO cycles have far-reaching effects. During an El Niño year, the northern jet stream—which is normally over Canada—splits and is drawn south over the United States. This pulls moist air from the Pacific and Gulf of Mexico inland, bringing intense storms and heavy rains from California across the midwestern states. The intervening La Niña years bring hot, dry weather to the same areas. Oregon, Washington, and British Columbia, on the other hand, tend to have warm, sunny weather in El Niño years rather than their usual rain. Droughts in Australia and Indonesia during El Niño episodes cause disastrous crop failures and forest fires, including one in Borneo in 1983 that burned 3.3 million ha (8 million acres).

Some climatologists believe that ENSO events are becoming stronger or more frequent because of global climate change.

There are signs that warm ocean-surface temperatures are spreading, which could contribute to El Niño strength or frequency. On the other hand, increased cloud cover over warmer oceans could raise global albedo, and strong convection currents generated by these storms could pump heat into the stratosphere. This might have an overall cooling effect and act as a safety valve for global warming.

9.3 Global Warming Is Happening

Many scientists regard anthropogenic (human-caused) global climate change to be the most important environmental issue of our times. The possibility that humans might alter world climate is not a new idea. John Tyndall measured the infrared absorption of various gases and described the greenhouse effect in 1859. In 1895, Svante Arrhenius, who subsequently received a Nobel Prize for his work in chemistry, predicted that CO_2 released by coal burning could cause global warming.

A scientific consensus is emerging

The first evidence that human activities are increasing atmospheric CO_2 came from an observatory on top of the Mauna Loa volcano in Hawaii. The observatory was established in 1957 as part of an International Geophysical Year, and was intended to provide data on air chemistry in a remote, pristine environment. Surprisingly, measurements showed CO_2 levels increasing about 0.5 percent per year, rising from 315 ppm in 1958 to 385 ppm in 2008 (fig. 9.9). This increase isn't a perfectly straight line, however. Because a majority of the world's land and vegetation are in the Northern Hemisphere, northern seasons dominate the signal. Every May, CO_2 levels drop slightly as plant growth on northern continents use CO_2 in photosynthesis. During the northern winter, levels rise again as respiration releases CO_2.

The **Intergovernmental Panel on Climate Change (IPCC)** brings together scientists and government representatives from 130 countries to review scientific evidence on the causes and likely effects of human-caused climate change. In 2007, the IPCC issued its fourth report. The result of 6 years of work by 2,500 scientists, the four volumes of the report represent a consensus by more than 90 percent of all the scientists working on climate change. This report has the strongest expression of certainty to date from the climate community. It says that global warming is "very likely" caused by humans. This implies a 90 percent probability that changes we are now seeing are caused by human actions. It's reported, however, that most delegates wanted the statement to say that human causation is "virtually certain," meaning a 99 percent probability. To achieve a consensus, the lead authors accepted the less definite wording. Most scientists, however, regard the evidence as "indisputable."

The estimated range of global temperature increase by the end of this century also has been increased in this report. The third IPCC report, published in 2001, calculated that average global surface temperatures would increase 1.4–5.8°C by 2100. The fourth report gives a variety of scenarios depending on population and economic growth, energy conservation and efficiency, and adoption (or lack thereof) of greenhouse gas controls. This gives a range of increases from 1.1–6.4°C (2–11.5°F) higher than now by 2100. According to the IPCC, the "best estimate" for temperature rise is now 1.8–4.0°C (3.2–7.8°F). To put that in perspective, the change in average global temperature between now and the middle of the last glacial period is estimated to be about 5°C.

Global warming could mean more than 1 million human deaths and hundreds of billions of dollars in costs by 2100. Extreme weather, including droughts, floods, heat waves, and hurricanes, is a major concern. These extremes have already increased significantly in the past decade, and are expected to get even worse in the future.

Sea levels are projected to rise 17–57 cm (7–23 in.) by the end of this century. If recent rapid melting of polar ice sheets and Greenland glaciers continues, it could add another 15 cm (6 in.) to that total. If nothing is done to halt greenhouse emissions, complete melting of Greenland's ice sheet would raise sea level by more than 6 m (nearly 20 ft). This would flood most of Florida, a broad swath of the Gulf Coast, most of Manhattan Island, Shanghai, Hong Kong, Tokyo, Kolkata, Mumbai, and about two-thirds of the other largest cities in the world.

Following the release of the IPCC report, 45 nations called for a new global environmental body to regulate greenhouse emissions, perhaps with policing powers to punish violators. Conspicuously absent from this initiative were the world's biggest emitters of these gases, including the United States, China, and India. These major emitters opposed mandatory cuts in greenhouse gas emissions. Voluntary programs, they maintain, are enough.

Greenhouse gases have many sources

Since preindustrial times atmospheric concentrations of CO_2, CH_4, and N_2O have climbed by over 31 percent, 151 percent, and 17 percent, respectively. Carbon dioxide is by far the most important cause of anthropogenic climate change (fig. 9.10a). Burning fossil fuels, making cement, burning forests and grasslands, and other human activities release more than 33 billion tons of CO_2 every year, on average, containing some 9 billion tons of carbon (fig. 9.10b). About 3 billion tons of this excess carbon is taken up by terrestrial ecosystems, and around 2 billion tons are absorbed by the oceans, leaving an annual atmospheric increase of some 4 billion tons per year. If current trends continue, CO_2 concentrations could reach about 500 ppm (approaching twice the preindustrial level of 280 ppm) by the end of the twenty-first century.

Although rarer than CO_2, methane absorbs 23 times as much infrared energy and is accumulating in the atmosphere about twice as fast as CO_2. Methane, the main component of natural gas, is released by ruminant animals, wet-rice paddies, coal mines, landfills, wetlands, and pipeline leaks. Reservoirs for hydroelectricity, usually promoted as a clean power source, also produce methane from rotting plants. Philip Fearnside, an ecologist at Brazil's National Institute for Amazon Research, calculates that decaying vegetation in the reservoir behind the Cura-Una dam in Para Province emits so much carbon dioxide and methane every year that it causes three and a half times as much global warming as

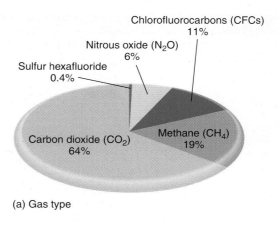

(a) Gas type

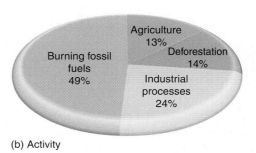

(b) Activity

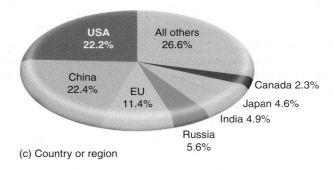

(c) Country or region

Figure 9.10 Percentage contributions to global warming by (a) gas type, (b) source activity, and (c) country or region.
Data from World Resources Institute and U.N. Statistics Division.

would generating the same amount of energy by burning fossil fuels.

Chlorofluorocarbons (CFCs) also store heat from infrared energy. CFC releases in developed countries have declined since many of their uses were banned, but increasing production in developing countries, such as China and India, remains a problem. Nitrous oxide (N_2O) is produced by burning organic material and by soil denitrification. As fig. 9.10a shows, CFCs and N_2O together are thought to account for only about 17 percent of human-caused global warming.

For many years, the United States was the world's largest source of greenhouse gases. With less than 5 percent of the world population, the United States released one-quarter or more of the global CO_2 emissions. In 2007, however, China was reported to

Table 9.1	Large Country per Capita CO_2 Emissions
Country	**CO_2 Production (Tons per Person)**
Australia	20.24
United States	20.14
Canada	19.24
Saudi Arabia	15.61
Belgium	13.10
Russia	11.88
Norway	11.40
Czech Republic	11.20
Germany	10.24
Slovakia	9.66
Japan	9.65
Spain	9.6
South Africa	9.56
United Kingdom	9.55

Source: International Energy Agency, 2007.

have passed the United States in total CO_2 emissions (fig. 9.10c). Rising affluence in China has fueled a rapidly growing demand for energy, the vast majority of which comes from coal. China is now building at least one large coal-burning power plant per week. Another large source of CO_2 in China is cement production. Worldwide, cement manufacturing accounts for 4 percent of all CO_2 emissions. Chinese cement companies, stimulated by the world's largest building boom, now produce nearly half of the world supply, and these plants are responsible for about 10 percent of all Chinese CO_2 emissions.

Although China now leads the world in total CO_2 releases, its per capita annual emissions are less than one-fifth those of the United States. India, with only one ton of CO_2 per person, has only one-twentieth as much as the United States. Australia currently has the highest per capita CO_2 emissions, while Canada is in third place, closely behind the United States. Table 9.1 lists emissions only for some of the largest countries with high emissions because there are some curious anomalies in the data. Oil-rich small countries, such as the Middle Eastern oil emirates, have very high per capita CO_2 output. Qatar, for example, produces more than three times as much CO_2 per person as Australia. And some island nations and city-states have even higher releases. Gibraltar, for example, is described by the International Energy Agency as releasing 155 tons of CO_2 per person, perhaps because of the large military presence. But the total output of these small populations is only a small contribution to global emissions.

To put carbon emissions in perspective, Africa, on average, produces just over one ton of CO_2 per person per year. The lowest emissions in the world are in Chad, where per capita production is only one-thousandth that of the United States. Some countries with high standards of living release relatively little CO_2. Sweden, for example, produces only 6.5 tons per person per year. Remarkably, Sweden, by adopting renewable energy sources and strict conser-

vation measures, has reduced its carbon emissions by 40 percent over the past 30 years, while also experiencing dramatic increases in both personal income and quality of life.

Perhaps the biggest question in environmental science today is whether China and India, which now have both the world's largest populations and are among the fastest growing economies, will follow the development path of Australia and the United States or that of Sweden and Switzerland. The world's future depends on how that question is answered.

Evidence of climate change is overwhelming

There is little doubt that our planet is warming. The American Geophysical Union, one of the nation's largest and most respected scientific organizations, says, "As best as can be determined, the world is now warmer than it has been at any point in the last two millennia, and, if current trends continue, by the end of the century it will likely be hotter than at any point in the last two million years." Some of the evidence for this warming includes the following:

- Over the last century the average global temperature has climbed about 0.6°C (1°F). Nineteen of the 20 warmest years in the past 150 have occurred since 1980. The hottest years globally in recorded history were 1998 and 2005.

- Polar regions have warmed much faster than the rest of the world. In Alaska, western Canada, and eastern Russia, average temperatures have increased as much as 4°C (7°F) over the past 50 years. Permafrost is melting; houses, roads, pipelines, sewage systems, and transmission lines are being damaged as the ground sinks beneath them. Trees are tipping over, and a spruce beetle infestation in Alaska (made possible by warmer winters) has killed 250 million spruce trees on the Kenai Peninsula. Coastal villages are having to relocate inland as ice breakup and increasingly severe storms erode the arctic shoreline.

- Arctic sea ice is only half as thick now as it was 30 years ago and the area covered by summer sea ice has decreased by more than 4 million km^2 (about half the size of the United States) in just three decades. By 2040, the Arctic Ocean could be totally ice-free in the summer, if present trends continue. This is bad news for polar bears, which depend on the ice to hunt seals. An aerial survey in 2005 found bears swimming across as much as 260 km (160 mi) of open water to reach the pack ice. After a major storm, dozens of bears were missing or spotted dead in the water. In 2008 the United States added polar bears to the endangered species list because of loss of arctic ice. Loss of sea ice is also devastating for Inuit people whose traditional lifestyle depends on ice for travel and hunting.

- Ice shelves on the Antarctic Peninsula are breaking up and disappearing rapidly. A recent survey found that 90 percent of the glaciers on the peninsula are now retreating an average of 50 m per year. The Greenland ice cap also is reported to be melting twice as fast as it did a few years ago. Because ice shelves are floating, they don't affect sea level when they melt. Greenland's massive ice cap, however, holds enough water to raise sea level by about 6 m (about 20 ft) if it all melts.

- Alpine glaciers everywhere are retreating rapidly. Mount Kilimanjaro has lost 85 percent of its famous ice cap since 1915. By 2015, all the permanent ice on the mountaintop is expected to be gone. In 1972, Venezuela had six glaciers; now it has only two. When Montana's Glacier National Park was created in 1910, it held some 150 glaciers. Now fewer than 30 dwindling glaciers remain (fig. 9.11). If current trends continue, all will have melted by 2030.

- So far, the oceans have been buffering the effects of our greenhouse emissions both by absorbing CO_2 directly and by storing heat. Deep-diving sensors show that the oceans are absorbing 0.85 watts per m^2 more than is radiated back to space. This absorption slows current warming, but also means that even if we reduce our greenhouse gas emissions today, it will take centuries to dissipate that stored heat. The higher levels of CO_2 being absorbed are acidifying the oceans, and could have adverse effects on sea life. Mollusks and corals, for example, have a more difficult time making calcium carbonate shells and skeletons at the lower pH.

- Sea level has risen worldwide approximately 15–20 cm (6–8 in.) in the past century. About one-quarter of this increase is ascribed to melting glaciers; roughly half is due to thermal expansion of seawater. If all of Antarctica were to melt, the resulting rise in sea level could be several hundred meters.

- Satellite images and surface measurements show that growing seasons are now as much as three weeks longer in a band across northern Eurasia and North America than they were 30 years ago (see fig. 1.7). New plants and animals, never before seen in the far north, are now extending their territories, while native species, such as musk ox, caribou, walrus, and seal populations are declining as climate changes.

- Droughts are becoming more frequent and widespread. In Africa, for example, droughts have increased about 30 percent since 1970. In 2005, the United Nations reported that 60 million people in 36 countries needed emergency food aid. Drought was the single largest cause of famine, although wars, economics, and internal politics always exacerbate these emergencies.

- Biologists report that many animals are breeding earlier or extending their range into new territory as the climate changes. In Europe and North America, for example, 57 butterfly species have either died out at the southern end of their range, or extended the northern limits, or both. Plants, also, are moving into new territories. Given enough time and a route for migration, many species may adapt to new conditions, but we now are forcing them to move much faster than many achieved at the end of the last ice age (fig. 9.12).

Figure 9.11 Alpine glaciers everywhere are retreating rapidly. These images show the Grinnell Glacier in 1914 and 1998. By 2030, if present melting continues, there will be no glaciers in Glacier National Park.

- The disappearance of amphibians, such as the beautiful golden toads from the cloud forests of Costa Rica or western toads from Oregon's Cascade Range, is thought to have been caused at least in part by changing weather patterns. Emperor and Adélie penguin populations have declined by half over the past 50 years as the ice shelves on which they depend for feeding and breeding disappear.

- Coral reefs worldwide are "bleaching," losing key algae and resident organisms, as water temperatures rise above 30°C (85°F). With reefs nearly everywhere threatened by pollution, overfishing, and other stressors, scientists worry that rapid climate change could be the final blow for many species in these complex, biologically rich ecosystems.

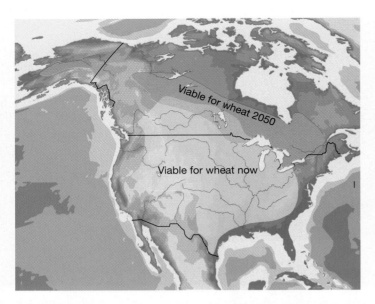

Figure 9.12 Most of the central United States is suitable for growing wheat now, but if current trends continue, the climatic conditions for wheat could be in central Canada in 2050.

Figure 9.13 More than 1 million people had to be evacuated from New Orleans in 2005 when a storm surge created by Hurricane Katrina breached the levees and allowed Lake Pontchartrain to flood into the city.

- Storms are becoming stronger and more damaging. The 2005 Atlantic storm season was the most severe on record, with 26 named tropical storms, twice as many as the average over the past 30 years. Some of this increased frequency could be a natural cycle, but the greater strength of these storms (an unprecedented 7 hurricanes reached level 5, the highest category) undoubtedly reflects higher sea surface temperatures. Hurricane Katrina, which demolished a 230,000 km² (90,000 mi²) area of the U.S. Gulf Coast (an area about the size of Great Britain) and flooded New Orleans (fig. 9.13) caused at least $200 billion in damage, the greatest storm loss in American history.

Global warming will be expensive

In 2006, Sir Nicholas Stern, former chief economist of the World Bank, issued a study on behalf of the British government on the costs of global climate change. It was one of the most strongly worded warnings to date from a government report. He said, "Scientific evidence is now overwhelming: climate change is a serious global threat, and it demands an urgent global response." If we don't act soon, Stern estimates, immediate costs of climate change will be at least 5 percent of the global GDP each year. If a wider range of risks is taken into account, the damage could equal 20 percent of the annual global economy. That would disrupt our economy and society on a scale similar to the great wars and economic depression of the first half of the twentieth century.

The report estimates that reducing greenhouse gas emissions now to avoid the worst impacts of climate change would cost only about 1 percent of the annual global GDP. That means that a dollar invested now could save us $20 later in this century. "We can't wait the five years it took to negotiate Kyoto," Sir Nicholas says, "We

simply don't have time." The actions we take—or fail to take—in the next 10 to 20 years will have a profound effect on those living in the second half of this century and in the next. Energy production, Stern suggests, will have to be at least 80 percent decarbonized by 2050 to stabilize our global climate.

As chapter 2 points out, global climate change is a moral and ethical issue as well as a practical one. Many religious leaders are joining with scientists and business leaders to campaign for measures to reduce greenhouse gas emissions. Ultimately, the people likely to suffer most from global warming are the poorest in Africa, Asia, and Latin America who have contributed least to the problem. Those of us in the richer countries will likely have resources to blunt problems caused by climate change, but residents of poorer countries will have fewer options. The Stern report says that without action, at least 200 million people could become refugees as their homes are hit by drought or floods. Furthermore, there's a question of intergenerational equity. What kind of world are we leaving to our children and grandchildren? What price will they pay if we fail to act?

The Stern review recommends four key elements for combating climate change. They are: (1) *emissions trading* to promote cost-effective emissions reductions, (2) *technology sharing* that would double research investment in clean energy technology and accelerate spread of that technology to developing countries, (3) *reduce deforestation,* which is a quick and highly cost-effective way to reduce emissions, and (4) *help poorer countries* by honoring pledges for development assistance to adapt to climate change.

The fourth IPCC report—issued in 2007—estimates the cost of preventing CO_2 doubling to be even less than Sir Nicholas calculates. The IPCC says it would cost only 0.12 percent of annual global GDP to reduce carbon emissions the 2 percent per year necessary to stabilize world climate. Former president Bill Clinton says that combating climate change doesn't have to mean economic hardship. It could be the biggest development stimulus since World War II, creating millions of jobs and saving trillions of dollars in foreign fuel imports.

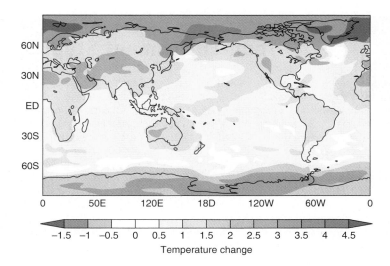

Figure 9.14 Predicted warming for carbon dioxide doubling in 2100. Note general warming of 1–2°C with much greater temperature rise in polar regions.
Source: Data from Dr. Tim Barnet, Scripps Institute of Oceanography, 2002.

A vigorous market for emissions trading already exists. In 2006, about 700 million tons of carbon equivalent credits were exchanged for some (U.S.) $3.5 billion. In 2008, trade is expected to be the equivalent of 4.8 billion tons of carbon reduction. Ultimately, this market may grow to $500 billion per year by 2050, and could represent an important source of development assistance for poorer countries. Market mechanisms for controlling climate change are discussed further in chapter 14.

Both wildlife and people will have to adapt

In many cases, both organisms and human infrastructure will not be able to move or adapt quickly enough to accommodate the rates at which climate is now changing. Tropical areas will probably not get much warmer than they are now, but some middle- and high-latitude areas could experience unaccustomed heat (fig. 9.14). Of course there could also be some good results from climate change. Residents of northern Canada, Siberia, and Alaska may enjoy warmer temperatures, longer growing seasons, and longer ice-free shipping seasons from their ports. The opening of the fabled Northwest Passage through the Arctic Ocean could cut thousands of kilometers off the sea route from London to Vladivostok. Stronger monsoons might bring welcome rains to Africa's Sahel, but we're already seeing extreme droughts in other parts of Africa that put millions of people at risk of food shortages.

Carbon dioxide is a fertilizer for plants, and higher CO_2 levels bring faster growth for many plant species. Sherwood Idso of the U.S. Department of Agriculture claims that increased CO_2 concentrations could trigger lush plant growth that would make crops abundant. But other researchers point out that the effects of elevated CO_2 in real-world situations may be very different from conditions in a greenhouse (see Exploring Science, p. 212).

Enough water is stored in glacial ice caps in Greenland and Antarctica to raise sea levels around 100 m (300 ft) if they all melt.

About one-third of the world's population now live in areas that would be flooded if that happens. Even the 75 cm (30 in.) sea-level rise expected by 2050 will flood much of south Florida, Bangladesh, Pakistan, and many other low-lying coastal areas. Most of the world's largest urban areas are on coastlines. Wealthy cities such as New York or London can probably afford to build dikes to keep out rising seas, but poorer cities such as Jakarta, Kolkata, or Manila might simply be abandoned as residents flee to higher ground. Several small island countries such as the Maldives, the Bahamas, Kiribati, and the Marshall Islands could become uninhabitable if sea levels rise a meter or more. The South Pacific nation of Tuvalu has already announced that it is abandoning its island homeland. All 11,000 residents will move to New Zealand, perhaps the first of many climate change refugees.

Insurance companies worry that the $2 trillion in insured property along U.S. coastlines is at increased risk from a combination of high seas and catastrophic storms. At least 87,000 homes in the United States within 150 m (500 ft) of a shoreline are in danger of coastal erosion or flooding in the next 50 years. Accountants warn that loss of land and structures to flooding and coastal erosion together with damage to fishing stocks, agriculture, and water supplies could raise worldwide insurance claims from about $50 billion, which they were in the 1970s, to more than $150 billion per year in 2010. Some of this increase in insurance claims is that more people are living in dangerous places, but extra-severe storms only exacerbate this problem.

Infectious diseases are likely to increase as the insects and rodents that carry them spread to new areas. Already we have seen diseases such as malaria, dengue fever, and West Nile virus appear in parts of North America where they had never before been reported. Coupled with the movement of hundreds of millions of environmental refugees and greater crowding as people are forced out of areas made uninhabitable by rising seas and changing climates, the spread of epidemic diseases will also increase.

An ominous consequence of arctic permafrost melting might be the release of vast stores of methane hydrate now locked in frozen ground and in sediments in the ocean floor. Together with the increased oxidation of high-latitude peat lands potentially caused by warmer and dryer conditions, release of these carbon stores could add as much CO_2 to the atmosphere as all the fossil fuels ever burned. We could trigger a disastrous positive feedback loop in which the effects of warming cause even more warming. On the other hand, increased ocean evaporation might intensify snowfall at high latitudes so that arctic glaciers and snow pack would increase rather than decrease. Clearly, we're disturbing a complex system and facing consequences that we don't fully understand.

9.4 The Kyoto Protocol Attempts to Slow Climate Change

One of the centerpieces of the 1992 United Nations Earth Summit meeting in Rio de Janeiro was the Framework Convention on Climate Change, which set an objective of stabilizing greenhouse gas emissions to reduce the threats of global warming. At a

Exploring
SCIENCE:

Carbon-Enrichment Studies

Figure 1 Student researchers collect samples for analysis at Cedar Creek's Free Air Carbon Enrichment study. The ring of striped poles blows CO2 over the prairie plants growing in experimental plots.

Among the greenhouse gases, none has been studied more than carbon dioxide (CO_2). As you've learned already in this chapter, CO_2 has been analyzed in 400,000-year-old air bubbles taken from glacial ice, measured atop Hawaii's Mauna Loa volcano since 1957, and implicated in a whole host of current climate changes. No one disputes that CO_2 levels in the troposphere increased from 280 to 380 ppm between 1750 and 2005. Few argue that it may exceed 500 ppm—nearly twice its preindustrial concentration—before the twenty-first century ends, but we're not sure what the effects of these high CO_2 levels may be on ecological processes.

Plants need CO_2 to carry out photosynthesis. We know that many plant species flourish in elevated CO_2 levels. That's why greenhouses often pipe in extra CO_2 to accelerate growth. Some horticulturists predict that doubling CO_2 atmospheric concentrations will produce such lush vegetation that the earth will return to a verdant Eden. Others contend that how plants behave in the ideal conditions of a greenhouse or growth chamber is very different from what happens in the real world. How can we know what to expect if current trends continue?

In 1997, Peter Reich and his colleagues began experiments at the University of Minnesota's Cedar Creek Natural History Area that attempt to mimic natural conditions more

faithfully than is possible in a growth chamber. At Cedar Creek, the researchers erected several large rings of towers from which CO_2 is released (fig. 1). Sensors track wind speeds and CO_2 concentrations within the ring, while a computer regulates gas flows to maintain 560 ppm throughout the experimental plot. Even when winds are swirling around the research site, the computer can adjust gas flows to keep CO_2 levels constant. Because the CO_2 isn't contained, these experiments are called Free Air Carbon Enrichment (FACE). Control plots have exactly the same design as the experimental ones except that the control towers blow plain air rather than CO_2-enriched air. The Cedar Creek Long-Term Ecological Research is an important center for the study of the relationship between biodiversity and prairie ecosystem stability, so the FACE research also included several different levels of native prairie species richness, as well as different fertilizer treatments, in experimental plots.

Dr. Reich and colleagues discovered that the response of plants to elevated CO_2 and nitrogen fertilization depended on how many species were in a test plot. In a paper published in 2001 in *Nature*, they showed that the amount of plant material (biomass) produced each year rose when CO_2 and nitrogen were added to the test plots. But test plots containing 9 or 16 plant species benefited much more than test plots with a single plant species (fig. 2). In a later paper in the *Proceedings of the National Academy of Sciences*, Dr. Reich and his team also demonstrated that, if test plots contained ecologically different plants (e.g., a legume and a summer-blooming grass), more biomass developed than in test plots with a single one of those functional plant groups. In other words, vegetation that contained a single plant species responded positively to rising carbon dioxide, but several plant species with different ecological functions responded much better, even without fertilization. The different species and functional groups compensated for each other in the face of a varying climate and environment.

Similar research is being done on forest species and agricultural crops. Peter Curtis, an Ohio State University researcher, writing in the journal *Phytologist*, summarized the results of dozens of FACE experiments on a number of species. In grain crop plants the total seed weight—and therefore food yield—increased by up to 25 percent with a doubling of CO_2 con-

centration, but in all species except legumes the nutritional value of those plants fell. In one study of tall fescue (*Festuca alatior*), one of the most important pasture grasses in the United States, the digestible protein that grazers could extract from the CO_2-stimulated grass actually fell below levels in existing forage, despite fertilization. So, while we may have luxuriant plant growth under elevated CO_2, it may be less valuable for certain uses than what we have today.

In the end, raising CO_2 through human activities represents a huge global experiment not unlike the FACE experiments, but with an important difference. If CO_2 reaches 500 ppm by 2100, we'll be stuck with those conditions—and their effects—for many years. What basic research tells us in the meantime about plant responses, including our important crop plants, will be important in understanding how to adapt to the challenges we face.

Dr. Reich says, "The idea that carbon dioxide will uniformly increase plant growth and therefore provide a large and predictable sink for carbon dioxide is naive at best. This type of model fails to consider the real complexity of the carbon dioxide fertilization effect. Species composition, community diversity, and the amount of nitrogen in the system interact to produce varying responses. In particular, reduced biodiversity and the widespread limited amount of nitrogen in the soil will reduce the CO_2 fertilization effect in terrestrial plant biomass."

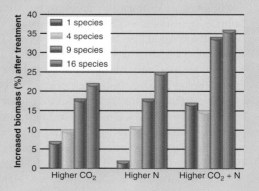

Figure 2 The Cedar Creek BioCON study demonstrates strength in numbers. Single species plots did not respond as well to CO_2 increases as multi-species plots.
Data from Peter Reich.

follow-up conference in Kyoto, Japan, in 1997, 160 nations agreed to roll back CO_2, methane, and nitrous oxide emissions about 5 percent below 1990 levels by 2012. Three other greenhouse gases, hydrofluorocarbons, perfluorocarbons, and sulfur hexafluoride, would also be reduced, although from what level was not decided. Known as the **Kyoto Protocol**, this treaty sets different limits for individual nations, depending on their output before 1990. Poorer nations, such as China and India, were exempted from emission limits to allow development to increase their standard of living. Wealthy countries created the problem, the poorer nations argue, and the wealthy should deal with it.

Although the United States took a lead role in negotiating a compromise at Kyoto that other countries could accept, President George W. Bush refused to honor U.S. commitments. Claiming that reducing carbon emissions would be too costly for the U.S. economy, he said, "We're going to put the interests of our own country first and foremost." The United States, which depends solely on voluntary efforts to limit greenhouse gas emissions, had a 2 percent rise in those releases in 2007. At this rate, the U.S. will be 25 percent above 1990 emissions by 2012.

Meanwhile, 126 countries have ratified the Kyoto Protocol. If the U.S. continues to refuse to comply, it could have severe ramifications on U.S. corporations engaged in international business. Thousands of American businesses—including most of the largest ones—fall into this category. Having to modify their products and practices for overseas markets while not doing so for domestic sales would be expensive and would put them at a disadvantage with competitors not subject to these costs. Moreover, because Kyoto is based on a global cap-and-trade program, companies that get in on it early will have an advantage—they can buy cheap emissions credits before the price gets bid up.

Many of the largest business conglomerates in America have joined environmental groups to call for strong national legislation to achieve significant reductions of greenhouse gas emissions. Those companies would prefer a single national standard rather than a jumble of conflicting local and state rules. Knowing that climate controls are inevitable, they'd rather know now how they'll have to adapt rather than wait until a crisis causes us to demand sudden, radical changes.

At a Kyoto follow-up meeting held in Bali in 2007, the United States finally acquiesced to global pressure and signed an action plan that commits all developed countries to adopt mitigation plans. U.S. negotiators refused, however, to promise specific reductions.

There has been a dramatic reversal in public understanding of global warming. Just a few years ago, about two-thirds of the American public believed that science of global climate change was highly uncertain. Now, all but a few recognize the importance of this issue. A great deal of credit for this change goes to former vice president Al Gore, whose movie, *An Inconvenient Truth*, showed the evidence to millions. Al Gore shared the Nobel Peace Prize with the IPCC in 2007.

Even the military is becoming concerned about global warming. In 2007, the U.S. Military Advisory Board said, "Climate change, national security, and energy dependence are a related set of global challenges that will lead to tensions even in stable regions of the world." Some international aid agencies point to the civil war and accompanying humanitarian crisis in the Darfur region of Sudan. These problems are rooted in drought and food shortages caused by changing weather patterns that have led to years of below normal rainfall and desertification. Global climate change may bring more such conflicts along with the millions of refugees and tragic suffering that we see now in the African Sahel.

In the next section we'll look at some of the ways we can reverse the process of global warming.

There are many ways we can control greenhouse emissions

Is the situation hopeless? Can we do anything to reverse the path we're on? No single step can reverse our release of greenhouse gases, but many simple policy changes could help stabilize our climate. One of the most encouraging analyses in this area is that of Steven Pacala and Robert Socolow at Princeton University. They say we already have the scientific, technical, and industrial know-how to stabilize our CO_2 emissions over the next half century, while still meeting the world's energy needs. In several recent publications, this group examines a portfolio of 14 policy options for reducing our CO_2 output. No single step can do the entire job by itself, they conclude, but the potential benefits of the entire portfolio are large enough that not every option must be used.

To avoid a doubling of atmospheric CO_2 we need to reduce our annual carbon emissions by about 8 billion tons by 2058. To make this easier to understand, the Princeton scientists divide the "stabilization triangle" between our current path and the flat emissions that we need in order to avoid doubling CO_2 levels into seven "wedges" (fig. 9.15 and table 9.2). Each wedge represents 1 GT

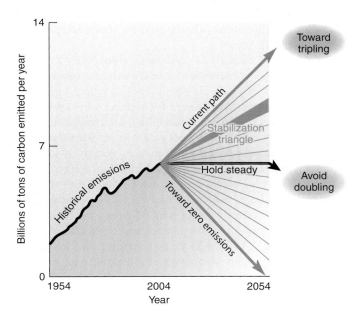

Figure 9.15 What would it take to stabilize our carbon emissions? Some authors argue that we already have the science and technology to do so. All we need is the political will.

Table 9.2	Actions to Reduce Global CO_2 Emissions by 1 Billion Tons over 50 Years
1.	Double the fuel economy for 2 billion cars from 30 to 60 mpg.
2.	Cut average annual travel per car from 10,000 to 5,000 miles.
3.	Improve efficiency in heating, cooling, lighting, and appliances by 25 percent.
4.	Update all building insulation, windows, and weather stripping to modern standards.
5.	Boost efficiency of all coal-fired power plants from 32 percent today to 60 percent (through co-generation of steam and electricity).
6.	Replace 800 large coal-fired power plants with an equal amount of gas-fired power (four times current capacity).
7.	Capture CO_2 from 800 large coal-fired, or 1,600 gas-fired, power plants and store it securely.
8.	Replace 800 large coal-fired power plants with an equal amount of nuclear power (twice the current level).
9.	Add 2 million 1 MW windmills (50 times current capacity).
10.	Generate enough hydrogen from wind to fuel a billion cars (4 million 1 MW windmills).
11.	Install 2,000 GW of photovoltaic energy (700 times current capacity).
12.	Expand ethanol production to 2 trillion liters per year (50 times current levels).
13.	Stop all tropical deforestation and replant 300 million ha of forest.
14.	Apply conservation tillage to all cropland (10 times current levels).

Source: Data from Pacala and Socolow, 2004.

Figure 9.16 Synthetic "trees" could capture CO_2 from the air. Each of these structures would be about 100 m (300 ft) tall and would have an array of metal slats 50 × 50 m across. Each of these trees could remove 90,000 tons of CO_2 per year from the air, an amount equal to the CO_2 produced by 15,000 cars. It would take a quarter of a million of them to remove all the CO_2 produced annually by humans.

(1 billion tons) of carbon emissions avoided in 2058, or a cumulative total of 25 billion tons over the next 50 years, compared to a "business as usual" scenario.

Because most of our CO_2 emissions come from fossil fuel combustion, energy conservation and a switch to renewable fuels probably are the first places we should look to reduce our climate problem. For example, 2 wedges (2 GT reduction in carbon emissions) could be accomplished by doubling our average fuel economy from the expected 30 miles per gallon in 2058 to 60 mpg, and by halving the number of miles driven each year by switching to walking, biking, or using mass transit. Simply installing the most efficient lighting and appliances available, along with improved insulation in buildings, could save another 2 GT of carbon emissions. Similarly, improving power plant efficiency and reducing the energy consumption of industrial processes could reduce carbon output one wedge (1 GT) in 50 years. Capturing and storing CO_2 released by power plants, methane wells, and other large sources could save another billion tons of carbon. Much of this CO_2 could be injected into oil wells to improve crude oil production.

As you can see, altogether these policy changes add up to considerably more than the 7 GT per year reduction we need to make by 2058 to avoid doubling atmospheric CO_2 concentrations. At least seven other immediately available options exist for reducing greenhouse gases by the equivalent of 1 GT of carbon. If we did all of them, we could reverse the present trajectory and move toward zero greenhouse emissions.

There also are many ways individuals can contribute to this effort (see What Can You Do? p. 215). Perhaps the easiest way to become carbon neutral is to simply pay someone to do it for you. There are many organizations you can pay to plant trees, or do other carbon storage activities to compensate for your carbon emissions. British Petroleum, for example, will compensate for the CO_2 emitted by a year's driving of the average car for about $40. You could make a guilt-free round-trip airplane flight from New York to Los Angeles for about $6.25. Or Carbonfund.org will offset for all the greenhouse gas emissions (both direct and indirect) for the average American for $99 per year.

There are some worries about these programs, however. In some places, planting trees is inappropriate. Fast-growing trees, such as the eucalyptus or pines often used in reforestation programs, can dry streams and springs and deplete soil nutrients. Furthermore, if the trees are grown for a few years and then harvested and burned, you haven't done anything for the planet. If you're going to finance tree planting, ask where the trees will be planted and how they'll be protected.

Progress is being made

Many countries are working to reduce greenhouse emissions. The United Kingdom, for example, had already rolled CO_2 emissions back to 1990 levels by 2000 and vowed to reduce them 60 percent by 2050. Britain already has started to substitute natural gas for coal, promote energy efficiency in homes and industry, and raise its

What Can You Do?

Reducing Individual CO₂ Emissions

Each of us can take steps to reduce global warming. While individually each change may have only a small impact, collectively they add up. Furthermore, most will save money in the long run and have other environmental and health benefits by reducing air pollution and resource consumption. The savings vary depending on where and how you live, but the following are averages for the United States.

	CO₂ Reduction (Pounds per Year)	Approximate Yearly Savings (U.S. $)
1. Keep your car tires at full pressure, avoid quick starts and stops, and drive within the speed limit.	1,100	$130
2. Carpool, walk, or take the bus once per week.	800	$100
3. Turn off the lights when you leave a room.	300	$10
4. Replace all your incandescent light bulbs with compact fluorescent bulbs.	100 each	$5 each
5. Raise your air conditioning 2°.	400	$20
6. Lower your furnace thermostat 2°.	550	$50
7. When heating or cooling, close doors and windows.	500	$20
8. When heating or cooling aren't needed, open doors and windows.	500	$20
9. Unplug TVs, DVD players, computers, and other instant-on electronics.	250	$20
10. Eat local, seasonal food.	250	variable
11. Take only five-minute showers.	250	$25
12. Take the train instead of flying 500 miles.	300	$100
13. Air dry your clothes.	700	$100
14. Replace your old car, truck, or SUV with one that gets at least 50 mpg.	6,000	$750
15. Defrost your refrigerator and keep coils and door seals clean.	700	$100

Source: Interfaith Power and Light, 2007.

already high gasoline tax. Plans are to "decarbonize" British society and to decouple GNP growth from CO₂ emissions. New carbon taxes are expected to lower CO₂ releases and trigger a transition to renewable energy over the next five decades. New Zealand Prime Minister Helen Clark pledged that her country will be the first to be "**carbon neutral**," that is, to reduce net greenhouse gas emissions to zero, although she didn't say when this will occur.

Germany, also, has reduced its CO₂ emissions at least 10 percent by switching from coal to gas and by encouraging energy efficiency throughout society. Atmospheric scientist Steve Schneider calls this a "no regrets" policy; even if we didn't need to stabilize our climate, many of these steps save money, conserve resources, and have other environmental benefits. Nuclear power also is being promoted as an alternative to fossil fuels. It's true that nuclear reactions don't produce greenhouse gases, but security worries and unresolved problems of how to store wastes safely make this option unacceptable to many people.

Many people believe renewable energy sources offer the best solution to climate problems. Chapter 12 discusses options for conserving energy and switching to renewable sources, such as solar, wind, geothermal, biomass, and fuel cells. Denmark, the world's leader in wind power, now gets 20 percent of its electricity from windmills. Plans are to generate half of the nation's electricity from offshore wind farms by 2030. Even China has promised to cut 10 percent of its CO₂ emissions per unit of economic output by shifting to renewable energy and conservation.

In addition to reducing its output, there are options for capturing and storing CO₂. Dr. Klaus S. Lackner of Columbia University has proposed building synthetic "trees" to capture CO₂ from the air (fig. 9.16). A calcium hydroxide solution trickling across the "leaves" would absorb CO₂. A structure with 2,500 m² of collecting area could remove 90,000 tons of CO₂ per year from the air (1,000 times more than a natural tree), an amount equal to the CO₂ produced by 15,000 cars. About 250,000 synthetic trees worldwide would be needed to soak up the 22 billion tons of CO₂ produced annually by burning fossil fuels. Critics claim that synthesizing the hydroxide needed for this reaction and disposing of the vast amount of calcium carbonate it would create may take more energy than the fuel that produced the CO₂ released. The first of these "trees" is now being built in Arizona.

Planting real trees can also be effective if they're allowed to mature to old-growth status or are made into products, such as window frames and doors, that will last for many years before they're burned or recycled. Nobel Prize-winner Wangari Maathai has started a campaign to plant one billion trees. Economists say this is much cheaper than fuel switching or other mitigation plans. Farmland also can serve as a carbon sink if farmers change their crop mixture and practice minimum till cultivation that keeps carbon in the soil.

Geoengineering may be necessary

Many people worry that efforts to reduce greenhouse emissions may be too little and too late. Despite decades of conferences and negotiations, emissions in most countries have continued to increase. It looks now as if we've already started to alter our global climate. Some scientists are beginning to argue that we need to take dramatic steps to adapt to climate change or to reverse warming through technological means. Among the adaptation steps might be to start massive transplanting of species to places where we think future climate will be suitable. For humans, we may need an enormous migration program away from coastal regions that will likely be flooded by rising seas.

The ocean fertilization schemes described in the opening case study for this chapter are an example of **geoengineering**, in which we attempt to deliberately alter global-scale ecological processes. Some other proposals for climate control include placing enormous mirrors in space to deflect sunlight away from the earth, or covering ice sheets in Greenland and Antarctica with reflective, insulating covers to prevent their melting. Others have suggested pumping a vast amount of sulfate aerosols into the atmosphere to shade the earth. This idea is modeled on the effects of volcanoes. When Mount Pinatubo erupted in the Philippines in 1991, it released 20 million tons of SO_2 and dust into the air, which cooled global temperatures about $0.5°C$ ($0.9°F$) for more than a year. All these schemes would be tremendously expensive and carry large risks of unforeseen consequences.

Proponents of ocean fertilization claim that it's just like planting trees except that it doesn't take up valuable land. It might even increase productivity in the ocean and help rebuild fish stocks reduced by overharvesting. On the other hand, we don't know what the ecological effects may be from massive ocean eutrophication. Perhaps a better use of algae is to grow them in ponds or other on-land containers. Cultivated next to a power plant, algae could absorb CO_2 from combustion while being kept warm with excess steam. Algae can double their biomass in just 24 hours. Some algae are as much as 50 percent oil that can be used as biodiesel. Where a hectare of land produces about 3,000 L of ethanol per year from corn, or about 600 L of biodiesel from soybeans, that same hectare devoted to algal culture could theoretically produce at least 45,000 L of biofuel annually.

Another way to store CO_2 is to inject it into geologic formations. Since 1996, Norway's Statoil has been pumping more than 1 million metric tons of CO_2 per year into an aquifer 1,000 m below the seafloor in the North Sea. The pressurized CO_2 enhances oil recovery. It also saves money because otherwise the company would have to pay a $50 per ton carbon tax on its emissions. Around the world, deep, briny aquifers could store a century worth of CO_2 output at current fossil fuel consumption rates. A number of companies have started, or are now planning, similar schemes.

Proponents of **carbon management**, as these various projects are called, argue that it may be cheaper to clean up fossil fuel effluents than to switch to renewable energy sources. By far, the easiest way to remove CO_2 from a coal-fired power plant is the integrated gasification combined cycle (IGCC) technology described in chapter 12. Utilities, eager to continue business as usual, are touting "clean-coal" techniques and carbon sequestration as the answer to global warming. These approaches are feasible. Geological formations around the world could hold hundreds of years' worth of CO_2 at current production levels. These formations, however, aren't necessarily near the sites where we now produce CO_2. Shipping billions of tons of CO_2 long distances could become a serious bottleneck. It may be far more efficient—not to mention more environmentally benign—to switch

Figure 9.17 While air quality is improving in many industrialized countries, newly developing countries have growing pollution problems. Xi'an, China, often has particulate levels above 300 µg/m³.

to local alternative energy sources or to ship energy (electricity) or carbon-neutral fuels (ethanol or hydrogen) to places where they're needed.

Most attention is focused on CO_2 because it lasts in the atmosphere, on average, for about 120 years. Methane (CH_4) and other greenhouse gases are much more powerful infrared absorbers but remain in the air for a much shorter time. NASA's Dr. James Hansen made a controversial suggestion, however, that the best short-term attack on warming might come by focusing not on carbon dioxide but on methane. Reducing gas pipeline leaks would conserve this valuable resource as well as help the environment.

Methane from many landfills, oil wells, and coal mines that once would have been burned or simply vented into the air is now being collected and used to generate electricity. Rice paddies are a major methane source because underwater, anaerobic (without oxygen) decomposition produces CH_4 instead of CO_2. Changing flooding schedules and fertilization techniques can reduce methane production. Finally, ruminant animals (such as cows, camels, and buffalo) create large amounts of methane from digestive systems. Modified diets can reduce these emissions.

Soot, while not mentioned in the Kyoto Protocol, might also be important in global warming. Dark, airborne particles absorb both ultraviolet and visible light, converting them to heat energy. According to some calculations, reducing soot from diesel engines, coal-fired generators, forest fires, and wood stoves might reduce net global warming by 40 percent within three to five years. Curbing soot emissions would also have beneficial health effects.

In the United States, more than 700 cities and 39 states have announced their own plans to combat global warming. And 450 college campuses have pledged to reduce greenhouse emissions. Some have promised to be carbon neutral by 2020. Some corporations are following suit. British Petroleum has set a goal of cutting CO_2 releases from all its facilities by 10 percent before 2010. Each of us can make a contribution in this effort. As Pacala and Socolow point out, simply driving less and buying high-mileage vehicles could save about 2 billion tons of carbon emissions over the next 50 years.

In the midst of all the debate about how serious the consequences of global climate change may or may not be, we need to remember that many of the proposed solutions are advantageous in their own right. Moving from fossil fuels to renewable energy sources such as solar or wind power, for example, would free us from dependence on foreign oil and would improve air quality. Planting trees makes the world a more pleasant place to live and provides habitat for wildlife. Making buildings more energy efficient and buying high-mileage vehicles saves money in the long run. Walking, biking, and climbing stairs are good for your health as well as reducing traffic congestion and energy consumption. Reducing waste, recycling, and other forms of sustainable living improve our environment in many ways in addition to helping fight climate change. It's important to focus on these positive effects rather than to look only at the gloom-and-doom scenarios for global climate catastrophes. As the Irish statesman and philosopher Edmund Burke said, "Nobody made a greater mistake than he who did nothing because he could do only a little."

9.5 Air Pollution

As the global warming debate shows, human activities can influence climate processes. In this section, we will discuss ways that human activities affect air quality and environmental processes on a more local scale.

According to the Environmental Protection Agency (EPA), Americans release some 147 million metric tons of air pollution (not counting carbon dioxide or wind-blown soil) each year. Worldwide emissions of these pollutants are around 2 billion metric tons per year. Even remote, pristine wilderness areas are now affected. Over the past 20 years, however, air quality has improved in most cities in Western Europe, North America, and Japan. Many young people might be surprised to learn that, a generation ago, most American cities were much dirtier than they are today. The EPA estimates that, since 1990, when regulation of the most hazardous materials began, air toxics emissions have been reduced more than 1 million tons per year. This is almost ten times the reductions achieved in the previous 20 years. Since the 1970s, the levels of major pollutants monitored by the EPA have decreased in the United States, despite population growth of more than 30 percent. Pollution reductions have resulted mainly from greater efficiency and pollution-control technologies in factories, power plants, and automobiles. Our success in controlling some of the most serious air pollutants gives us hope for similar progress in other environmental problems.

While developed countries have been making progress, however, air quality in the developing world has been getting much worse. Especially in the burgeoning megacities of rapidly industrializing countries (chapter 14), air pollution often exceeds World Health Organization standards by large margins. In Lahore, Pakistan, and Xi'an, China, for instance, airborne dust, smoke, and dirt often are ten times higher than levels considered safe for human health (fig. 9.17).

Studies of air pollutants over southern Asia reveal that a 3 km (2 mi) thick toxic cloud of ash, acids, aerosols, dust, and photochemical reactants covers the entire Indian subcontinent for much of the year. Nobel laureate Paul Crutzen estimates that up to 2 million people in India alone die each year from atmospheric pollution. Produced by forest fires, the burning of agricultural wastes, and dramatic increases in the use of fossil fuels, the Asian smog layer cuts the amount of solar energy reaching the earth's surface beneath it by up to 15 percent. Meteorologists suggest that the cloud—80 percent of which is human-made—could disrupt monsoon weather patterns and cut rainfall over northern Pakistan, Afghanistan, western China, and central Asia by up to 40 percent.

When this "Asian Brown Cloud" drifts out over the Indian Ocean at the end of the monsoon season, it cools sea temperatures and may be changing El Niño/Southern Oscillation patterns in the Pacific Ocean as well. As UN Environment Programme Executive Director Klaus Töpfer said, "There are global implications because a pollution parcel like this, which stretches 3 km high, can travel half way round the globe in a week."

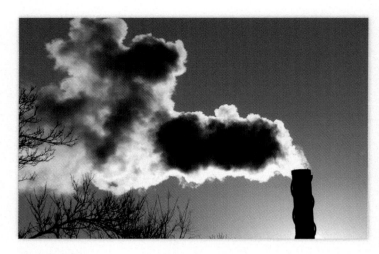

Figure 9.18 Primary pollutants are released directly from a source into the air. A point source is a specific location of highly concentrated discharge, such as this smokestack.

We have different ways to describe pollutants

In the United States, principal pollutants are identified and regulated by the Clean Air Act of 1970. A **point source** is a smokestack or some other concentrated pollution origin (fig. 9.18). **Primary pollutants** are released in a harmful form. Secondary pollutants, by contrast, become hazardous after reactions in the air. **Photochemical oxidants** (compounds created by reactions driven by solar energy) and atmospheric acids are probably the most important secondary pollutants. **Fugitive**, or **nonpoint-source, emissions** are those that do not go through a smokestack. By far

the most massive example of this category is dust from soil erosion, strip mining, rock crushing, and building construction (and destruction). Leaking valves and pipe joints contribute as much as 90 percent of the hydrocarbons and volatile organic chemicals emitted from oil refineries and chemical plants.

Conventional, or **criteria, pollutants** are a group of seven major pollutants (sulfur dioxide, carbon monoxide, particulates, volatile organic compounds, nitrogen oxides, ozone, and lead) that contribute the largest volume of air-quality degradation and are considered the most serious threat of all air pollutants to human health and welfare. Figure 9.19 shows the main sources

Figure 9.19 Anthropogenic sources of six of the primary "criteria" air pollutants in the United States.
Source: UNEP, 1999.

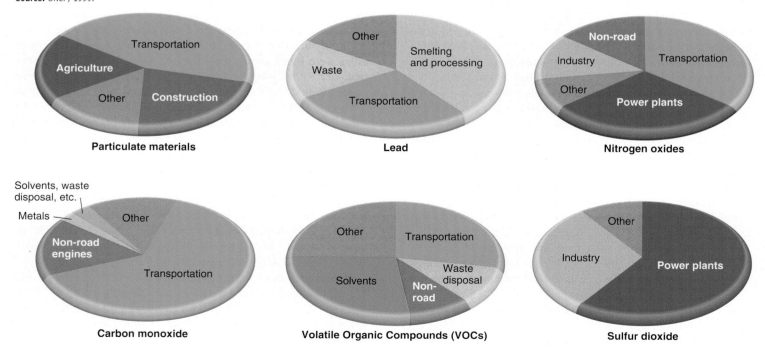

Particulate materials

Lead

Nitrogen oxides

Carbon monoxide

Volatile Organic Compounds (VOCs)

Sulfur dioxide

Table 9.3	Photochemical Oxidant Production
Steps	**Photochemical Products**
1. NO + VOC →	NO_2 (nitrogen dioxide)
2. NO_2 + UV sunlight →	NO + O (nitric oxide + atomic oxygen)
3. O + O_2 →	O_3 (ozone)
4. NO_2 + VOC →	PAN (peroxyacetyl nitrate)

of six of these criteria pollutants. Since 1970 the Clean Air Act has mandated that the EPA set allowable limits for concentrations of these pollutants in **ambient air** (the air around us), especially in cities.

The EPA also monitors **unconventional pollutants**, compounds that are produced in less volume than conventional pollutants but that are especially toxic or hazardous. Among these are asbestos, benzene, beryllium, mercury, polychlorinated biphenyls (PCBs), and vinyl chloride. Most of these materials have no natural source in the environment (to any great extent) and are, therefore, only anthropogenic in origin.

Aesthetic degradation is another important form of pollution. Noise, odors, and light pollution may not be life threatening, but they reduce the quality of our lives. In most urban areas, it is difficult or impossible to see stars in the sky at night because of dust in the air and stray light from buildings, outdoor advertising, and streetlights.

Sources and problems of major pollutants

Most conventional pollutants are produced primarily from burning fossil fuels, especially in coal-powered electric plants and in cars and trucks, as well as in processing natural gas and oil. Others, especially sulfur and metals, are by-products of mining and manufacturing processes. Of the 188 air toxics listed in the Clean Air Act, about two-thirds are volatile organic compounds (VOCs), and most of the rest are metal compounds. In this section we will discuss the characteristics and origin of the major pollutants.

Sulfur dioxide (SO_2) is a colorless, corrosive gas that damages both plants and animals. Once in the atmosphere, it can be further oxidized to sulfur trioxide (SO_3), which reacts with water vapor or dissolves in water droplets to form sulfuric acid (H_2SO_4), a major component of acid rain. Sulfur dioxide and sulfate ions are probably second only to smoking as causes of air-pollution–related health damage. Sulfate particles and droplets also reduce visibility in the United States by as much as 80 percent.

Nitrogen oxides (NO_x) are highly reactive gases formed when combustion initiates reactions between atmospheric nitrogen and oxygen. The initial product, nitric oxide (NO), oxidizes further in the atmosphere to nitrogen dioxide (NO_2), a reddish-brown gas that gives photochemical smog its distinctive color. Because these gases convert readily from one form to the other, the general term NO_x is used to describe these gases. Nitrogen oxides combine with water to form nitric acid (HNO_3), which is also a major component of acid precipitation. Excess nitrogen in water is causing eutrophi-

cation of inland waters and coastal seas. It may also encourage the growth of weedy species that crowd out native plants.

Carbon monoxide (CO) is less common but more dangerous than the principal form of atmospheric carbon, carbon dioxide (CO_2). CO is a colorless, odorless, but highly toxic gas produced mainly by incomplete combustion of fuel (coal, oil, charcoal, wood, or gas). CO inhibits respiration in animals by binding irreversibly to hemoglobin in blood. In the United States, two-thirds of the CO emissions are created by internal combustion engines in transportation. Land-clearing fires and cooking fires also are major sources. About 90 percent of the CO in the air is consumed in photochemical reactions that produce ozone.

Particulate material includes dust, ash, soot, lint, smoke, pollen, spores, algal cells, and many other suspended materials. Aerosols, or extremely minute particles or liquid droplets suspended in the air, are included in this class. Particulates often are the most apparent form of air pollution, since they reduce visibility and leave dirty deposits on windows, painted surfaces, and textiles. Breathable particles smaller than 2.5 μm are among the most dangerous of this group because they can damage lung tissues. Asbestos fibers and cigarette smoke are among the most dangerous respirable particles in urban and indoor air because they are carcinogenic.

Volatile organic compounds (VOCs) are organic (carbon containing) gases. Plants, bogs, and termites are the largest sources of VOCs, especially isoprenes (C_5H_8), terpenes ($C_{10}H_{15}$), and methane (CH_4). These volatile hydrocarbons are generally oxidized to CO and CO_2 in the atmosphere.

More dangerous synthetic organic chemicals, such as benzene, toluene, formaldehyde, vinyl chloride, phenols, chloroform, and trichloroethylene, are released by human activities. Principal sources are incompletely burned fuels from vehicles, power plants, chemical plants, and petroleum refineries. These chemicals play an important role in the formation of photochemical oxidants.

Photochemical oxidants are products of secondary atmospheric reactions driven by solar energy (table 9.3). One of the most important of these reactions involves formation of singlet (atomic) oxygen by splitting nitrogen dioxide (NO_2). This atomic oxygen then reacts with another molecule of O_2 to make ozone (O_3). Although ozone is important in the stratosphere, in ambient air it is highly reactive and damages vegetation, animal tissues, and building materials. Ozone's acrid, biting odor is a distinctive characteristic of photochemical smog.

Lead and other toxic elements

Toxic metals and halogens are chemical elements that are toxic when concentrated and released in the environment. Principal metals of concern are lead, mercury, arsenic, nickel, beryllium, cadmium, thallium, uranium, cesium, and plutonium. Halogens (fluorine, chlorine, bromine, and iodine) are highly reactive toxic elements. Most of these materials are mined and used in manufacturing. Metals commonly occur as trace elements in fuels, especially coal.

Lead and mercury are widespread neurotoxins that damage the nervous system. By some estimates, 20 percent of all inner-city children suffer some degree of developmental retardation from high environmental lead levels. Long-range transport of lead and mercury through the air is causing bioaccumulation in remote aquatic ecosystems, such as arctic lakes and seas. Chlorine is a toxic halogen widely used in bleach, plastics, and other products. Methyl bromide (a powerful fungicide used in agriculture) and **chlorofluorocarbons** (propellants and refrigerants) are also implicated in ozone depletion.

Indoor air can be more dangerous than outdoor air

We have spent a considerable amount of effort and money to control the major outdoor air pollutants, but we have only recently become aware of the dangers of indoor air pollutants. The U.S. EPA has found that indoor concentrations of toxic air pollutants are often higher than outdoors. Furthermore, people generally spend more time inside than out and therefore are exposed to higher doses of these pollutants. In some cases, indoor air in homes has chemical concentrations that would be illegal outside or in the workplace. Under some circumstances, compounds such as chloroform, benzene, carbon tetrachloride, formaldehyde, and styrene can be 70 times higher in indoor air than in outdoor air. Molds, pathogens, and other biohazards also represent serious indoor pollutants.

Figure 9.20 Some 2.5 billion people, mainly women and children, spend hours each day in poorly ventilated kitchens and living spaces where carbon monoxide, particulates, and cancer-causing hydrocarbons often reach dangerous levels.

Cigarette smoke is without doubt the most important air contaminant in developed countries in terms of human health. The U.S. surgeon general has estimated that 400,000 people die each year in the United States from emphysema, heart attacks, strokes, lung cancer, or other diseases caused by smoking. These diseases are responsible for 20 percent of all mortality in the United States, or four times as much as infectious agents. Total costs for early deaths and smoking-related illnesses are estimated to be $100 billion per year. Eliminating smoking probably would save more lives than any other pollution-control measure.

In the less-developed countries of Africa, Asia, and Latin America, where such organic fuels as firewood, charcoal, dried dung, and agricultural wastes make up the majority of household energy, smoky, poorly ventilated heating and cooking fires represent the greatest source of indoor air pollution (fig. 9.20). The World Health Organization (WHO) estimates that 2.5 billion people—more than one-third of the world's population—are adversely affected by pollution from this source. In particular, women and small children spend long hours each day around open fires or unventilated stoves in enclosed spaces.

9.6 Interactions Between Climate Processes and Air Pollution

Physical processes in the atmosphere transport, concentrate, and disperse air pollutants. To comprehend the global effects of air pollution, it is necessary to understand how climate processes interact with pollutants. Global warming in which pollutants are altering the earth's energy budget, is the best-known case of interaction between anthropogenic pollutants and the atmosphere. In this section we will survey other important climate-pollution interactions.

Air pollutants can travel far

Dust and fine aerosols can be carried great distances by the wind. Pollution from the industrial belt between the Great Lakes and the Ohio River Valley regularly contaminates the Canadian Maritime Provinces and sometimes can be traced as far as Ireland. Similarly, dust storms from China's Gobi and Takla Makan Deserts routinely close schools, factories, and airports in Japan and Korea, and often reach western North America. In one particularly severe dust storm in 1998, chemical analysis showed that 75 percent of the particulate pollution in Seattle, Washington, air came from China. Similarly, dust from North Africa regularly crosses the Atlantic and contaminates the air in Florida and the Caribbean Islands (fig. 9.21). This dust can carry pathogens and is thought to be the source of diseases attacking Caribbean corals. Soil scientists estimate that 3 billion tons of sand and dust are blown around the world every year.

Increasingly sensitive monitoring equipment has begun to reveal industrial contaminants in places usually considered among the cleanest in the world. Samoa, Greenland, and even Antarctica and the North Pole all have heavy metals, pesticides, and radioactive elements in their air. Since the 1950s, pilots flying in the high

Figure 9.21 A massive dust storm extends more than 1,600 km (1,000 mi) from the coast of western Sahara and Morocco. Storms such as this can easily reach the Americas, and they have been linked both to the decline of coral reefs in the Caribbean and to the frequency and intensity of hurricanes formed in the eastern Atlantic Ocean.

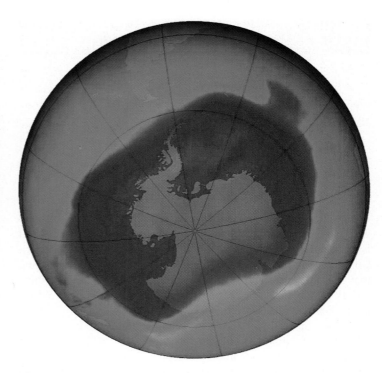

Figure 9.22 In 2006, stratospheric ozone was depleted over an area (*dark, irregular circle*) that covered 29.5 million km², or more than the entire Antarctic continent. Although CFC production is declining, this was the largest area ever recorded.

Arctic have reported dense layers of reddish-brown haze clouding the arctic atmosphere. Aerosols of sulfates, soot, dust, and toxic heavy metals, such as vanadium, manganese, and lead, travel to the pole from the industrialized parts of Europe and Russia.

Circulation of the atmosphere tends to transport contaminants toward the poles. Volatile compounds evaporate from warm areas, travel through the atmosphere, then condense and precipitate in cooler regions. Over several years, contaminants migrate to the coldest places, generally at high latitudes, where they bioaccumulate in food chains. Whales, polar bears, sharks, and other top carnivores in polar regions have been shown to have dangerously high levels of pesticides, metals, and other hazardous air pollutants in their bodies. The Inuit people of Broughton Island, well above the Arctic Circle, have higher levels of polychlorinated biphenyls (PCBs) in their blood than any other known population, except victims of industrial accidents. Far from any source of this industrial by-product, these people accumulate PCBs from the flesh of fish, caribou, and other animals they eat.

Stratospheric ozone is declining

Long-range pollution transport and the chemical reactions of atmospheric gases and pollution produce the phenomenon known as the ozone hole (fig. 9.22). The ozone "hole," really a thinning of ozone concentrations in the stratosphere, was discovered in 1985 but has probably been developing since at least the 1960s.

Chlorine-based aerosols, such as chlorofluorocarbons (CFCs), are the principal agents of ozone depletion. Nontoxic, nonflammable, chemically inert, and cheaply produced, CFCs were extremely useful as industrial gases and in refrigerators, air conditioners, styrofoam insulation, and aerosol spray cans for many years. From the 1930s until the 1980s, CFCs were used all over the world and widely dispersed through the atmosphere.

Although ozone is a pollutant in the ambient air, ozone in the stratosphere is valuable because it absorbs much of the ultraviolet (UV) radiation entering the atmosphere. UV radiation harms plant and animal tissues, including the eyes and the skin. A 1 percent loss of ozone could result in about a million extra human skin cancers per year worldwide. Excessive UV exposure could reduce agricultural production and disrupt ecosystems. Scientists worry, for example, that high UV levels in Antarctica could reduce populations of plankton, the tiny floating organisms that form the base of a food chain that includes fish, seals, penguins, and whales in Antarctic seas.

Antarctica's exceptionally cold winter temperatures (−85° to −90°C) help break down ozone. During the long, dark, winter months, strong winds known as the circumpolar vortex circle the pole. These winds isolate antarctic air and allow stratospheric temperatures to drop low enough to create ice crystals at high altitudes—something that rarely happens elsewhere in the world. Ozone and chlorine-containing molecules are absorbed on the surfaces of these ice particles. When the sun returns in the spring,

Table 9.4	Stratospheric Ozone Destruction by Chlorine Atoms and UV Radiation
Step	**Products**
1. $CFCl_3$ (chlorofluorocarbon) + UV energy	$CFCl_2$ + Cl
2. $Cl + O_3$	$ClO + O_2$
3. O_2 + UV energy	2O
4. $ClO + 2O$	$O_2 + Cl$
5. Return to step 2	

it provides energy to liberate chlorine ions, which readily bond with ozone, breaking it down to molecular oxygen (table 9.4). It is only during the antarctic spring (September through December) that conditions are ideal for rapid ozone destruction. During that season, temperatures are still cold enough for high-altitude ice crystals, but the sun gradually becomes strong enough to drive photochemical reactions.

As the antarctic summer arrives, temperatures moderate somewhat, the circumpolar vortex breaks down, and air from warmer latitudes mixes with antarctic air, replenishing ozone concentrations in the ozone hole. Slight decreases worldwide result from this mixing, however. Ozone re-forms naturally, but not nearly as fast as it is destroyed. Since the chlorine atoms are not themselves consumed in reactions with ozone, they continue to destroy ozone for years, until they finally precipitate or are washed out of the air. Almost every year since it was discovered, the antarctic ozone hole has grown. In 2000 the region of ozone depletion covered 29.8 million km² (about the size of North America).

There are signs of progress

The discovery of stratospheric ozone losses brought about a remarkably quick international response. In 1987 an international meeting in Montreal, Canada, produced the Montreal Protocol, the first of several major international agreements on phasing out most use of CFCs by 2000. As evidence accumulated, showing that losses were larger and more widespread than previously thought, the deadline for the elimination of all CFCs (halons, carbon tetrachloride, and methyl chloroform) was moved up to 1996, and a $500 million fund was established to assist poorer countries in switching to non-CFC technologies. Fortunately, alternatives to CFCs for most uses already exist. The first substitutes are hydrochlorofluorocarbons (HCFCs), which release much less chlorine per molecule. Eventually, scientists hope to develop halogen-free molecules that work as well and are no more expensive than CFCs.

There is some evidence that the CFC ban is already having an effect. CFC production in most industrialized countries has fallen sharply since 1988 (fig. 9.23), and CFCs are now being removed from the atmosphere more rapidly than they are being added. In 50 years or so, stratospheric ozone levels are expected to be back to normal. Unfortunately, there may be a downside to stopping ozone destruction. Ozone is a potent greenhouse gas. Some models suggest that lower ozone levels have been offsetting the effects

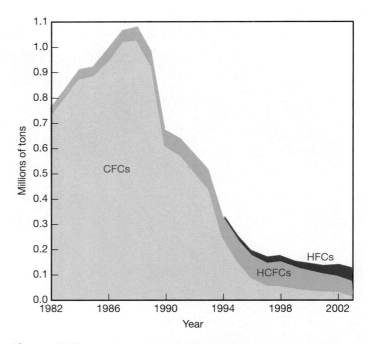

Figure 9.23 The Montreal Protocol has been remarkably successful in eliminating CFC production. The remaining HFC and HCFC use is primarily in developing countries, such as China and India.

of increased CO_2. When ozone is restored, global warming may be accelerated. As so often is the case, when we disturb one environmental factor, we affect others as well.

Cities create dust domes, smog, and heat islands

Temperature inversions can concentrate dangerous levels of pollutants within cities. Normally, air temperatures decrease with elevation above the earth's surface. An inversion reverses these conditions, with cool, dense air lying below a warmer, lighter layer. In these very stable conditions, convection currents cannot develop, and there is no mechanism for dispersing pollutants. The most stable inversion conditions are usually created by rapid night-time cooling in a valley or basin where air movement is restricted. Los Angeles is a classic example of an area with conditions that create temperature inversions and photochemical smog. Mountains surround the city on three sides, reducing wind movement; the climate is sunny; and heavy traffic creates high pollution levels. Skies are generally clear at night, allowing rapid radiant heat loss, and the ground cools quickly. Surface air layers are cooled by contact with the cool ground surface, while upper layers remain relatively warm. Density differences slow vertical mixing. As long as the atmosphere is still and stable, pollutants accumulate near the ground where they are produced and where we breathe them.

Morning sunlight initiates photochemical oxidation in the concentrated aerosols and gaseous chemicals in the inversion layer. A brown haze of hazardous chemicals, especially ozone and nitrogen dioxide, quickly develops. Although recent air quality regulations

(a)

(b)

Figure 9.24 In 1975, acid precipitation from the copper-nickel smelters (tall stacks in background) had killed all the vegetation and charred the pink granite bedrock black for a large area around Sudbury, Ontario (a). By 2005, a scrubby forest was growing again around Sudbury, but the rock surfaces remain burned black (b).

have helped tremendously, on summer days, ozone concentrations in the Los Angeles basin still can reach unhealthy levels.

Heat islands and **dust domes** occur in cities even without inversion conditions. With their low albedo, concrete and brick surfaces in cities absorb large amounts of solar energy. A lack of vegetation or water results in very slight evaporation (latent heat production); instead, available solar energy is turned into heat. As a result, temperatures in cities are frequently 3° to 5°C (5° to 9°F) warmer than in the surrounding countryside, a condition known as an urban heat island. Tall buildings create convective updrafts that sweep pollutants into the air. Stable air masses created by this heat island over the city concentrate pollutants in a dust dome.

9.7 Effects of Air Pollution

Air pollution is equally serious for ecosystem health and for human health. In this section we will review the most important effects of air pollution.

Polluted air is unhealthy

Consequences of breathing dirty air include increased probability of heart attacks, respiratory diseases, and lung cancer. This can mean as much as a five- to ten-year decrease in life expectancy if you live in the worst parts of Los Angeles or Baltimore, for example. Of course, the intensity and duration of your exposure, as well as your age and general health, are extremely important: you are much more likely to be at risk if you are very young, very old, or already suffering from some respiratory or cardiovascular disease. Bronchitis and emphysema are common chronic conditions resulting from air pollution. The U.S. Office of Technology Assessment estimates that 250,000 people suffer from pollution-related bronchitis and emphysema in the United States, and some 50,000 excess deaths each year are attributable to complications of these diseases, which are probably second only to heart attack as a cause of death.

Conditions are often much worse in developing countries. The United Nations estimates that at least 1.3 billion people around the world live in areas where the air is dangerously polluted. In many parts of the former Soviet Union, for example, respiratory ailments, cardiovascular diseases, lung cancer, infant mortality, and miscarriages are as much as 50 percent higher than in countries with cleaner air. And in China, city dwellers are four to six times more likely than country folk to die of lung cancer. The World Health Organization estimates that 4 million people die each year from diseases exacerbated by air pollution.

How does air pollution cause these health effects? Because they are strong oxidizing agents, sulfates, SO_2, NO_x, and O_3 irritate and damage delicate tissues in the eyes and lungs. Fine, suspended particulate materials penetrate deep into the lungs, causing irritation, scarring, and even tumor growth. Heart stress results from impaired lung functions. Carbon monoxide binds to hemoglobin, reducing oxygen flow to the brain. Headaches, dizziness, and heart stress result. Lead also binds to hemoglobin, damaging critical neurons in the brain and resulting in mental and physical impairment and developmental retardation.

Plants are sensitive to pollutants

In the early days of industrialization, fumes from furnaces, smelters, refineries, and chemical plants often destroyed vegetation and created desolate, barren landscapes around mining and manufacturing centers. The copper-nickel smelter at Sudbury, Ontario, is a spectacular example. Starting in 1886, open-bed roasting was used to purify sulfide ores of nickel and copper. The resulting sulfur dioxide and sulfuric acid destroyed almost all plant life within about 30 km (18.6 mi) of the smelter. Rains washed away the exposed soil, leaving a barren moonscape of blackened bedrock (fig. 9.24). Recently, emission controls have been introduced, and the environment is beginning to recover, although exposed rock is still black, and the forest cover is mostly small and rather spindly.

Certain combinations of environmental factors have **synergistic effects** in which the injury caused by exposure to two factors together is more than the sum of exposure to each factor individually. For instance, white pine seedlings exposed to subthreshold concentrations of ozone and sulfur dioxide individually do not suffer any visible injury. If the same concentrations of pollutants are given together, however, visible damage occurs.

Pollutant levels too low to produce visible symptoms of damage may still have important effects. Field studies show that yields in some crops, such as soybeans, may be reduced as much as 50 percent by currently existing levels of oxidants in ambient air. Some plant pathologists suggest that ozone and photochemical oxidants are responsible for as much as 90 percent of agricultural, ornamental, and forest losses from air pollution. The total costs of this damage may be as much as $10 billion per year in North America alone.

Smog and haze reduce visibility

We have only recently realized that pollution affects rural areas as well as cities. Even supposedly pristine places such as our national parks are suffering from air pollution. Grand Canyon National Park, where maximum visibility used to be 300 km (185 mi), is now so smoggy on some winter days that visitors can't see the opposite rim only 20 km (12.5 mi) across the canyon. Mining operations, smelters, and power plants (some of which were moved to the desert to improve air quality in cities such as Los Angeles) are the main culprits. Huge regions are affected by pollution. A gigantic "haze blob" as much as 3,000 km (about 2,000 mi) across covers much of the eastern United States in the summer, cutting visibility as much as 80 percent. People become accustomed to these conditions and don't realize that the air once was clear. Studies indicate, however, that, if all human-made sources of air pollution were shut down, the air would clear up in a few days, and there would be about 150-km (90-mi) visibility nearly everywhere, rather than the 15 km to which we have become accustomed.

Acid deposition has many effects

Acid precipitation, the deposition of wet, acidic solutions or dry, acidic particles from the air, became widely recognized as a pollution problem only in the last 20 years. But the concept has been recognized since the 1850s. We describe acidity in terms of pH, with substances below pH 7 being acidic (chapter 2). Normal, unpolluted rain generally has a pH of about 5.6 due to carbonic acid created when rainwater reacts with CO_2 in the air. Downwind of industrial areas, rainfall acidity can reach levels below pH 4.3, more than ten times as acidic as normal rain. Acid fog, snow, mist, and dew can deposit damaging acids on plants, in water systems, and on buildings. Furthermore, fallout of dry sulfate and nitrate particles can account for as much as half of the acidic deposition in some areas.

A vigorous program of pollution control has been undertaken by Canada, the United States, and several European countries since the widespread recognition of acid rain. SO_2 and NO_x emissions

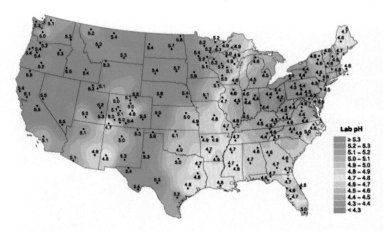

Figure 9.25 Acid precipitation over the United States, 2000.
Source: Data from National Acid Deposition Program, 2001.

from power plants have decreased dramatically over the past three decades over much of Europe and eastern North America as a result of pollution-control measures. However, rain falling in these areas remains acidic (fig. 9.25). Apparently, alkaline dust that would once have neutralized acids in air has been depleted by years of acid rain and is now no longer effective.

Aquatic effects of acid deposition

Aquatic ecosystems in Scandinavia were among the first discovered to be damaged by acid precipitation. Prevailing winds from Germany, Poland, and other parts of Europe deliver acids generated by industrial and automobile emissions—principally H_2SO_4 and HNO_3. The thin, acidic soils and nutrient-poor lakes and streams in the mountains of southern Norway and Sweden have been severely affected by this acid deposition. Most noticeable is the reduction of trout, salmon, and other game fish, whose eggs and fry die below pH 5. Aquatic plants, insects, and invertebrates also suffer. Many lakes in Sweden are now so acidic that they will no longer support game fish or other sensitive aquatic organisms. Large parts of Europe and eastern North America have also been damaged by acid precipitation.

Forest damage

Since the early 1980s observers have documented forest declines apparently related to acid damage in both Europe and North America. A detailed 1980 ecosystem inventory on Camel's Hump Mountain in Vermont found that seedling production, tree density, and viability of spruce-fir forests at high elevations had declined about 50 percent in 15 years. On Mount Mitchell in North Carolina, nearly all the trees above 2,000 m (6,000 ft) are losing needles, and about half are dead (fig. 9.26). Damage has been reported throughout Europe, from the Netherlands to Switzerland, as well as in China and the states of the former Soviet Union. In 1985 West German foresters estimated that about half the total forest area in West Germany (more than 4 million ha) was declining. The loss to the forest industry is estimated to be about 1 billion DM (deutsche marks) per year.

Figure 9.26 A Fraser fir forest on Mount Mitchell, North Carolina, killed by acid rain, insect pests, and other stressors.

Figure 9.27 Atmospheric acids, especially sulfuric and nitric acids, have almost completely eaten away the face of this medieval statue. Each year, the total losses from air pollution damage to buildings and materials amount to billions of dollars.

High-elevation forests are most severely affected. Mountain tops often have thin, acidic soils under normal conditions, with little ability to buffer, or neutralize, acid precipitation. Acidic fog and mist frequently rest on mountaintops, lengthening plants' exposure to acidity. Mountaintops also catch more rain and snow than lowlands, so they are more exposed to acidic precipitation.

The most visible mechanism of forest decline is direct damage to plant tissues and seedlings. Nutrient availability in forest soils is also depleted, however. Acids can also dissolve and mobilize toxic metals, such as aluminum, and weakened trees become susceptible to diseases and insect pests.

Buildings and monuments

In cities throughout the world, air pollution is destroying some of the oldest and most glorious buildings and works of art. Smoke and soot coat buildings, paintings, and textiles. Acids dissolve limestone and marble, destroying features and structures of historic buildings (fig. 9.27). The Parthenon in Athens, the Taj Mahal in Agra, the Coliseum in Rome, medieval cathedrals in Europe, and the Washington Monument in Washington, D.C., are slowly dissolving and flaking away because of acidic fumes in the air. On a more mundane level, air pollution and acid precipitation corrode steel in reinforced concrete, weakening buildings, roads, and bridges. Limestone, marble, and some kinds of sandstone flake and crumble. The Council on Environmental Quality estimates that U.S. economic losses from architectural damage caused by air pollution amount to about $4.8 billion in direct costs and $5.2 billion in property-value losses each year.

9.8 Air Pollution Control

"Dilution is the solution to pollution" was one of the early approaches to air pollution control. Tall smokestacks were built to send emissions far from the source, where they became uniden-

tifiable and largely untraceable. But dispersed and diluted pollutants are now the source of some of our most serious pollution problems. We are finding that there is no "away" to which we can throw our waste products.

The most effective pollution-control strategy is to minimize production

Since most air pollution in the developed world is associated with transportation and energy production, the most effective strategy would be conservation: reducing electricity consumption, insulating homes and offices, and developing better public transportation could all greatly reduce air pollution in the United States, Canada, and Europe. Alternative energy sources, such as wind and solar power, produce energy with little or no pollution, and these and other technologies are becoming economically competitive (chapter 12). In addition to conservation, pollution can be controlled by technological innovation.

Particulate removal involves filtering air emissions. Filters trap particulates in a mesh of cotton cloth, spun glass fibers, or asbestos-cellulose. Industrial air filters are generally giant bags 10 to 15 m long and 2 to 3 m wide. Effluent gas is blown through the bag, much like the bag on a vacuum cleaner. Every few days or weeks, the bags are opened to remove the dust cake. Electrostatic precipitators are the most common particulate controls in power plants. Ash particles pick up an electrostatic surface charge as they pass between large electrodes (fig. 9.28). The electrically charged particles then precipitate (collect) on an oppositely charged collecting plate. These precipitators consume a large amount of electricity, but

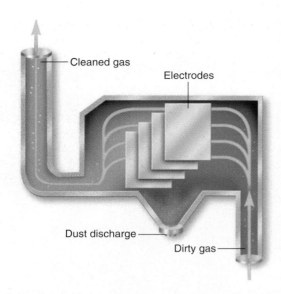

Figure 9.28 An electrostatic precipitator traps particulate material. Particles are given an electrical charge, which makes them attracted to electrically charged plates. Effluent passes between these plates, leaving particles behind, on its way to the smoke stack.

maintenance is relatively simple, and collection efficiency can be as high as 99 percent. The ash collected by both of these techniques is a solid waste (often hazardous due to the heavy metals and other trace components of coal or other ash source) and must be buried in landfills or other solid waste disposal sites.

Sulfur removal is important because sulfur oxides are among the most damaging of all air pollutants in terms of human health and ecosystem viability. Switching from soft coal with a high sulfur content to low-sulfur coal is the surest way to reduce sulfur emissions. High-sulfur coal is frequently politically or economically expedient, however. In the United States, Appalachia, a region of chronic economic depression, produces mostly high-sulfur coal. In China, much domestic coal is rich in sulfur. Switching to cleaner oil or gas would eliminate metal effluents as well as sulfur. Cleaning fuels is an alternative to switching. Coal can be crushed, washed, and gasified to remove sulfur and metals before combustion. This improves heat content and firing properties but may replace air pollution with solid waste and water pollution problems; furthermore, these steps are expensive.

Sulfur can also be removed to yield a usable product instead of simply a waste disposal problem. Elemental sulfur, sulfuric acid, and ammonium sulfate can all be produced using catalytic converters to oxidize or reduce sulfur. Markets have to be reasonably close and fly ash contamination must be reduced as much as possible for this procedure to be economically feasible.

Nitrogen oxides (NO_x) can be reduced in both internal combustion engines and industrial boilers by as much as 50 percent by carefully controlling the flow of air and fuel. Staged burners, for example, control burning temperatures and oxygen flow to prevent formation of NO_x. The catalytic converter on your car uses platinum-palladium and rhodium catalysts to remove up to 90 percent of NO_x, hydrocarbons, and carbon monoxide at the same time.

Hydrocarbon controls mainly involve complete combustion or the control of evaporation. Hydrocarbons and volatile organic compounds are produced by incomplete combustion of fuels or by solvent evaporation from chemical factories, paints, dry cleaning, plastic manufacturing, printing, and other industrial processes. Closed systems that prevent escape of fugitive gases can reduce many of these emissions. In automobiles, for instance, positive crankcase ventilation (PCV) systems collect oil that escapes from around the pistons and unburned fuel and channels them back to the engine for combustion. Controlling leaks from industrial valves, pipes, and storage tanks can have a significant impact on air quality. Afterburners are often the best method for destroying volatile organic chemicals in industrial exhaust stacks.

Clean air legislation is controversial

Throughout history, countless ordinances have prohibited emission of objectionable smoke, odors, and noise. Air pollution traditionally has been treated as a local problem, however. The Clean Air Act of 1963 was the first national legislation in the United States aimed at air pollution control. The act provided federal grants to states to combat pollution but was careful to preserve states' rights to set and enforce air quality regulations. It soon became obvious that some pollution problems cannot be solved on a local basis.

In 1970, an extensive set of amendments essentially rewrote the Clean Air Act. These amendments identified the "criteria" pollutants discussed earlier in this chapter, and established primary and secondary standards for ambient air quality. Primary standards are intended to protect human health, while secondary standards are set to protect materials, crops, climate, visibility, and personal comfort.

Since 1970 the Clean Air Act has been modified, updated, and amended many times. The most significant amendments were in the 1990 update. Amendments have involved acrimonious debate, with bills sometimes languishing in Congress from one session to the next because of disputes over burdens of responsibility and cost and definitions of risk. A 2002 report concluded that simply by enforcing existing clean air legislation, the United States could save at least another 6,000 lives per year and prevent 140,000 asthma attacks.

Throughout its history the Clean Air Act has been controversial. Victims of air pollution demand more protection; industry and special interest groups complain that controls are too expensive.

One of the most contested aspects of the act is the "new source review," which was established in 1977. This provision was originally adopted because industry argued that it would be intolerably expensive to install new pollution-control equipment on old power plants and factories that were about to close down anyway. Congress agreed to "grandfather" or exempt existing equipment from new pollution limits with the stipulation that when they were upgraded or replaced, more stringent rules would apply. The result was that owners kept old facilities operating precisely because they were exempted from pollution control. In fact, corporations poured millions into aging power plants and factories, expanding their capacity rather than build new ones. Thirty years later, most of those grandfathered plants are still going strong, and continue to be among the biggest contributors to smog and acid rain.

The Clinton administration attempted to force utilities to install modern pollution control on old power plants when they replaced or repaired equipment. President Bush, however, said that determining which facilities are new, and which are not, represented a cumbersome and unreasonable imposition on industries. The EPA subsequently announced it would abandon new source reviews, depending instead on voluntary emissions controls and a trading program for air pollution allowances.

Environmental groups generally agree that cap-and-trade (which sets maximum amounts for pollutants, and then lets facilities facing costly cleanup bills to pay others with lower costs to reduce emissions on their behalf) has worked well for sulfur dioxide. When trading began in 1990, economists estimated that eliminating 10 million tons of sulfur dioxide would cost $15 billion per year. Left to find the most economical ways to reduce emissions, however, utilities have been able to reach clean air goals for one-tenth that price. A serious shortcoming of this approach is that while trading has resulted in overall pollution reduction, some local "hot spots" remain where owners have found it cheaper to pay someone else to reduce pollution than to do it themselves. Knowing that the average person is enjoying cleaner air isn't much comfort if you're living in one of the persistently dirty areas.

9.9 Current Conditions and Future Prospects

Although the United States has not yet achieved the Clean Air Act goals in many parts of the country, air quality has improved dramatically in the last decade in terms of the major large-volume pollutants. For 23 of the largest U.S. cities, the number of days each year in which air quality reached the hazardous level is down 93 percent from a decade ago. Of 97 metropolitan areas that failed to meet clean air standards in the 1980s, 41 are now in compliance. For many cities, this is the first time they met air quality goals in 20 years.

There have been some notable successes and some failures. The EPA estimates that between 1970 and 1998, lead fell 98 percent, SO_2 declined 35 percent, and CO shrank 32 percent. Filters, scrubbers, and precipitators on power plants and other large stationary sources are responsible for most of the particulate and SO_2 reductions. Catalytic converters on automobiles are responsible for most of the CO and O_3 reductions.

The only conventional "criteria" pollutants that have not dropped significantly are particulates and NO_x. Because automobiles are the main source of NO_x, cities, such as Nashville, Tennessee, and Atlanta, Georgia, where pollution comes largely from traffic, still have serious air quality problems. Rigorous pollution controls are having a positive effect on Southern California air quality. Los Angeles, which had the dirtiest air in the nation for decades, wasn't even in the top 20 polluted cities in 2007.

Particulate matter (mostly dust and soot) is produced by agriculture, fuel combustion, metal smelting, concrete manufacturing, and other activities. Industrial cities, such as Baltimore, Maryland, and Baton Rouge, Louisiana, also have continuing problems. Eighty-five other urban areas are still considered nonattainment regions. In spite of these local failures, however, 80 percent of the United States now meets the National Ambient Air Quality Standards. This improvement in air quality is perhaps the greatest environmental success story in our history.

Air pollution remains a problem in many places

The outlook is not so encouraging in other parts of the world. The major metropolitan areas of many developing countries are growing at explosive rates to incredible sizes (chapter 14), and environmental quality is abysmal in many of them. Mexico City remains notorious for bad air. Pollution levels exceed WHO health standards 350 days per year, and more than half of all city children have lead levels in their blood high enough to lower intelligence and retard development. Mexico City's 131,000 industries and 2.5 million vehicles spew out more than 5,500 tons of air pollutants daily. Santiago, Chile, averages 299 days per year on which suspended particulates exceed WHO standards of 90 mg/m^3.

While China is making efforts to control air and water pollution (chapter 1), many of China's 400,000 factories have no air pollution controls. Experts estimate that home coal burners and factories emit 10 million tons of soot and 15 million tons of sulfur dioxide annually and that emissions have increased rapidly over the past 20 years. Seven of the ten cities in the world with the worst air quality are in China. Sheyang, an industrial city in northern China, is thought to have the world's worst continuing particulate problem, with peak winter concentrations over 700 mg/m^3 (nine times U.S. maximum standards). Airborne particulates in Sheyang exceed WHO standards on 347 days per year. It's estimated that air pollution is responsible for 400,000 premature deaths every year in China. Beijing, Xi'an, and Guangzhou also have severe air pollution problems. The high incidence of cancer in Shanghai is thought to be linked to air pollution.

There are signs of hope

Not all is pessimistic, however. There have been some spectacular successes in air pollution control. Sweden and West Germany (countries affected by forest losses due to acid precipitation) cut their sulfur emissions by two-thirds between 1970 and 1985. Austria and Switzerland have gone even further, regulating even motorcycle emissions. The Global Environmental Monitoring System (GEMS) reports declines in particulate levels in 26 of 37 cities worldwide. Sulfur dioxide and sulfate particles, which cause acid rain and respiratory disease, have declined in 20 of these cities.

Figure 9.29 Air quality in Delhi, India, has improved dramatically since buses, auto-rickshaws, and taxis were required to switch from liquid fuels to compressed natural gas. This is one of the most encouraging success stories in controlling pollution in the developing world.

Figure 9.30 Cubatao, Brazil, was once considered one of the most polluted cities in the world. Better environmental regulations and enforcement along with massive investments in pollution-control equipment have improved air quality significantly.

Even poor countries can control air pollution. Delhi, India, for example was once considered one of the world's ten most polluted cities. Visibility often was less than 1 km on smoggy days. Health experts warned that breathing Delhi's air was equivalent to smoking two packs of cigarettes per day. Pollution levels exceeded World Health Organization standards by nearly five times. Respiratory diseases were widespread, and the cancer rate was significantly higher than surrounding rural areas. The biggest problem was vehicle emissions, which contributed about 70 percent of air pollutants (industrial emissions made up 20 percent, while burning of garbage and firewood made up most of the rest).

In the 1990s, catalytic converters were required for automobiles, and unleaded gasoline and low-sulfur diesel fuel were introduced. In 2000, more than private automobiles were required to meet European standards, and in 2002, more than 80,000 buses, auto-rickshaws, and taxis were required to switch from liquid fuels to compressed natural gas (fig. 9.29). Sulfur dioxide and carbon monoxide levels have dropped 80 percent and 70 percent, respectively, since 1997. Particulate emissions dropped by about 50 percent. Residents report that the air is dramatically clearer and more healthy. Unfortunately, rising prosperity, driven by globalization of information management, has doubled the number of vehicles on the roads, threatening this progress. Still, the gains made in New Delhi are encouraging for people everywhere.

Twenty years ago, Cubatao, Brazil, was described as the "Valley of Death," one of the most dangerously polluted places in the world. A steel plant, a huge oil refinery, and fertilizer and chemical factories churned out thousands of tons of air pollutants every year that were trapped between onshore winds and the uplifted plateau on which São Paulo sits (fig. 9.30). Trees died on the surrounding hills. Birth defects and respiratory diseases were

alarmingly high. Since then, however, the citizens of Cubatao have made remarkable progress in cleaning up their environment. The end of military rule and restoration of democracy allowed residents to publicize their complaints. The environment became an important political issue. The state of São Paulo invested about $100 million and the private sector spent twice as much to clean up most pollution sources in the valley. Particulate pollution was reduced 75 percent, ammonia emissions were reduced 97 percent, hydrocarbons that cause ozone and smog were cut 86 percent, and sulfur dioxide production fell 84 percent. Fish are returning to the rivers, and forests are regrowing on the mountains. Progress is possible! We hope that similar success stories will be obtainable elsewhere.

Conclusion

We appear to be at a tipping point in our attitudes toward climate change. Where, a few years ago, most people in America thought that this was just a wild theory of a small group of crazy scientists, we've gone to a widespread acceptance that the science of global warming is indisputable. The surprising success of former Vice President Al Gore's documentary, *An Inconvenient Truth*, has been a powerful force in this social transformation. Perhaps more important is that many people can actually see tangible proof in their own lives that something strange is happening to our climate.

But is it too late to do anything? Most scientists believe there is time, if we act quickly and effectively, to avoid the worst consequences of global climate change. It's encouraging to know that we don't need a technological miracle to save us from this impending catastrophe. We have the ability to dramatically reduce our greenhouse gas emissions right now. And it will probably save money and create jobs to do so.

The success of the Montreal Protocol in eliminating CFCs is a landmark in international cooperation on an environmental program. While the stratospheric ozone hole continues to grow because of global warming effects and the residual chlorine in the air released decades ago, we expect the ozone depletion to end in about 50 years. This is one of the few global environmental threats that has had such a rapid and successful resolution. Let's hope that others will follow.

Progress in reducing local pollution in developing countries, such as Brazil and India, also is encouraging. Problems that once seemed overwhelming can be overcome. In some cases, it requires lifestyle changes or different ways of doing things to bring about progress, but as the Chinese philosopher Lao Tsu wrote, "A journey of a thousand miles must begin with a single step."

Practice Quiz

1. What are the "stabilization wedges" suggested by Pacala and Socolow at Princeton University (see table 9.2)? How many wedges do we need to accomplish to flatten our CO_2 emissions?

2. What is the *greenhouse effect*, and how does it work?

3. Why are we worried about greenhouse gases?

4. What is the *thermohaline ocean conveyor* and what is happening to it?

5. Describe the El Niño/Southern Oscillation.

6. What gas, action, and country make the largest contribution to global warming?

7. What has been the greatest air pollution control success in the United States since 1970?

8. Define *primary air pollutant, secondary air pollutant, photochemical oxidant, point source*, and *fugitive emissions*.

9. What is destroying stratospheric ozone, and where does this happen?

10. What is the "new source review"?

Critical Thinking and Discussion Questions

Apply the principles you have learned in this chapter to discuss these questions with other students.

1. El Niño is a natural climate process, but some scientists suspect it is changing because of global warming. What sort of evidence would you look for to decide if El Niño has gotten stronger recently?

2. One of the problems with the Kyoto Protocol and with the Clean Air Act is that economists and scientists define problems differently and have contrasting priorities. How would an economist and an ecologist explain disputes over the Kyoto Protocol differently?

3. Economists and scientists often have difficulty reaching common terms for defining and solving issues such as the Clean Air Act renewal. How might their conflicting definitions be reshaped to make the discussion more successful?

4. Why do you think controlling greenhouse gases is such a difficult problem? List some of the technological, economic, political, emotional, and other factors involved. Whose responsibility is it to reduce our impacts on climate?

5. Air pollution often originates in one state or country but causes health and crop damage in other areas. For example, mercury from Midwestern power plants is harming plants, water, and health in eastern states. How should states, or countries, negotiate the costs of controlling these pollutants?

 # Data Analysis: Graphing Air Pollution Control

Reduction of acid-forming air pollutants in Europe is an inspiring success story. The first evidence of ecological damage from acid rain came from disappearance of fish from Scandinavian lakes and rivers in the 1960s. By the 1970s, evidence of air pollution damage to forests in northern and central Europe alarmed many people. International agreements, reached since the mid-1980s have been highly successful in reducing emissions of SO_2 and NO_x as well as photochemical oxidants, such as O_3. The graph on page 230 shows reductions in SO_2 emissions in Europe between 1990 and 2002. The light blue area shows actual SO_2 emissions. Blue represents changes due to increased nuclear and renewable energy. Orange shows reductions due to energy conservation. Green shows

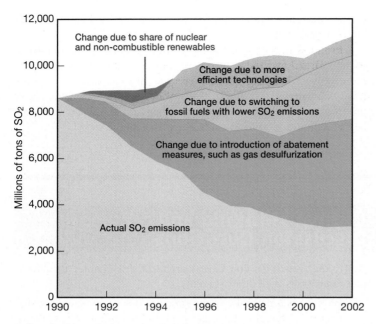

Sulfur dioxide emission reductions in Europe, 1990–2002.

improvement from switching to low-sulfur fuels. Purple shows declines due to increased abatement measures (flue gas scrubbers). The upper boundary of each area indicates what emissions would have been without pollution control.

1. How much have actual SO_2 emissions declined since 1990?

2. How much lower were SO_2 emissions in 2002 than they would have been without pollution control (either in percentage or actual amount)?

3. What percentage of this reduction was due to abatement measures, such as flue gas scrubbers?

4. What percent was gained by switching to low-sulfur fuels?

5. How much did energy conservation contribute?

6. What happened to nuclear power?

For Additional Help in Studying This Chapter, please visit our website at www.mhhe.com/cunningham5e. You will find practice quizzes, key terms, answers to end of chapter questions, additional case studies, an extensive reading list, and Google Earth™ mapping quizzes.

Surface levels of Georgia's Lake Lanier dropped to record low levels as the southeastern United States experienced its worst drought in a century. Towns were forced to ration water as dwindling supplies threatened industry, agriculture, and wildlife.

10 Water: Resources and Pollution

Learning Outcomes

After studying this chapter, you should be able to answer the following questions:

- Where does our water come from? How do we use it?
- Where and why do water shortages occur?
- How can we increase water supplies? What are some costs of these methods?
- How can *you* conserve water?
- What is *water pollution*? What are its sources and effects?
- Why are sewage treatment and clean water important in developing countries?
- How can we control water pollution?

I tell you gentlemen; you are piling up a heritage of conflict and litigation of water rights, for there is not sufficient water to supply the land.

–John Wesley Powell

Adjusting to Drought

Normally, the southeastern United States is one of the wettest parts of the country. Average annual rainfall in Georgia, for example, is about 130 cm (50 in.) or nearly four times as much as Utah. For the past several years, however, the region has been gripped by the most severe drought in more than 100 years. Most of the southern Appalachians have received less than one-quarter of the usual precipitation. In 2007, Georgia Governor Sonny Perdue declared an emergency for most of his state. An exceptional drought—the most severe level designated by the U.S. Weather Bureau—had reduced lake surfaces, stream flows, and drinking water supplies to their lowest levels in recorded history.

Governor Perdue prohibited outdoor watering, car washing (except for commercial operations), and other inessential activities, while he urged residents to conserve water wherever possible. He also asked the Army Corps of Engineers, which regulates flow from headwater lakes into the Chattahoochee River, to hold water back to preserve drinking supplies for Atlanta and other cities. Some smaller Georgia towns are now completely out of water and are having to truck in drinking supplies. They face a very uncertain future if rains don't return soon.

Florida and Alabama, which lie downstream on the Chattahoochee, cried foul, arguing that Georgia was trying to grab water needed for agriculture, industry, domestic supplies, and endangered species survival in their states. They said that Georgia had allowed unrestrained urban growth while failing to control excessive consumption or plan for potential emergencies. Atlanta, they point out, is one of the few metropolitan areas in the country—some others are Los Angeles, Denver, Phoenix, and Las Vegas—without a permanent water supply.

This emergency reopened a decades-old debate about water allocation in the Apalachicola, Chattahoochee, and Flint (ACF) rivers, which are shared by Georgia, Alabama, and Florida (fig. 10.1). For years it has been recognized that these rivers don't have enough water for all the competing uses within their combined watershed. The situation is exacerbated by pollution from farms, industry, and urban areas that make the water unsuitable for many uses. The U.S. Environmental Protection Agency named the 100 km of the Chattahoochee south of Atlanta one of the five most polluted rivers in the United States. Millions of dollars in fines were levied on the city to encourage pollution control. Atlanta diluted contaminates by releasing more clean water from reservoirs upstream, but claimed that little could be done to prevent much of the pollution from the large and rapidly growing city.

Florida's Apalachicola Bay, where the combined rivers flow into the Gulf, is particularly threatened by what's happening upstream. Pollution, reduced river flow, and increasing salinity in the estuary jeopardize the bay's multimillion dollar per year fishing and tourism industries. The U.S. Fish and Wildlife Service has demanded that more water be released into the rivers to protect several species of endangered mussels and sturgeon that live and spawn in the Apalachicola estuary. In 2002, Georgia, Alabama, and Florida agreed on a plan to reallocate water within the AFC basin, but the subsequent drought has given all three states an opening to challenge the settlement.

Fish and Wildlife officials, meanwhile, have requested more water to protect the endangered species. Governor Perdue called this request downright dangerous. "How can the government care more about mussels and sturgeons than humans?" he asked. Meanwhile, Senator Richard Shelby of Alabama slipped language into a massive end-of-year spending bill, blocking any change in water-sharing in the region. Some Georgians have suggested that water be imported from the Tennessee River, which still has an abundant supply, but other states oppose intrabasin transfers.

Water shortages have long been a problem in the western United States, with growing cities in California and neighboring states struggling to acquire enough water without destroying aquatic ecosystems. But like Atlanta, many growing metropolitan areas in other regions are increasingly competing with other users for water. Around the world, cities are experiencing water shortages, while pollution makes the water we do have less useful. The United Nations warns that water supplies are likely to become one of the most pressing environmental issues of the twenty-first century. By 2025, two-thirds of all humans could be living in countries where water resources are inadequate. In this chapter, we'll look at the sources of our fresh water, what we do with it, and how we might protect its quality and extend its usefulness.

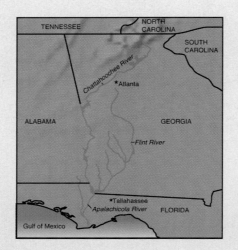

Figure 10.1 Rising in the mountains of northern Georgia, the Chattahoochee flows south through Atlanta before joining the Flint to form the Apalachicola River.

10.1 Water Resources

Water is a marvelous substance—flowing, swirling, seeping, constantly moving from sea to land and back again. It shapes the earth's surface and moderates our climate. Water is essential for life. It is the medium in which all living processes occur (chapter 2). Water dissolves nutrients and distributes them to cells, regulates body temperature, supports structures, and removes waste products. About 60 percent of your body is water. You could survive for weeks without food, but only a few days without water.

Water also is needed for agriculture, industry, transportation, and a host of other human uses. In short, clean freshwater is one of our most vital natural resources.

The hydrologic cycle constantly redistributes water

The water we use cycles endlessly through the environment. The total amount of water on our planet is immense—more than 1,404 million km³ (370 billion billion gal) (table 10.1). This water evaporates from moist surfaces, falls as rain or snow, passes through living organisms, and returns to the ocean in a process known as the **hydrologic cycle** (see fig. 2.21). Every year, about 500,000 km³, or a layer 1.4 m thick, evaporates from the oceans. More than 90 percent of that moisture falls back on the ocean. The 47,000 km³ carried onshore joins some 72,000 km³ that evaporate from lakes, rivers, soil, and plants to become our annual, renewable freshwater supply. Plants play a major role in the hydrologic cycle, absorbing groundwater and pumping it into the atmosphere by transpiration (transport plus evaporation). In tropical forests, as much as 75 percent of annual precipitation is returned to the atmosphere by plants.

Solar energy drives the hydrologic cycle by evaporating surface water, which becomes rain and snow. Because water and sunlight are unevenly distributed around the globe, water resources are very uneven. At Iquique in the Chilean desert, for instance, no rain has fallen in recorded history. At the other end

Table 10.1	Units of Water Measurement
One cubic kilometer (km³) equals 1 billion cubic meters (m³), 1 trillion liters, or 264 billion gal.
One acre-foot is the amount of water required to cover an acre of ground 1 ft deep. This is equivalent to 325,851 gal, or 1.2 million liters, or 1,234 m³, approximately the amount consumed annually by a family of four in the United States.
One cubic foot per second of river flow equals 28.3 liters per second, or 449 gal per minute.

of the scale, 22 m (72 ft) of rain was recorded in a single year at Cherrapunji in India. Figure 10.2 shows broad patterns of precipitation around the world. Most of the world's rainiest regions are tropical, where heavy rainy seasons occur, or in coastal mountain regions. Deserts occur on every continent just outside the tropics (the Sahara, the Namib, the Gobi, the Sonoran, and many others). Rainfall is also slight at very high latitudes, another high-pressure region.

Mountains also influence moisture distribution. The windward sides of mountain ranges, including the Pacific Northwest and the flanks of the Himalayas, are typically wet and have large rivers; on the leeward sides of mountains, in areas known as the rain shadow, dry conditions dominate, and water can be very scarce. The windward side of Mount Waialeale on the island of Kauai, for example, is one of the wettest places on earth, with an

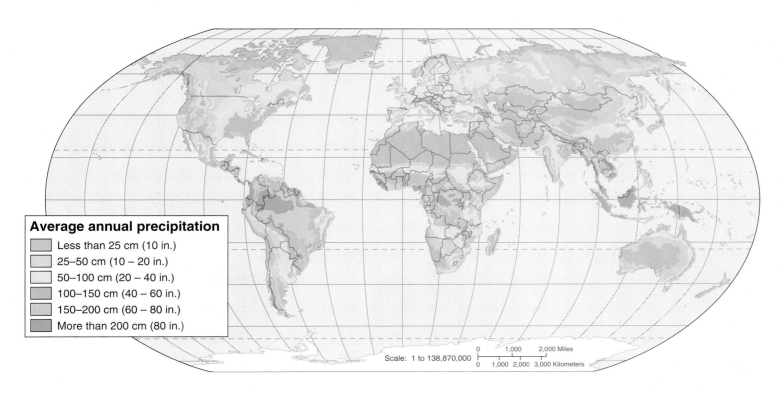

Average annual precipitation

- Less than 25 cm (10 in.)
- 25–50 cm (10 – 20 in.)
- 50–100 cm (20 – 40 in.)
- 100–150 cm (40 – 60 in.)
- 150–200 cm (60 – 80 in.)
- More than 200 cm (80 in.)

Scale: 1 to 138,870,000

0   1,000   2,000 Miles
0 1,000 2,000 3,000 Kilometers

Figure 10.2 Average annual precipitation. Note wet areas that support tropical rainforests occur along the equator, while the major world deserts occur in zones of dry, descending air between 20° and 40° north and south.

Table 10.2 Earth's Water Compartments

Compartment	Volume (1,000 km³)	Percent of Total Water	Average Residence Time
Total	1,386,000	100	2,800 years
Oceans	1,338,000	96.5	3,000 to 30,000 years*
Ice and snow	24,364	1.76	1 to 100,000 years*
Saline groundwater	12,870	0.93	Days to thousands of years*
Fresh groundwater	10,530	0.76	Days to thousands of years*
Fresh lakes	91	0.007	1 to 500 years*
Saline lakes	85	0.006	1 to 1,000 years*
Soil moisture	16.5	0.001	2 weeks to 1 year*
Atmosphere	12.9	0.001	1 week
Marshes, wetlands	11.5	0.001	Months to years
Rivers, streams	2.12	0.0002	1 week to 1 month
Living organisms	1.12	0.0001	1 week

Source: Data from UNEP, 2002.

*Depends on depth and other factors.

annual rainfall around 1,200 cm (460 in.). The leeward side, only a few kilometers away, has an average yearly rainfall of only 46 cm (18 in.).

An important consideration in water availability is whether there is rainfall year-round, especially in the growing season. Another question is whether hot, dry weather evaporates available moisture. These factors help determine the amount and variety of biological activity in a place (chapter 5).

10.2 Major Water Compartments

The distribution of water often is described in terms of interacting compartments in which water resides, sometimes briefly and sometimes for eons (table 10.2). The length of time water typically stays in a compartment is its **residence time**. On average, a water molecule stays in the ocean for about 3,000 years, for example, before it evaporates and starts through the hydrologic cycle

again. Nearly all the world's water is in the oceans (fig. 10.3). Oceans play a crucial role in moderating the earth's temperature, and over 90 percent of the world's living biomass is contained in the oceans. What we mainly need, though, is fresh water. Of the 2.4 percent that is fresh, most is locked up in glaciers or in groundwater. Amazingly, only about .02 percent of the world's water is in a form accessible to us and to other organisms that rely on fresh water.

Groundwater stores most fresh, liquid water

Originating as precipitation that percolates into layers of soil and rock, groundwater makes up the largest compartment of liquid, fresh water. The groundwater within 1 km of the surface is more than 100 times the volume of all the freshwater lakes, rivers, and reservoirs combined.

Plants get moisture from a relatively shallow layer of soil containing both air and water, known as the **zone of aeration**

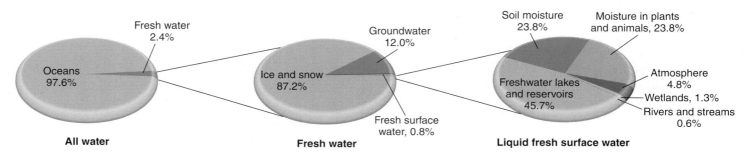

Figure 10.3 Less than 1 percent of fresh water, and less than 0.02 percent of all water, is fresh, liquid surface water on which terrestrial life depends.
Source: U.S. Geological Survey.

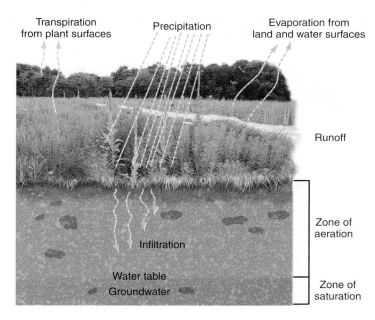

Figure 10.4 Precipitation that does not evaporate or run off over the surface percolates through the soil in a process called infiltration. The upper layers of soil hold droplets of moisture between air-filled spaces. Lower layers, where all spaces are filled with water, make up the zone of saturation, or groundwater.

(fig. 10.4). Depending on rainfall amount, soil type, and surface topography, the zone of aeration may be a few centimeters or many meters deep. Lower soil layers, where all soil pores are filled with water, make up the **zone of saturation**, the source of water in most wells; the top of this zone is the **water table**.

Geologic layers that contain water are known as **aquifers**. Aquifers may consist of porous layers of sand or gravel or of cracked or porous rock. Below an aquifer, relatively impermeable layers of rock or clay keep water from seeping out at the bottom. Instead, water seeps more or less horizontally through the porous layer. Depending on geology, it can take from a few hours to several years for water to move a few hundred meters through an aquifer. If impermeable layers lie above an aquifer, pressure can develop within the water-bearing layer. Pressure in the aquifer can make a well flow freely at the surface. These free-flowing wells and springs are known as *artesian* wells or springs.

Areas where surface water filters into an aquifer are **recharge zones** (fig. 10.5). Most aquifers recharge extremely slowly, and road and house construction or water use at the surface can further slow recharge rates. Contaminants can also enter aquifers through recharge zones. Urban or agricultural runoff in recharge zones is often a serious problem. About 2 billion people—approximately one-third of the world's population—depend on groundwater for drinking and other uses. Every year 700 km^3 are withdrawn by humans, mostly from shallow, easily polluted aquifers.

Rivers, lakes, and wetlands cycle quickly

Fresh, flowing surface water is one of our most precious resources. Rivers contain a minute amount of water at any one time. Most rivers would begin to dry up in weeks or days if they were not constantly replenished by precipitation, snowmelt, or groundwater seepage.

We compare the size of rivers in terms of **discharge**, or the amount of water that passes a fixed point in a given amount of time. This is usually expressed as liters or cubic feet of water per second. The 16 largest rivers in the world carry nearly half of all surface runoff on the earth, and a large fraction of that occurs in a single river, the Amazon, which carries 10 times as much water as the Mississippi (table 10.3).

Lakes contain nearly 100 times as much water as all rivers and streams combined, but much of this water is in a few of the world's largest lakes. Lake Baikal in Siberia, the Great Lakes of North America, the Great Rift Lakes of Africa, and a few other

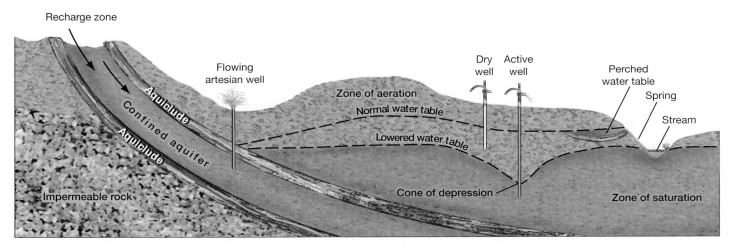

Figure 10.5 An aquifer is a porous or cracked layer of rock. Impervious rock layers (aquicludes) keep water within a confined aquifer. Pressure from uphill makes an artesian well flow freely. Pumping can create a cone of depression, which leaves shallower wells dry.

Table 10.3 World's Ten Largest Rivers

River	Location	Annual Discharge (m³/second) *
Amazon	Brazil, Peru	175,000
Orinoco	Venezuela, Colombia	45,300
Congo	Congo	39,200
Yangtze	Tibet, China	28,000
Brahmaputra	South Asia	19,000
Mississippi	United States	18,400
Mekong	Southeast Asia	18,300
Paraná	Paraguay, Argentina	18,000
Yenisey	Russia	17,200
Lena	Russia	16,000

Source: Data from World Resource Institute.

*1 m³ = 264 gal.

lakes contain vast amounts of water, not all of it fresh. Worldwide, lakes are almost as important as rivers in terms of water supplies, food, transportation, and settlement.

Wetlands—bogs, swamps, wet meadows, and marshes—play a vital and often unappreciated role in the hydrologic cycle. Their lush plant growth stabilizes soil and holds back surface runoff, allowing time for infiltration into aquifers and producing even, year-long stream flow. When wetlands are disturbed, their natural water-absorbing capacity is reduced, and surface waters run off quickly, resulting in floods and erosion during the rainy season and low stream flow the rest of the year.

The atmosphere is one of the smallest compartments

The atmosphere contains only 0.001 percent of the total water supply, but it is the most important mechanism for redistributing water around the world. An individual water molecule resides in the atmosphere for about ten days, on average. Some water evaporates and falls within hours. Water can also travel halfway around the world before it falls, replenishing streams and aquifers on land.

10.3 Water Availability and Use

Clean, fresh water is essential for nearly every human endeavor. Collectively, we now appropriate more than half of all the freshwater in the world. Perhaps more than any other environmental factor, the availability of water determines the location and activities of humans on the earth. **Renewable water supplies** are resources that are replenished regularly—mainly surface water and shallow groundwater. Renewable water is most plentiful in the tropics, where rainfall is heavy, followed by midlatitudes, where rainfall is regular.

Many countries experience water scarcity and stress

As you can see in figure 10.2, South America, West Central Africa, and South and Southeast Asia all have areas of very high rainfall. Brazil and the Democratic Republic of Congo, because they have high precipitation levels and large land areas, are among the most water-rich countries on earth. Canada and Russia, which are both very large, also have large annual water supplies. The highest per capita water supplies generally occur in countries with wet climates and low population densities. Iceland, for example, has about 160 million gallons per person per year. In contrast, Bahrain, where temperatures are extremely high and rain almost never falls, has essentially no natural fresh water. Almost all of Bahrain's water comes from imports and desalinized seawater. Egypt, in spite of the fact that the Nile River flows through it, has only about 11,000 gallons of water annually per capita, or about 15,000 times less than Iceland.

Periodic droughts create severe regional water shortages (fig. 10.6). Droughts are most common and often most severe in semiarid zones where moisture availability is the critical factor in determining plant and animal distribution. Undisturbed ecosystems often survive extended droughts with little damage, but introduction of domestic animals and agriculture disrupts native vegetation and undermines natural adaptations to low moisture levels.

Droughts are often cyclic, and land-use practices exacerbate their effects. In the United States, the cycle of drought seems to be about 30 years. There were severe dry years in the 1870s, 1900s, 1930s, 1950s, and 1970s. The worst of these in economic and social terms were the 1930s. Poor soil conservation practices and a series of dry years in the Great Plains combined to create the

Figure 10.6 Droughts can cause severe hardships for both humans and wildlife. Will droughts be more serious and more frequent as a consequence of global warming?

"dust bowl." Wind stripped topsoil from millions of hectares of land, and billowing dust clouds turned day into night. Thousands of families were forced to leave farms and migrate to cities.

In the first decade of the twenty-first century, much of the western United States has been exceptionally dry. And now, as the opening case study for this chapter shows, the Southeast, which normally has a wet climate, has been experiencing catastrophic drought (fig. 10.7). Perhaps this is simply an example of ordinary climatic fluctuations, but many climatologists predict that global warming (chapter 9) will cause more frequent and more severe droughts in many places. Higher temperatures cause more evaporation and dry out the ground. Rising thermals block the flow of

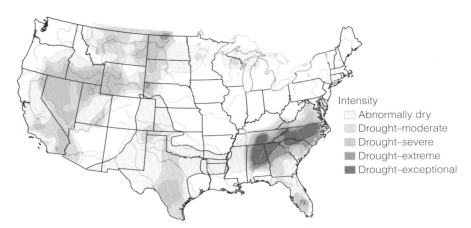

Intensity
- ☐ Abnormally dry
- ☐ Drought–moderate
- ☐ Drought–severe
- ■ Drought–extreme
- ■ Drought–exceptional

Figure 10.7 Much of the southeastern United States is experiencing the most severe drought in more than a century. This has created a controversy among many water users.

moist air from coastal regions, and the rain that does fall often evaporates before it reaches the earth.

The United States government projects that 36 states will have water shortages by 2012. Increased temperatures, disturbed weather patterns, population growth, urban sprawl, waste, and wasteful uses all will contribute to these shortages. The effects on water supplies may well be the most serious consequences of global climate change.

Agriculture is our greatest water user

In contrast to energy resources, which usually are consumed when used, water can be used over and over if it is not too badly contaminated. Water **withdrawal** is the total amount of water taken from a water body. Much of this water could be returned to circulation in a reusable form. Water **consumption**, on the other hand, is loss of water due to evaporation, absorption, or contamination.

The natural cleansing and renewing functions of the hydrologic cycle replace the water we need if natural systems are not overloaded or damaged. Water is a renewable resource, but renewal takes time. The rate at which many of us now use water may make it necessary to conscientiously protect, conserve, and replenish our water supply.

Water use has been increasing about twice as fast as population growth over the past century. Water withdrawals are expected to continue to grow as more land is irrigated to feed an expanding population (fig. 10.8). Conflicts increase as different countries, economic sectors, and other stakeholders compete for the same, limited water supply. Water wars may well be the major source of hostilities in the twenty-first century.

Worldwide, agriculture claims about 70 percent of total water withdrawal, ranging from 93 percent of all water used in India to only 4 percent in Kuwait, which cannot afford to spend its limited water on crops. In many developing countries and in parts of the United States, the most common type of irrigation is to simply flood the whole field or run water in rows between crops. As much as half the water can be lost through evaporation or seepage from unlined irrigation canals bringing water to fields. Sprinklers are more efficient in distributing water, but they are more costly and energy intensive. Water-efficient drip irrigation can save significant amounts of water but currently is used on only about 1 percent of the world's croplands.

Industry uses about one-fourth of water withdrawals worldwide. Some European countries use 70 percent of water for industry; less-industrialized countries use as little as 5 percent. Cooling water for power plants is by far the largest single industrial use of water, typically accounting for 50 to 75 percent of industrial withdrawal. A rapidly growing demand for water is for biofuel production. It currently takes 4 to 5 liters of water to produce 1 liter of ethanol. Water shortages may limit a switch to biofuels.

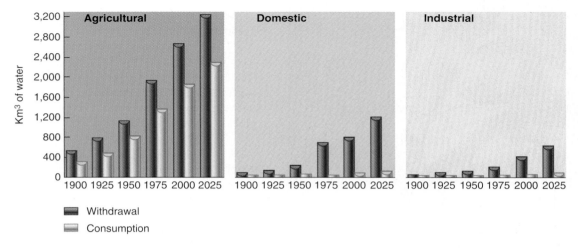

Figure 10.8 Growth of water withdrawal and consumption, by sector, with projected levels to 2025.
Source: UNEP, 2002.

Domestic, or household, water use accounts for only about 6 percent of world water use. This includes water for drinking, cooking, and washing. The amount of water used per household varies enormously, however, depending on a country's wealth. The United Nations reports that people in developed countries consume on average, about ten times more water daily than those in developing nations. Poorer countries can't afford the infrastructure to obtain and deliver water to citizens. Inadequate water supplies, on the other hand, prevent agriculture, industry, sanitation, and other developments that reduce poverty.

10.4 Freshwater Shortages

Clean drinking water and basic sanitation are necessary to prevent communicable diseases and to maintain a healthy life. For many of the world's poorest people, one of the greatest environmental threats to health remains the continued use of polluted water. The United Nations estimates that at least a billion people lack access to safe drinking water and 2.5 billion don't have adequate sanitation. These deficiencies result in hundreds of millions of cases of water-related illness and more than 5 million deaths every year. As populations grow, more people move into cities, and agriculture and industry compete for increasingly scarce water supplies, water shortages are expected to become even worse.

By 2025 two-thirds of the world's people will be living in water-stressed countries—defined by the United Nations as consumption of more than 10 percent of renewable freshwater resources. One of the United Nations Millennium goals is to reduce by one-half the proportion of people without reliable access to clean water and improved sanitation.

There have been many attempts to enhance local supplies and redistribute water. Towing icebergs from Antarctica has been proposed, and creating rain in dry regions has been accomplished, with mixed success, by cloud seeding—distributing condensation nuclei in humid air to help form raindrops. Desalination is locally

important: in the arid Middle East, where energy and money are available but water is scarce, desalination is sometimes the principal source of water. Some American cities, such as San Diego, also depend partly on energy-intensive desalination.

Many people lack access to clean water

The World Health Organization considers an average of 1,000 m^3 (264,000 gal) per person per year to be a necessary amount of water for modern domestic, industrial, and agricultural uses. Some 45 countries, most of them in Africa or the Middle East, cannot meet the minimum essential water needs of all their citizens. In some countries, the problem is access to *clean* water. In Mali, for example, 88 percent of the population lacks clean water; in Ethiopia, it is 94 percent. Rural people often have less access to clean

Figure 10.9 Village water supplies in Ghana.

water than do city dwellers. Causes of water shortages include natural deficits, overconsumption by agriculture or industry, and inadequate funds for purifying and delivering good water.

More than two-thirds of the world's households have to fetch water from outside the home (fig. 10.9). This is heavy work, done mainly by women and children and sometimes taking several hours a day. Improved public systems bring many benefits to these poor families.

Availability does not always mean affordability. A typical poor family in Lima, Peru, for instance, uses one-sixth as much water as a middle-class American family but pays three times as much for it. If they followed government recommendations to boil all water to prevent cholera, up to one-third of the poor family's income could be used just in acquiring and purifying water.

Investments in rural development have brought significant improvements in recent years. Since 1990, nearly 800 million people—about 13 percent of the world's population—have gained access to clean water. The percentage of rural families with safe drinking water has risen from less than 10 percent to nearly 75 percent.

Groundwater supplies are being depleted

Groundwater provides nearly 40 percent of the fresh water for agricultural and domestic use in the United States. Nearly half of all Americans and about 95 percent of the rural population depend on groundwater for drinking and other domestic purposes. Overuse of these supplies dries up wells, natural springs, and even groundwater-fed wetlands, rivers, and lakes. Pollution of aquifers through dumping of contaminants on recharge zones, leaks through abandoned wells, or deliberate injection of toxic wastes can make this valuable resource unfit for use.

In many areas of the United States, groundwater is being withdrawn from aquifers faster than natural recharge can replace it. On a local level, this causes a cone of depression in the water table. On a broader scale, heavy pumping can deplete a whole aquifer. The Ogallala Aquifer underlies eight Great Plains states from Texas to North Dakota. This porous bed of sand, gravel, and sandstone once held more water than all the freshwater lakes, streams, and rivers on the earth. Excessive pumping for irrigation has removed so much water that wells have dried up in many places, and farms, ranches, even whole towns are being abandoned. Recharging many such aquifers will take thousands of years. Using "fossil" water like this is essentially water mining. For all practical purposes, these aquifers are nonrenewable resources.

Water withdrawal also allows aquifers to collapse. Subsidence, or sinking of the ground surface, follows. The San Joaquin Valley in California has sunk more than 10 m in the past 50 years because of excessive groundwater pumping. Where aquifers become compressed, recharge becomes impossible.

Another consequence of aquifer depletion is saltwater intrusion. Along coastlines and in areas where saltwater deposits are left from ancient oceans, overuse of freshwater reservoirs often allows saltwater to intrude into aquifers used for domestic and agricultural purposes.

Diversion projects redistribute water

Dams and canals are a foundation of civilization because they store and redistribute water for farms and cities. Many great civilizations have been organized around large-scale canal systems, including ancient empires of Sumeria, Egypt, and the Inca. As modern dams and water diversion projects have grown in scale and number, though, their environmental costs have raised serious questions about efficiency, costs, and the loss of river ecosystems.

More than half of the world's 227 largest rivers have been dammed or diverted. Of the 50,000 large dams in the world, 90 percent were built in the twentieth century. Half of those are in China, and China continues to build and plan dams on its remaining rivers. Dams are justified in terms of flood control, water storage, and electricity production. However, the costs of relocating villages, as well as lost fishing, farming, and water losses to evaporation, are enormous. Economically speaking, at least one-third of the world's large dams should never have been built. The largest water diversion project in the world is now being built in China (see What Do You Think? p. 241).

One of the most disastrous diversion histories is that of the Aral Sea. Situated in arid Central Asia, on the border of Kazakhstan and Uzbekistan, the Aral Sea is a shallow inland sea fed by rivers from distant mountains. Starting in the 1950s, the Soviet Union began diverting these rivers to water cotton fields and rice fields. Gradually, the Aral Sea has evaporated, leaving vast, toxic salt flats (fig. 10.10). The economic value of the cotton and rice has probably never met the cost of lost fisheries, villages, and health.

Recently some river flow has been restored to the "Small Aral" or northern lobe of the once-great sea. Water levels have risen 8 m and native fish are being reintroduced. It's hoped that one day commercial fishing may be resumed. The fate of the larger, southern remnant is more uncertain. There may never be enough water to refill it, and if there were, the toxins left in the lake bed could make it unusable anyway.

Questions of justice often surround dam projects

While dams provide hydroelectric power and water to distant cities, local residents often suffer economic and cultural losses. In some cases, dam builders have been charged with using public money to increase the value of privately held farmlands, as well as encouraging inappropriate farming and urban growth in arid lands.

On India's Narmada River, a proposed series of 30 dams has incited bitter protest. Many of the 1 million villagers and tribal people being displaced by this project have engaged in civil disobedience and mass protests, which have gone on for the past 15 years or more. In neighboring Nepal, there are fears that earthquakes will cause dams to collapse, leading to catastrophic flooding.

Canada's James Bay project, built by Hydro-Quebec to sell electricity to New York markets, diverted three major rivers flowing west into Hudson Bay and flooded more than 10,000 km^2 (4,000 m^2) of forest and tundra. In 1984, 10,000 caribou drowned while trying to follow ancient migration routes across newly flooded land. Native

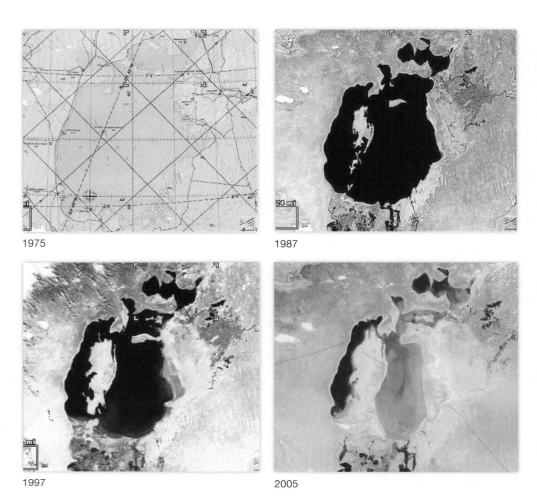

1975

1987

1997

2005

Figure 10.10 For 30 years, rivers feeding the Aral Sea have been diverted to irrigate cotton and rice fields. The Aral Sea has lost more than 60 percent of its water. Dust storms from remaining salt flats now contaminate irrigated fields.

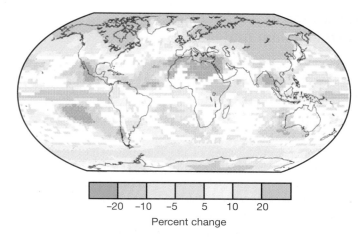

Percent change

Figure 10.11 Relative changes in precipitation (in percentage) for the period 2090–2099 compared to 1980–1989, predicted by the Intergovernmental Panel on Climate Change.
Source: IPCC, 2007.

Cree communities have been devastated by the loss of hunting and fishing sites. In addition, mercury leaching out of rocks in the newly submerged lands has entered the food chain, and many residents have shown signs of food poisoning.

Would you fight for water?

Many environmental scientists have warned that water shortages could lead to wars between nations. *Fortune* magazine wrote "water will be to the 21st century what oil was to the 20th." With one-third of all humans living in areas with water stress now, the situation could become much worse as population grows and climate change dries up some areas and brings more severe storms to others. Already we've seen skirmishes—if not outright warfare—over insufficient water. In Kenya, for instance, nomadic tribes have fought over dwindling water and grazing. An underlying cause of the genocide now occurring in the Darfur region of Sudan is water scarcity. When rain was plentiful, Arab pastoralists and African farmers coexisted peacefully. Drought—perhaps caused by global warming—has upset that truce. The hundreds of thousands who have fled to Chad could be considered climate refugees as well as war victims.

Although they haven't usually risen to the level of war, there have been at least 37 military confrontations in the past 50 years in which water has been at least one of the motivating factors. Thirty of those conflicts have been between Israel and its neighbors. India, Pakistan, and Bangladesh also have confronted each other over water rights, and Turkey and Iraq threatened to send their armies to protect access to the water in the Tigris and Euphrates Rivers. Water can even be used as a weapon. Saddam Hussein cut off water flow into the massive Iraq marshes as a way of punishing his enemies among the Marsh Arabs. Drying of the marshes drove 140,000 people from their homes and destroyed a unique way of life. It also caused severe ecological damage to what is regarded by many as the original Garden of Eden.

Public anger over privatization of the public water supply in Bolivia sparked a revolution that overthrew the government in 2000. Water sales are already a $400-billion-a-year business. Multinational corporations are moving to take control of water systems in many countries. Who owns water and how much they are able to charge for it could become the question of the century. Investors are now betting on scarce water resources by buying future water rights. One Canadian water company, Global Water Corporation, puts it best: "Water has moved from being an endless commodity that may be taken for granted to a rationed necessity that may be taken by force."

WHAT DO YOU THINK?

China's South-to-North Water Diversion

Water is inequitably distributed in China. In the south, torrential monsoon rains cause terrible floods. A 1931 flood on the Yangtze displaced 56 million people and killed 3.7 million (the worst natural disaster in recorded history). Northern and western China, on the other hand, are too dry, and getting drier. At least 200 million Chinese live in areas without sufficient fresh water. The government has warned that unless new water sources are found soon, many of those people (including the capital Beijing, with roughly 20 million residents) will have to be moved. But where could they go? Southern China has water, but doesn't need more people.

The solution, according to the government, is to transfer some of the extra water from south to north. A gargantuan project is now underway to do just that. Work has begun to build three major canals to carry water from the Yangtze River to northern China. Ultimately, it's planned to move 45 billion m^3 per year (more than twice the flow of the Colorado River through the U.S. Grand Canyon) up to 1,600 km (1,000 mi) north. The initial cost estimate of this scheme is about 400 billion yuan (roughly U.S. $62 billion), but it could easily be twice that much.

The eastern route uses the Grand Canal, built by Zhou and Sui emperors 1,500 years ago across the coastal plain between Shanghai and Beijing. This project is already operational. It's relatively easy to pump water through the existing waterways, but they're so polluted by sewage and industrial waste that northern cities—even though they're desperately dry—are reluctant to accept this water.

The central route will draw water from the reservoir behind the nearly completed Three Gorges Dam on the Yangtze. Part of the motivation for building this controversial dam and flooding the historic Three Gorges (see related story "Three Gorges Dam" at www.mhhe.com/cunningham5e) was to provide energy and raise the river level for the South-to-North project. This middle canal will cross several major mountain ranges and dozens of rivers, including the Han and the Yellow Rivers. Currently, work is in progress on raising the Danjiangkou Dam and enlarging its reservoir as part of this route. This will displace some 200,000 people, but planners say it's worthwhile to benefit a thousand times as many. It's hoped this segment will be finished by 2020.

The western route is the most difficult and expensive. It would tunnel through rugged mountains, across aqueducts, and over deep canyons for more than 250 km (160 mi), from the upper Yangtze to the Yellow River,

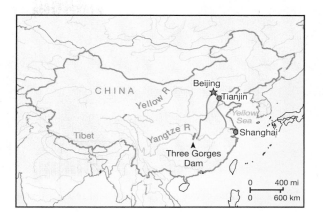

With the Yellow River nearly depleted by overuse, northern China now plans canals (*red*) to deliver Yangtze water to Beijing.

where they both spill off the Tibetan Plateau. This phase won't be finished until at least 2050. If global warming melts all Tibet's glaciers, however, it may not be feasible anyway.

Planners have waited a lifetime to see this project move forward. Revolutionary leader Mao Zedong proposed it 50 years ago. Environmental scientists worry, however, that drawing down the Yangtze will worsen pollution problems (already exacerbated by the Three Gorges Dam), dry up downstream wetlands, and possibly even alter ocean circulation and climate along China's eastern coast. Although southern China has too much water during the rainy season, even there cities face water shortages because of rapidly growing populations and severe pollution problems. At least half of all major Chinese rivers are too polluted for human consumption. Drawing water away from the rivers on which millions rely only makes pollution problems worse.

What do you think? Are there other ways that China could adapt to uneven water distribution? If you were advising the Chinese government, what safeguards would you recommend to avoid unexpected consequences from the gargantuan project?

Freshwater shortages may become much worse in many places because of global climate change. In 2007, the Intergovernmental Panel on Climate Change (IPCC) issued its fourth report on climate change. Figure 10.11 shows a summary of predictions from several models on likely changes in global precipitation for the period 2090–2099 compared to 1980–1999. White areas are where less than two-thirds of the models agree on likely outcomes; stippled areas are where more than 90 percent of the models agree. How does this map compare to figure 10.2? Which areas do you think are most likely to suffer from water shortages by the end of this century? If you lived in one of those areas, would you fight for water?

10.5 Water Management and Conservation

Watershed management and conservation are often more economical and environmentally sound ways to prevent flood damage and store water for future use than building huge dams and reservoirs. A **watershed**, or catchment, is all the land drained by a stream or river. It has long been recognized that retaining vegetation and groundcover in a watershed helps hold back rainwater and lessens downstream floods. In 1998 Chinese officials acknowledged that unregulated timber cutting upstream on the Yangtze

Figure 10.12 By using native plants in a natural setting, residents of Phoenix save water and fit into the surrounding landscape.

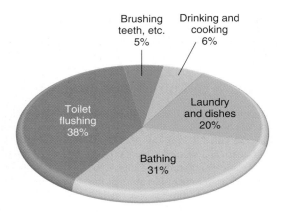

Figure 10.13 Typical household water use in the United States.
Source: Data from U.S. Environmental Protection Agency, 2004.

contributed to massive floods that killed 30,000 people. Similarly, after disastrous floods in the upper Mississippi Valley in 1993, it was suggested that, rather than allowing residential, commercial, or industrial development on floodplains, these areas should be reserved for water storage, aquifer recharge, wildlife habitat, and agriculture. Further discussion of flooding hazards can be found in chapter 11.

Sound farming and forestry practices can reduce runoff. Retaining crop residue on fields reduces flooding, and minimizing plowing and forest cutting on steep slopes protects watersheds. Effects of deforestation on weather and water supplies are discussed in chapter 6. Wetlands conservation preserves natural water storage capacity and aquifer recharge zones. A river fed by marshes and wet meadows tends to run consistently clear and steady, rather than in violent floods.

A series of small dams on tributary streams can hold back water before it becomes a great flood. Ponds formed by these dams provide useful wildlife habitat and stock-watering facilities. They also catch soil where it could be returned to the fields. Small dams can be built with simple equipment and local labor, eliminating the need for massive construction projects and huge dams.

In 1998 U.S. Forest Service chief Mike Dombeck, announced a major shift in his agency's priorities. "Water," he said, "is the most valuable and least appreciated resource the national forests provide. More than 60 million people in 33 states obtain their drinking water from national forest lands. Protecting watersheds is far more economically important than logging or mining, and will be given the highest priority in forest planning."

Everyone can help conserve water

We could probably save as much as half of the water we now use for domestic purposes without great sacrifice or serious changes in our lifestyles. Simple steps, such as taking shorter showers, fixing leaks, and washing cars, dishes, and clothes as efficiently as possible, can go a long way toward forestalling the water shortages that many authorities predict. Isn't it better to adapt to more conserva-

tive uses now when we have a choice than to be forced to do it by scarcity in the future?

Conserving appliances, such as low-volume shower heads and efficient dishwashers, can reduce water consumption greatly (see What Can You Do? p. 243). If you live in an arid part of the country, you might consider whether you really need a lush, green lawn that requires constant watering, feeding, and care. Planting native groundcover in a "natural lawn" or developing a rock garden or landscape in harmony with the surrounding ecosystem can be both ecologically sound and aesthetically pleasing (fig. 10.12). Cultivated lawns, golf courses, and parks in the United States use more water, fertilizer, and pesticides per hectare than any other kind of land.

Toilets are our greatest domestic water user (fig. 10.13). Usually each flush uses several gallons of water to dispose of a few ounces of waste. On average, each person in the United States uses about 50,000 L (13,000 gal) of drinking-quality water annually to flush toilets. Low-flush toilets and composting toilets can drastically reduce this water use. Low-flow shower heads can reduce our second-largest household water use.

These steps are so important that a number of cities (including Los Angeles, Orlando, Austin, and Phoenix) ordered that water-saving toilets, showers, and faucets be installed in all new buildings. The motivation was two-fold: to relieve overburdened sewer systems and to conserve water.

Significant amounts of water also can be reclaimed and recycled. California uses more than 555 million m³ (450,000 acre-feet) of recycled water annually. That's equivalent to about two-thirds of the water consumed by Los Angeles every year.

Efficiency is reducing water use in many areas

Growing recognition that water is a precious and finite resource has changed policies and encouraged conservation across the United States. Despite a growing population, the United States is now saving some 144 million l (38 million gal) per day—a tenth the volume of Lake Erie—compared with per capita consumption rates of 20 years ago. With 37 million more people in the United

Figure 10.14 Sewer outfalls, industrial effluent pipes, acid draining out of abandoned mines, and other point sources of pollution are generally easy to recognize. Pollution control laws have made this sight less common today than it once was.

States now than in 1980, we get by with 10 percent less water. New requirements for water-efficient fixtures and low-flush toilets in many cities help conserve water on the home front. More efficient irrigation methods on farms also are a major reason for the downward trend. New sprinkler systems have small spray heads just a foot or so above the plant tops and apply water much more directly. Even better is drip irrigation, which applies water directly to plant roots. California and Florida farmers currently water about 500,000 ha with this technique.

Charging a higher proportion of real costs to users of public water projects has helped encourage conservation, and so have water marketing policies that allow prospective users to bid on water rights. Both the United States and Australia have had effective water pricing and allocation policies that encourage the most socially beneficial uses and discourage wasteful water uses. It will be important, as water markets develop, to be sure that environmental, recreational, and wildlife values are not sacrificed to the lure of high-bidding industrial and domestic users.

Several countries with desperate water shortages, including Singapore, Australia, and parts of the United States, are using recycled water for drinking. Former Australian Environment Minister Malcolm Turnbull said, "It may sound yucky, but we're not getting rain; we've got no choice."

10.6 Water Pollution

Any physical, biological, or chemical change in water quality that adversely affects living organisms or makes water unsuitable for desired uses can be considered pollution. There are natural sources of water contamination, such as poison springs, oil seeps, and sedimentation from erosion, but here we will focus primarily on human-caused changes that affect water quality or usability.

What Can You Do?

Saving Water and Preventing Pollution

Each of us can conserve much of the water we use and avoid water pollution in many simple ways.

- Don't flush every time you use the toilet. Take shorter showers, and shower instead of taking baths.

- Don't let the faucet run while brushing your teeth or washing dishes. Draw a basin of water for washing and another for rinsing dishes. Don't run the dishwasher when it's half full.

- Use water-conserving appliances: low-flow showers, low-flush toilets, and aerated faucets.

- Fix leaking faucets, tubs, and toilets. A leaky toilet can waste 50 gal per day. To check your toilet, add a few drops of dark food coloring to the tank and wait 15 minutes. If the tank is leaking, the water in the bowl will change color.

- Put a brick or full water bottle in your toilet tank to reduce the volume of water in each flush.

- Dispose of used motor oil, household hazardous waste, batteries, and so on responsibly. Don't dump anything down a storm sewer that you wouldn't want to drink.

- Avoid using toxic or hazardous chemicals for simple cleaning or plumbing jobs. A plunger or plumber's snake will often unclog a drain just as well as caustic acids or lye. Hot water and soap can accomplish most cleaning tasks.

- If you have a lawn, use water, fertilizer, and pesticides sparingly. Plant native, low-maintenance plants that have low water needs.

- If possible, use recycled (gray) water for lawns, house plants, and car washing.

Pollution includes point sources and nonpoint sources

Pollution control standards and regulations usually distinguish between point and nonpoint pollution sources. Factories, power plants, sewage treatment plants, underground coal mines, and oil wells are classified as **point sources** because they discharge pollution from specific locations, such as drain pipes, ditches, or sewer outfalls (fig. 10.14). These sources are discrete and identifiable, so they are relatively easy to monitor and regulate. It is generally possible to divert effluent from the waste streams of these sources and treat it before it enters the environment.

In contrast, **nonpoint sources** of water pollution are diffuse, having no specific location where they discharge into a particular body of water. They are much harder to monitor and regulate than point sources because their origins are hard to identify. Nonpoint sources include runoff from farm fields and feedlots, golf courses, lawns and gardens, construction sites, logging areas, roads, streets, and parking lots. While point sources may be fairly uniform and predictable throughout the year, nonpoint sources are often highly

Table 10.4 Major Categories of Water Pollutants

Category	Examples	Sources
Cause Ecosystem Disruption		
1. Oxygen-demanding wastes	Animal manure, plant residues	Sewage, agricultural runoff, paper mills, food processing
2. Plant nutrients	Nitrates, phosphates, ammonium	Agricultural and urban fertilizers, sewage, manure
3. Sediment	Soil, silt	Land erosion
4. Thermal changes	Heat	Power plants, industrial cooling
Cause Health Problems		
1. Pathogens	Bacteria, viruses, parasites	Human and animal excreta
2. Inorganic chemicals	Salts, acids, caustics, metals	Industrial effluents, household cleansers, surface runoff
3. Organic chemicals	Pesticides, plastics, detergents, oil, gasoline	Industrial, household, and farm use
4. Radioactive materials	Uranium, thorium, cesium, iodine, radon	Mining and processing of ores, power plants, weapons production, natural sources

episodic. The first heavy rainfall after a dry period may flush high concentrations of gasoline, lead, oil, and rubber residues off city streets, for instance, while subsequent runoff may be much cleaner.

Perhaps the ultimate in diffuse, nonpoint pollution is atmospheric deposition of contaminants carried by air currents and precipitated into watersheds or directly onto surface waters as rain, snow, or dry particles. The Great Lakes, for example, have been found to be accumulating industrial chemicals, such as PCBs (polychlorinated biphenyls) and dioxins, as well as agricultural toxins, such as the insecticide toxaphene, that cannot be accounted for by local sources alone. The nearest sources for many of these chemicals are sometimes thousands of kilometers away.

Biological pollution includes pathogens and waste

Although the types, sources, and effects of water pollutants are often interrelated, it is convenient to divide them into major categories for discussion (table 10.4). Here, we look at some of the important sources and effects of different pollutants.

Pathogens

The most serious water pollutants in terms of human health worldwide are pathogenic (disease-causing) organisms (chapter 8). Among the most important waterborne diseases are typhoid, cholera, bacterial and amoebic dysentery, enteritis, polio, infectious hepatitis, and schistosomiasis. Malaria, yellow fever, and filariasis are transmitted by insects that have aquatic larvae. Altogether, at least 25 million deaths each year are blamed on these water-related diseases. Nearly two-thirds of the mortalities of children under 5 years old are associated with waterborne diseases.

The main source of these pathogens is untreated or improperly treated human wastes. Animal wastes from feedlots or fields near waterways and food processing factories with inadequate waste treatment facilities also are sources of disease-causing organisms.

In developed countries, sewage treatment plants and other pollution-control techniques have reduced or eliminated most of the worst sources of pathogens in inland surface waters. Furthermore, drinking water is generally disinfected by chlorination, so epidemics of waterborne diseases are rare in these countries. The United Nations estimates that 90 percent of the people in developed countries have adequate (safe) sewage disposal, and 95 percent have clean drinking water.

The situation is quite different in less-developed countries, where billions of people lack adequate sanitation and access to clean drinking water. Conditions are especially bad in remote, rural areas, where sewage treatment is usually primitive or nonexistent and purified water is either unavailable or too expensive to obtain. The World Health Organization estimates that 80 percent of all sickness and disease in less-developed countries can be attributed to waterborne infectious agents and inadequate sanitation.

Detecting specific pathogens in water is difficult, time-consuming, and costly, so water quality is usually described in terms of concentrations of **coliform bacteria**—any of the many types that live in the colon, or intestines, of humans and other animals. The most common of these is *Escherichia coli* (or *E. coli*). These bacteria occur naturally in our intestines. Other bacteria such as *Shigella*, *Salmonella*, or *Listeria*, can also cause serious, even fatal, illness. If any coliform bacteria are present in a water sample, infectious pathogens are usually assumed to be present also. Therefore, the Environmental Protection Agency (EPA) considers water with any coliform bacteria at all to be unsafe for drinking.

Biological oxygen demand

The amount of oxygen dissolved in water is a good indicator of water quality and of the kinds of life it will support. An oxygen content above 6 parts per million (ppm) will support game fish and other desirable forms of aquatic life. At oxygen levels below 2 ppm, water will support mainly worms, bacteria, fungi, and other detritus feeders and decomposers. Oxygen is added to water by diffusion from the air, especially when turbulence and mixing rates are high, and by photosynthesis of green plants, algae, and cyanobacteria. Turbulent, rapidly flowing water is constantly aerated, so it often recovers quickly from oxygen-depleting processes.

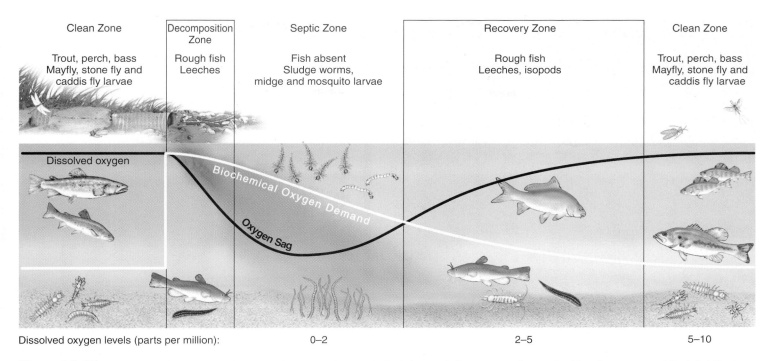

Clean Zone	Decomposition Zone	Septic Zone	Recovery Zone	Clean Zone
Trout, perch, bass Mayfly, stone fly and caddis fly larvae	Rough fish Leeches	Fish absent Sludge worms, midge and mosquito larvae	Rough fish Leeches, isopods	Trout, perch, bass Mayfly, stone fly and caddis fly larvae

Dissolved oxygen

Biochemical Oxygen Demand

Oxygen Sag

Dissolved oxygen levels (parts per million):   0–2   2–5   5–10

Figure 10.15 Oxygen sag downstream of an organic source. A great deal of time and distance may be required for the stream and its inhabitants to recover.

Oxygen is removed from water by respiration and chemical processes that consume oxygen. Because oxygen is so important in water, **dissolved oxygen (DO)** levels are often measured to compare water quality in different places.

Adding organic materials, such as sewage or paper pulp, to water stimulates activity and oxygen consumption by decomposers. Consequently, **biochemical oxygen demand (BOD)**, or the amount of dissolved oxygen consumed by aquatic microorganisms, is another standard measure of water contamination. Alternatively, chemical oxygen demand (COD) is a measure of all organic matter in water.

Downstream from a point source, such as a municipal sewage plant discharge, a characteristic decline and restoration of water quality can be detected either by measuring DO content or by observing the types of flora and fauna that live in successive sections of the river. The oxygen decline downstream is called the **oxygen sag** (fig. 10.15). Upstream from the pollution source, oxygen levels support normal populations of clean-water organisms. Immediately below the source of pollution, oxygen levels begin to fall as decomposers metabolize waste materials. Rough fish, such as carp, bullheads, and gar, are able to survive in this oxygen-poor environment, where they eat both decomposer organisms and the waste itself.

Farther downstream, the water may become so oxygen depleted that only the most resistant microorganisms and invertebrates can survive. Eventually, most of the nutrients are used up, decomposer populations are smaller, and the water becomes oxygenated once again. Depending on the volumes and flow rates of the effluent plume and the river receiving it, normal communities may not appear for several miles downstream.

Plant nutrients and cultural eutrophication

Water clarity (transparency) is affected by sediments, chemicals, and the abundance of plankton organisms; clarity is a useful measure of water quality and water pollution. Rivers and lakes that have clear water and low biological productivity are said to be **oligotrophic** (*oligo* = little + *trophic* = nutrition). By contrast, **eutrophic** (*eu* + *trophic* = well-nourished) waters are rich in organisms and organic materials. Eutrophication, an increase in nutrient levels and biological productivity, often accompanies successional changes (chapter 5) in lakes. Tributary streams bring in sediments and nutrients that stimulate plant growth. Over time, ponds and lakes often fill in, becoming marshes or even terrestrial biomes. The rate of eutrophication depends on water chemistry and depth, volume of inflow, mineral content of the surrounding watershed, and biota of the lake itself.

Human activities can greatly accelerate eutrophication, an effect called **cultural eutrophication**. Cultural eutrophication is mainly caused by increased nutrient input into a water body. Increased productivity in an aquatic system sometimes can be beneficial. Fish and other desirable species may grow faster, providing a welcome food source. Often, however, eutrophication produces "blooms" of algae or thick growths of aquatic plants stimulated by elevated phosphorus or nitrogen levels (fig. 10.16). Bacterial populations then increase, fed by larger amounts of organic matter. The water often becomes cloudy, or turbid, and has unpleasant tastes and odors. Cultural eutrophication can accelerate the "aging" of a water body enormously over natural rates. Lakes and reservoirs that normally might exist for hundreds or thousands of years can be filled in a matter of decades.

Figure 10.16 Eutrophic lake. Nutrients from agriculture and domestic sources have stimulated growth of algae and aquatic plants. This reduces water quality, alters species composition, and lowers the lake's recreational and aesthetic values.

Figure 10.17 Mercury contamination is the most common cause of impairment of U.S. rivers and lakes. Forty-five states have issued warnings about eating locally caught freshwater fish. Long-lived, top predators are especially likely to bioaccumulate toxic concentrations of mercury.

Eutrophication also occurs in marine ecosystems, especially in nearshore waters and partially enclosed bays or estuaries. Partially enclosed seas, such as the Black, Baltic, and Mediterranean Seas, tend to be in especially critical condition. During the tourist season, the coastal population of the Mediterranean, for example, swells to 200 million people. Eighty-five percent of the effluents from large cities go untreated into the sea. Beach pollution, fish kills, and contaminated shellfish result. Extensive "dead zones" often form where rivers dump nutrients into estuaries and shallow seas (see Exploring Science, p. 247). A federal study of the condition of U.S. coastal waters found that 28 percent of estuaries are impaired for aquatic life, and 80 percent of all coastal water is in fair to poor condition.

Marine animals in hypoxic zones die not only because of depleted oxygen, but also because of high concentrations of harmful organisms, including toxic algae, pathogenic fungi, and parasitic protists. Excessive nutrients support blooms of these deadly aquatic microorganisms in polluted nearshore waters. Red tides—and other colors, depending on the species involved—have become increasingly common where nutrients and wastes wash down rivers. (See related story "A Flood of Pigs" at www.mhhe.com/cunningham5e.)

Inorganic pollutants include metals, salts, and acids

Some toxic inorganic chemicals are naturally released into water from rocks by weathering processes (chapter 11). Humans accelerate the transfer rates in these cycles thousands of times above natural background levels by mining, processing, using, and discarding minerals.

Among the chemicals of greatest concern are heavy metals, such as mercury, lead, tin, and cadmium. Supertoxic elements, such as selenium and arsenic, also have reached hazardous levels in some waters. Other inorganic materials, such as acids, salts, nitrates, and chlorine, that are nontoxic at low concentrations may become concentrated enough to lower water quality and adversely affect biological communities.

Metals

Many metals, such as mercury, lead, cadmium, and nickel, are highly toxic in minute concentrations. Because metals are highly persistent, they accumulate in food chains and have a cumulative effect in humans.

Currently the most widespread toxic metal contamination in North America is mercury released from incinerators and coal-burning power plants. Transported through the air, mercury precipitates in water supplies, where it bioconcentrates in food webs to reach dangerous levels in top predators. As a general rule, Americans are warned not to eat more than one meal of fish per week (fig. 10.17). Top marine predators, such as shark, swordfish, bluefin tuna, and king mackerel, tend to have especially high mercury content. Pregnant women and small children should avoid these species entirely. Public health officials estimate that 600,000 American children now have mercury levels in their bodies high enough to cause mental and developmental problems, while one woman in six in the United States has blood-mercury concentrations that would endanger a fetus.

Mine drainage and leaching of mining wastes are serious sources of metal pollution in water. A survey of water quality in eastern Tennessee found that 43 percent of all surface streams and lakes and more than half of all groundwater used for drinking supplies were contaminated by acids and metals from mine drainage. In some cases, metal levels were 200 times higher than what is considered safe for drinking water.

Exploring SCIENCE:

Studying the Gulf Dead Zone

In the 1980s shrimp boat crews noticed that certain locations off the Gulf Coast of Louisiana were emptied of all aquatic life. Since the region supports shrimp, fish, and oyster fisheries worth $250 to $450 million per year, these "dead zones" were important to the economy as well as to the Gulf's ecological systems. In 1985, Nancy Rabelais, a scientist working with Louisiana Universities Marine Consortium, began mapping areas of low oxygen concentrations in the Gulf waters. Her results, published in 1991, showed that vast areas, just above the floor of the Gulf, had oxygen concentration less than 2 parts per million (ppm), a level that eliminated all animal life except primitive worms. Healthy aquatic systems usually have about 10 ppm dissolved oxygen. What caused this hypoxic (oxygen-starved) area to develop?

Rabelais and her team tracked the phenomenon for several years, and it became clear that the dead zone was growing larger over time, that poor shrimp harvests coincided with years when the zone was large, and that the size of the dead zone, which ranges from 5,000 to 20,000 km² (about the size of New Jersey), depended on rainfall and runoff rates from the Mississippi River. Excessive nutrients, mainly nitrogen, from farms and cities far upstream on the Mississippi River, were the suspected culprit.

How did Rabelais and her team know that nutrients were the problem? They noticed that each year, 7–10 days after large spring rains in the agricultural parts of the upper Mississippi watershed, oxygen concentrations in the Gulf drop from 5 ppm to below 2 ppm. These rains are known to wash soil, organic debris, and last year's nitrogen-rich fertilizers from farm fields. The scientists also knew that saltwater ecosystems normally have little available nitrogen, a key nutrient for algae and plant growth. Pulses of agricultural runoff were followed by a profuse growth of algae and phytoplankton (tiny floating plants). Such a burst of biological activity produces an excess of dead plant cells and

fecal matter that drifts to the seafloor. Shrimp, clams, oysters, and other filter feeders normally consume this debris, but they can't keep up with the sudden flood of material. Instead, decomposing bacteria in the sediment break down the debris, and they consume most of the available dissolved oxygen as well. Putrefying sediments also produce hydrogen sulfide, which further poisons the water near the seafloor.

In well-mixed water bodies, as in the open ocean, oxygen from upper layers of water is frequently mixed into lower water layers. Warm, protected water bodies are often stratified, however, as abundant sunlight keeps the upper layers warmer, and less dense, than lower layers. Denser lower layers cannot mix with upper layers unless strong currents or winds stir water.

Many enclosed coastal waters, including Chesapeake Bay, Long Island Sound, the Mediterranean Sea, and the Black Sea, tend to be stratified and suffer hypoxic conditions that destroy bottom and near-bottom communities. There are about 200 dead zones around the world, and the number has doubled each decade since dead zones were first observed in the 1970s. The Gulf of Mexico is second in size behind a 100,000 km² dead zone in the Baltic Sea.

Can dead zones recover? Yes. Water is a forgiving medium, and organisms use nitrogen quickly. In 1996 in the Black Sea region,

farmers in collapsing communist economies cut their nitrogen applications by half out of economic necessity; the Black Sea dead zone disappeared, while farmers saw no drop in their crop yields. In the Mississippi watershed, farmers can afford abundant fertilizer, and they fear they can't afford to risk underfertilizing. Because of the great geographic distance between the farm states and the Gulf, Midwestern states have been slow to develop an interest in the dead zone. At the same time, concentrated feedlot production of beef and pork is rapidly increasing, and feedlot runoff is the fastest growing, and least regulated, source of nutrient enrichment in rivers.

In 2001, federal, state, and tribal governments forged an agreement to cut nitrogen inputs by 30 percent and reduce the size of the dead zone to 5,000 km². This agreement represented astonishingly quick research and political response to scientific results, but it doesn't appear to be enough. Computer models suggest that it would take a 40–45 percent reduction in nitrogen to achieve the 5,000 km² goal.

Human activities have increased the flow of nitrogen reaching U.S. coastal waters by four to eight times since the 1950s. Phosphorus, another key nutrient, has tripled. This case study shows how water pollution can connect far-distant places, such as Midwestern farmers and Louisiana shrimpers.

The Mississippi River drains 40 percent of the coterminous United States, including the most heavily farmed states. Nitrogen fertilizer produces a summer "dead zone" in the Gulf of Mexico.

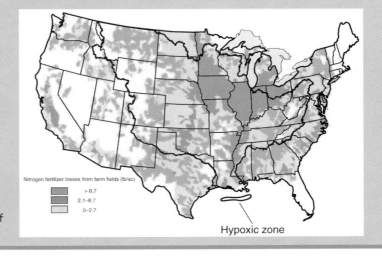

Nitrogen fertilizer losses from farm fields (lb/ac)
- > 8.7
- 2.1–8.7
- 0–2.7

Hypoxic zone

Nonmetallic salts

Some soils contain high concentrations of soluble salts, including toxic selenium and arsenic (see related story "Arsenic in Drinking Water" at www.mhhe.com/cunningham5e). You have probably heard of poison springs and seeps in the desert, where percolating groundwater brings these compounds to the surface. Irrigation and drainage of desert soils can mobilize these materials on a larger

scale and result in serious pollution problems, as in Kesterson Marsh in California, where selenium poisoning killed thousands of migratory birds in the 1980s.

Salts, such as sodium chloride (table salt), that are nontoxic at low concentrations also can be mobilized by irrigation and concentrated by evaporation, reaching levels that are toxic for plants and animals. Salinity levels in the Colorado River and surrounding farm fields have become so high in recent years that millions of hectares of valuable croplands have had to be abandoned. In northern states, millions of tons of sodium chloride and calcium chloride are used to melt road ice in the winter. Leaching of road salts into surface waters has a devastating effect on some aquatic ecosystems.

Acids and bases

Acids are released as by-products of industrial processes, such as leather tanning, metal smelting and plating, petroleum distillation, and organic chemical synthesis. Coal mining is an especially important source of acid water pollution. Sulfur compounds in coal react with oxygen and water to make sulfuric acid. Thousands of kilometers of streams in the United States have been acidified by acid mine drainage, some so severely that they are essentially lifeless.

Acid precipitation (chapter 9) also acidifies surface-water systems. In addition to damaging living organisms directly, these acids leach aluminum and other elements from soil and rock, further destabilizing ecosystems.

Organic chemicals include pesticides and industrial substances

Thousands of different natural and synthetic organic chemicals are used in the chemical industry to make pesticides, plastics, pharmaceuticals, pigments, and other products that we use in everyday life. Many of these chemicals are highly toxic (chapter 8). Exposure to very low concentrations (perhaps even parts per quadrillion, in the case of dioxins) can cause birth defects, genetic disorders, and cancer. Some can persist in the environment because they are resistant to degradation and toxic to organisms that ingest them.

The two principal sources of toxic organic chemicals in water are (1) improper disposal of industrial and household wastes and (2) pesticide runoff from farm fields, forests, roadsides, golf courses, and private lawns. The EPA estimates that about 500,000 metric tons of pesticides are used in the United States each year. Much of this material washes into the nearest waterway, where it passes through ecosystems and may accumulate in high levels in nontarget organisms. The bioaccumulation of DDT in aquatic ecosystems was one of the first of these pathways to be understood (chapter 8). Dioxins and other chlorinated hydrocarbons (hydrocarbon molecules that contain chlorine atoms) have been shown to accumulate to dangerous levels in the fat of salmon, fish-eating birds, and humans and to cause health problems similar to those resulting from toxic metal compounds.

Hundreds of millions of tons of hazardous organic wastes are thought to be stored in dumps, landfills, lagoons, and underground tanks in the United States (chapter 13). Many, perhaps most, of these sites have leaked toxic chemicals into surface waters, groundwater, or both. The EPA estimates that about 26,000 hazardous waste sites will require cleanup because they pose an imminent threat to public health, mostly through water pollution.

Is bottled water safer?

It has become trendy to drink bottled water. Every year, Americans buy about 28 billion bottles of water at a cost of about $15 bil-

lion with the mistaken belief that it's safer than tap water. Worldwide, some 160 billion liters (42 billion gallons) of bottled water are consumed annually. Public health experts say that municipal water is often safer than bottled water because most large cities test their water supplies every hour for up to 25 different chemicals and pathogens, while the requirements for bottled water are much less rigorous. About one-quarter of all bottled water in the United States is simply reprocessed municipal water, and much of the rest is drawn from groundwater aquifers, which may or may not be safe. A recent survey of bottled water in China found that two-thirds of the samples tested had dangerous levels of pathogens and toxins.

Though the plastics used for bottling water are easily recycled, 80 percent of the bottles purchased in the United States end up in a landfill (the recycling rate is even poorer in most other countries). Overall, the average energy cost to make the plastic, fill the bottle, transport it to market, and then deal with the waste would be "like filling up a quarter of every bottle with oil," says water-expert Peter Gleick. Furthermore, it takes 3 to 5 times as much water to make the bottles as they hold. In blind tasting tests, most adults either can't tell the difference between municipal and bottled water, or they actually prefer municipal water. Furthermore, water that's been sitting in plastic bottles for weeks or months can leach out plasticizers and other toxic chemicals.

In most cases, bottled water is expensive, wasteful, and often less safe than most municipal water. Drink tap water and do a favor for your environment, your budget, and, possibly, your health.

Sediment and heat also degrade water

Sediment is a natural and necessary part of river systems. Sediment fertilizes floodplains and creates fertile deltas. But human activities, chiefly farming and urbanization, greatly accelerate erosion and increase sediment loads in rivers. Silt and sediment are considered the largest source of water pollution in the United States, being responsible for 40 percent of the impaired river miles in EPA water quality surveys. Cropland erosion contributes about 25 billion metric tons of soil, sediment, and suspended solids to world surface waters each year. Forest disturbance, road building, urban construction sites, and other sources add at least 50 billion additional tons.

This sediment fills lakes and reservoirs, obstructs shipping channels, clogs hydroelectric turbines, and makes purification of drinking water more costly. Sediments smother gravel beds in which insects take refuge and fish lay their eggs. Sunlight is blocked, so that plants cannot carry out photosynthesis, and oxygen levels decline. Murky, cloudy water also is less attractive for swimming, boating, fishing, and other recreational uses (fig. 10.18). Sediment washed into the ocean clogs estuaries and coral reefs.

Thermal pollution, usually effluent from cooling systems of power plants or other industries, alters water temperature. Raising or lowering water temperatures from normal levels can adversely affect water quality and aquatic life. Water temperatures are usually much more stable than air temperatures, so aquatic organisms tend to be poorly adapted to rapid temperature changes. Lowering the temperature of tropical oceans by even 1° can be lethal to

Figure 10.18 Sediment and industrial waste flow from this drainage canal into Lake Erie.

some corals and other reef species. Raising water temperatures can have similar devastating effects on sensitive organisms. Oxygen solubility in water decreases as temperatures increase, so species requiring high oxygen levels are adversely affected by warming water.

Humans also cause thermal pollution by altering vegetation cover and runoff patterns. Reducing water flow, clearing stream-side trees, and adding sediment all make water warmer and alter the ecosystems in a lake or stream.

Warm-water plumes from power plants often attract fish and birds, which find food and refuge there, especially in cold weather. This artificial environment can be a fatal trap, however. Florida's manatees, an endangered mammal, are attracted to the abundant food supply and warm water in power plant thermal plumes. Often they are enticed into spending the winter much farther north than they normally would. On several occasions, a midwinter power plant breakdown has exposed a dozen or more of these rare animals to a sudden, deadly thermal shock.

10.7 Water Quality Today

Surface-water pollution is often both highly visible and one of the most common threats to environmental quality. In more developed countries, reducing water pollution has been a high priority over the past few decades. Billions of dollars have been spent on control programs, and considerable progress has been made. Still, much remains to be done.

The 1972 Clean Water Act protects our water

Like most developed countries, the United States and Canada have made encouraging progress in protecting and restoring water quality in rivers and lakes over the past 40 years. In 1948 only about one-third of Americans were served by municipal sewage systems, and most of those systems discharged sewage without any treat-

ment or with only primary treatment (the bigger lumps of waste are removed). Most people depended on cesspools and septic systems to dispose of domestic wastes.

Areas of progress

The 1972 Clean Water Act established a National Pollution Discharge Elimination System (NPDES), which requires an easily revoked permit for any industry, municipality, or other entity dumping wastes in surface waters. The permit requires disclosure of what is being dumped and gives regulators valuable data and evidence for litigation. As a consequence, only about 10 percent of our water pollution now comes from industrial and municipal point sources. One of the biggest improvements has been in sewage treatment.

Since the Clean Water Act was passed in 1972, the United States has spent more than $180 billion in public funds and perhaps ten times as much in private investments on water pollution control. Most of that effort has been aimed at point sources, especially to build or upgrade thousands of municipal sewage treatment plants. As a result, nearly everyone in urban areas is now served by municipal sewage systems, and no major city discharges raw sewage into a river or lake except as overflow during heavy rainstorms.

This campaign has led to significant improvements in surface-water quality in many places. Fish and aquatic insects have returned to waters that formerly were depleted of life-giving oxygen. Swimming and other water-contact sports are again permitted in rivers, in lakes, and at ocean beaches that once were closed by health officials.

The Clean Water Act goal of making all U.S. surface waters "fishable and swimmable" has not been fully met, but currently the EPA reports that 91 percent of all monitored river miles and 88 percent of all assessed lake acres are suitable for their designated uses. This sounds good, but you have to remember that not all water bodies are monitored. Furthermore, the designated goal for some rivers and lakes is merely to be "boatable." Water quality doesn't have to be very high to allow boating. Even in "fishable" rivers and lakes, there isn't a guarantee that you can catch anything other than rough fish, such as carp or bullheads, nor can you be sure that what you catch is safe to eat. Even with billions of dollars of investment in sewage treatment plants, elimination of much of the industrial dumping and other gross sources of pollutants, and a general improvement in water quality, the EPA reports that 21,000 water bodies still do not meet their designated uses. According to the EPA, an overwhelming majority of the American people—almost 218 million—live within 16 km (10 mi) of an impaired water body.

In 1998 a new regulatory approach to water quality assurance was instituted by the EPA. Rather than issue standards on a river by river approach or factory-by-factory permit discharge, the focus was changed to watershed-level monitoring and protection. Some 4,000 watersheds are now monitored for water quality (fig. 10.19). You can find information about your watershed at www.epa.gov/owow/tmdl/. The intention of this program is to give the public more and better information about the health of their watersheds. In addition, states can have greater flexibility as they identify impaired water bodies and set priorities, and new tools can be used to achieve goals. States are required to identify waters not

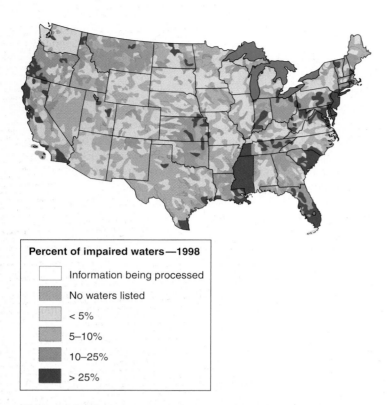

Percent of impaired waters—1998

- ☐ Information being processed
- No waters listed
- < 5%
- 5–10%
- 10–25%
- > 25%

Figure 10.19 Percent of impaired U.S. rivers in the contiguous 48 states by watershed in 1998.
Source: Data from U.S. Environmental Protection Agency, 1999.

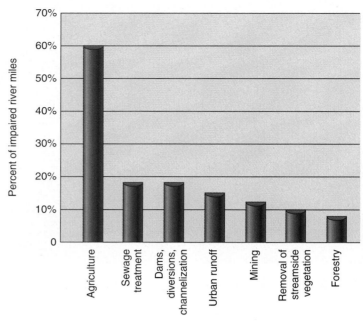

Figure 10.20 Percentage of impaired river miles in the United States by source of damage. Totals add up to more than 100 percent because one river can be affected by many sources.
Source: Data from USDA and Natural Resources Conservation Service, *America's Private Land: A Geography of Hope*, USDA.

meeting water quality goals and to develop **total maximum daily loads (TMDL)** for each pollutant and each listed water body. A TMDL is the amount of a particular pollutant that a water body can receive from both point and nonpoint sources. It considers seasonal variation and includes a margin of safety.

Currently, all 56 U.S. states and territories have submitted TMDL lists, and the EPA has approved most of them. Of the 5.6 million km of rivers monitored, only 480,000 km fail to meet their clean water goals. Similarly, of 40 million lake ha, only 12.5 percent (in about 20,000 lakes) fail to meet their goal. To give states more flexibility in planning, the EPA has proposed new rules that include allowances for reasonably foreseeable increases in pollutant loadings to encourage "Smart Growth." In the future, TMDLs also will include load allocations from all nonpoint sources, including air deposition and natural background levels.

An encouraging example of improved water quality is seen in Lake Erie. Although widely regarded as "dead" in the 1960s, the lake today is promoted as the "walleye capital of the world." Bacteria counts and algae blooms have decreased more than 90 percent since 1962. Water that once was murky brown is now clear. Interestingly, part of the improved water quality is due to immense numbers of exotic zebra mussels, which filter the lake water very efficiently. Swimming is now officially safe along 96 percent of the lake's shoreline. Nearly 40,000 nesting pairs of double-crested cormorants nest in the Great Lakes region, up from only about 100 in the 1970s.

Canada's 1970 Water Act has produced comparable results. Seventy percent of all Canadians in towns over 1,000 population are now served by some form of municipal sewage treatment. In Ontario, the vast majority of those systems include tertiary treatment. After ten years of controls, phosphorus levels in the Bay of Quinte in the northeast corner of Lake Ontario have dropped nearly by half, and algal blooms that once turned waters green are less frequent and less intense than they once were. Elimination of mercury discharges from a pulp and paper mill on the Wabigoon-English River system in western Ontario has resulted in a dramatic decrease in mercury contamination. Twenty years ago this mercury contamination was causing developmental retardation in local residents. Extensive flooding associated with hydropower projects has raised mercury levels in fish to dangerous levels elsewhere, however.

Remaining problems

The greatest impediments to achieving national goals in water quality in both the United States and Canada are sediment, nutrients, and pathogens, especially from nonpoint discharges of pollutants. These sources are harder to identify and to reduce or treat than are specific point sources. About three-fourths of the water pollution in the United States comes from soil erosion, fallout of air pollutants, and surface runoff from urban areas, farm fields, and feedlots. In the United States, as much as 25 percent of the 46,800,000 metric tons (52 million tons) of fertilizer spread on farmland each year is carried away by runoff (fig. 10.20).

Cattle in feedlots produce some 129,600,000 metric tons (144 million tons) of manure each year, and the runoff from these sites is rich in viruses, bacteria, nitrates, phosphates, and other contaminants. A single cow produces about 30 kg (66 lbs) of manure per day. Some feedlots have 100,000 animals with no provision for capturing or treating runoff water. Imagine drawing your drinking water downstream from such a facility. Pets also can be a problem. It is estimated that the wastes from about a half million dogs in New York City are disposed of primarily through storm sewers and therefore do not go through sewage treatment.

Loading of both nitrates and phosphates in surface water have decreased from point sources but have increased about four-fold since 1972 from nonpoint sources. Fossil fuel combustion has become a major source of nitrates, sulfates, arsenic, cadmium, mercury, and other toxic pollutants that find their way into water. Carried to remote areas by atmospheric transport, these combustion products now are found nearly everywhere in the world. Toxic organic compounds, such as DDT, PCBs, and dioxins, also are transported long distances by wind currents.

Developing countries often have serious water pollution

Japan, Australia, and most of Western Europe also have improved surface-water quality in recent years. Sewage treatment in the wealthier countries of Europe generally equals or surpasses that in the United States. Sweden, for instance, serves 98 percent of its population with at least secondary sewage treatment (compared with 70 percent in the United States), and the other 2 percent have primary treatment. Poorer countries have much less to spend on sanitation. Spain serves only 18 percent of its population with even primary sewage treatment. In Ireland, it is only 11 percent, and in Greece, less than 1 percent of the people have even primary treatment. Most of the sewage, both domestic and industrial, is dumped directly into the ocean.

The fall of the "iron curtain" in 1989 revealed appalling environmental conditions in much of the former Soviet Union and its satellite states in eastern and central Europe. The countries closest geographically and socially to western Europe, the Czech Republic, Hungary, East Germany, and Poland, have made massive investments and encouraging progress toward cleaning up environmental problems. Parts of Russia itself, however, along with former socialist states in the Balkans and Central Asia, remain some of the most polluted places on earth. In Russia, for example, only about half the tap water is fit to drink. In cities like St. Petersburg, even boiling and filtering isn't enough to make municipal water safe.

As we saw earlier in this chapter, at least 200 million Chinese live in areas without sufficient fresh water. Sadly, pollution makes much of the limited water unusable (fig. 10.21). It's estimated that 70 percent of China's surface water is unsafe for human consumption, and that the water in half the country's major rivers is so contaminated that it's unsuited for any use, even agriculture. The situation in Shanxi Province exemplifies the problems of water pollution in China. An industrial powerhouse, in the north-central part of the country, Shanxi has about one-third of China's known coal resources and currently produces about two-thirds of

the country's energy. In addition to power plants, major industries include steel mills, tar factories, and chemical plants.

Economic growth has been pursued in recent decades at the expense of environmental quality. According to the Chinese Environmental Protection Agency, the country's ten worst polluted cities are all in Shanxi. Factories have been allowed to exceed pollution discharges with impunity. For example, 3 million tons of wastewater is produced every day in the province with two-thirds of it discharged directly into local rivers without any treatment. Locals complain that the rivers, which once were clean and fresh, now run black with industrial waste. Among the 26 rivers in the province, 80 percent were rated Grade V (unfit for any human use) or higher in 2006. More than half the wells in Shanxi are reported to have dangerously high arsenic levels. Many of the 85,000 reported public protests in China in 2006 involved complaints about air and water pollution.

However, there are some encouraging pollution-control stories. In 1997 Minamata Bay in Japan, long synonymous with mercury poisoning, was declared officially clean again. Another important success is found in Europe, where one of its most important rivers has been cleaned up significantly through international cooperation. The Rhine, which starts in the rugged Swiss Alps and winds 1,320 km through five countries before emptying through a Dutch delta into the North Sea, has long been a major commercial artery into the heart of Europe. More than 50 million people live in its catchment basin, and nearly 20 million get their drinking water from the river or its tributaries. By the 1970s, the Rhine had become so polluted that dozens of fish species disappeared and swimming was discouraged along most of its length.

Efforts to clean up this historic and economically important waterway began in the 1950s, but a disastrous fire at a chemical warehouse near Basel, Switzerland, in 1986 provided the impetus for major changes. Through a long and sometimes painful series

Figure 10.21 Half of the water in China's major rivers is too polluted to be suitable for any human use. Although the government has spent billions of yuan in recent years, dumping of industrial and domestic waste continues at dangerous levels.

Figure 10.22 Ditches in this Haitian slum serve as open sewers, into which all manner of refuse and waste are dumped. The health risks of living under these conditions are severe.

of international conventions and compromises, land-use practices, waste disposal, urban runoff, and industrial dumping have been changed and water quality has significantly improved. Oxygen concentrations have gone up five-fold since 1970 (from less than 2 mg/l to nearly 10 mg/l, or about 90 percent of saturation) in long stretches of the river. Chemical oxygen demand has fallen five-fold during the same period, and organochlorine levels have decreased as much as ten-fold. Many species of fish and aquatic invertebrates have returned to the river. In 1992, for the first time in decades, mature salmon were caught in the Rhine.

The less-developed countries of South America, Africa, and Asia have even worse water quality than do the poorer countries of Europe. Sewage treatment is usually either totally lacking or woefully inadequate. In urban areas, 95 percent of all sewage is discharged untreated into rivers, lakes, or the ocean. Low technological capabilities and little money for pollution control are made even worse by burgeoning populations, rapid urbanization, and the shift of much heavy industry (especially the dirtier ones) from developed countries where pollution laws are strict to less-developed countries where regulations are more lenient.

Appalling environmental conditions often result from these combined factors (fig. 10.22). Two-thirds of India's surface waters are contaminated sufficiently to be considered dangerous to human health. The Yamuna River in New Delhi has 7,500 coliform bacteria per 100 ml (37 times the level considered safe for swimming in the United States) *before* entering the city. The coliform count increases to an incredible 24 *million* cells per 100 ml as the river leaves the city! At the same time, the river picks up some 20 million liters of industrial effluents every day from New Delhi. It's

no wonder that disease rates are high and life expectancy is low in this area. Only 1 percent of India's towns and cities have any sewage treatment, and only eight cities have anything beyond primary treatment.

In Malaysia, 42 of 50 major rivers are reported to be "ecological disasters." Residues from palm oil and rubber manufacturing, along with heavy erosion from logging of tropical rainforests, have destroyed all higher forms of life in most of these rivers. In the Philippines, domestic sewage makes up 60 to 70 percent of the total volume of Manila's Pasig River. Thousands of people use the river not only for bathing and washing clothes but also as their source of drinking and cooking water.

Groundwater is especially hard to clean up

About half the people in the United States, including 95 percent of those in rural areas, depend on underground aquifers for their drinking water. This vital resource is threatened in many areas by overuse and pollution and by a wide variety of industrial, agricultural, and domestic contaminants. For decades it was widely assumed that groundwater was impervious to pollution because soil would bind chemicals and cleanse water as it percolated through. Springwater or artesian well water was considered to be the definitive standard of water purity, but that is no longer true in many areas.

One of the serious sources of groundwater pollution throughout the United States is MTBE (methyl tertiary butyl ether), a suspected carcinogen added to gasoline to reduce carbon monoxide and ozone in urban air. Aquifers across the United States have been contaminated—mainly from leaking underground storage tanks at gas stations. In one U.S. Geological Survey (USGS) study, 27 percent of shallow urban wells tested contained MTBE. The additive is being phased out, but plumes of tainted water will continue to move through aquifers for decades to come. Liability for this contamination is a highly contentious issue.

The EPA estimates that every day some 4.5 trillion l (1.2 trillion gal) of contaminated water seep into the ground in the United States from septic tanks, cesspools, municipal and industrial landfills and waste disposal sites, surface impoundments, agricultural fields, forests, and wells (fig. 10.23). The most toxic of these are probably waste disposal sites. Agricultural chemicals and wastes are responsible for the largest total volume of pollutants and area affected. Because deep underground aquifers often have residence times of thousands of years, many contaminants are extremely stable once underground. It is possible, but expensive, to pump water out of aquifers, clean it, and then pump it back.

In farm country, especially in the Midwest's corn belt, fertilizers and pesticides commonly contaminate aquifers and wells. Herbicides such as atrazine and alachlor are widely used on corn and soybeans and show up in about half of all wells in Iowa, for example. Nitrates from fertilizers often exceed safety standards in rural drinking water. These high nitrate levels are dangerous to infants (nitrates combine with hemoglobin in the blood and result in "blue-baby" syndrome).

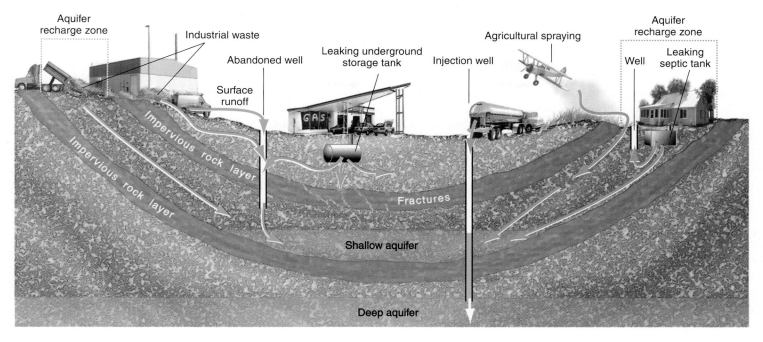

Figure 10.23 Sources of groundwater pollution. Septic systems, landfills, and industrial activities on aquifer recharge zones leach contaminants into aquifers. Wells provide a direct route for injection of pollutants into aquifers.

Every year, epidemiologists estimate that around 1.5 million Americans fall ill from infections caused by fecal contamination. In 1993, for instance, a pathogen called cryptosporidium got into the Milwaukee public water system, making 400,000 people sick and killing at least 100 people. The total costs of these diseases amount to billions of dollars per year. Preventative measures, such as protecting water sources and aquifer recharge zones and updating treatment and distribution systems, would cost far less.

Ocean pollution has few controls

Although we don't use ocean waters directly, ocean pollution is serious and one of the fastest-growing water pollution problems. Coastal bays, estuaries, shoals, and reefs are often overwhelmed by pollution. Dead zones and poisonous algal blooms are increasingly widespread. Toxic chemicals, heavy metals, oil, sediment, and plastic refuse affect some of the most attractive and productive ocean regions. The potential losses caused by this pollution amount to billions of dollars each year. In terms of quality of life, the costs are incalculable.

Discarded plastic flotsam and jetsam are becoming a ubiquitous mark of human impact on the oceans. Since plastic is lightweight and nonbiodegradable, it is carried thousands of miles on ocean currents and lasts for years. Even the most remote beaches of distant islands are likely to have bits of polystyrene foam containers or polyethylene packing material that were discarded half

a world away. It has been estimated that 6 million metric tons of plastic bottles, packaging material, and other litter are tossed from ships every year into the ocean, where they ensnare and choke seabirds, mammals, and even fish (fig. 10.24).

Oil pollution affects beaches and open seas around the world. Oceanographers estimate that between 3 million and 6 million metric tons of oil are discharged into the world's oceans each year from oil tankers, fuel leaks, intentional discharges of fuel oil, and coastal industries. About half of this amount is due to maritime transport. Of this portion, most is not from dramatic, headline-making accidents such as the 1989 *Exxon Valdez* spill in

Figure 10.24 A deadly necklace. Marine biologists estimate that cast-off nets, plastic beverage yokes, and other packing residue kill hundreds of thousands of birds, mammals, and fish each year.

Alaska but, rather, from routine, open-sea bilge pumping and tank cleaning. These activities are illegal but very common.

The transport of huge quantities of oil creates opportunities for major oil spills through a combination of human and natural hazards. Military conflict in the Middle East destabilizes shipping routes. More important, drilling and transport in stormy seas cause spills. Plans to drill for oil along the seismically active California and Alaska coasts have been controversial because of the damage that spills could cause to these biologically rich coastal ecosystems.

Fortunately, awareness of ocean pollution is growing. Oil spill cleanup technologies and response teams are improving, although most oil is eventually decomposed by natural bacteria. Efforts are growing to control waste plastic. Many states now require that six-pack yokes be made of biodegradable or photodegradable plastic, limiting their longevity as potential killers. International concern about ocean ship waste is increasing, and some shipping companies have been prosecuted and fined for dumping fuel oil. Beach pollution—mainly plastic debris, but also sewage waste, oil, and chemical contaminants—is becoming more common, but it is also more frequently reported in the mainstream media. Volunteer efforts are helping to reduce beach pollution locally: in one day, volunteers in Texas gathered more than 300 tons of plastic refuse from Gulf Coast beaches.

10.8 Pollution Control

The cheapest and most effective way to reduce pollution is to avoid producing it or releasing it in the first place. Eliminating lead from gasoline has resulted in a widespread and significant decrease in the amount of lead in U.S. surface waters. Studies have shown that as much as 90 percent less road deicing salt can be used in many areas without significantly affecting the safety of winter roads. Careful handling of oil and petroleum products can greatly reduce the amount of water pollution caused by these materials. Although we still have problems with persistent chlorinated hydrocarbons spread widely in the environment, the banning of DDT and PCBs in the 1970s has resulted in significant reductions in levels in wildlife.

Industry can reduce pollution by recycling or reclaiming materials that otherwise might be discarded in the waste stream. These approaches usually have economic as well as environmental benefits. Companies can extract valuable metals and chemicals and sell them, instead of releasing them as toxic contaminants into the water system. Both markets and reclamation technologies are improving as awareness of these opportunities grows. In addition, modifying land use is an important component of reducing pollution.

Nonpoint sources are often harder to control than point sources

Farmers have long contributed a huge share of water pollution, especially in the developed world. Increasingly, though, farmers are finding ways to save money and water quality at the same time. Soil conservation practices on farmlands (chapter 7) maintain soil fertility, as well as protect water quality. Precise application of fertilizer, irrigation water, and pesticides saves money and reduces water contamination. Preserving wetlands that act as natural processing facilities for removing sediment and contaminants helps protect surface and groundwaters.

In urban areas, reducing waste that enters storm sewers is essential. It is getting easier for city residents to recycle waste oil and to properly dispose of paint and other household chemicals that they once dumped into storm sewers or the garbage. Urbanites can also minimize use of fertilizers and pesticides. Regular street sweeping greatly reduces nutrient loads (from decomposing leaves and debris) in rivers and lakes. Runoff can also be diverted away from streams and lakes. Many cities are separating storm sewers and municipal sewage lines to avoid overflow during storms.

A good example of the problems of watershed management is seen in Chesapeake Bay, America's largest estuary. Once fabled for its abundant oysters, crabs, shad, striped bass, and other valuable fisheries, the bay had deteriorated seriously by the early 1970s. Citizens' groups, local communities, state legislatures, and the federal government together established an innovative pollution-control program that made the bay the first estuary in America targeted for protection and restoration.

Among the principal objectives of this plan is reducing nutrient loading through land-use regulations in the bay's six watershed states to control agricultural and urban runoff. Pollution-prevention measures, such as banning phosphate detergents, also are important, as are upgrading wastewater treatment plants and improving compliance with discharge and filling permits. Efforts are underway to replant thousands of hectares of sea grasses and to restore wetlands that filter out pollutants. Since the 1980s, annual phosphorous discharges into Chesapeake Bay have dropped 40 percent. Nitrogen levels, however, have remained constant or have even risen in some tributaries. Although progress has been made, the goals of reducing both nitrogen and phosphate levels by 40 percent and restoring viable fish and shellfish populations are still decades away. Still, as EPA Administrator Carol Browner says, it demonstrates the "power of cooperation" in environmental protection. (See related story "Watershed Protection in the Catskills" at www.mhhe.com/cunningham5e.)

Human waste degrades naturally in low concentrations

As we have already seen, human and animal wastes usually create the most serious health-related water pollution problems. More than 500 types of disease-causing (pathogenic) bacteria, viruses, and parasites can travel from human or animal excrement through water.

Natural processes

In the poorer countries of the world, most rural people simply go out into the fields and forests to relieve themselves, as they have

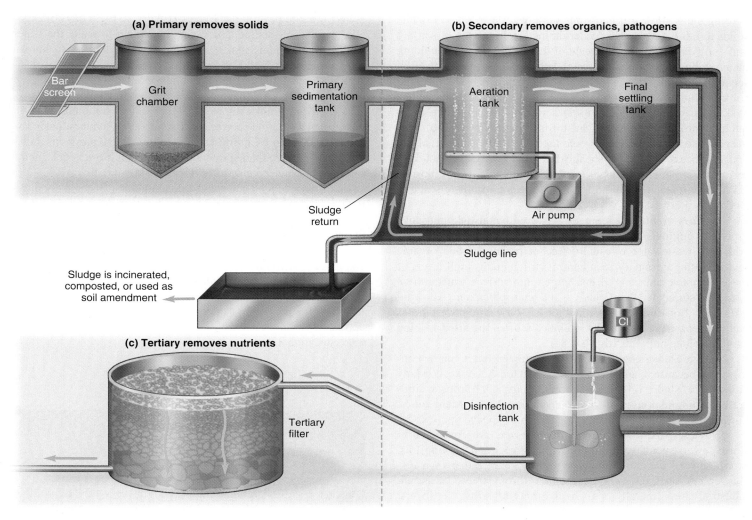

(a) Primary removes solids

Bar screen

Grit chamber

Primary sedimentation tank

(b) Secondary removes organics, pathogens

Aeration tank

Final settling tank

Sludge return

Air pump

Sludge line

Sludge is incinerated, composted, or used as soil amendment

(c) Tertiary removes nutrients

Tertiary filter

Disinfection tank

Cl

Figure 10.25 Activated sludge wastewater treatment. (a) Primary treatment removes only solids and suspended sediment. (b) Secondary treatment, through aeration of activated sludge, followed by sludge removal and chlorination of effluent, kills pathogens and removes most organic material. (c) During tertiary treatment, passage through a trickling bed evaporator and/or a tertiary filter further removes inorganic nutrients, oxidizes any remaining organics, and reduces effluent volume.

always done. Where population densities are low, natural processes eliminate wastes quickly, making this a relatively safe method of sanitation. The high population densities of cities, however, make this practice unworkable. Even major cities of many less-developed countries have serious problems with sewage disposal.

Where intensive agriculture is practiced—especially in wet rice paddy farming in Asia—it has long been customary to collect "night soil" (human and animal waste) to be spread on the fields as fertilizer. This waste is a valuable source of plant nutrients, but it is also a source of disease-causing pathogens in the food supply.

Until about 50 years ago, most rural American families and quite a few residents of towns and small cities depended on a pit toilet, or "outhouse," for waste disposal. Untreated wastes tended to seep into the ground, however, and pathogens sometimes contaminated drinking water. The development of septic tanks and properly constructed drain fields represented a considerable

improvement in public health. Septic systems allow solids to settle in a tank, where bacteria decompose them; liquids percolate through soil, where soil bacteria presumably purify them. Where population densities are not too high, this can be an effective method of waste disposal. With urban sprawl, however, groundwater pollution often becomes a problem.

Municipal treatment has three levels of quality

Over the past 100 years, sanitary engineers have developed ingenious and effective municipal wastewater treatment systems to protect human health, ecosystem stability, and water quality. This topic is an important part of pollution control, and is a principal responsibility of every municipal government.

Primary treatment physically separates large solids from the waste stream with screens and settling tanks (fig. 10.25a).

Settling tanks allow grit and some dissolved (suspended) organic solids to fall out as sludge. Water drained from the top of settling tanks still carries up to 75 percent of the organic matter, including many pathogens. These pathogens and organics are removed by **secondary treatment**, in which aerobic bacteria break down dissolved organic compounds. In secondary treatment, effluent is aerated, often with sprayers or in an aeration tank, in which air is pumped through the microorganism-rich slurry (fig. 10.25b). Fluids can also be stored in a sewage lagoon, where sunlight, algae, and air process waste more cheaply but more slowly. Effluent from secondary treatment processes is usually disinfected with chlorine, UV light, or ozone to kill harmful bacteria before it is released to a nearby waterway.

Tertiary treatment removes dissolved metals and nutrients, especially nitrates and phosphates, from the secondary effluent. Although wastewater is usually free of pathogens and organic material after secondary treatment, it still contains high levels of these inorganic nutrients. If discharged into surface waters, these nutrients stimulate algal blooms and eutrophication. Allowing effluent to flow through a wetland or lagoon can remove nitrates and phosphates. Alternatively, chemicals often are used to bind and precipitate nutrients (fig. 10.25c).

Sewage sludge can be a valuable fertilizer, but it can be controversial because it can contain metals and toxic chemicals. Many cities spread sludge on farms and forest lands, and European cities are beginning to convert it to methane (natural gas). Many cities, however, incinerate or landfill sludge, both expensive options.

In many American cities, sanitary sewers are connected to storm sewers, which carry runoff from streets and parking lots. Storm sewers are routed to the treatment plant rather than discharged into surface waters because runoff from streets, yards, and industrial sites generally contains a variety of refuse, fertilizers, pesticides, oils, rubber, tars, lead (from gasoline), and other undesirable chemicals. Unfortunately heavy storms often overload the system, especially where the system is old and already overburdened. As a result, large volumes of raw sewage and toxic surface runoff are dumped directly into receiving waters. To prevent this overflow, cities are spending hundreds of millions of dollars to separate storm and sanitary sewers.

Constructed wetlands are low-cost treatment

A number of alternative treatment systems have been developed. One of the most attractive is **constructed wetlands**, artificial marshes designed to filter and decompose waste. Arcata, California, for instance, needed an expensive sewer plant upgrade. Instead, the city transformed a 65-ha garbage dump into a series of ponds and marshes that serve as a simple, low-cost, waste treatment facility (chapter 2). Arcata saved millions of dollars and improved the environment simultaneously. The marsh is a haven for wildlife and has become a prized recreation area for the city. Eventually, the purified water flows into Humboldt Bay, where marine life flourishes.

Similar wetland waste treatment systems are now operating in many developing countries. Effluent from these operations can be used to irrigate crops or raise fish for human consumption if care

is taken first to destroy pathogens. Usually 20 to 30 days of exposure to sun, air, and aquatic plants is enough to make the water safe. These systems make an important contribution to human food supplies. A 2,500 ha waste-fed aquaculture facility in Kolkata (Calcutta), for example, supplies about 7,000 metric tons of fish annually to local markets.

The World Bank estimates that more than 3 billion people will be without sanitation services by the year 2030 under a business-as-usual scenario. With investments in innovative programs, however, sanitation could be provided to about half those people, and a great deal of misery and suffering could be avoided.

Waste treatment can even be done indoors. Ocean Arks International in Falmouth, Massachusetts, has been developing holistic systems for water purification that are combinations of different plants and animals, including algae, rooted aquatic plants, clams, snails, and fish, each chosen to provide a particular service in a contained environment. Technically, the water that has flowed through such a system is drinkable, although few people feel comfortable doing so. More often, the final effluent is used to flush toilets or for irrigation. Called ecological engineering, this novel approach can save resources and money, and it can serve as a valuable educational tool (fig. 10.26).

Figure 10.26 In-house wastewater treatment in Oberlin College's Environmental Studies building. Constructed wetlands outside, and tanks inside, allow plants to filter water and remove nutrients.

Conventional treatment misses new pollutants

Countless additional organic compounds enter our water unmonitored. In 2002, the USGS released the first-ever study of pharmaceuticals and hormones in streams. Scientists sampled 130 streams, looking for 95 contaminants, including antibiotics, natural and synthetic hormones, detergents, plasticizers, insecticides, and fire retardants (fig. 10.27). All these substances were found, usually in low concentrations. One stream had 38 of the compounds tested. Drinking water standards exist for only 14 of the 95 substances. A similar study found the same substances in groundwater, which is much harder to clean than surface waters. What are the effects of these widely used chemicals, on our environment or on people consuming the water? Nobody knows. This study is a first step toward filling huge gaps in our knowledge about their distribution, though.

Remediation can involve containment, extraction, or plants

Just as there are many sources of water contamination, there are many ways to clean it up. New developments in environmental engineering are providing promising solutions to many water pollution problems. Containment methods keep dirty water from spreading. Chemicals can be added to toxic wastewater to precipitate, immobilize, or solidify contaminants. Many pollutants can be destroyed or detoxified by chemical reactions that oxidize, reduce, neutralize, hydrolyze, precipitate, or otherwise change their chemical composition. Where chemical techniques are ineffective, physical methods may work. Solvents and other volatile organic compounds, for instance, can be stripped from solution by aeration and then burned in an incinerator. (See related story "International Accord to Clean Up the Rhine River" at www.mhhe.com/cunningham5e.)

Often, living organisms can clean contaminated water effectively and inexpensively. We call this **bioremediation**. Restored wetlands, for instance, along stream banks or lake margins can effectively filter out sediment and remove pollutants. Some plants are very efficient at taking up heavy metals and organic contaminants. Bioremediation offers exciting and inexpensive alternatives to conventional cleanup.

Lowly duckweed (*Lemna* sp.), the green scum you often see covering the surface of eutrophic ponds, grows fast and can remove large amounts of organic nutrients from water. Large duckweed lagoons are being used as inexpensive, low-tech, sewage treatment plants in developing countries. The duckweed can also be harvested and used as feed, fuel, or fertilizer. Up to 35 percent of its dry mass is protein—about twice as much as alfalfa, a popular animal feed.

10.9 Water Legislation

Water pollution control has been among the most broadly popular and effective of all environmental legislation in the United States. It has not been without controversy, however. Table 10.5 describes some of the most important water legislation in the United States.

The Clean Water Act was ambitious, popular, and largely successful

Passage of the U.S. Clean Water Act of 1972 was a bold, bipartisan step that made clean water a national priority. Along with the Endangered Species Act and the Clean Air Act, this is one of the most significant and effective pieces of environmental legislation ever passed by the U.S. Congress. It also is an immense and complex law, with more than 500 sections regulating everything from urban runoff, industrial discharges, and municipal sewage treatment to land-use practices and wetland drainage.

The ambitious goal of the Clean Water Act was to return all U.S. surface waters to "fishable and swimmable" conditions. For point sources, the act requires discharge permits and use of the best practicable control technology (BPT). For toxic substances, the act sets national goals of best available, economically achievable technology (BAT) and zero discharge goals for 126 priority toxic pollutants. As discussed earlier, these regulations have had a positive effect on water quality. While

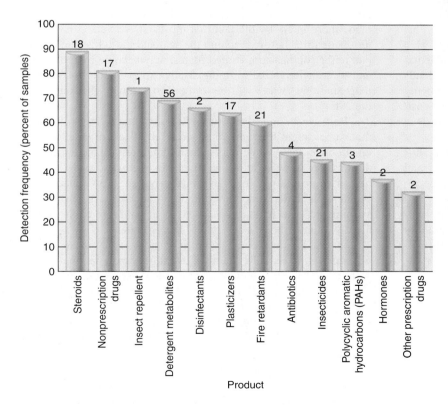

Figure 10.27 Detection frequency of organic, wastewater contaminants in a recent USGS survey. Maximum concentrations in water samples are shown above the bars in micrograms per liter. Dominant substances included DEET insect repellent, caffeine, and triclosan, which comes from antibacterial soaps.

Table 10.5   Some Important Water Quality Legislation

1. *Federal Water Pollution Control Act (1972).* Establishes uniform nationwide controls for each category of major polluting industries.

2. *Marine Protection Research and Sanctuaries Act (1972).* Regulates ocean dumping and established sanctuaries for protection of endangered marine species.

3. *Ports and Waterways Safety Act (1972).* Regulates oil transport and the operation of oil-handling facilities.

4. *Safe Drinking Water Act (1974).* Requires minimum safety standards for every community water supply. Among the contaminants regulated are bacteria, nitrates, arsenic, barium, cadmium, chromium, fluoride, lead, mercury, silver, and pesticides; radioactivity and turbidity also are regulated. This act also contains provisions to protect groundwater aquifers.

5. *Resource Conservation and Recovery Act (RCRA) (1976).* Regulates the storage, shipping, processing, and disposal of hazardous wastes and sets limits on the sewering of toxic chemicals.

6. *Toxic Substances Control Act (TOSCA) (1976).* Categorizes toxic and hazardous substances, establishes a research program, and regulates the use and disposal of poisonous chemicals.

7. *Comprehensive Environmental Response, Compensation, and Liability Act (CERCLA) (1980)* and *Superfund Amendments and Reauthorization Act (SARA) (1984).* Provide for sealing, excavation, or remediation of toxic and hazardous waste dumps.

8. *Clean Water Act (1985)* (amending the *1972 Water Pollution Control Act).* Sets as a national goal the attainment of "fishable and swimmable" quality for all surface waters in the United States.

9. *London Dumping Convention (1972).* Calls for an end to all ocean dumping of industrial wastes, tank-washing effluents, and plastic trash. The United States is a signatory to this international convention.

---

not yet swimmable or fishable everywhere, surface-water quality in the United States has significantly improved on average over the past quarter century. Perhaps the most important result of the act has been investment of $54 billion in federal funds and more than $128 billion in state and local funds for municipal sewage treatment facilities.

Opponents of federal regulation have tried repeatedly to weaken or eliminate the Clean Water Act. They regard restriction of their "right" to dump toxic chemicals and wastes into wetlands and waterways to be an undue loss of freedom. They resent being forced to clean up municipal water supplies and call for cost/benefit analysis that places greater weight on economic interests in all environmental planning.

Supporters of the Clean Water Act would like to see a shift away from an "end-of-the-pipe" focus on effluent removal and more attention to changing industrial processes, so that toxic substances aren't produced in the first place. Many people also would like to see stricter enforcement of existing regulations, mandatory minimum penalties for violations, more effective community right-to-know provisions, and increased powers for citizen lawsuits against polluters.

Conclusion

Water is a precious resource. As human populations grow and climate change affects rainfall patterns, water is likely to become even more scarce in the future. Already, about 2 billion people live in water-stressed countries (where there are inadequate supplies to meet all demands), and at least half those people don't have access to clean drinking water. Depending on population growth rates and climate change, it's possible that by 2050 there could be 7 billion people (about 60 percent of the world population) living in areas with water stress or scarcity. Conflicts over water rights are becoming more common between groups within countries and between neighboring countries that share water resources. This is made more likely by the fact that most major rivers cross two or more countries before reaching the sea, and droughts, such as the one in the southeastern United States, may become more frequent and severe with global warming. Many experts agree with *Fortune* magazine that "water will be to the 21st century what oil was to the 20th."

Forty years ago, rivers in the United States were so polluted that some caught fire while others ran red, black, orange, or other unnatural colors with toxic industrial wastes. Many cities still dumped raw sewage into local rivers and lakes, so that warnings had to be posted to avoid any bodily contact. We've made huge progress since that time. Not all rivers and lakes are "fishable or swimmable," but federal, state, and local pollution controls have greatly improved our water quality in most places.

In rapidly developing countries, such as China and India, water pollution remains a serious threat to human health and ecosystem well-being. It will take a massive investment to correct this growing problem. But there are relatively low-cost solutions to many pollution issues. The example of Arcata's constructed wetland for ecological sewage treatment shows us that we can find low-tech, inexpensive ways to reduce pollution. Living machines for water treatment in individual buildings or communities also offer hope for better ways to treat our wastes. Perhaps you can use the information you've learned by studying environmental science to make constructive suggestions for your own community.

1. Describe the path a molecule of water might follow through the hydrologic cycle from the ocean to land and back again.

2. About what percent of the world's water is liquid, fresh, surface water that supports most terrestrial life (see fig. 10.3)?

3. What is an *aquifer*? How does water get into an aquifer? Explain the idea of an *artesian well* and a *cone of depression*.

4. What is the difference between water *withdrawal* and *consumption*? Which sector of water use (see fig. 10.8) consumes most globally? Overall, has water use increased in the past century? Has efficiency increased or decreased in the three main use sectors?

5. Describe at least one example of the environmental costs of water diversion from rivers to farms or cities.

6. Explain the difference between point and nonpoint pollution. Which is harder to control? Why?

7. Why are nutrients considered pollution? Explain the ideas of *eutrophication* and an *oxygen sag* (see fig. 10.15).

8. Describe primary, secondary, and tertiary water treatment.

9. What are some sources of groundwater contamination? Why is groundwater pollution such a difficult problem?

10. Why are nutrients important factors in water pollution and eutrophication? What two elements are the most important nutrients in water pollution?

Apply the principles you have learned in this chapter to discuss these questions with other students.

1. What changes might occur in the hydrologic cycle if our climate were to warm or cool significantly?

2. Why does it take so long for deep ocean waters to circulate through the hydrologic cycle? What happens to substances that contaminate deep ocean water or deep aquifers in the ground?

3. How precise do you think the estimate is that 1.5 billion people lack access to safe drinking water? If, in 2025, two-thirds of the world's population live in countries with severe water shortages, what might be the number of people who lack access to clean water? Why is this number ambiguous?

4. Do you think that water pollution is worse now than it was in the past? What considerations go into a judgment such as this? How do your personal experiences influence your opinion?

5. What additional information would you need to make a judgment about whether conditions are getting better or worse? How would you weigh different sources, types, and effects of water pollution?

6. Under what conditions might sediment in water or cultural eutrophication be beneficial? How should we balance positive and negative effects?

Data Analysis: Graphing Global Water Stress and Scarcity

According to the United Nations, **water stress** is when annual water supplies drop below 1,700 m³ per person. **Water scarcity** is defined as annual water supplies below 1,000 m³ per person. More than 2.8 billion people in 48 countries will face either water stress or scarcity conditions by 2025. Of these countries, 40 are expected to be in West Asia or Africa. By 2050, far more people could be facing water shortages, depending both on population projections and scenarios for water supplies based on global warming and consumption patterns. The following graph in this box shows an estimate for water stress and scarcity in 1995 together with three possible scenarios (high, medium, and low population projections) for 2050. You'll remember from chapter 4 that according to the 2004 UN population revision, the low projection for 2050 is about 7.6 billion, the medium projection is 8.9 billion, and the high projection is 10.6 billion.

1. What are the combined numbers of people who could experience water stress and scarcity under the low, medium, and high scenarios in 2050?

2. What proportion (percentage) of 7.6 billion, 8.9 billion, and 10.6 billion would this be?

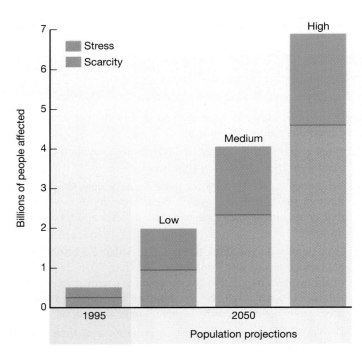

Global water stress and scarcity

3. How does the percentage of the population in these two categories vary in the three estimates?

4. Why is the proportion of people in the scarce category so much larger in the high projection?

5. How many liters are in 1,000 m³? How many gallons?

6. How does 1,000 m³ compare to the annual consumption by the average family of four in the United States? (Hint: Look at table 10.1 and the table of units of measurement conversions at the end of this book).

7. Why isn't the United States (as a whole) considered to be water stressed?

For Additional Help in Studying This Chapter, please visit our website at www.mhhe.com/cunningham5e. You will find practice quizzes, key terms, answers to end of chapter questions, additional case studies, an extensive reading list, and Google Earth™ mapping quizzes.

An aerial view of the Jonah Field in the Upper Green River Basin shows some of the thousands of well pads connected by a spider web of service roads and pipelines spread across the landscape. This gas field lies across the annual migration route of 50,000 antelope, the greatest mammal migration in the lower 48 states.

11 Environmental Geology and Earth Resources

Learning Outcomes

After studying this chapter, you should be able to answer the following questions:

- What is coal-bed methane? Why is it both important and controversial?
- How do continental and oceanic crust differ?
- Why do volcanoes occur?
- What are some of the environmental and social costs of mining and oil- and gas-drilling?
- What are some solutions to those costs?
- How can we reduce our consumption of geologic resources?
- Why is mass wasting a problem?

When we heal the earth, we heal ourselves.

–David Orr

Coal-Bed Methane: A Clean Fuel or a Dirty Business?

Of the many resources we extract from the earth, fossil fuels are among the most valuable. Coal-bed methane, from expansive coal deposits mainly in Western states, is one of the newest and most controversial of these energy sources. Methane (CH_4) is the main component of natural gas. It occurs naturally within coal deposits, formed within ancient swamp muck and peat bogs, as well as by the same geologic heat and pressure that compressed that ancient muck and peat into rock-hard coal.

Traditionally, methane has been a hazardous mine gas that can suffocate miners or explode, or it has been released to the atmosphere, where it contributes to global warming. Most coal-bed methane is held in place by pressure from overlying aquifers (rock formations saturated with water). To release the gas, water is pumped from the aquifer to the surface. As water pressure is released, methane escapes from the coal seam. It rises to the surface with the water and is captured. All that groundwater carries salts and minerals dissolved from the surrounding rock, mainly sodium, barium, bicarbonates, sulfate, boron, and iron. Although none of these is toxic at low levels, they harm both animals and plants at high concentrations.

The volume of water pumped to release coal-bed methane is phenomenal. A typical coal-bed methane well produces 75,000 liters of water per day. Wells are usually distributed thickly across the landscape (see opening photo), multiplying the volume of water produced in a watershed. Much of this water is allowed to seep into groundwater and shallow aquifers or is released into natural waterways. The increased water volume erodes stream banks, and the elevated salt levels poison fish, wildlife, and aquatic ecosystems. In arid Western states, natural streams and groundwater are also needed for irrigating farms and pastures. In some areas, groundwater is contaminated. In others, water tables have fallen as coal-bed aquifers are drawn down, and farmers and ranchers lose their wells and springs. Wetlands also dry up as groundwater falls, and wildlife is left stranded. "It may be a clean fuel," says one rancher, "but it's a dirty business."

The sheer extent and density of operations also make coal-bed methane controversial. In Wyoming's Powder River Basin, energy companies have already installed nearly 14,000 wells and have proposed 39,000 more. Eventually, this area could contain as many as 140,000 wells, together with a sprawling network of roads, pipelines, compressor stations, and wastewater pits. The Green River Basin and the San Juan Basin, with three to five times as much potential gas and oil as the Powder River Basin, may see even greater environmental damage (fig. 11.1).

In 2002 the U.S. Environmental Protection Agency (EPA) gave its worst possible rating to the Environmental Impact Statement (EIS) for the proposed Powder River wells because of concerns over wastewater disposal. Nevertheless, the Bush administration approved the plan and ordered federal land managers across the Rockies to look for ways to remove or reduce environmental restrictions on gas drilling. This order came in spite of a federal study finding that 63 percent of the natural gas in the region was completely open to drilling and only 12 percent was completely protected.

An unlikely coalition of ranchers, hunters, anglers, conservationists, water users, and renewable energy activists have banded together to fight coal-bed gas extraction, calling on Congress to protect private property rights, to enforce the Clean Water Act, and to conserve sensitive public lands. Lifelong Republicans have banded together with tree-hugging environmentalists to protect their way of life.

On the other hand, oil and gas companies argue that they are contributing to American energy independence. As long as we choose to drive large vehicles, commute long distances, and heat and cool large houses, we will need fossil fuels. Methane is the cleanest-burning fossil fuel, producing far less particulates, sulfur oxides, and other pollutants than coal. Brine from the gas wells may be disagreeable, but it isn't nearly as toxic and long-lasting as nuclear waste. The dense networks of roads and drilling pads are unsightly and hazardous to wildlife, but they're better than strip-mining.

Expect to see many new disputes over fossil fuel extraction in the decades ahead. Geologists estimate that at least 346 trillion ft³ of "technically recoverable" natural gas and 62 billion barrels of petroleum occurs in the five intermountain basins stretching from Montana to New Mexico. Extracting these would involve disturbing vast stretches of country, but it could also provide us with a 15-year supply of gas at present rates of use, and at least four times as much oil as the most optimistic estimates for the Arctic National Wildlife Refuge. The total value of methane and petroleum liquids from the Rocky Mountains could be as much as $200 billion over the next decade. Beyond this, 13 percent of the lower 48 states has some coal, and extractable methane occurs with many of these deposits. Vast deposits of oil-saturated sand and shale also underlie the forests and plains of Alberta and neighboring provinces. The debates over extracting these reserves will not end soon.

Fossil fuels are just one of the essential geologic resources on which we depend. In this chapter we'll examine geologic resources and consider why they're distributed as they are. We will also consider some of the environmental and social costs of extracting them, as well as ways those costs can be reduced.

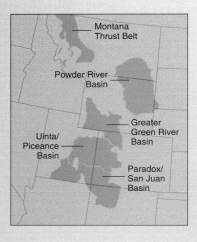

Figure 11.1 Coal-bed methane deposits occur in five intermountain basins in the western United States.

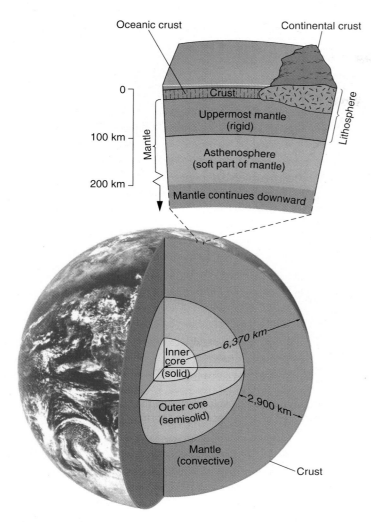

Figure 11.2 Earth's cross section. Slow convection in the mantle causes the thin, brittle crust to move.

Table 11.1	Eight Most Common Chemical Elements (Percent) in Whole Earth and Crust		
Whole Earth		**Crust**	
Iron	33.3	Oxygen	45.2
Oxygen	29.8	Silicon	27.2
Silicon	15.6	Aluminum	8.2
Magnesium	13.9	Iron	5.8
Nickel	2.0	Calcium	5.1
Calcium	1.8	Magnesium	2.8
Aluminum	1.5	Sodium	2.3
Sodium	0.2	Potassium	1.7

11.1 Earth Processes Shape Our Resources

Every one of us shares the benefits of geologic resources. Right now you are probably wearing several geologic products: plastics, including glasses and synthetic fabric, are made from oil; iron, copper, and aluminum mines produced your snaps, zippers, and the screws in your glasses; silver, gold, and diamond mines may have produced your jewelry. All of us also share responsibility for the environmental and social devastation that often results from mining and drilling. Fortunately, there are many promising solutions to reduce these costs, including recycling and alternative materials. The question is whether voters will demand that we use these technologies—and whether consumers will share the costs of responsible production.

Why are these resources distributed as they are? To understand how and where earth resources are created, we must examine the earth's structure and the processes that shape it.

Earth is a dynamic planet

Although we think of the ground under our feet as solid and stable, the earth is a dynamic and constantly changing structure. Titanic forces inside the earth cause continents to split, move apart, and then crash into each other in slow but inexorable collisions.

The earth is a layered sphere. The **core**, or interior, is composed of a dense, intensely hot mass of metal—mostly iron—thousands of kilometers in diameter (fig. 11.2). Solid in the center but more fluid in the outer core, this immense mass generates the magnetic field that envelops the earth.

Surrounding the molten outer core is a hot, pliable layer of rock called the **mantle**. The mantle is much less dense than the core because it contains a high concentration of lighter elements, such as oxygen, silicon, and magnesium.

The outermost layer of the earth is the cool, lightweight, brittle rock **crust**. The crust below oceans is relatively thin (8–15 km), dense, and young (less than 200 million years old) because of constant recycling. Crust under continents is relatively thick (25–75 km), light, and as old as 3.8 billion years, with new material being added continually. It also is predominantly granitic, while oceanic crust is mainly dense basaltic rock. Table 11.1 compares the composition of the whole earth (dominated by the dense core) and the crust.

Tectonic processes reshape continents and cause earthquakes

The huge convection currents in the mantle are thought to break the overlying crust into a mosaic of huge blocks called **tectonic plates** (fig. 11.3). These plates slide slowly across the earth's surface like wind-driven ice sheets on water, in some places breaking up into smaller pieces, in other places crashing ponderously into each other to create new, larger landmasses. Ocean basins form where continents crack and pull apart. The Atlantic Ocean, for example, is growing slowly as Europe and Africa move away from the Americas. **Magma** (molten rock) forced up through the cracks forms new oceanic crust that piles up underwater in **mid-ocean ridges**. Creating the largest mountain range in the world,

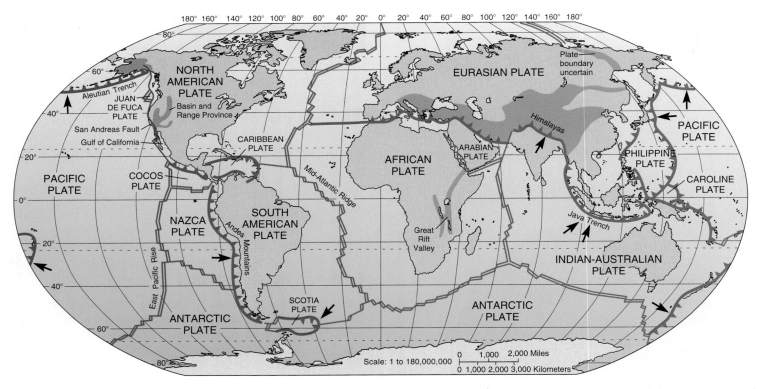

Figure 11.3 Map of tectonic plates. Plate boundaries are dynamic zones, characterized by earthquakes, volcanism, and the formation of great rifts and mountain ranges. Arrows indicate direction of subduction where one plate is diving beneath another. These zones are sites of deep trenches in the ocean floor and high levels of seismic and volcanic activity.

Sources: Data from U.S. Department of the Interior and U.S. Geological Survey.

these ridges wind around the earth for 74,000 km (46,000 mi) (see fig. 11.3). Although concealed from our view, this jagged range boasts higher peaks, deeper canyons, and sheerer cliffs than any continental mountains. Slowly spreading from these fracture zones, ocean plates push against continental plates.

Earthquakes are caused by grinding and jerking as plates slide past each other. Mountain ranges like those on the west coast of North America and in Japan are pushed up at the margins of colliding continental plates. The Himalayas are still rising as the Indian subcontinent grinds slowly into Asia. Southern California is slowly sailing north toward Alaska. In about 30 million years, Los Angeles will pass San Francisco, if both still exist by then.

When an oceanic plate collides with a continental landmass, the continental plate usually rides up over the seafloor, while the oceanic plate is **subducted**, or pushed down into the mantle, where it melts and rises back to the surface as magma (fig. 11.4). Deep ocean trenches mark these subduction zones, and volcanoes form where the magma erupts through vents and fissures in the overlying crust. Trenches and volcanic mountains ring the Pacific Ocean rim

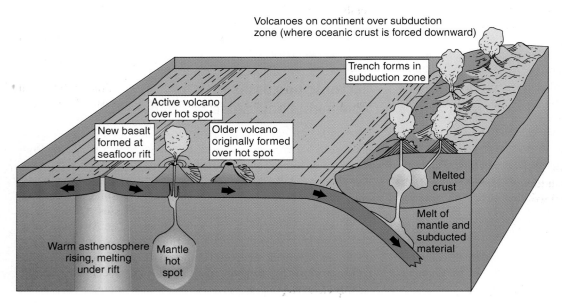

Volcanoes on continent over subduction zone (where oceanic crust is forced downward)

Trench forms in subduction zone

Active volcano over hot spot

New basalt formed at seafloor rift

Older volcano originally formed over hot spot

Melted crust

Melt of mantle and subducted material

Warm asthenosphere rising, melting under rift

Mantle hot spot

Figure 11.4 Plate tectonic movement. Where thin, oceanic plates diverge, upwelling magma forms mid-ocean ridges. A chain of volcanoes, such as the Hawaiian Islands, may form as plates pass over a hot spot. Where plates converge, melting can cause volcanoes, such as the Cascades.

Figure 11.5 Pangaea, an ancient supercontinent of 200 million years ago, combined all the world's continents in a single landmass. Continents have combined and separated repeatedly.

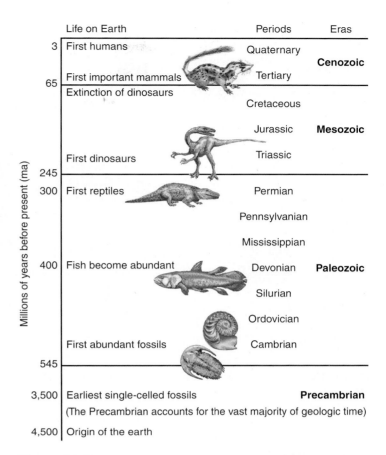

Figure 11.6 Periods and eras in geologic time, and major lifeforms that mark some periods.

from Indonesia to Japan to Alaska and down the west coast of the Americas, forming a so-called ring of fire where oceanic plates are being subducted under the continental plates. This ring is the source of more earthquakes and volcanic activity than any other region on the earth.

Over millions of years, continents can drift long distances. Antarctica and Australia once were connected to Africa, for instance, somewhere near the equator and supported luxuriant forests. Geologists suggest that several times in the earth's history most or all of the continents have gathered to form supercontinents, which have ruptured and re-formed over hundreds of millions of years (fig. 11.5). The redistribution of continents has profound effects on the earth's climate and may help explain the periodic mass extinctions of organisms marking the divisions between many major geologic periods (fig. 11.6).

11.2 Minerals and Rocks

A **mineral** is a naturally occurring, inorganic solid with a specific chemical composition and a specific internal crystal structure. A mineral is solid; therefore, ice is a mineral (with a distinct composition and crystal structure), but liquid water is not. Similarly molten lava is not crystalline, although it generally hardens to create distinct minerals. Metals (such as iron, copper, aluminum, or gold) come from mineral ores, but once purified, metals are no longer crystalline and thus are not minerals. Depending on the conditions in which they were formed, mineral crystals can be

microscopically small, such as asbestos fibers, or huge, such as the tree-size selenite crystals recently discovered in a Chihuahua, Mexico mine.

A **rock** is a solid, cohesive aggregate of one or more minerals. Within the rock, individual mineral crystals (or grains) are mixed together and held firmly in a solid mass. The grains may be large or small, depending on how the rock was formed, but each grain retains its own unique mineral qualities. Each rock type has a characteristic mixture of minerals, grain sizes, and ways in which the grains are mixed and held together. Granite, for example, is a mixture of quartz, feldspar, and mica crystals. Rocks with a granite-like mineral content but much finer crystals are called rhyolite; chemically similar rocks with large crystals are called pegmatite.

The rock cycle creates and recycles rocks

Although rocks appear hard and permanent, they are part of a relentless cycle of formation and destruction. They are crushed, folded, melted, and recrystallized by dynamic processes related to those that shape the large-scale features of the earth's crust. We call this cycle of creation, destruction, and metamorphosis the **rock cycle** (fig. 11.7). Understanding something of how this

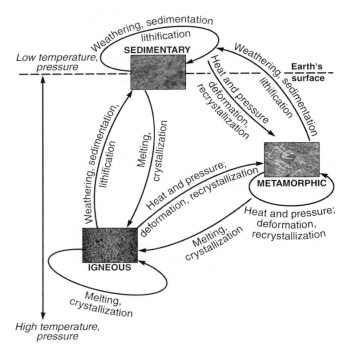

Figure 11.7 The rock cycle includes a variety of geologic processes that can transform any rock.

cycle works helps explain the origin and characteristics of different types of rocks.

There are three major rock classifications: igneous, metamorphic, and sedimentary. **Igneous rocks** (from *igni*, the Latin word for fire) are solidified from hot, molten magma or lava. Most rock in the earth's crust is igneous. Magma extruded to the surface from volcanic vents cools quickly to make finely crystalline rocks, such as basalt, rhyolite, or andesite. Magma that cools slowly in subsurface chambers or is intruded between overlying layers makes coarsely crystalline rocks, such as gabbro (rich in iron and silica) or granite (rich in aluminum and silica), depending on the chemical composition of the magma.

Metamorphic rocks form from the melting, contorting, and recrystallizing of other rocks. Deep in the ground, tectonic forces squeeze, fold, heat, and recrystallize solid rock. Under these conditions, chemical reactions can alter both the composition and the structure of the component minerals. Metamorphic rocks are classified by their chemical composition and by the degree of recrystallization: some minerals form only under extreme pressure and heat (diamonds or jade, for example); others form under more moderate conditions (graphite or talc). Some common metamorphic rocks are marble (from limestone), quartzite (from sandstone), and slate (from mudstone and shale). Metamorphic rocks often have beautiful colors and patterns left by the twisting and folding that created them.

Sedimentary rocks are formed when loose grains of other rocks are consolidated by time and pressure. Sandstone, for example, is solidified from layers of sand, and mudstone consists of extremely hardened mud and clay. Tuff is formed from volcanic

ash, and conglomerates are aggregates of sand and gravel. Some sedimentary rocks develop from crystals that precipitate out of extremely salty water. Rock salt, made of the mineral halite, is ground up to produce ordinary table salt (sodium chloride). Salt deposits often form when a body of saltwater dries up, leaving salt crystals behind. Limestone is a rock composed of cemented remains of marine organisms. You can often see the shapes of shells and corals in a piece of limestone. Sedimentary formations often have distinctive layers that show different conditions when they were laid down. Erosion can reveal these layers and inform us of their history (fig. 11.8).

Weathering and sedimentation

Most crystalline rocks are extremely hard and durable, but exposure to air, water, changing temperatures, and reactive chemical agents slowly breaks them down in a process called **weathering** (fig. 11.9). Mechanical weathering is the physical breakup of rocks into smaller particles without a change in chemical composition of the constituent minerals. You have probably seen the results of mechanical weathering in rounded rocks in rivers or on shorelines, smoothed by constant tumbling in waves or currents. On a larger scale, mountain valleys are carved by rivers and glaciers.

Chemical weathering is the selective removal or alteration of specific minerals in rocks. This alteration leads to weakening and disintegration of rock. Among the more important chemical weathering processes are oxidation (combination of oxygen with an element to form an oxide or a hydroxide mineral) and hydrolysis (hydrogen atoms from water molecules combine with other chemicals to form acids). The products of these reactions are more susceptible to both mechanical weathering and dissolving in water. For instance, when carbonic acid (formed when rainwater absorbs CO_2) percolates through porous limestone layers in the ground, it dissolves the rock and creates caves.

Figure 11.8 Eons of erosion in Arizona's Grand Canyon reveals many sedimentary layers, and some deep metamorphic rocks.

Figure 11.9 Weathering slowly reduces an igneous rock to loose sediment. Here, exposure to moisture expands minerals in the rock, and frost may also force the rock apart.

Table 11.2	Primary Uses of Some Major Metals
Metal	**Use**
Aluminum	Packaging foods and beverages (38%), transportation, electronics
Chromium	High-strength steel alloys
Copper	Building construction, electric and electronic industries
Iron	Heavy machinery, steel production
Lead	Leaded gasoline, car batteries, paints, ammunition
Manganese	High-strength, heat-resistant steel alloys
Nickel	Chemical industry, steel alloys
Platinum group	Automobile catalytic converters, electronics, medical uses
Gold	Medical, aerospace, electronic uses; accumulation as monetary standard
Silver	Photography, electronics, jewelry

Particles of rock loosened by wind, water, ice, and other weathering forces are carried downhill, downwind, or downstream until they come to rest again in a new location. The deposition of these materials is called **sedimentation**. Water, wind, and glaciers deposit particles of sand, clay, and silt far from their source. Much of the American Midwest, for instance, is covered with hundreds of meters of sedimentary material left by glaciers (till, or rock debris deposited by glacial ice), wind (loess, or fine dust deposits), river deposits of sand and gravel, and ocean deposits of sand, silt, clay, and limestone.

11.3 Economic Geology and Mineralogy

Economic mineralogy is the study of minerals that are valuable for manufacturing and trade. Most economic minerals are metal ores, minerals with unusually high concentrations of metals. Lead, for example, often comes from the mineral galena (PbS), and copper comes from sulfide ores, such as bornite (Cu_5FeS_4). Nonmetallic geologic resources include graphite, feldspar, quartz crystals, diamonds, and other crystals that are valued for their usefulness or beauty. Metals have been so important in human affairs that major epochs of human history are commonly known by their dominant materials and the technology involved in using those materials (Stone Age, Bronze Age, Iron Age, etc.). The mining, processing, and distribution of these materials have broad implications for both our culture and our environment. Most economically valuable crustal resources exist everywhere in small amounts; the important thing is to find them concentrated in economically recoverable levels.

Public policy in the United States has encouraged mining on public lands as a way of boosting the economy and utilizing nat-

ural resources. Today these policies are controversial, and there are efforts to recover public revenue from these publicly owned resources. (See related story "Should We Revise the 1872 Mining Law?" at www.mhhe.com/cunningham5e.)

Metals are essential to our economy

Metals are malleable substances that are useful and valuable because they are strong, relatively light, and can be reshaped for many purposes. The availability of metals and the methods to extract and use them have determined technological developments, as well as economic and political power for individuals and nations.

The metals consumed in greatest quantity by world industry include iron (740 million metric tons annually), aluminum (40 million metric tons), manganese (22.4 million metric tons), copper and chromium (8 million metric tons each), and nickel (0.7 million metric tons). Most of these metals are consumed in the United States, Western Europe, Japan, and China. They are produced primarily in South America, South Africa, and Russia (fig. 11.10). It is easy to see how these facts contribute to a worldwide mineral trade network that has become crucially important to the economic and social stability of all nations involved. Table 11.2 shows the primary uses of these metals.

Nonmetal mineral resources include gravel, clay, glass, and salts

Nonmetal minerals constitute a broad class that covers resources from gemstones to sand, gravel, salts, limestone, and soils. Sand and gravel production for road and building construction comprise by far the greatest volume and dollar value of all nonmetal mineral resources and a far greater volume than all metal ores. Sand and gravel are used mainly in brick and concrete construction, in

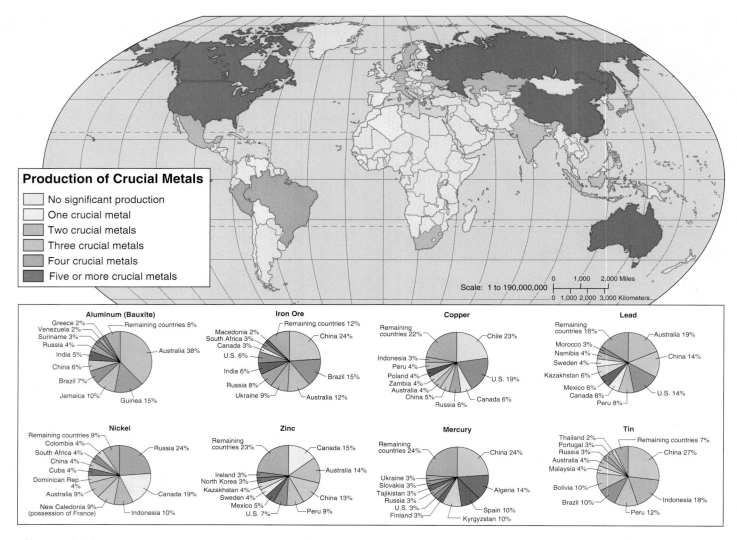

Figure 11.10 World production of metals most essential for an industrial economy. Principal consumers are the United States, Western Europe, Japan, and China.

paving, as loose road filler, and for sandblasting. High-purity silica sand is our source of glass. These materials usually are retrieved from surface pit mines and quarries, where they were deposited by glaciers, winds, or ancient oceans.

Limestone, like sand and gravel, is mined and quarried for concrete and crushed for road rock. It also is cut for building stone, pulverized for use as an agricultural soil additive that neutralizes acidic soil, and roasted in lime kilns and cement plants to make plaster (hydrated lime) and cement.

Evaporites (materials deposited by evaporation of chemical solutions) are mined for halite, gypsum, and potash. These are often found at or above 97 percent purity. Halite, or rock salt, is used for water softening and ice melting on winter roads in some northern areas. Refined, it is a source of table salt. Gypsum (calcium sulfate) now makes our plaster wallboard, but it has been used to cover walls ever since the Egyptians plastered their frescoed tombs along the Nile River some 5,000 years ago. Potash is an evaporite composed of a variety of potassium chlorides and

potassium sulfates. These highly soluble potassium salts have long been used as a soil fertilizer.

Sulfur deposits are mined mainly for sulfuric acid production. In the United States, sulfuric acid use amounts to more than 200 lbs per person per year, mostly because of its use in industry, car batteries, and some medicinal products.

Durable, highly valuable, and easily portable, gemstones and precious metals have long been a way to store and transport wealth. Unfortunately, these valuable materials also have bankrolled despots, criminal gangs, and terrorism in many countries. In recent years, brutal civil wars in Africa have been financed—and often motivated by—gold, diamonds, tantalum ore, and other high-priced commodities. In 2004 investigators discovered that members of the terrorist group Al Qaida used diamonds purchased in Liberia to finance the September 11, 2001 attacks in the United States. Much of this illegal trade ends up in the $100 billion per year global jewelry trade, two-thirds of which sells in the United States. Many people who treasure a diamond ring or a gold wedding band as a symbol of love and devotion are unaware that it may have been obtained through slave labor, torture, and environmentally destructive mining and processing methods. Civil rights organizations are campaigning to require better documentation of the origins of gems and precious metals to prevent their use as financing for crimes against humanity (see related story "Conflict Diamonds" at www.mhhe.com/cunningham5e).

In 2004 a group of Nobel Peace Laureates called on the World Bank to overhaul its policies on lending for resource extractive industries. "War, poverty, climate change, and ongoing violations of human rights—all of these scourges are all too often linked to the oil and mining industries," wrote Archbishop Desmond Tutu, winner of the 1984 Nobel Peace Prize for helping eliminate apartheid in South Africa. In response, the World Bank appointed an Extractive Industries Review headed by former Indonesian Environment Minister Emil Salim. Its final report agreed with many concerns raised by environmental and community organizations.

The earth provides almost all our fuel

Modern society functions largely on energy produced from geologic deposits of oil, coal, and natural gas. Nuclear energy, which runs on uranium, makes up a large portion of our electricity resources. Energy production from these sources is discussed in chapter 12. Oil, coal, and gas are organic, created over millions of years as extreme heat and pressure transformed the remains of ancient organisms. They are not minerals, because they have no crystalline structure, but they can be considered part of economic mineralogy because they are such important geologic resources. In addition to providing energy, oil is the source material for plastics, and natural gas is used to make agricultural fertilizers.

The search for these organic deposits is one of the most important parts of economic mineralogy (see Exploring Science, p. 270). Debates over exploitation and ownership of these resources play important roles in national and international politics. The Persian Gulf War of 1990 was fought over control of vast underground oil deposits, as has been Russia's war in Chechnya. Debates over exploiting potential oil reserves in the Alaskan National Wildlife Refuge (ANWR) have raged in the United States for decades. (See related stories "Exploiting Oil in ANWR" at www.mhhe.com/cunningham5e.) Recently, Canada's oil shale and tar sands have become another source of oil (chapter 12).

11.4 Environmental Effects of Resource Extraction

Each of us depends daily on geologic resources mined or pumped from sites around the world. We use scores of metals and minerals, many of which we've never even heard of, in our lights, computers, watches, fertilizers, and cars. Extracting and purifying these resources can have severe environmental and social consequences. The most obvious effect of mining and well drilling is often the disturbance or removal of the land surface. Farther-reaching effects,

Figure 11.11 Thousands of abandoned mines on public lands poison streams and groundwater with acid, metal-laced drainage. This old mine in Montana drains into the Blackfoot River, the trout stream featured in Norman Maclean's book *A River Runs through It.*

though, include air and water pollution. The EPA lists more than 100 toxic air pollutants, from acetone to xylene, released from U.S. mines and wells every year. Nearly 80,000 metric tons of particulate matter (dust) and 11,000 tons of sulfur dioxide are released from nonmetal mining alone. Chemical- and sediment-runoff pollution is a major problem in many local watersheds.

Gold and other metals are often found in sulfide ores that produce sulfuric acid when exposed to air and water. In addition, metal elements often occur in very low concentrations—10 to 20 parts per billion may be economically extractable for gold, platinum, and other metals. Consequently, vast quantities of ore must be crushed and washed to extract metals. Cyanide, mercury, and other toxic substances are used to chemically separate metals from the minerals that contain them, and these substances can easily contaminate lakes and streams. Further, a great deal of water is used in washing crushed ore with cyanide and other solutions. In arid Nevada, the USGS estimates that mining consumes about 230,000 m^3 (60 million gal) per day. After use in ore processing, much of this water contains sulfuric acid, arsenic, heavy metals, and other contaminants. Mine runoff leaking into lakes and streams can damage or destroy aquatic ecosystems (fig. 11.11).

Exploring SCIENCE: Prospecting with New Technology

How do geologists discover oil and gas deposits? Oil and gas are among the most lucrative of earth resources, but they are hidden deep in the ground, sometimes thousands of meters deep. The main method for finding these hidden reservoirs is seismic exploration and 3-D imaging.

Seismology is the study of how waves of energy (seismic waves) move through the earth. Seismic exploration involves producing, and then detecting, seismic pulses. The way these seismic waves move through the earth tells a great deal about the density of the rocks, where layers of differing density meet, where the earth is fractured, and where hidden pockets of rock are saturated with oil or gas.

To carry out seismic exploration, geologists set out an array of "geophones" to record vibrations in the earth, and then produce a series of intense vibrations along a survey line. Initially, dynamite blasts were used to create seismic waves in the earth. Now "thumper trucks" are usually used (fig. 1). Thumper trucks don't actually thump; instead, they lower a plate to the ground, press on the plate with the truck's 30-ton mass, and then produce low-frequency vibrations that travel through the earth. This process is repeated every 100 m or so, along survey lines that may run many kilometers across the landscape. As the vibrations bounce off of rock layers in the ground, seismic waves travel back to the geophones. The geophones record how long the waves took to travel—thus how dense the rocks were—and send the data to a central computer.

By compiling the data from many geophones, and from hundreds or thousands of vibration events, geologists can reconstruct the shapes of features in the ground. Originally, these shapes were mapped two-dimensionally, along a single track of the truck or dynamite blasts. As computers have improved, they have become able to process the terabytes of data needed to map a grid of survey lines. Using a grid of survey lines, geologists can create three-dimensional models of structures in the earth (fig. 2—3-D image). Increasingly, geologists re-survey sites after pumping begins, to detect how and where oil and gas are depleted. This approach is called "4-dimensional imaging" because repeat visits add a time dimension to the process.

These methods have greatly reduced the number of test wells needed to explore for gas and oil, and they have allowed oil and gas companies to locate small, deep, or unconventional deposits they might never have found with other methods.

Local conservation groups, especially in Wyoming and Utah, however, still object to the damage thumper trucks leave on the landscape. In the dry, open country of Western states, soils and vegetation are easily disturbed and recover very slowly. The deep ruts left by the 30-ton trucks, as they crawl cross-country up hills and across gullies, usually in trains of four or five trucks together, will be visible for many decades, or possibly centuries, to come. Similarly, in the delicate tundra of the Arctic, networks of survey tracks disappear slowly. In the mountainous rainforests of Ecuador, meanwhile, felling trees for survey lines has become a dominant cause of deforestation.

Increasingly, oil and gas exploration are moving offshore, as land-based resources become harder to extract. Most new discoveries are hundreds of meters below the sea, as well as thousands of meters below the seafloor. For undersea exploration, ships carry powerful airguns. Blasts from the airguns produce sound waves that travel into the seafloor, and reflected wavelenghts are picked up by sensors towed behind the ship. Although ships leave no tracks, marine biologists worry about the effects of noise on whales and other marine mammals.

Seismic imaging is one of the huge recent advances in the science of finding and extracting fossil fuels. It has reduced the damage of test wells but also made exploration feasible in wild and roadless areas previously thought useless to oil companies. How do you think we should manage exploration in public lands, as new technology makes it easier? How would your answers differ if you were a rancher in Wyoming, a Senator from Wyoming, or someone from a distant state?

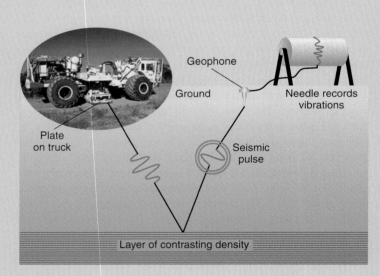

Figure 1

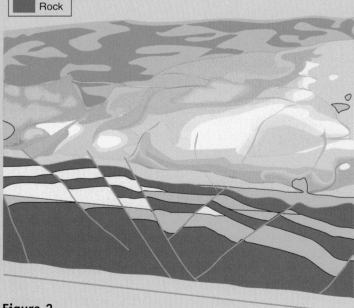

Gas
Oil
Water
Rock

Figure 2

Mining has enormous environmental effects

There are many techniques for extracting geologic materials. The most common methods are open-pit mining, strip-mining, and underground mining. An ancient method of accumulating gold, diamonds, and coal is placer mining, in which pure nuggets are washed from stream sediments. Since the California gold rush of 1849, placer miners have used water cannons to blast away hillsides. This method, which chokes stream ecosystems with sediment, is still used in Alaska, Canada, and many other regions. Another ancient, and much more dangerous, method is underground mining. Ancient Roman, European, and Chinese miners tunneled deep into tin, lead, copper, coal, and other mineral seams. Mine tunnels occasionally collapse, and natural gas in coal mines can explode. Water seeping into mine shafts also dissolves toxic minerals. Contaminated water seeps into groundwater; it is also pumped to the surface, where it enters streams and lakes.

In underground coal mines, another major environmental risk is fires. Hundreds of coal mines smolder in the United States, China, Russia, India, South Africa, and Europe. The inaccessibility and size of these fires make many impossible to extinguish or control. One mine fire in Centralia, Pennsylvania, has been burning since 1962; control efforts have cost at least $40 million, but the fire continues to expand. China, which depends on coal for much of its heating and electricity, has hundreds of smoldering mine fires; one has been burning for 400 years. According to a recent study from the International Institute for Aerospace Survey in the Netherlands, these fires consume up to 200 million tons of coal every year and emit as much carbon dioxide as all the cars in the United States. Toxic fumes containing mercury, arsenic, selenium, radon, and other hazardous emissions are also released from these fires.

Open-pit mines are used to extract massive beds of metal ores and other minerals. The size of modern open pits can be hard to comprehend. The Bingham Canyon mine, near Salt Lake City, Utah, is 800 m (2,640 ft) deep and nearly 4 km (2.5 mi) wide at the top. More than 5 billion tons of copper ore and waste material have been removed from the hole since 1906. A chief environmental challenge of open-pit mining is that groundwater accumulates in the pit. In metal mines, a toxic soup results. No one yet knows how to detoxify these lakes, which endanger wildlife and nearby watersheds. (See related story "Death in a Mine Pit" at www.mhhe.com/cunningham5e.)

Half the coal used in the United States comes from strip mines. Since coal is often found in expansive, horizontal beds, the entire land surface can be stripped away to cheaply and quickly expose the coal. The overburden, or surface material, is placed back into the mine, but usually in long ridges called spoil banks. Spoil banks are very susceptible to erosion and chemical weathering. Since the spoil banks have no topsoil (the complex organic mixture that supports vegetation—see chapter 7), revegetation occurs very slowly.

The 1977 federal Surface Mining Control and Reclamation Act (SMCRA) requires better restoration of strip-mined lands, especially where mines replaced prime farmland. Since then, the record of strip-mine reclamation has improved substantially. Complete mine restoration is expensive, often more than $10,000 per hectare. Restoration is also difficult because the developing soil is usually acidic and compacted by the heavy machinery used to reshape the land surface.

Bitter controversy has grown recently over mountaintop removal, a coal mining method practiced mainly in Appalachia. Long, sinuous ridge-tops are removed by giant, 20-story-tall shovels to expose horizontal beds of coal (fig. 11.12). Up to 215 m (700 ft) of ridge-top is pulverized and dumped into adjacent river valleys. The debris can be laden with selenium, arsenic, coal, and other toxic substances. At least 900 km (560 mi) of streams have been buried in West Virginia alone. Environmental lawyers have sued to stop the destruction of streams, arguing that it violates the Clean Water Act (chapter 10). In response, the Bush administration issued a "clarification" of the Clean Water Act, rewriting the law to allow stream filling by any sort of mining or industrial debris, rather than forbidding any stream destruction. Environmentalists charge that the new wording subverts the Clean Water Act, giving a green light to all sorts of stream destruction. West Virginians are deeply divided about mountaintop removal because they live in affected stream valleys but also depend on a coal economy.

The Mineral Policy Center in Washington, D.C., estimates that 19,000 km (12,000 mi) of rivers and streams in the United States are contaminated by mine drainage. The EPA estimates that cleaning up impaired streams, along with 550,000 abandoned mines in the United States, may cost $70 billion. Worldwide, mine closing and rehabilitation costs are estimated in the trillions of dollars. Because of the volatile prices of metals and coal, many mining companies have gone bankrupt before restoring mine sites, leaving the public responsible for cleanup. In 2002 more than 500 leading mine executives and their critics convened at the Global Mining Initiative, a meeting in Toronto aimed at improving the sustainability of mining. Executives acknowledged that in the future they will increasingly be held liable for environmental damages, and they said they were seeking ways to improve the industry's social and environmental record. Jay Hair, secretary general of the International Council on Mining and Metals, stated that "environmental protection and social responsibility are important" and that mining companies were interested in participating in sustainable development. Mine executives also recognized that, increasingly, big cleanup bills will cut into company values and stock prices. Finding creative ways to keep mines cleaner from the start will make good economic sense, even if it's not easy to do.

Processing contaminates air, water, and soil

Metals are extracted from ores by heating or by using chemical solvents. Both processes release large quantities of toxic materials that can be even more environmentally hazardous than mining. **Smelting**—roasting ore to release metals—is a major source of air pollution. One of the most notorious examples of ecological devastation from smelting is a wasteland near Ducktown, Tennessee. In the mid-1800s, mining companies began excavating the rich copper deposits in the area. To extract copper from the ore, they built huge, open-air wood fires, using timber from the surrounding forest. Dense clouds of sulfur dioxide released from sulfide ores poisoned the vegetation and acidified the soil over a 50 mi^2 (13,000 ha) area. Rains washed the soil off the denuded land, creating a barren moonscape.

Sulfur emissions from Ducktown smelters were reduced in 1907 after Georgia sued Tennessee over air pollution. In the 1930s the Tennessee Valley Authority (TVA) began treating the soil and replanting trees to cut down on erosion. Recently, upwards of $250,000 per year has been spent on this effort. While the trees and other plants are still spindly and feeble, more than two-thirds of the area is considered "adequately" covered with vegetation. Similarly, smelting of copper-nickel ore in Sudbury, Ontario, a century ago caused widespread ecological destruction that is slowly being repaired following pollution-control measures (see fig. 9.24).

Chemical extraction is used to dissolve or mobilize pulverized ore, but it uses and pollutes a great deal of water. A widely used method is **heap-leach extraction**, which involves piling crushed ore in huge heaps and spraying it with a dilute alkaline-cyanide solution (fig. 11.13). The solution percolates through the pile and dissolves gold. The gold-containing solution is then pumped to a processing plant that removes the gold by electrolysis. A thick clay pad and plastic liner beneath the ore heap is supposed to keep

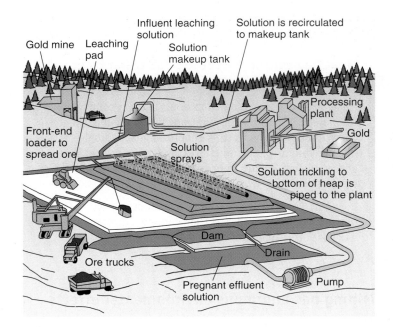

Figure 11.13 In a heap-leach operation, huge piles of low-grade ore are heaped on an impervious pad and sprayed continuously with a cyanide solution. As the leaching solution trickles through the crushed ore, it extracts gold and other precious metals. The "pregnant" effluent solution is then pumped to a processing plant, where metals are extracted and purified. This technique is highly profitable but carries large environmental risks.
Source: Data from George Laycock, *Audubon Magazine*, vol. 91(7), July 1989.

the poisonous cyanide solution from contaminating surface or groundwater, but leaks are common.

Once all the gold is recovered, mine operators may simply walk away from the operation, leaving vast amounts of toxic effluent in open ponds behind earthen dams. A case in point is the Summitville Mine near Alamosa, Colorado. After extracting $98 million in gold, the absentee owners declared bankruptcy in 1992, abandoning millions of tons of mine waste and huge, leaking ponds of cyanide. The Environmental Protection Agency may spend more than $100 million trying to clean up the mess and keep the cyanide pool from spilling into the Alamosa River.

11.5 Conserving Geologic Resources

Conservation offers great potential for extending our supplies of economic minerals and reducing the effects of mining and processing. The advantages of conservation are significant: less waste to dispose of, less land lost to mining, and less consumption of money, energy, and water resources.

Recycling saves energy as well as materials

Some waste products already are being exploited, especially for scarce or valuable metals. Aluminum, for instance, must be extracted from bauxite by electrolysis, an expensive, energy-

Table 11.3	Energy Requirements in Producing Various Materials from Ore and Raw Source Materials	
	Energy Requirement (MJ/kg)[1]	
Product	New	From Scrap
Glass	25	25
Steel	50	26
Plastics	162	n.a.[2]
Aluminum	250	8
Titanium	400	n.a.[2]
Copper	60	7
Paper	24	15

Source: Data from E. T. Hayes, *Implications of Materials Processing*, 1997.
[1]Megajoules per kilogram.
[2]Not available.

intensive process. Recycling waste aluminum, such as beverage cans, on the other hand, consumes one-twentieth of the energy of extracting new aluminum. Today, nearly two-thirds of all aluminum beverage cans in the United States are recycled, up from only 15 percent 20 years ago. The high value of aluminum scrap ($650 a ton versus $60 for steel, $200 for plastic, $50 for glass, and $30 for paperboard) gives consumers plenty of incentive to deliver their cans for collection. Recycling is so rapid and effective that half of all the aluminum cans now on a grocer's shelf will be made into another can within two months. Table 11.3 shows the energy cost of extracting other materials.

Platinum, the catalyst in automobile catalytic exhaust converters, is valuable enough to be regularly retrieved and recycled from used cars (fig. 11.14). Other commonly recycled metals are gold,

Figure 11.14 The richest metal source we have—our mountains of scrapped cars—offers a rich, inexpensive, and ecologically beneficial resource that can be "mined" for a number of metals.

silver, copper, lead, iron, and steel. The last four are readily available in a pure and massive form, including copper pipes, lead batteries, and steel and iron auto parts. Gold and silver are valuable enough to warrant recovery, even through more difficult means.

Despite its potential, however, recycling rates remain modest (fig. 11.15). Can you think of some explanations for these rates?

While total U.S. steel production has fallen in recent decades—largely because of inexpensive supplies from new and efficient Japanese steel mills—a new type of mill subsisting entirely on a readily available supply of scrap/waste steel and iron is a growing industry. **Minimills**, which remelt and reshape scrap iron and steel, are smaller and cheaper to operate than traditional integrated mills that perform every process from preparing raw ore to finishing iron and steel products. Minimills produce steel at between $225 and $480 per metric ton, while steel from integrated mills costs $1,425 to $2,250 per metric ton, on average. The energy cost is also lower in minimills: 5.3 million BTU/ton of steel, compared with 16.08 million BTU/ton in integrated mill furnaces. Minimills now produce about half of all U.S. steel. Recycling is slowly increasing as raw materials become more scarce and wastes become more plentiful.

New materials can replace mined resources

Mineral and metal consumption can be reduced by new materials or new technologies developed to replace traditional uses. This is a long-standing tradition; for example, bronze replaced stone technology, and iron replaced bronze. More recently, the introduction of plastic pipe has decreased our consumption of copper, lead, and steel pipes. In the same way, the development of fiber-optic technology and satellite communication reduces the need for copper telephone wires.

Iron and steel have been the backbone of heavy industry, but we are now moving toward other materials. One of our primary uses for iron and steel has been machinery and vehicle parts. In automobile production, steel is being replaced by polymers

Figure 11.15 Rates of metal recycling remain modest for most metals even though recycling saves energy and the environment.
Data Source: U.S. Geological Survey.

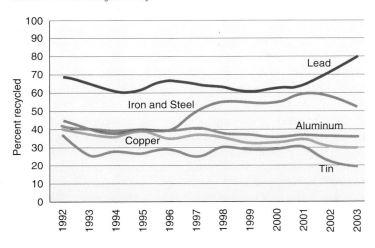

(long-chain organic molecules similar to plastics), aluminum, ceramics, and new high-technology alloys. All of these reduce vehicle weight and cost, while increasing fuel efficiency. Some of the newer alloys that combine steel with titanium, vanadium, or other metals wear much better than traditional steel. Ceramic engine parts provide heat insulation around pistons, bearings, and cylinders, keeping the rest of the engine cool and operating efficiently. Plastics and glass fiber–reinforced polymers are used in body parts and some engine components.

Electronics and communications (telephone) technology, once major consumers of copper and aluminum, now use ultra-high-purity glass cables to transmit pulses of light, instead of metal wires carrying electron pulses. Once again, this technology has been developed for its greater efficiency and lower cost, but it also affects consumption of our most basic metals.

11.6 Geologic Hazards

Earthquakes, volcanoes, floods, and landslides are normal earth processes, events that have made our earth what it is today. However, when they affect human populations, their consequences can be among the worst and most feared disasters that befall us.

Earthquakes are frequent and deadly hazards

The 2004 earthquake and tsunami in Banda Aceh, Indonesia, stunned the world, killed over 230,000 people, and caused damage as far away as Africa. Less than a year later, an earthquake in Pakistan killed 80,000 people. **Earthquakes** are sudden movements in the earth's crust that occur along faults (planes of weakness), where one rock mass slides past another one. When movement along faults occurs gradually and relatively smoothly, it is called creep or seismic slip and may be undetectable to the casual observer. When friction prevents rocks from slipping easily, stress builds up until it is finally released with a sudden jerk. The point on a fault at which the first movement occurs during an earthquake is called the epicenter.

Earthquakes have always seemed mysterious, sudden, and violent, coming without warning and leaving ruined cities and dislocated landscapes in their wake. Cities such as Kobe, Japan, or Mexico City, parts of which are built on soft landfill or poorly consolidated soil, usually suffer the greatest damage from earthquakes (fig. 11.16). Water-saturated soil can liquify when shaken. Buildings sometimes sink out of sight or fall down like a row of dominoes under these conditions (see related stories on earthquakes at www.mhhe.com/cunningham5e).

Earthquakes frequently occur along the edges of tectonic plates, especially where one plate is being subducted, or pushed down, beneath another. Earthquakes also occur in the centers of continents, however. In fact, one of the largest earthquakes ever recorded in North America was one of an estimated magnitude 8.8 that struck the area around New Madrid, Missouri, in 1812. Fortunately, few people lived there at the time, and the damage was minimal.

Figure 11.16 Earthquakes are most devastating where building methods cannot withstand shaking. The 2008 earthquake in China killed more than 60,000 people and left 5 million homeless.

Modern contractors in earthquake zones attempt to prevent damage and casualties by constructing buildings that can withstand tremors. The primary methods used are heavily reinforced structures, strategically placed weak spots in the building that can absorb vibration from the rest of the building, and pads or floats beneath the building on which it can shift harmlessly with ground motion.

One of the most notorious effects of earthquakes is the **tsunami**, Japanese for "harbor wave." These giant sea swells, such as the one originating in Banda Aceh, Indonesia in 2005, can move at 1,000 kph (600 mph), or faster, away from the center of an earthquake. When these swells approach the shore, they can create breakers as high as 65 m (nearly 200 ft). Tsunamis also can be caused by underwater volcanic explosions or massive seafloor slumping.

Volcanoes eject deadly gases and ash

Volcanoes and undersea magma vents are the sources of most of the earth's crust. Over hundreds of millions of years, gaseous emissions from these sources formed the earth's earliest oceans and atmosphere. Many of the world's fertile soils are weathered

Figure 11.17 Lava and ashflow spill down the slopes of Mayon volcano in the Philippines in this September 23, 1984, image. Because more than 73,000 people evacuated danger zones, this eruption caused no casualties.

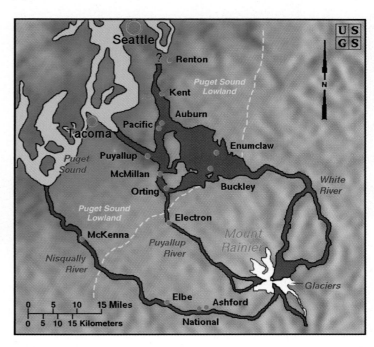

Figure 11.18 Volcanic activity at Mount Rainier has produced at least 12 large mudflows in the past 6,000 years (*brown areas*). Future flows could threaten large populations now in the area.
Source: Data from T. W. Sisson, USGS Open File Report 95-642.

volcanic materials. Volcanoes have also been an ever present threat to human populations (fig. 11.17). One of the most famous historic volcanic eruptions was that of Mount Vesuvius in southern Italy, which buried the cities of Herculaneum and Pompeii in A.D. 79. The mountain had been showing signs of activity before it erupted, but many citizens chose to stay and take a chance on survival. On August 24, the mountain buried the two towns in ash. Thousands were killed by the dense, hot, toxic gases that accompanied the ash flowing down from the volcano's mouth. It continues to erupt from time to time.

Nuees ardentes (French for "glowing clouds") are deadly, denser-than-air mixtures of hot gases and ash like those that inundated Pompeii and Herculaneum. Temperatures in these clouds may exceed 1,000°C, and they move at more than 100 kph (60 mph). *Nuees ardentes* destroyed the town of St. Pierre on the Caribbean island of Martinique on May 8, 1902. Mount Pelee released a cloud of *nuees ardentes* that rolled down through the town, killing between 25,000 and 40,000 people within a few minutes. All of the town's residents died except for a single prisoner being held in the town dungeon.

Disastrous mudslides are also associated with volcanoes. The 1985 eruption of Nevado del Ruíz, 130 km (85 mi) northwest of Bogotá, Colombia, caused mudslides that buried most of the town of Armero and devastated the town of Chinchina. An estimated 25,000 people were killed. Heavy mudslides also accompanied the eruption of Mount St. Helens in Washington in 1980. Sediments mixed with melted snow destroyed roads, bridges, and property, but because of sufficient advance warning, there were few casualties. Geologists worry that similar mudflows from an eruption at Mount Rainier would threaten much larger populations (fig. 11.18).

Volcanic eruptions often release large volumes of ash and dust into the air. Mount St. Helens expelled 3 km³ of dust and ash, causing ash fall across much of North America. This was only a

minor eruption. An eruption in a bigger class of volcanoes was that of Tambora in Indonesia in 1815, which expelled 175 km³ of dust and ash, more than 58 times that of Mount St. Helens. These dust clouds circled the globe and reduced sunlight and air temperatures enough so that 1815 was known as the year without a summer.

It is not just a volcano's dust that blocks sunlight. Sulfur emissions from volcanic eruptions combine with rain and atmospheric moisture to produce sulfuric acid (H_2SO_4). Droplets of H_2SO_4 interfere with solar radiation and can significantly cool the world climate. In 1991 Mount Pinatubo in the Philippines emitted 20 million tons of sulfur dioxide aerosols, which remained in the stratosphere for two years. This thin haze cooled the entire earth by 1°C for about two years. It also caused a 10 to 15 percent reduction in stratospheric ozone, allowing increased ultraviolet light to reach the earth's surface.

Floods are part of a river's land-shaping processes

Like earthquakes and volcanoes, **floods** are normal events that cause damage when people get in the way. As rivers carve and shape the landscape, they build broad **floodplains**, level expanses that are periodically inundated. Large rivers, such as the Mississippi, can have huge floodplains. Many cities have been built on these flat, fertile plains, which are so convenient to the river. Floodplains that flood very irregularly may appear safe for many years, but eventually, most floodplains do flood. The severity of floods can be described by the height of water

Figure 11.19 The overflowing Cedar River filled most of downtown Cedar Rapids, Iowa during the 2008 floods. More than 40,000 people were forced from their homes in Iowa, and economic losses were estimated to be in the billions of dollars. Are human-caused environmental changes partly to blame for the severity of disasters such as this?

above the normal stream banks or by how frequently a similar event normally occurs—on average—for a given area. Note that these are statistical averages over long time periods. A "10-year flood" would be expected to occur once in every ten years; a "100-year flood" would be expected to occur once every century. But two 100-year floods can occur in successive years or even in the same year.

Among direct natural disasters, floods take the largest number of human lives. A flood on the Yangtze River in China in 1931 killed 3.7 million people, making it the most deadly natural disaster in recorded history. In another flood on China's Yellow River in 1959, about two million people died, mostly due to famine and disease. Torrential rains in June 2008 caused massive flooding across the American Midwest. Many cities in Iowa, Wisconsin, Illinois, and Indiana experienced their highest water levels in more than a century. In Cedar Rapids, Iowa, for example, almost the entire downtown was inundated by the overflowing Cedar River (fig. 11.19).

The 2008 floods in the Mississippi River basin caused billions of dollars in property damage. The biggest economic loss from floods is usually not the buildings and property they wash away, but rather the contamination they cause. Virtually everything floodwaters touch in a house—carpets, furniture, drapes, electronics, even drywall and insulation—must be removed and discarded because of the sewage, toxic chemicals, farm waste, dead animals, and smelly mud carried by the water. In some cases, sediment left behind by floods has completely buried whole towns.

In 2008, more than 2 million ha (5 million acres) of rich farmland was inundated in the heart of America's corn (maize) and soybean producing area. The USDA estimates that this will result in loss of 200 million bushels (about 5 million metric tons) of corn and 40 million bushels of soybeans at a time when these commodities are already at record high prices. These losses are likely to exacerbate worldwide food shortages.

Are these floods related to global climate change? Many climate scientists predict that global warming will cause more extreme weather events, including both severe droughts in some places and more intense rainfall in others. In addition, many other human activities increase both the severity and frequency of floods. Covering the land with hardened surfaces, such as roads, parking lots, and building roofs, reduces water infiltration into the soil and speeds the rate of runoff into streams and lakes. Clearing forests for agriculture and destroying natural wetlands also increases both the volume and rate of water discharge after a storm. In Iowa, for example, at least 99 percent of the natural wetlands that existed before settlement have been filled for farmland and urban development.

Even more than development, though, flood-control structures have separated floodplains from rivers. Levees and flood walls are built to contain water within riverbanks, and river channels are dredged and deepened to allow water to recede faster. Every flood-control structure simply transfers the problem downstream, however. The water has to go somewhere. If it doesn't soak into the ground upstream, it will simply exacerbate floods somewhere downstream—leading to more levee development, and then to more flooding farther downstream, and so on.

Flood control

More than $25 billion of river-control systems have been built on the Mississippi and its tributaries. These systems have protected many communities over the past century. In the major floods of 1993, however, this elaborate system helped turn a large flood into a major disaster. Deprived of the ability to spill out over floodplains, the river is pushed downstream to create faster currents and deeper floods until eventually a levee gives way somewhere. Hydrologists calculate that the floods of 1993 were about 3 m (10 ft) higher than they would have been, given the same rainfall in 1900 before the flood-control structures were in place.

Under current rules, the government is obligated to finance most levees and flood-control structures. Many people think that it would be much better to spend this money to restore wetlands, replace groundcover on water courses, build check dams on small streams, move buildings off the floodplain, and undertake other nonstructural ways of reducing flood danger. According to this view, floodplains should be used for wildlife habitat, parks, recreation areas, and other uses not susceptible to flood damage.

The National Flood Insurance Program administered by the Federal Emergency Management Agency (FEMA) was intended to aid people who cannot buy insurance at reasonable rates, but its effects have been to encourage building on the floodplains by making people feel that, whatever happens, the government will take care of them. Many people would like to relocate homes and businesses out of harm's way after the recent floods or to improve them so they will be less susceptible to flooding, but owners of damaged property can collect only if they rebuild in the same place and in the same way as before. This perpetuates problems rather than solves them.

Mass wasting includes slides and slumps

Gravity constantly pulls downward on every material everywhere on earth. Hillsides, beaches, even relatively flat farm fields

Figure 11.20 Mass wasting includes the collapse of unstable hill slopes, such as this one in Laguna Beach, California. Land clearing and building can accelerate this natural process.

can lose material to erosion. Often water helps mobilize loose material, and catastrophic slumping, beach erosion, and gully development can occur in a storm. A general term for downhill slides of earth is "mass wasting."

Landslides are sudden collapses of hillsides. In the United States alone, landslides and related mass wasting cause over $1 billion in property damage every year. When unconsolidated sediments on a hillside are saturated by a storm or exposed by logging, road building, or house construction, slopes are especially susceptible to sudden landslides (fig. 11.20).

Often people are unaware of the risks they face by locating on or under unstable hillsides. Sometimes they simply ignore clear and obvious danger. In southern California, where land prices are high, people often build expensive houses on steep hills and in narrow canyons. Most of the time, this dry environment appears quite stable, but in fact, steep hillsides slip and slump frequently. Especially when soil is exposed or rainfall is heavy, mudslides and debris flows can destroy whole neighborhoods. In developing countries, mudslides have buried entire villages in minutes. **Soil creep**, on the other hand, moves material inexorably downhill at an imperceptibly slow pace.

Erosion destroys fields and undermines buildings

Gullying is the development of deep trenches on relatively flat ground. Especially on farm fields, which have a great deal of loose soil unprotected by plant roots, rainwater running across the surface can dig deep gullies. Sometimes land becomes useless for farming because gullying is so severe and because erosion has removed the fertile topsoil. Agricultural soil erosion has been described as an invisible crisis. Erosion has reduced the fertility of millions of acres of prime farmland in the United States alone.

Beach erosion occurs on all sandy shorelines because the motion of the waves is constantly redistributing sand and other sediments. One of the world's longest and most spectacular sand beaches runs down the Atlantic coast of North America from New England to Florida and around the Gulf of Mexico. Much of this beach lies on some 350 long, thin **barrier islands** that stand between the mainland and the open sea. Behind these barrier islands lie shallow bays or brackish lagoons fringed by marshes or swamps.

Early inhabitants recognized that exposed, sandy shores were hazardous places to live, and they settled on the bay side of barrier islands or as far upstream on coastal rivers as was practical. Modern residents, however, place a high value on living where they have an ocean view and ready access to the beach. And they assume that modern technology makes them immune to natural forces. The most valuable and prestigious property is closest to the shore. Over the past 50 years, more than 1 million acres of estuaries and coastal marshes have been filled to make way for housing or recreational developments.

Construction directly on beaches and barrier islands can cause irreparable damage to the whole ecosystem. Normally fragile vegetative cover holds the shifting sand in place. Damaging this vegetation with construction, building roads, and breaching dunes with roads can destabilize barrier islands. Storms then wash away beaches or even whole islands. A single severe storm in 1962 caused $300 million in property damage along the east coast of the United States and left hundreds of beach homes tottering into the sea (fig. 11.21). FEMA estimates that 25 percent of all coastal homes in the United States will have the ground washed out from under them by 2060.

Cities and individual property owners often spend millions of dollars to protect beaches from erosion. Sand is dredged from the ocean floor or hauled in by the truckload, only to wash away again in the next storm. Building artificial barriers, such as groins or jetties, can trap migrating sand and build beaches in one area, but they often starve downstream beaches and make erosion there even worse (fig. 11.22).

As is the case for inland floodplains, government policies often encourage people to build where they probably shouldn't. Subsidies for road building and bridges, support for water and sewer projects, tax exemptions for second homes, flood insurance, and disaster relief are all good for the real estate and construction businesses but invite people to build in risky places. Flood insurance typically costs $300 per year for $80,000 of coverage. In 1998 FEMA paid out $40 billion in claims, 80 percent of which

Figure 11.21 Winter storms have eroded the beach and undermined the foundations of homes on this barrier island. Breaking through protective dunes to build such houses damages sensitive plant communities and exposes the whole island to storm sand erosion. Coastal zone management attempts to limit development on fragile sites.

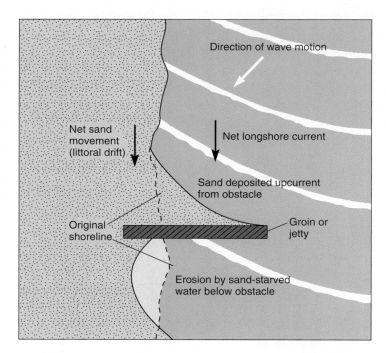

Figure 11.22 Groins are structures built perpendicular to a beach to slow erosion and trap sand. They build up beaches toward prevailing wind and wave direction but starve beaches and increase erosion downstream. Dashed line indicates original shoreline.

were flood-related. Settlement usually requires that structures be rebuilt exactly where and as they were before. There is no restriction on how many claims can be made, and policies are rarely canceled, no matter what the risk. Some beach houses have been rebuilt—at public expense—three times in a decade. The General Accounting Office found that 2 percent of federal flood policies were responsible for 30 percent of the claims.

The Coastal Barrier Resources Act of 1982 prohibited federal support, including flood insurance, for development on sensitive islands and beaches. In 1992, however, the U.S. Supreme Court ruled that ordinances forbidding floodplain development amount to an unconstitutional "taking," or confiscation, of private property.

Conclusion

Coal-bed methane is one of many important resources produced by geologic forces. The study of geology has allowed us to predict where these resources can be found. Seismic exploration is one of many advanced techniques used for exploration. Oil and coal are also earth energy resources on which we depend. The extraction of earth resources often carries severe environmental costs, however, including water contamination, habitat destruction, and air pollution.

The earth's surface is shaped by shifting pieces of crust, which split and collide slowly but continuously. Earthquakes, volcanoes, and mountains occur on plate margins. Rocks are composed of minerals, and rocks can be described according to their origins in

molten material (igneous rocks), eroded or deposited sediments (sedimentary rocks), or materials transformed by heat and pressure deep in the earth (metamorphic rocks).

Earth resources, including oil, gas, and coal, are the foundation of our economy. Materials like sand, gravel, gypsum, and salt are inexpensive, but they are essential and valuable resources for construction. Metals are expensive to extract, but they are extremely valuable because they are ductile (bendable) yet strong and because they carry electricity (e.g., copper). Gemstones are notable mainly because of their aesthetic value. Diamonds and other fine gems, which are easy to transport, have been used to fund wars, including activities of al-Qaida in 2001.

Earth resources can be recycled, saving money, energy, and environmental quality. Recycling aluminum consumes one-twentieth of the energy to extract new aluminum, for example, and recycling copper takes about one-eighth as much energy. We can also save energy and resources by replacing traditional materials with newer, more efficient ones. Fiber-optic communication lines have replaced much of our copper wiring, adding speed and efficiency while saving copper use.

Geologic hazards are an important cause of property loss, and of mortality in some areas. Volcanoes and earthquakes are spectacular but irregular disasters; floods and erosion occur frequently in many areas, especially on floodplains, beaches, and other shorelines. These events are natural but are responsible for enormous damage to cities, houses, farmland, and other property.

Practice Quiz

1. How does tectonic plate movement create ocean basins, mid-ocean ridges, and volcanoes?

2. What is the "ring of fire"?

3. Describe the processes and components of the rock cycle (fig. 11.7).

4. What is the difference between metals and nonmetal mineral resources?

5. What is a *mineral* and a *rock*? Why are pure metals not minerals?

6. Which countries are the single greatest producers of our major metals? (See fig. 11.10.)

7. Describe some of the mining, processing, and drilling methods that can degrade water or air quality.

8. Compare the different mining methods of underground, open-pit, strip, and placer mining, as well as mountaintop removal.

9. What resources, aside from minerals themselves, can be saved by recycling?

10. Give an example of resource substitution.

11. Describe the most deadly risks of volcanoes.

12. What is *mass wasting*? Give three examples and explain why they are a problem.

13. Why is building on barrier islands risky?

14. What is a *floodplain*? Why is building on floodplains controversial?

15. Describe the processes of chemical weathering and mechanical weathering. How do these processes contribute to the recycling of rocks?

16. The Mesozoic period begins and ends with the appearance and disappearance of dinosaurs. What fossils mark the other geologic eras? (See fig. 11.6.)

17. Describe how seismic exploration works in finding oil and gas deposits. What are its costs and benefits?

Critical Thinking and Discussion Questions

Apply the principles you have learned in this chapter to discuss these questions with other students.

1. Understanding and solving the environmental problems of mining are basically geologic problems, but geologists need information from a variety of environmental and scientific fields. What are some of the other sciences (or disciplines) that could contribute to solving mine contamination problems?

2. Geologists are responsible for identifying and mapping mineral resources. But mineral resources are buried below the soil and covered with vegetation. How do you suppose geologists in the field find clues about the distribution of rock types?

3. If you had an igneous rock with very fine crystals and one with very large crystals, which would you expect to have formed deep in the ground, and why?

4. Heat and pressure tend to help concentrate metal ores. Explain why such ores might often occur in mountains such as the Andes in South America.

5. The idea of tectonic plates shifting across the earth's surface is central to explanations of geologic processes. Why is this idea still called the "theory" of plate tectonic movement?

6. Geologic evidence from fossils and sediments provided important evidence for past climate change. What sorts of evidence in the rocks and landscape around you suggest that the place where you live once looked much different than it does today?

Data Analysis: Exploring Recent Earthquakes

Go to the USGS Earthquake Center (http://earthquake.usgs.gov/eqcenter/). This is a rich source of global earthquake information.

1. Look at the USA map of earthquakes. Where are the greatest concentrations of earthquakes? Can you explain why those locations have lots of earthquakes?

2. Find the World Lists of earthquakes, and click on Magnitude 5+. Where was the largest recent earthquake? What was its magnitude? Its depth? Are listed quakes clustered in a few regions? Where are those regions?

3. Still looking at the world list of large earthquakes, make a graph showing the frequency of different magnitudes for listed earthquakes: On a piece of paper, draw a line with earthquake magnitude classes, then mark an X for each earthquake over the appropriate size class, like this:

```
X
X      X           X
5.0    5.5    6.0   6.5    7.0    7.5  8.0+
```

This kind of graph is called a frequency distribution graph, or a histogram. Would you describe earthquake sizes as mostly small, evenly distributed, mostly large, or mostly in the middle of the range?

Understanding Volcanoes

USGS Hawaii Volcano Observatory is one of the most important places for studying and understanding volcanoes. Visit the observatory's website at http://hvo.wr.usgs.gov/.

1. Which of the volcanoes discussed at this website is the largest in the world? Which is the most active?

2. Why are earthquakes discussed on this volcano website?

3. What are the major hazards to people from these volcanoes?

For a good collection of volcano images, go to the USGS Cascades Volcano Observatory at http://vulcan.wr.usgs.gov/Photo/framework.html.

Evaluating Erosion on Farmland

The Natural Resources Conservation Service (NRCS), part of the U.S. Department of Agriculture, is the agency that monitors agricultural resources and conditions. Among its data-gathering efforts, the NRCS produces a Natural Resources Inventory (NRI), with maps of farmland conditions across the mainland United States. Visit the NRCS/NRI website at www.nrcs.usda.gov/technical/land/erosion.html.

Find the list of erosion maps. Look at several of the maps. First identify the meaning of the colors. Then identify the major concentration of high and low erosion rates.

1. Where is wind erosion worst? Where is water erosion worst?

2. Where are the most acres of highly erodible cropland? What kind of physical features might occur there?

For Additional Help in Studying This Chapter, please visit our website at www.mhhe.com/cunningham5e. You will find practice quizzes, key terms, answers to end of chapter questions, additional case studies, an extensive reading list, and Google Earth™ mapping quizzes.

Windmills supply all the electricity used on Denmark's Ærø Island, while solar and biomass energy provide space heating and vehicle fuel. Altogether, 100 percent of the island's energy comes from renewable sources.

12 Energy

Learning Outcomes

After studying this chapter, you should be able to answer the following questions:

- What are our dominant sources of energy?
- What is peak oil production? Why is it hard to evaluate future oil production?
- How important is coal in domestic energy production?
- What are the environmental effects of coal burning? Is clean coal possible?
- How do nuclear reactors work? What are some of their advantages and disadvantages?
- What are our main renewable forms of energy?
- Could solar, wind, hydropower, and other renewables eliminate the need for fossil fuels?
- What are photovoltaic cells, and how do they work?
- What are biofuels? What are arguments for and against their use?

We are not only responsible for what we do, but also for what we do not do.

–Moliere

Renewable Energy Islands

Denmark has substantial oil and gas supplies under the North Sea, but the Danes have chosen to wean themselves away from dependence on fossil fuels. Currently the world leader in renewable energy, Denmark now gets 20 percent of its power from solar, wind, and biomass. Some parts of this small, progressive country have moved even further toward sustainability. One of the most inspiring examples of these efforts are the small islands of Samsø and Ærø, which now get 100 percent of their energy from renewable sources.

Samsø and Ærø lie between the larger island of Zealand (home to Copenhagen) and the Jutland Peninsula. The islands are mostly agricultural. Together, they have an area of about 200 km² (77 mi²) and a population of about 12,000 people. In 1997, Samsø and Ærø were chosen in a national competition to be renewable energy demonstration projects. The first step in energy independence is conservation. As you'll learn in this chapter, Denmark uses roughly half as much energy per person as the United States, although by most measures the Danes have a higher standard of living than most Americans. Danish energy conservation is achieved with high-efficiency appliances, superior building insulation,

high-mileage vehicles, and other energy-saving measures. Most homes are clustered in small villages, both to save agricultural land and to facilitate district heating. Living closely together also makes having a private automobile less necessary.

Some 30 large wind generators provide 100 percent of Samsø and Ærø's electricity. Two-thirds of these windmills are located offshore, and are publicly owned. The 11 onshore wind turbines are mostly privately owned, but a share of the profits is used to finance other community energy projects. Space heating accounts for about one-third of the energy consumption on the islands. District heating systems provide most of this energy. Several large solar collector arrays supply about half the hot water for space heating and household use (fig. 12.1). Biomass-based (straw, wood chips, manure) systems supply the remainder of the island's heating needs. Some of this biomass comes from energy crops (fast-growing elephant grass and hybrid poplars, for example, are grown on marginal farmland), while the rest comes from agricultural waste. Biodiesel (primarily from rapeseed oil) fuels farm tractors and ferries, while most passenger vehicles are electric.

Geothermal pumps supplement the solar water heaters, and in one village a recently closed landfill produces methane that is used to run a small electric generator. Nuclear power is considered an unacceptable option in Denmark and doesn't feature in current energy plans. Samsø and Ærø have won numerous prizes and awards for their pioneering conversion to renewable energy. Over the past 20 years, as a result of other projects like those on Samsø and Ærø, both Denmark's fossil fuel consumption and their greenhouse gas emissions have remained constant. All of us could learn from their example.

Many other countries, both in Europe and elsewhere in the world, are turning to renewable energy to reduce their dependence on environmentally damaging and politically unstable fossil fuels. In this chapter, we'll look at world energy resources as well as how we currently obtain the energy we use, and what our options are for finding environmentally and socially sustainable ways to meet our energy needs.

Figure 12.1 A 19,000 m² array of solar water heaters provides space heating for the town of Marstal on Ærø Island.

12.1 Energy Resources and Uses

Energy sources drive our economy today, and many of our most important questions in environmental science have to do with energy resources—from air pollution, climate change, and mining impacts to technological innovations in alternative energy sources. On the islands of Ærø and Samsø, the Danish government chose to address these issues by investing in renewable energy sources. Throughout Denmark, the state has subsidized wind power research, so that now Danes are world leaders in this fast-growing industry.

Most of us think about energy choices chiefly in terms of how much different sources cost. But energy policies can foster innovation or maintain conventional strategies. Landmark rules in the 2007 United States energy bill, for example, required a quadrupling of ethanol production from biomass by 2020, in just 13 years. In

combination with agricultural subsidies, ethanol receives supports of about $7 billion per year, or about $2 per gallon. The same energy bill established the first new standards for vehicle efficiency in over 30 years. This kind of rule may make it possible for you to buy a car that gets 60–70 miles per gallon of gas. That kind of efficiency hadn't been available in the United States (although it has been available elsewhere) since the 1970s, with the exception of hybrid cars. Thus, national energy policies can influence how much you pay at the pump, as well as national dependence on imported energy. U.S. energy policies give the lion's share of funding to oil and gas development ($13 billion in tax incentives in 2007), nuclear energy ($25 billion in underwriting), coal, and gas. Additional, smaller tax breaks and federal research funding have subsidized wind, solar, and other alternative energy forms. High energy prices also drive innovation, which can lead to newer, cheaper, more sustainable sources of

Table 12.1 Energy Units

1 joule (J) = work needed to accelerate 1 kg 1 m/sec² for 1 m (or 1 amp/sec flowing through 1 ohm resistance)

1 watt (W) = 1 J per second

1 kilowatt hour (kWh) = 1,000 W exerted for 1 hour (or 3.6 million J)

1 megawatt (MW) = 1 million (10^6) W

1 gigajoule (GJ) = 1 billion (10^9) J

1 standard barrel (bbl) of oil = 42 gal (160 liters)

Table 12.2 Energy Uses

Uses	kWh/year*
Computer	100
Television	125
100 W light bulb	250
15 W fluorescent bulb	40
Dehumidifier	400
Dishwasher	600
Electric stove/oven	650
Clothes dryer	900
Refrigerator	1100

*Averages shown; actual rates vary greatly.

Source: U.S. Department of Energy.

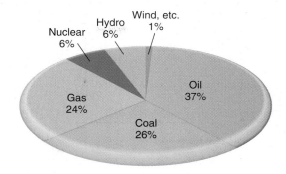

Figure 12.2 Worldwide commercial energy consumption. This does not include energy collected for personal use or traded in informal markets.
Source: Data from British Petroleum, 2006.

energy. In this chapter we'll examine our different energy options, how we use them, and how they influence environmental systems.

How do we measure energy?

To understand the magnitude of energy use, it is helpful to know the units used to measure it. **Work** is the application of force over distance, and we measure work in **joules** (table 12.1). **Energy** is the capacity to do work. **Power** is the rate of energy flow or the rate of work done: for example, one **watt** (W) is one joule per second. If you use a 100-watt light bulb for 10 hours, you have used 1,000 watt-hours, or one kilowatt-hour (kWh). Most American households use about 11,000 kWh per year (table 12.2).

Fossil fuels supply most of our energy

Fossil fuels (petroleum, natural gas, and coal) now provide about 87 percent of all commercial energy in the world (fig. 12.2). Renewable sources—solar, wind, geothermal, and hydroelectricity—make up about 7 percent of our commercial power (but hydro accounts for almost all of that). Excluded from these figures are energy sources not traded commercially. Wood biomass provides the primary energy source for at least a billion people in developing countries. This is an important energy source for poor people but can be a serious cause of forest destruction (chapter 6). Residential solar water heating and cogeneration (producing electricity from waste heat) are also important, and under-used, sources.

Nuclear power is roughly equal to hydroelectricity worldwide (about 6 percent of all commercial energy), but it makes up about 20 percent of all electric power in more-developed countries, such as Canada and the United States. We have enough nuclear fuel to produce power for a long time, and it has the benefit of not contributing to global warming (chapter 9), but as we discuss later in this chapter, safety concerns make this option unacceptable to most people.

Perhaps the most important facts about fossil fuel consumption are that we in the 20 richest countries consume nearly 80 percent of the natural gas, 65 percent of the oil, and 50 percent of the coal produced each year. Although we make up less than one-fifth of the world's population, we use more than one-half of the commercial energy supply. The United States, for instance, constitutes only about 4.5 percent of the world's population but consumes about one-quarter of all fossil fuels.

How much energy do you use every year? Most of us don't think about it much, but maintaining the lifestyle we enjoy requires an enormous energy input. On average, each person in the United States and Canada uses more than 300 gigajoules (GJ) (equivalent to about 60 barrels of oil) per year. By contrast, in some of the poorest countries of the world, such as Ethiopia, Nepal, and Bhutan, each person generally consumes less than 1 GJ per year. This means that each of us consumes, on average, almost as much energy in a single day as a person in one of these countries consumes in a year.

Clearly, energy consumption is linked to the comfort and convenience of our lives (fig. 12.3). Those of us in the richer countries enjoy many amenities not available to most people in the world. The link is not absolute, however. Several European countries, including Sweden, Denmark, and Finland, have higher standards of living than does the United States by almost any measure but use about half as much energy.

How do we use energy?

The largest share (32.6 percent) of the energy used in the United States is consumed by industry (fig. 12.4). Mining, milling, smelting, and forging of primary metals consume about one-quarter of

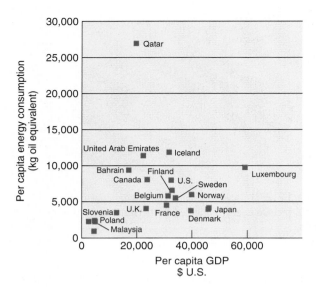

Figure 12.3 Per capita energy consumption and GDP. In general, income and standard of living increase with energy availability, but some energy-rich countries, such as Qatar, use vast amounts of energy without improving human development very much. Some countries, such as Norway, Denmark, and Finland, have a much higher human development index than the United States while consuming only about half as much energy. Because of the scale of this graph, low-income, low-energy countries aren't visible.
Source: Data from UNDP, 2004.

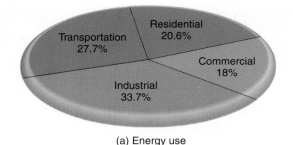

(a) Energy use

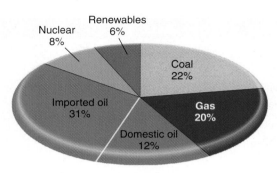

(b) U.S. energy consumption

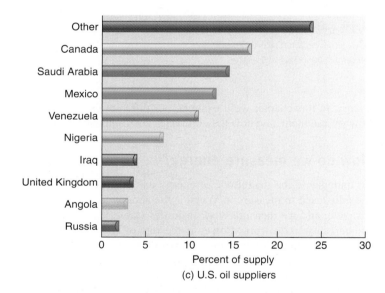

(c) U.S. oil suppliers

Figure 12.4 Fossil fuels supply 85 percent of the energy used in the United States, and oil (three-quarters of it imported) makes up half that amount. Canada is now the largest oil supplier for the United States, and industry is the largest energy-use sector. Transportation, however, uses the vast bulk of all oil.
Source: Data from U.S. Department of Energy, 2006.

that industrial energy share. The chemical industry is the second largest industrial user of fossil fuels, but only half of its use is for energy generation. The remainder is raw material for plastics, fertilizers, solvents, lubricants, and hundreds of thousands of organic chemicals in commercial use. The manufacture of cement, glass, bricks, tile, paper, and processed foods also consumes large amounts of energy.

Residential and commercial customers use roughly 37 percent of the primary energy consumed in the United States, mostly for space heating, air conditioning, lighting, and water heating.

Transportation consumes about 27 percent of all energy used in the United States each year. About 98 percent of that energy comes from petroleum products refined into gasoline and diesel fuel, and the remaining 2 percent is provided by natural gas and electricity. Almost three-quarters of all transport energy is used by motor vehicles. Nearly 3 trillion passenger miles and 600 billion ton miles of freight are carried annually by motor vehicles in the United States. About 75 percent of all freight traffic in the United States is carried by trains, barges, ships, and pipelines, but because they are very efficient, they use only 12 percent of all transportation fuel.

Producing and transporting energy also consumes and wastes energy. About half of all the energy in primary fuels is lost during conversion to more useful forms, while being shipped to the site of end use, or during use. Electricity is generally promoted as a clean, efficient source of energy because, when it is used to run a resistance heater or an electrical appliance, almost 100 percent of

its energy is converted to useful work and no pollution is given off. What happens before then, however? We often forget that huge amounts of pollution are released during mining and burning of the coal that fires power plants. Furthermore, nearly two-thirds of the energy in the coal that generated that electricity was lost in thermal conversion in the power plant. About 10 percent more is lost during transmission and stepping down to household voltages.

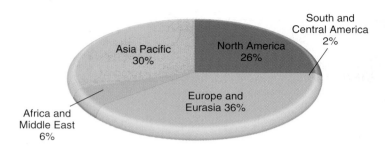

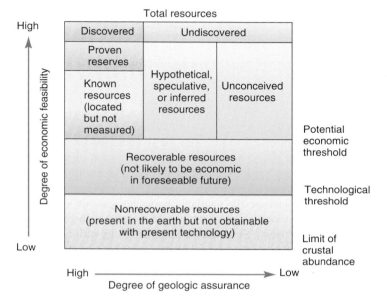

Figure 12.5 Proven-in-place coal reserves by region, 2006.
Source: British Petroleum, 2006.

Figure 12.6 Categories of natural resources according to economic and technological feasibility, as well as geologic assurance.

Figure 12.7 Some giant mining machines stand as tall as a 20-story building and can scoop up thousands of cubic meters of rock per hour. Note the pick-up trucks in the machine's shadow.

Oil supplies over 40 percent of U.S. energy demands and over 99 percent of fuel for cars and trucks, according to the U.S. Department of Energy (fig. 12.4). Coal-fired power plants supply most of our electrical energy. Impurities in coal, especially sulfur and mercury, are also our most important source of air pollution. Natural gas is a cleaner energy source, and 90 percent of new power plants in the next 20 years will probably burn natural gas.

Coal resources are vast

World coal deposits are vast, ten times greater than conventional oil and gas resources combined, and one-quarter of global coal deposits are in the United States (fig. 12.5). Coal seams can be 100 m thick and can extend across tens of thousands of square kilometers that were vast, swampy forests in prehistoric times. The total resource is estimated to be 10 trillion metric tons. If all this coal could be extracted, and if coal consumption continued at present levels, this would amount to several thousand years' supply. At present rates of consumption, these **proven-in-place reserves**—those explored and mapped but not necessarily economic at today's prices—will last about 200 years. Proven reserves are generally a small fraction of a total resource (fig. 12.6).

Do we really want to use all of the coal? Coal mining is a dirty, dangerous activity. Underground mines are notorious for cave-ins, explosions, and lung diseases, such as black-lung suffered by miners. Surface mines (called strip mines, where large machines strip off overlying sediment to expose coal seams) are cheaper and generally safer for workers than tunneling, but leave huge holes where coal has been removed and vast piles of discarded rock and soil (fig. 12.7).

An especially damaging technique employed in Appalachia is called mountaintop removal. Typically, the whole top of a mountain

Natural gas is our most efficient fuel. Only 10 percent of its energy content is lost in shipping and processing, since it moves by pipelines and usually needs very little refining. Ordinary gas-burning furnaces are about 75 percent efficient, and high-economy furnaces can be as much as 95 percent efficient. Because natural gas has more hydrogen per carbon atom than oil or coal, it produces about half as much carbon dioxide—and therefore contributes half as much to global warming—per unit of energy.

12.2 Fossil Fuels

Fossil fuels are organic (carbon-based) compounds derived from decomposed plants, algae, and other organisms buried in rock layers for hundreds of millions of years. Most of the richest deposits date to about 286 million to 360 million years ago (the Mississippian, Pennsylvanian, and Permian periods: see fig. 11.6), when the earth's climate was much warmer and wetter than it is now.

ridge is scraped off to access buried coal. The waste rock pushed down into the nearest valley buries forests, streams, houses, and farms (see chapter 11 for further discussion of this issue). Mine reclamation is now mandated in the United States, but efforts often are only partially successful.

Coal comes in a variety of forms with varying chemical composition, hardness, and energy content. Lignite, the softest of this family, is not much more concentrated than peat and provides about 2,800 kWh/ metric ton. Subbituminous and bituminous coal have energy contents between 5,000 and 10,000 kWh/ton. Anthracite, which is the hardest coal, can be 96 percent carbon and produce over 10,000 kWh/ton.

Coal burning releases large amounts of air pollution. Coal often contains toxic impurities, such as mercury, arsenic, chromium, and lead, which are released into the air when coal is burned. Every year the roughly one billion tons of coal burned in the United States (83 percent for electric power generation) releases 18 million metric tons of sulfur dioxide (SO_2), 5 million metric tons of nitrogen oxides (NO_x), 4 million metric tons of airborne particulates, 600,000 metric tons of hydrocarbons and carbon monoxide, 40 tons of mercury, and close to a trillion metric tons of carbon dioxide (CO_2). This is about three-quarters of the SO_2, one-third of the NO_x, and about half of the industrial CO_2 released in the United States each year. Sulfur and nitrogen oxides combine with water in the air to form sulfuric and nitric acid, making coal burning the largest single source of acid rain in many areas (chapter 9).

New plants can be clean

Because coal causes so much air pollution, a great deal of effort has been invested in developing clean coal plants. While the initial cost of these plants is higher than older technology, they can pay for themselves over time. One of these systems is **integrated gasification combined cycle (IGCC)**, a technology that could produce zero-emissions electricity from coal. Power plants using this system could generate electricity while capturing and permanently storing carbon dioxide and other pollutants. An IGCC plant has been operating successfully for the past decade just outside of Tampa, Florida. Every day, the Polk power plant converts 2,400 tons of coal into 250 megawatts (MW) of electricity, or enough power for about 100,000 homes. Unlike conventional coal-fired power plants, an IGCC doesn't actually burn the coal. It converts the coal into gas and then burns the gas in a turbine (fig. 12.8). To do this, the coal is first ground into a fine powder and mixed with water to create a slurry. The slurry is pumped at high pressure into a gasification chamber, where it mixes with 96 percent pure oxygen, and is heated to 1,370°C (2,500°F). The coal

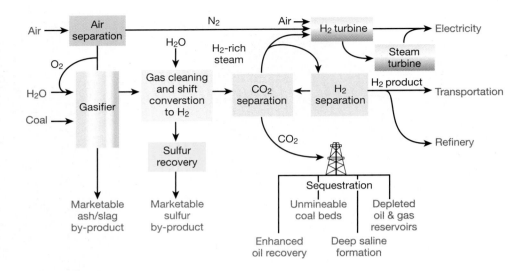

Figure 12.8 Clean coal technology could contribute to energy independence, while also reducing our greenhouse gas emissions.

doesn't burn; instead it reacts with the oxygen and breaks down into a variety of gases, mostly hydrogen and carbon dioxide. The gases are cooled, separated, and converted into easily managed forms.

After purification, the synthetic hydrogen gas (or syngas) is pumped to the combustion turbine, which spins a huge magnet to produce electricity. Superheated gases from the turbine are fed into a steam generator that drives another turbine to produce more electrical current. Combining these two turbines makes an IGCC about 15 percent more efficient than a normal coal-fired power plant. Perhaps even better is that the hydrogen gas could power fuel cells if they become commercially feasible.

Contaminants, such as sulfur dioxide (SO_2), ash, and mercury, that often go up the smokestack in a normal coal-burning plant, are captured and sold to make the IGCC cleaner and more economical. Sulfur is marketed as fertilizer; ash and slag are sold to cement companies. Mercury removal is an important public health benefit. All the slurry water is recycled to the gasifier; there is no waste water and very little solid waste. Because of these efficiencies, the Polk plant produces the cheapest electricity in the whole Tampa system. It doesn't now capture carbon dioxide, because it isn't required to, but it could easily do so. If we had CO_2 emission limits, IGCC plants could either pump it into deep wells for storage, or use it to enhance oil and natural gas recovery.

Because the Polk plant is so successful, Tampa Electric now plans to build another, even larger IGCC plant. Unfortunately this progressive outlook isn't typical of the whole power industry. Of the 80 or so new coal-fired power plants planned for construction over the next decade, only ten are slated to be IGCC, largely because of construction costs. While an IGCC can be very economical to operate, it costs 15 to 20 percent more to build than a conventional design. If industries were required to either sequester their CO_2 or pay a tax for not doing so, clean coal technology

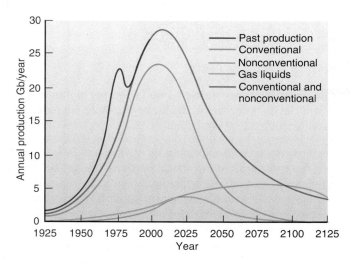

Figure 12.9 Worldwide production of crude oil with predicted Hubbert production. Gb = billion barrels.
Source: Jean Laherrère, www.hubbertpeak.org.

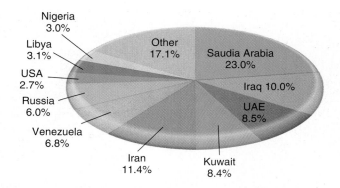

Figure 12.10 Proven oil reserves. Ten countries account for nearly 84 percent of all known recoverable oil.
Source: Data from British Petroleum, 2006.

would be much more attractive, and our contributions to global warming would be far lower. Although the Tampa plant is the only one in the United States, Japan has about 18 IGCC plants.

China and India, both of which have very large coal resources, now burn about half of all coal mined annually in the world. Both of these countries plan to increase coal production greatly in the next few decades to fuel their rapidly growing economies. If they do so, we could have run-away global climate change due to the enormous amount of CO_2 they will emit. We desperately need to persuade them to switch to renewable energy sources.

Have we passed peak oil?

In the 1940s Dr. M. King Hubbert, a Shell Oil geophysicist, predicted that oil production in the United States would peak in the 1970s, based on estimates of U.S. reserves at the time. Hubbert's predicted peak was correct, and subsequent calculations have estimated a similar peak in global oil production in about 2005–2010 (fig. 12.9). While global production has not yet slowed, many oil experts expect that we will pass this peak in the next few years.

About half of the world's original 4 trillion bbl (600 billion metric tons) of liquid oil are thought to be ultimately recoverable. (The rest is too diffuse, too tightly bound in rock formations, or too deep to be extracted.) Of the 2 trillion recoverable barrels, roughly 1.15 trillion bbl are in proven reserves (defined in fig. 12.6). We have already used more than 0.5 trillion bbl—almost half of proven reserves—and the remainder is expected to last 40 years at current consumption rates of 28.5 billion bbl per year. Middle Eastern countries have more than half of world supplies (fig. 12.10).

Consumption rates continue to climb, however, both in developed regions such as North America and Europe, and in the fast-growing economies such as China, India, and Brazil. China's energy demands have more than tripled in the past 35 years (much

of this energy is used to produce goods for the U.S. and European markets), and China anticipates another doubling of energy demands in the next 15 years. Much of this energy will come from domestic coal and hydropower, but it is clear that competition is growing for global oil and gas supplies. Competition has already raised oil prices, from around $15 per barrel in 1993 to more than $150 per barrel in 2008.

The exact shape and timing of the curves in fig. 12.9 depend on many factors. Global economic growth will increase consumption rates, but economic downturns will lead to conservation. Development of new extraction technologies has extended supplies, as have unconventional resources such as tar sands. These unconventional sources become economically recoverable when prices are high. More technological innovation is also supported by high prices. At the same time, high prices encourage conservation. The cost of insulating your house is easier to justify when heating and cooling costs are high. Companies can afford to experiment with new designs for light bulbs, cars, and consumer electronics when they are confident consumers want them.

Domestic oil supplies are limited

The United States has used more than half of its technically recoverable petroleum resources. About 30.7 billion bbl are proven in place. At 2008 U.S. rates of consumption (32 million bbl/day), that's enough for about 2.5 years, if we were to stop all imports. Opening the Arctic National Wildlife Refuge to drilling, an option hotly disputed in recent years, would add between 4 and 10 billion barrels, or enough for another 4–10 months, according to the U.S. Geological Survey.

Other U.S. regions with potential for new oil discoveries include portions of the continental shelf on the coast of California, the Arctic Ocean, and the Grand Banks, all of which provide important wildlife habitat and fishery habitat. Concern about environmental damage from drilling, and oil spills like that of the *Exxon Valdez* in Alaska's Prince William Sound, make many observers worry about exploitation in such sensitive areas

Figure 12.11 High-efficiency "smart" cars have been available for many years in Europe. Getting the equivalent of 60 mpg, they produce far less pollution than a typical American car. They are easy to maneuver in crowded city streets, and two can park in a standard parking space. (US models get about 36 mpg.)

(see related story, "Oil Drilling in the Arctic," at www.mhhe.com/cunningham5e). The huge amounts of oil used and transported around the world result in very serious oil and water pollution. Oceanographers estimate that between 3 million and 6 million metric tons of oil is discharged into the world's rivers and oceans every year. About half of this comes from oil tanker accidents and from dumping of oily ballast water. The rest is mostly deliberate dumping of used motor oil. Many people think a storm sewer is a good place to discard waste oil and don't realize that the sewer probably leads directly to a lake, river, or seashore.

Conservation can make a big difference in stretching oil supplies. Transportation accounts for over 40 percent of U.S. oil use, and oil (refined to produce gasoline or diesel fuel) provides over 90 percent of energy in transportation. So vehicle efficiency can have a substantial influence in our overseas oil dependence (fig. 12.11; see Active Learning, at right). Household heating accounts for another 25 percent, so steps such as lowering thermostats and weather-stripping doors are also important.

Oil shales and tar sands contain huge amounts of petroleum

Estimates of our recoverable oil supplies usually don't account for the very large potential from unconventional resources. The World Energy Council estimates that oil shales, tar sands, and other unconventional deposits contain ten times as much oil as liquid petroleum reserves. **Tar sands** are composed of sand and shale particles coated with bitumen, a viscous, tar-like mixture of long-chain hydrocarbons. Shallow tar sands are excavated and mixed with hot water and steam to extract the bitumen, then fractionated to make useful products. For deeper deposits, superheated steam is injected to melt the bitumen, which can then be pumped to the sur-

face, like liquid crude. Once the oil has been retrieved, it still must be cleaned and refined to be useful. This costly, energy-intensive extraction becomes economically justified when oil prices rise above about $50 per barrel.

Canada and Venezuela have the world's largest and most accessible tar sand resources. Canadian deposits in northern Alberta are estimated to be equivalent to 1.7 trillion bbl of oil, and Venezuela has nearly as much. Together, these deposits are three times as large as all conventional liquid oil reserves. By 2010 Alberta is expected to increase its flow to 2 million bbl per day, or twice the maximum projected output of the Arctic National Wildlife Refuge

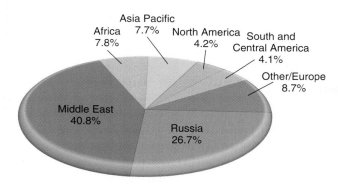

Figure 12.12 Proven natural gas reserves by region, 2002.
Source: Data from British Petroleum, 2006.

Figure 12.13 As domestic supplies of natural gas dwindle, the United States is turning increasingly to shipments of liquefied gas in specialized ships, such as this one at an Australian terminal. An explosion of one of these ships would release about as much energy as a medium-size atomic bomb.

(ANWR). Furthermore, because Athabascan tar sand beds are 40 times larger and much closer to the surface than ANWR oil, the Canadian resource will last longer and may be cheaper to extract. Canada is already the largest supplier of oil to the United States, having surpassed Saudi Arabia in 2000.

There are severe environmental costs, however, in producing this oil. A typical plant producing 125,000 bbl of oil per day creates about 15 million m³ of toxic sludge, releases 5,000 tons of greenhouse gases, and consumes or contaminates billions of liters of water each year. Surface mining in Canada could destroy millions of hectares of boreal forest. Native Cree, Chipewyan, and Metis people worry about effects on traditional ways of life if forests are destroyed and wildlife and water are contaminated. Many Canadians dislike becoming an energy colony for the United States, and environmentalists argue that investing billions of dollars to extract this resource simply makes us more dependent on fossil fuels.

Oil shales are fine-grained sedimentary rock rich in solid organic material called kerogen. Like tar sands, the kerogen can be heated, liquefied, and pumped out like liquid crude oil. Oil shale beds up to 600 m (1,800 ft) thick underlie much of Colorado, Utah, and Wyoming. If these deposits could be extracted at a reasonable price and with acceptable environmental impacts, they might yield the equivalent of several trillion barrels of oil. Mining and extraction of oil shale use vast amounts of water (a scarce resource in the arid western United States) and create enormous quantities of waste. The rock matrix expands when heated, resulting in two or three times the volume of waste as was dug out of the ground. Billions of dollars were spent in the 1980s on pilot projects to produce synthetic oil. When oil prices dropped, these schemes were abandoned. With rapidly rising crude oil prices in recent years, interest in oil shale has rekindled. In 2005, the Bureau of Land Management began selling leases once again for pilot projects in oil shale recovery.

Natural gas

Natural gas is the world's third largest commercial fuel, making up 24 percent of global energy consumption. Because natural gas produces only half as much CO_2 as an equivalent amount of coal, substitution could help reduce global warming (chapter 9).

Russia has 26.7 percent of known natural gas reserves (mostly in Siberia and the Central Asian republics) and accounts for about 35 percent of all global production. Both Eastern and Western Europe depend on gas from these wells. Figure 12.12 shows the distribution of proven natural gas reserves in the world.

The total ultimately recoverable natural gas resources in the world are estimated to be 10,000 trillion ft³, corresponding to about 80 percent as much energy as the recoverable reserves of crude oil. The proven world reserves of natural gas are 6,200 trillion ft³ (176 trillion metric tons). Because gas consumption rates are only about half of those for oil, current gas reserves represent roughly a 60-year supply at present usage rates. Proven reserves in the United States are about 185 trillion ft³, or 3 percent of the world total. This is about a ten-year supply at current rates of consumption. Known reserves are more than twice as large.

Large amounts of methane are released from coal deposits. The Rocky Mountain front in Colorado, Wyoming, and Montana could have 10 percent of the total world methane supply. Proposals for up to hundreds of thousands of coal-bed methane wells in Wyoming and Utah, for example, have raised concerns about water pollution and land damage (see opening case study, chapter 11). Other states could face similar problems.

World consumption of natural gas is growing by about 2.2 percent per year, considerably faster than either coal or oil. Much of this increase is in the developing world, where concerns about urban air pollution encourage the switch to cleaner fuel. Gas can be shipped easily and economically through buried pipelines. The United States has been fortunate to have abundant gas resources accessible by an extensive pipeline system. It is difficult and dangerous, however, to ship and store gas between continents. To make the process economic, gas is compressed and liquefied. At −160°C (−260°F) the liquid takes up about one-six-hundredth the volume of gas. Special refrigerated ships transport liquefied natural gas (LNG) (fig. 12.13). LNG weighs only about half as much

Figure 12.14 Two nuclear reactors (domes) at the San Onofre Nuclear Generating Station sit between the beach and Interstate 5, the major route between Los Angeles and San Diego.

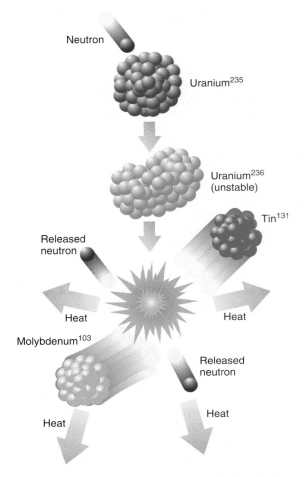

Figure 12.15 The process of nuclear fission is carried out in the core of a nuclear reactor. In the sequence shown here, the unstable isotope uranium-235 absorbs a neutron and splits to form tin-131 and molybdenum-103. Two or three neutrons are released per fission event and continue the chain reaction. The total mass of the reaction product is slightly less than the starting material. The residual mass is converted to energy (mostly heat).

as water, so the ships are very buoyant. Finding sites for terminals to load and unload these ships is difficult. Many cities are unwilling to accept the risk of an explosion of the volatile cargo. A fully loaded LNG ship contains about as much energy as a medium-size atomic bomb. Furthermore, huge amounts of seawater are used to warm and re-gasify the LNG. This can have deleterious effects on coastal ecology. However, in 2005, the U.S. Senate voted to eliminate state or local veto power in siting LNG terminals.

12.3 Nuclear Power

In 1953 President Dwight Eisenhower presented his "Atoms for Peace" speech to the United Nations. He announced that the United States would build nuclear-powered electrical generators to provide clean, abundant energy. He predicted that nuclear energy would fill the deficit caused by predicted shortages of oil and natural gas. It would provide power "too cheap to meter" for continued industrial expansion of both the developed and the developing world. Today there are about 440 reactors in use worldwide, 104 of these in the United States. Half of the U.S. plants (52) are more than 30 years old and are thus approaching the end of their expected operational life. Cracking pipes, leaking valves, and other parts increasingly require repair or replacement as a plant ages. Nuclear power now amounts to about 8 percent of U.S. energy supply (1 percent more than the world average). All of it is used to generate electricity.

Rapidly increasing construction costs, safety concerns, and the difficulty of finding permanent storage sites for radioactive waste have made nuclear energy less attractive than promoters expected in the 1950s. Of the 140 reactors on order in 1975, 100 were subsequently canceled. The costs of decommissioning old reactors is a serious concern, because deconstructing a worn-out plant may

cost ten times as much as building it in the first place. Unit 1 of the San Onofre plant near San Diego was closed down in 1992 (fig. 12.14). Deconstructing the plant required sawing through radioactive, meter-thick steel-reinforced walls of the containment chamber, a time-consuming and costly process that has been projected to cost $600 million. In 2008 the plant was still in the process of deconstruction.

The nuclear power industry has been campaigning for greater acceptance, arguing that reactors don't release greenhouse gases that cause global warming. About half the existing nuclear reactors in the United States have had their licenses renewed for an additionl 20 years beyond their original 40-year design, and several new plants are now on the drawing board. A number of prominent environmentalists are now promoting nuclear power as a solution to global climate change. Professor Robert Socolow of Princeton calculates that we would need to double the total global nuclear capacity (to about 700 large power plants) to eliminate 1 billion tons of annual CO_2 emissions. However, worries about accidents and susceptiblity to terrorist attacks make many people fearful of this solution.

How do nuclear reactors work?

The most commonly used fuel in nuclear power plants is U^{235}, a naturally occurring radioactive isotope of uranium. Uranium ore must be purified to a concentration of about 3 percent U^{235}, enough to sustain a chain reaction in most reactors. The uranium is then formed into cylindrical pellets slightly thicker than a pencil and about 1.5 cm long. Although small, these pellets pack an amazing amount of energy. Each 8.5 g pellet is equivalent to a ton of coal or 4 bbl of crude oil.

The pellets are stacked in hollow metal rods approximately 4 m long. About 100 of these rods are bundled together to make a **fuel assembly**. Thousands of fuel assemblies containing about 100 tons of uranium are bundled in a heavy steel vessel called the reactor core. Radioactive uranium atoms are unstable—that is, when struck by a high-energy subatomic particle called a neutron, they undergo **nuclear fission** (splitting), releasing energy and more neutrons. When uranium is packed tightly in the reactor core, the neutrons released by one atom will trigger the fission of another uranium atom and the release of still more neutrons (fig. 12.15). Thus, a self-sustaining **chain reaction** is set in motion, and vast amounts of energy are released.

The chain reaction is moderated (slowed) in a power plant by a neutron-absorbing cooling solution that circulates between the fuel rods. In addition, **control rods** of neutron-absorbing material, such as cadmium or boron, are inserted into spaces between fuel assemblies to shut down the fission reaction or are withdrawn to allow it to proceed. Water or some other coolant is circulated between the fuel rods to remove excess heat. The greatest danger in one of these complex machines is a cooling system failure. If the pumps fail or pipes break during operation, the nuclear fuel quickly overheats, and a "meltdown" can result that releases deadly radioactive material. Although nuclear power plants cannot explode like a nuclear bomb, the radioactive releases from a worst-case disaster, such as the meltdown of the Chernobyl reactor in the Soviet Ukraine in 1986, are just as devastating as a bomb. (See related story on Chernobyl at www.mhhe.com/cunningham5e.)

Nuclear reactor design

Seventy percent of the world's nuclear plants are pressurized water reactors (PWR). Water circulates through the core, absorbing heat as it cools the fuel rods (fig. 12.16). This primary cooling water is heated to 317°C (600°F) and reaches a pressure of 2,235 psi. It then is pumped to a steam generator, where it heats a secondary water-cooling loop. Steam from the secondary loop drives a high-speed turbine generator that produces electricity. Both the reactor vessel and the steam generator are contained in a thick-walled, concrete-and-steel containment building that prevents radiation from escaping and is designed to withstand high pressures and temperatures in case of accidents.

Overlapping layers of safety mechanisms are designed to prevent accidents, but these fail-safe controls make reactors very expensive and very complex. A typical nuclear power plant has 40,000 valves, compared with only 4,000 in a fossil fuel–fired plant of similar size. In some cases, the controls are so complex that they confuse operators and cause accidents, rather than prevent them. Under normal operating conditions, however, a PWR releases very little radioactivity and is probably less dangerous for nearby residents than a coal-fired power plant.

The Chernobyl plant that exploded and burned in Ukraine in 1986 used a graphite cooling design. Graphite has a high capacity for both capturing neutrons and dissipating heat. Designers claimed that these reactors could not possibly run out of control; unfortunately, they were proven wrong. The small cooling tubes are quickly blocked by steam if the cooling system fails, and the graphite core burns when exposed to air. Burning graphite in the Chernobyl nuclear plant made the fire much more difficult to control than it might have been in another reactor design. Radioactive particles spread across much of northern Europe before fires were put out. Fortunately, such catastrophes have been rare.

We lack safe storage for radioactive waste

One of the most difficult problems associated with nuclear power is the disposal of wastes produced during mining, fuel production, and reactor operation. How these wastes are managed may ultimately be the overriding obstacle to nuclear power.

Enormous piles of mine wastes and abandoned mill tailings in uranium-producing countries represent another serious waste

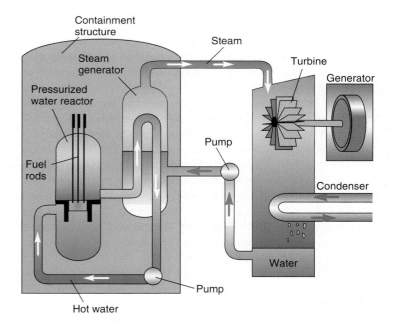

Figure 12.16 Pressurized water nuclear reactor. Water is super-heated and pressurized as it flows through the reactor core. Heat is transferred to nonpressurized water in the steam generator. The steam drives the turbogenerator to produce electricity.

Courtesy of Northern States Power Company. Minneapolis, MN.

Figure 12.17 Spent fuel is being stored temporarily in large, aboveground "dry casks" at many nuclear power plants.

disposal problem. Production of 1,000 tons of uranium fuel typically generates 100,000 tons of tailings and 3.5 million liters of liquid waste. There now are approximately 200 million tons of radioactive waste in piles around mines and processing plants in the United States. This material is carried by the wind or washes into streams, contaminating areas far from its original source. Canada has even more radioactive mine waste on the surface than does the United States.

In addition to the leftovers from fuel production, there are about 100,000 tons of low-level waste (contaminated tools, clothing, building materials, etc.) and about 15,000 tons of high-level (very radioactive) wastes in the United States. The high-level wastes consist mainly of spent fuel rods from commercial nuclear power plants and assorted wastes from nuclear weapons production. For the past 20 years, spent fuel assemblies from commercial reactors have been stored in deep, water-filled pools at the power plants. These pools were originally intended only as temporary storage until the wastes were shipped to reprocessing centers or permanent disposal sites.

With internal waste storage pools now full but neither reprocessing nor permanent storage available, a number of utility companies are beginning to store nuclear waste in large, metal dry casks placed outside power plants (fig. 12.17). These projects are meeting with fierce opposition from local residents, who fear the casks will leak. Most nuclear power plants are built near rivers, lakes, or seacoasts. Extremely toxic radioactive materials could spread quickly over large areas if leaks occur. A hydrogen gas explosion and fire in 1997 in a dry storage cask at Wisconsin's Point Beach nuclear plant intensified opponents' suspicions about this form of waste storage.

In 1987 the U.S. Department of Energy announced plans to build the first high-level waste repository on a barren desert ridge near Yucca Mountain, Nevada. After 15 years of research and $4 billion in exploratory drilling, it still isn't certain that the site is safe (see related story on Yucca Mountain at www.mhhe.com/cunningham5e). Intensely radioactive wastes are to be buried deep in the ground, where it is hoped that they will remain unexposed to groundwater and earthquakes for the thousands of years required for the radioactive materials to decay to a safe level. Although the area is very dry now, we can't be sure that it will always remain that way. Total costs now are expected to be at least $35 billion. Although the facility was supposed to open in 1998, the earliest possible date is now 2010.

Some nuclear experts believe that **monitored, retrievable storage** would be a much better way to handle wastes. This method involves holding wastes in underground mines or secure surface facilities where they can be watched. If canisters begin to leak, they could be removed for repacking. Safeguarding the wastes would be expensive, and the sites might be susceptible to wars and terrorist attacks. We might need a perpetual priesthood of nuclear guardians to ensure that the wastes are never released into the environment.

12.4 Energy Conservation

One of the best ways to avoid energy shortages and to relieve environmental and health effects of our current energy technologies is simply to use less (see What Can You Do? above). Much of the energy we consume is wasted. Our ways of using energy are so inefficient that most potential energy in fuel is lost as waste heat, becoming a form of environmental pollution. Conservation involves technology innovation as well as changes in behavior, but we have met these challenges in the past. Oil price shocks in the 1970s led to rapid improvements in industrial and household energy use. Although population and GDP have continued to grow since then, the **energy intensity**, or amount of energy needed to provide goods and services has declined (fig. 12.18). In response to federal regulations and high gasoline prices, automobile gas-mileage averages in the United States more than doubled from 13 mpg in 1975 to 28.8 mpg in 1988. Unfortunately, the oil glut and falling fuel prices of the 1990s discouraged further conservation. By 2004 the average slipped to only 20.4 mpg.

Much more could be done. High-efficiency automobiles are already available. Low-emission, hybrid gas-electric vehicles get up to 30.3 km/liter (72 mpg) on the highway (see related story on hybrid engines at www.mhhe.com/cunningham5e). Amory B. Lovins of the Rocky Mountain Institute in Colorado estimates that raising the average fuel efficiency of the U.S. car and light-truck fleet by 1 mpg would cut oil consumption about 295,000 bbl per day. In one year, this would equal the total amount the U.S. Department of the Interior hopes to extract from the Arctic National Wildlife Refuge in Alaska.

Many improvements in domestic energy efficiency have occurred in recent decades. Today's average new home uses one-half the fuel required in a house built in 1974, but much more can be done. Reducing air infiltration is usually the cheapest, quickest, and most effective way of saving energy because it is the largest source of losses in a typical house. It doesn't take much skill or investment to caulk around doors, windows, foundation joints, electrical outlets, and other sources of air leakage. Mechanical ventilation is needed to prevent moisture buildup in tightly sealed homes. Household energy losses can be reduced by one-half to three-fourths by using better insulation, installing double- or triple-glazed windows, purchasing thermally efficient curtains or window coverings, and sealing cracks and loose joints.

Green building can cut energy costs by half

Innovations in "green" building have been stirring interest in both commercial and household construction. Much of the innovation has occurred in large commercial structures, which have larger budgets—and more to save through efficiency—than most homeowners have. Elements of green building are evolving rapidly, but they include extra insulation in walls and roofs, coated windows to keep summer heat out and winter heat in, and recycled materials, which save energy in production. Orienting windows toward the sun, or providing roof overhangs for shade, are important for comfort as well as for saving money. Conservative water systems reduce water heating and water waste. Green roofs, planted with living vegetation and soil, provide insulation, slow the rate of rainfall runoff into sewers and rivers, provide wildlife habitat, and absorb carbon dioxide. Many large buildings now sport vegetated roofs, including Chicago's city hall and San Francisco's new California Academy of Sciences building in Golden Gate Park, which has 2.5 acres of native plantings on its roof. Around the world, cities are developing green building guidelines or codes that encourage these steps in new construction.

New houses can also be built with extra-thick, superinsulated walls and roofs. Windows can be oriented to let in sunlight, and eaves can be used to provide shade. Double-glazed windows that have internal reflective coatings and that are filled with an inert gas (argon or xenon) have an insulation factor of R11, the same as a standard 4-inch-thick insulated wall, or ten times as efficient as a single-pane window (fig 12.19). Superinsulated houses now being built in Sweden require 90 percent less energy for heating and cooling than the average American home.

Improved industrial design has also cut our national energy budget. More efficient electric motors and pumps, new sensors and control devices, advanced heat-recovery systems, and material recycling have reduced industrial energy requirements significantly. In the early 1980s, U.S. businesses saved $160 billion per year through conservation. When oil prices collapsed, however, many businesses returned to wasteful ways.

Cities can make surprising contributions to energy conservation. New York City has become a leader in this effort, replacing 11,000 traffic signals with more-efficient LEDs (light-emitting diodes), and 180,000 old refrigerators with energy-saving models. Ann Arbor, Michigan, replaced 1,000 streetlights with LED models. These lights saved the city over $80,000 in the first year, and will pay for themselves in just over two years.

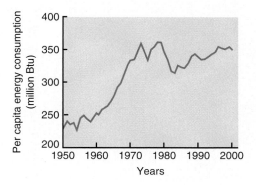

Figure 12.18 Per capita energy consumption in the United States rose rapidly in the 1960s. Price shocks in the 1970s encouraged conservation. Although GDP continued to grow in the 1980s and 90s, higher efficiency kept per capita consumption relatively constant. **Source:** Data from U.S. Department of Energy.

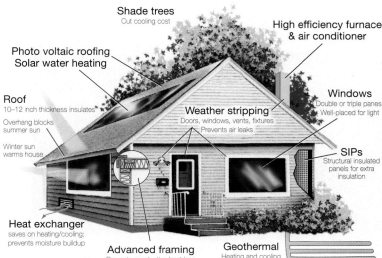

Figure 12.19 Energy-efficient building can lower energy costs dramatically. Many features can be added to older structures. New buildings that start with energy-saving features (such as SIPs or advanced framing) can save even more money.

Cogeneration makes electricity from waste heat

One of the fastest growing sources of new energy is **cogeneration**, the simultaneous production of both electricity and steam or hot water in the same plant. By producing two kinds of useful energy in the same facility, the net energy yield from the primary fuel is increased from 30–35 percent to 80–90 percent. In 1900 half the electricity generated in the United States came from plants that also provided industrial steam or district heating. As power plants became larger, dirtier, and less acceptable as neighbors, they were forced to move away from their customers. Waste heat from the turbine generators became an unwanted pollutant to be disposed of in the environment. Furthermore, long transmission lines, which are unsightly and lose up to 20 percent of the electricity they carry, became necessary.

By the 1970s, cogeneration had fallen to less than 5 percent of our power supplies, but interest in this technology is growing. District heating systems are being rejuvenated, and the EPA estimates that cogeneration could produce almost 20 percent of US electrical use, or the equivalent of 400 coal-fired plants.

12.5 Energy from Biomass

Plants capture immense amounts of solar energy by storing it in the chemical bonds of plant cells. Firewood is our original source of fuel. As recently as 1850, wood supplied 90 percent of the fuel used in the United States. For more than a billion people in developing countries, burning biomass remains the principal energy source for heating and cooking. An estimated 1,500 m³ of fuelwood is gathered each year globally. This amounts to half of all wood harvested. In urban areas of developing countries, wood is often sold in the form of charcoal (fig. 12.20). Wood gathering and charcoal burning are important causes of forest depletion in many rural areas—although commercial logging and conversion to farms and plantations are more rapid and widespread causes of forest loss globally. In some countries, such as Mauritania, Rwanda, and Sudan, firewood demand is ten times the sustainable yield.

In developed countries, where we depend on fossil fuels for most energy, wood burning is a minor heat source. Inefficient burning in stoves and fireplaces makes wood burning an important source of air pollution, especially soot and hydrocarbons, in some areas. However, biomasss burning provides an important fuel source for many medium-sized power plants, which produce steam for both heating and electricity. Many of these plants burn waste material such as urban tree clippings, which makes them an efficient, local, carbon-neutral source of energy (fig. 12.21).

Ethanol and biodiesel can contribute to fuel supplies

Biofuels, ethanol and biodiesel, are by far the biggest recent news in biomass energy. Globally, production of these two fuels is booming, from Brazil (which uses sugarcane) to Southeast Asia (oil palm

Figure 12.20 A charcoal market in Ghana. Firewood and charcoal provide the main fuel for billions of people. Forest destruction results in wildlife extinction, erosion, and water loss.

Figure 12.21 This district heating and cooling plant supplies steam heat, air conditioning, and electricity to three-quarters of downtown St. Paul, Minnesota, at about twice the efficiency of a remote power plant. It burns mainly urban tree trimming and, thus, is carbon neutral.

fruit) to the United States and Europe (corn, soybeans, rape seed). In the United States, both farm policies and energy policies have promoted biofuel crops. The energy bill passed by the U.S. Congress in 2007 required a four-fold increase in ethanol production, from 9 billion to 36 billion gallons (34 billion to 136 billion L) per year, in just 13 years. This rule provided an enormous boost to corn growers in the Midwest, where falling corn prices have plagued farmers for decades. In 2007 alone, the U.S. ethanol industry grew by 40 percent, corn production grew by 25 percent, and corn prices rose to historic highs above $5 per bushel. Still more important, the bill requires development of biofuels from inedible, woody parts of whole plants (cellulose), rather than the corn kernels used in most U.S. ethanol production. This change is important because corn is a relatively low-efficiency source of biomass (fig. 12.22), but most research and development have focused, until recently, on corn (see What Do You Think? p. 296).

Small amounts of ethanol have been added to gasoline for years, because oxygen-rich ethanol molecules help gasoline burn (oxidize) more completely. Ethanol helps reduce carbon monoxide (CO), an important pollutant, by converting it to carbon dioxide (CO_2). Most vehicles can burn up to 10 percent ethanol without damaging engines; newer "flexible fuel" vehicles, however, can burn up to 100 percent ethanol. Ethanol, the same alcohol used in beverages, is made by adding yeast to a liquid mix of water and ground grain, then fermenting it to produce alcohol.

Biodiesel, derived from organic oils, can be burned in normal diesel engines, and it can be much cheaper to produce than ethanol because it requires no fermentation. Just about anything organic, from turkey entrails and cow dung to soybeans, can be used as a source.

Will biofuels eliminate the need for other fuels? No. If all farmland in the United States were converted to corn-ethanol pro-duction, we would produce just a portion of the gasoline we use in a year. New strategies in conservation, as well as alternatives in fuel production, are needed to reduce our dependence on petroleum fuels.

Grasses and algae could grow fuel

Expanding markets for ethanol have spurred research into new fuel sources. Switchgrass (*Panicum virgatum*), a tall grass native to the Great Plains, has been one focus of attention. Switchgrass is a perennial grass, with deep roots that store carbon (and thus capture atmospheric greenhouse gases). Perennial roots also hold soil in place, unlike annual corn crops, which require constant disturbance of the soil surface. Because it doesn't require plowing and planting each year, a crop like switchgrass takes far less fuel than does corn. Consequently, switchgrass produces a 540 percent **net energy yield** (5.4 times as much energy produced as was used to grow it), according to a 2008 study published by the National Academy of Sciences. This compares to an approximate 25 percent net energy gain from corn-based ethanol. Switchgrass can also be grown on marginal lands unsuitable for crop planting (fig. 12.23).

Some studies suggest that mixed fields of perennial, native grasses could provide biofuels and wildlife habitat at the same time. While switchgrass is a native plant, a monoculture of this one species is little different for wildlife than other monocultures. One

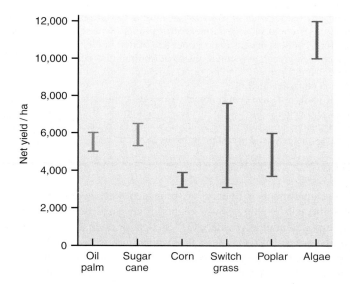

Figure 12.22 Proven biofuel sources include oil palms, sugarcane, and corn grain (maize). Other experimental sources may produce better yields, however (red = tropical plants; blue = temperate plants).
Source: Data from E. Marris, 2006. *Nature* 444: 670–678.

Figure 12.23 Experimental trials suggest that switchgrass and other perennial plants can have a high energy yield. Grasses can grow on marginal lands and minimize soil loss.

WHAT DO YOU THINK?

Can Biofuels Be Sustainable?

Biofuels (ethanol refined from plant material, and diesel fuel made from vegetable oils or animal fats) are thought by many people to be the answer to both our farm crisis and our fuel needs. Brazil is the current world leader in crop-based ethanol production. In 2006, Brazil refined some 18 billion l (4.7 billion gal) of ethanol, mostly from sugar cane waste called bagasse). By 2012, Brazil expects to double this output and export billions of liters of biofuels to other countries. Almost all new vehicles in Brazil now have flexfuel engines. Most motorists mix their own combination (about 80 percent ethanol is most common) right at the pump. They are motivated to do so, because ethanol is about 40 percent cheaper than gasoline. Rapidly increasing soy and corn production in Brazil (chapter 7) could also provide vegetable oils for diesel fuel.

The United States is currently second in world ethanol production. In 2006, the U.S. produced 15.2 billion l of ethanol (4 billion gal), mostly from shelled corn. This number rose to 36 billion l (9 billion gal) in 2007. Like Brazil, the U.S. hopes to double this yield by 2012. Almost all American states now require small amounts of oxygenated fuel, such as ethanol, to improve air quality. Some states mandate higher ethanol content as a way of reducing oil dependence and supporting farmers. The highest of these is in Minnesota, which requires 20 percent ethanol in gasoline, the highest concentration thought to be feasible without modifying ordinary engines.

But do these crops represent a net energy gain? Does it take more fossil fuel energy to grow, harvest, and process crop-based biofuels than you get back in the finished product? Experts hotly debate this question. For 15 years, David Pimental from Cornell University has repeatedly published calculations showing a net energy loss in biofuels. In 2005, he was joined by Tad Patzek from the University of California–Berkeley, in claims that it takes 29 percent more energy to refine ethanol from corn than it yields. Soy-based biodiesel and cellulosic ethanol are equally inefficient, these authors maintain.

On the other side of this debate are scientists, such as Bruce Dale of Michigan State University and John Sheehan from the U.S. National Renewable Energy Laboratory, who argue that biofuels produced by modern techniques represent a positive energy return. They say that Pimental and Patzek used outdated data and unreasonable assumptions in making their estimates. Dramatic improvements in farming productivity, coupled with much greater efficiency in ethanol fermentation now yield about 50 percent more energy in ethanol from corn, and about 60 percent more energy in ethanol from cellulose crops than in the inputs, they calculate. A major difference between the two sides in this debate is process energy for fermentation and purification. Pimental and Patzek assume this energy

Alcohol, mostly from sugarcane waste, is cheaper than either gasoline or diesel fuel in Brazil. All passenger vehicles in Brazil are required to use at least 20 percent ethanol.

would come from fossil fuels, but Dale and Sheehan argue that fermentation waste can be burned to make manufacturing self-sufficient, as is already being done in Brazil.

Another difference in the calculations between these two sides is how to account for crop inputs. Pimental and Patzek include not only the fuel to cultivate and harvest crops, but also the energy to manufacture the equipment used to grow and process the crops. Dale and Sheehan view this as an unfair handicap on biofuels. How far back do you carry the input? Do you also include the energy to feed and house the farmers who grow the crops, and the workers who make the machinery? Using Pimental and Patzek's formulas, Dale calculates that ethanol and soy diesel are much more energy efficient than gasoline (–39 percent) and coal-based electricity (–235 percent). By 2025, according to Dale and Sheehan, biofuels crops could provide farmers with profits of more than $5 billion per year, could save consumers about $20 billion in fuel costs, and could reduce U.S. greenhouse emissions by 1.7 billion tons per year—equal to more than 80 percent of transportation-related emissions and 22 percent of total emissions in 2002.

Clearly we have many choices in energy supplies, but weighing their costs and benefits inevitably involves assumptions. If you were making the decisions about energy resources, whose estimates would you use? Which sources would you support? Why?

---

study from Minnesota found that mixed prairie grasses provided a biomass yield, and potential ethanol yield, comparable to switchgrass but with more drought resistance, because in a mixed field, different species flourish under different weather conditions.

Algae could be an extremely efficient source of oil, or biodiesel, although this source remains experimental. Researchers have found strains of algae that grow rapidly under hot and saline conditions, producing abundant lipids (oils) that could be converted to biodiesel. Algae could be grown with recycled water, in containment ponds or chambers built on land that cannot be farmed. The

U.S. Department of Energy has proposed that algae ponds could be built near existing power plants. Flue gases bubbled through these ponds could provide CO_2 for algae growth, while preventing the escape of CO_2 to the atmosphere. Thus, algae could be a very cheap form of carbon capture. Although this technology has not been developed on a large scale, it shows great promise (see fig. 12.22).

Many other potential sources of biomass exist, including urban sewage, waste products from meat packing plants, orange peels from Florida citrus growers, and sawdust from lumber mills.

New plants built for these sources are among the more than 100 facilities now in development.

Effects on food and environment are uncertain

Will biofuel production affect food costs? Yes, but the seriousness of the problem depends on where you look. Since a $3 box of cereal contains only about one penny's worth of corn, a doubling of corn prices shouldn't affect the price of corn flakes much. Corn and soy costs, however, can make up 40–50 percent of meat prices. Many low-income Americans can ill afford steep price increases, but most people in wealthy countries spend only about 10 percent of their income on food. So most people could absorb somewhat higher food costs.

In developing countries, more than 50 percent of household income may be spent on food. Higher costs of cooking oil and grain can be devastating for family budgets. Regions dependent on food aid have seen supplies dwindling and prices rising. (For some farmers in developing countries, however, who have been struggling to compete with cheap imported corn from the United States, rising grain prices may be a blessing.) Gary Becker, a Nobel laureate in economics, calculates that a 30 percent rise in food prices would reduce living standards in rich countries by about 3 percent; in developing countries, living standards would drop by 20 percent.

Although ethanol and biodiesel are renewable fuels, they are not necessarily friendly to the environment. That depends on what kinds of plants are used and where they are grown. Algae or mixed prairie grasses could provide a sustainable, wildlife-friendly feed source with almost no soil erosion or water pollution, but these fuel sources are still hypothetical and experimental. In many areas, biofuel production has led to intensive cropping on erodible soils, as well as rapid clearing of grasslands and forests for crop fields. Accelerating soil erosion and plummeting biodiversity result. In Indonesia, conversion to oil palm plantations has become an important threat to primary rainforest habitat. In Brazil, grasslands and forests have been replaced with soy and sugarcane. Fertilizer-intensive crops, including corn and sugarcane, also increase nutrient runoff in rivers. Water shortages are also a concern. With current technology, 3 to 6 liters of water is needed to produce 1 liter of ethanol. In many farming states, there isn't enough water for both agriculture and food production. Plans for some new processing facilities have been scaled back because of water shortages.

These concerns are causing the European Union to consider new laws that ban biofuels grown with unsustainable practices. Whether these rules succeed in promoting good practices, and whether they spread to other regions, remains to be seen.

Methane from biomass is efficient and clean

Just about any organic waste, but especially sewage and manure, can be used to produce methane. Methane gas, the main component of natural gas, is produced when anaerobic bacteria (bacteria living in an oxygen-free, enclosed tank) digest organic matter (fig. 12.24). The main by-product of this digestion, CH_4, has no

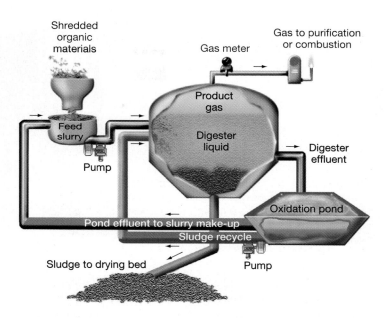

Figure 12.24 Continuous unit for converting organic material to methane by anaerobic fermentation. One kilogram of dry organic matter will produce 1–1.5 m^3 of methane, or 2,500–3,600 million calories per metric ton.

oxygen atoms because no oxygen was available in digestion. But this molecule oxidizes, or burns, easily, producing CO_2 and H_2O (water vapor). Consequently, methane is a clean, efficient fuel. Today, as more cities struggle to manage urban sewage and feedlot manure, methane could be a rich source of energy. In China, more than 6 million households use methane, also known as biogas, for cooking and lighting. Two large municipal facilities in Nanyang, China, provide fuel for more than 20,000 families.

Methane is a promising resource, but it has not been adopted as widely as it could be. Gas is harder to store and transport than liquid fuels like ethanol, and low prices for natural gas and other fuels have reduced incentives for building methane production systems. However, concerns about greenhouse gases may lead to further development, because methane is a powerful agent of atmospheric warming (chapter 9). Especially around livestock facilities, such as poultry or hog barns, large lagoons of liquid manure release a constant flow of methane to the atmosphere. These lagoons are also a threat to water bodies, because they occasionally overflow. But they could be a rich source of energy, which would save money as well as reducing atmospheric impacts. City sewage treatment plants and landfills also offer rich, and mostly untapped, potential for methane generation.

12.6 Wind and Solar Energy

In Denmark's efforts to reduce dependence on oil imports (opening case study), wind power has been the principal focus, followed by solar thermal (heat) systems. Relative to other alternative sources, wind is cheap and available almost everywhere. Although wind

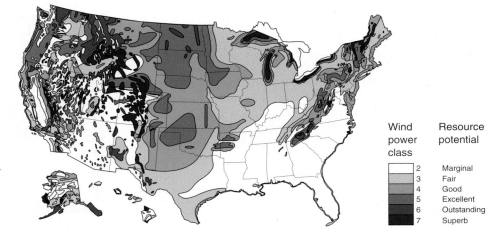

Figure 12.25 United States wind resource map. Mountain ranges and areas of the High Plains have the highest wind potential, but much of the country has a fair to good wind supply.
Source: Data from U.S. Department of Energy.

Wind power class	Resource potential
2	Marginal
3	Fair
4	Good
5	Excellent
6	Outstanding
7	Superb

turbines are highly visible, they have a small footprint, so they don't displace farming and other land uses. How people feel about the visibility of a wind farm depends on whether they are enthusiastic about energy alternatives, whether it earns money for their community, and which particular view is obstructed by the turbines.

Solar energy can be converted to heat (thermal energy), as well as electricity. The sun is an almost inconceivably rich source of energy. The average amount that reaches the earth's surface is some 10,000 times greater than all commercially sold energy used each year. However, this energy is diffuse and low in intensity. Innovations in recent years have produced new strategies for concentrating solar energy, to make it useful for more purposes.

Wind energy is our fastest growing renewable

Although Denmark remains the world leader in wind energy, the United States and other countries are catching up rapidly. Wind generation capacity in the United States grew by 20 percent in 2006 and by another 45 percent in 2007. Those increases brought the level of installed resources to 17 million MW, enough for almost 5 million homes. Wind power emerged as more than 1 percent of total U.S. energy demand in 2007. Texas is the leader in U.S. wind power, with just over one-third of installed capacity, followed by California, Minnesota, Iowa, and Washington.

Potential for wind development is almost inconceivably large. The World Meteorological Organization has estimated that wind could produce about 50 times the total capacity of all nuclear power plants now in operation. Interest in wind has grown as conventional fuel prices have risen. In the 1980s the state of California was the world's top wind power producer, with 90 percent of all existing wind power generators. Some 17,000 windmills marched across windy mountain ridges at Altamont, Tehachapi, and San Gorgonio Passes. Poor management, technical flaws, and overdependence on subsidies, however, led to bankruptcy of major corporations, including Kenetech, once the largest turbine producer in the United States. Now Danish, German, and Japanese wind machines are capturing the rapidly growing world market. Wind technology is now Denmark's second-largest export, employing 20,000 people and bringing in about $1 billion (U.S.) per year.

The 50,000 MW of global installed wind power currently in operation worldwide demonstrates the economy of wind turbines. Theoretically up to 60 percent efficient, windmills typically produce about 35 percent efficiency under field conditions. Where conditions are favorable, wind power is now cheaper than any other new energy source, with electric prices as low as 3¢ per kilowatt-hour in places with steady winds averaging at least 24 kph (15 mph). Large areas of western North America meet this requirement (fig. 12.25). Some energy experts have called North America's Great Plains the Saudi Arabia of wind power.

The World Energy Council predicts that wind could account for 200,000 MW of electricity by 2020, depending on how seriously politicians take global warming and how many uneconomical nuclear reactors go offline. One thousand megawatts meets the energy needs of about 50,000 typical U.S. households or is equivalent to about 6 million bbl of oil. Shell Oil suggests that half of all the world's energy could be wind and solar generated by the middle of this century.

In places like Denmark, which has winds above 16 mph an average of 245 days per year, windmills have long been recognized as a valuable energy source for pumping water and grinding grain. Germany, with 17,000 MW of installed capacity, now gets one-third of its electricity from wind power, and is the world leader in this technology. Spain is second with 10,000 MW, and the United States is third with 9,200 MW. The World Energy Council predicts that wind generating capacity could grow another tenfold by 2020.

Wind farms are large concentrations of wind generators producing commercial electricity (fig. 12.26). California continues to lead the United States in wind power, with 2,096 MW installed capacity in 2006, or about one-fifth of the U.S. total. Texas with 1,294 MW is second in installed capacity. Minnesota, Iowa, Oregon, Washington, Wyoming, Colorado, and New Mexico all have between 250 and 650 MW of installed wind power.

Do wind farms have any negative impacts? They generally occupy places with wind and weather too severe for residential or other development. Most wind farms are too far from residential areas to be heard or seen. But they do interrupt the view in remote, isolated places and destroy the sense of isolation and natural

Figure 12.26 Renewable energy sources, such as wind, solar energy, geothermal power, and biomass crops, could eliminate our dependence on fossil fuels and prevent extreme global climate change, if we act quickly.

beauty. Bird and bat kills have been a concern for some wind farms. Recent studies have shown that risks for birds may be less than for bats. Radio towers and lighted office buildings, both of which shine bright lights at night, the peak time for bird migration, are probably more important causes of bird mortality. Careful placement of wind farms outside of migration corridors and the addition of warning devices can reduce mortality greatly.

Wouldn't wind power take up a huge land area if we were to depend on it for a major part of our energy supply? As table 12.3 shows, the actual space taken up by towers, roads, and other structures on a wind farm is only about one-third as much as would be consumed by a coal-fired power plant or solar thermal energy system to generate the same amount of energy over a 30-year period. Furthermore, the land under windmills is more easily used for grazing or farming than is a strip-mined coal field or land under solar panels. Farmers are finding wind energy to be a lucrative crop. A single tower sitting on 0.1 ha (0.25 acre) can pay $100,000 per year.

When a home owner or community invests independently in wind generation, the same question arises as with solar energy: what should be done about energy storage when electricity production exceeds use? Besides the storage methods mentioned earlier, many private electricity producers believe the best use for excess electricity is to sell it to the public utility grid. In states that allow net energy pricing, you sell excess power from wind or solar systems back to a local utility. Ideally, the utility pays you something close to its average wholesale price. The 1978 Public Utilities Regulatory Policies Act required utilities to buy power generated by small hydro, wind, cogeneration (simultaneous production of useful heat and electricity), and other privately owned technologies at a fair price. Not all utilities yet comply, but some—notably in California, Oregon, Maine, and Vermont—are purchasing significant amounts of private energy.

Table 12.3   Jobs and Land Required for Alternative Energy Sources

Technology	Land Use (m² per gigawatt-hour for 30 years)	Jobs (per terawatt-hour per year)
Coal	3,642	116
Photovoltaic	3,237	175
Solar thermal	3,561	248
Wind	1,335	542

Source: Data from Lester R. Brown et al., *Saving the Planet*, 1991. W. W. Norton & Co., Inc.

Solar energy is diffuse but abundant

The sun is a giant nuclear furnace in space, constantly bathing our planet with a free energy supply. Solar heat drives winds and the hydrologic cycle. All biomass, as well as fossil fuels and our food (both of which are derived from biomass), results from conversion of light energy (photons) into chemical bond energy by photosynthetic bacteria, algae, and plants.

The average amount of solar energy reaching the earth's surface is some 10,000 times all the commercial energy used each year. However, this tremendous infusion of energy comes in a form that, until recently, has been too diffuse and low in intensity to be used except for environmental heating and photosynthesis. Figure 12.27 shows solar energy levels over the United States for typical summer and winter days.

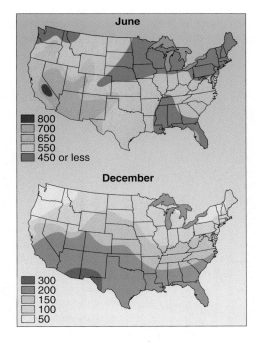

Figure 12.27 Average daily solar radiation in the United States in June and December. One langley, the unit for solar radiation, equals 1 cal/cm² of earth surface (3.69 Btu/ft²).
Source: Data from National Weather Bureau, U.S. Department of Commerce.

Passive solar absorbs heat; active solar pumps heated fluids

The opening case study shows the value of solar energy for water heating. In the case of the Danish islands of Ærø and Samsø, water is pumped through tubes and heated, then distributed to household radiators, or stored in massive heat-storage chambers underground. Heating water or other fluids is perhaps the most efficient way to use solar energy. In most American households, water heaters account for about 15 percent of household energy budgets. In countries where energy is more expensive, as in southern Europe, rooftop solar water heaters are standard household equipment. In Greece, Italy, Israel, and other countries, up to 70 percent of domestic hot water comes from solar collectors. In California, some 650,000 homes have solar water heating. Sometimes a rooftop unit only warms water, and an electric or gas water heater raises the temperature further. But even this contribution can make a big difference in energy use.

Massive heat storage has been used to capture solar energy for thousands of years. Thick adobe, stone, or dung walls absorb daytime heat and release it gradually at night. This **passive solar absorption** has been updated in modern homes with massive, heat-absorbing floors and walls, or with glass-walled "sun spaces" on the south side of a building.

Active solar systems like those in Ærø generally pump a heat-absorbing fluid medium (air, water, or an antifreeze solution) through a relatively small collector, rather than passively collecting heat in a stationary medium, such as masonry. Active collectors can be located adjacent to or on top of buildings, rather than being built into the structure.

A flat, black surface sealed with a double layer of glass makes a good solar collector. A fan circulates air over the hot surface and into the house through ductwork of the type used in standard forced-air heating. Alternatively, water can be pumped through the collector to pick up heat for space heating or to provide hot water.

Sunshine doesn't reach us all the time, of course. How can solar energy be stored for times when it is needed? A number of options are available. In a climate where sunless days are rare and seasonal variations are small, a small, insulated water tank is a good solar energy storage system. In areas where clouds block the sun for days at a time or where energy must be stored for winter use, a large, insulated bin containing a heat-storing mass, such as stone, water, or clay, provides solar energy storage. During the summer months, a fan blows the heated air from the collector into the storage medium. In the winter, a similar fan at the opposite end of the bin blows the warm air into the house. During the summer, the storage mass is cooler than the outside air, and it helps cool the house by absorbing heat. During the winter, it is warmer and acts as a heat source by radiating stored heat. In many areas, six or seven months' worth of thermal energy can be stored in 10,000 gal of water or 40 tons of gravel, about the amount of water in a small swimming pool or the gravel in two average-size dump trucks.

Solar thermal energy can also be used to generate electricity. These systems use parabolic mirrors (fig. 12.28) to heat oil or other fluids to temperatures as high as 400°C. The hot fluid is pumped to a central plant, where it heats water to produce steam that runs a turbine. California's Mojave Desert has had solar thermal facilities for decades and now has over 300 MW of installed capacity. A new facility scheduled to open in 2009 will produce 500 MW of electricity. Meanwhile, a new plant has been built near Las Vegas, Nevada, to provide 64 MW, enough for 15,000 homes.

Photovoltaic cells generate electricity directly

Although thermal energy is easier to collect from the sun, electrical energy is useful for more purposes. **Photovoltaic cells (PVCs)** have developed over the past 50 years, gradually becoming cheaper, more durable, and more widely used. Today they are competitive in cost with conventional energy sources. In 2007–2008, the U.S. Air Force installed the largest-ever photovoltaic array, with 70,000 panels distributed on 140 acres (55 ha) of Nevada desert at Nellis Air Force Base. The installation is designed to produce 14 MW of electricity, 25 percent of the base's needs, and save $1 million per year in energy costs. On a corporate scale, Google has installed 1.6 MW of solar panels at its headquarters, which should supply 30 percent of energy use. The energy saved should pay for the panel installation in 7–8 years—when tax incentives are included in calculations—after which the panels will provide cost-free electricity for the remainder of their 30-year life span.

As new types of PVCs are developed, and as the market grows, the costs should come down still more. New methods such as thin films (as opposed to the thicker traditional silicon PVCs) should be cheaper because they use less materials. Researchers have been trying to improve efficiency of new photovoltaic technologies, as well.

Photovoltaic cells capture solar energy and convert it directly to electrical current by separating electrons from their parent atoms and accelerating them across a one-way electrostatic barrier formed by the junction between two different types of

Figure 12.28 Parabolic mirrors focus sunlight on steam-generating tubes at this power plant in the California desert.

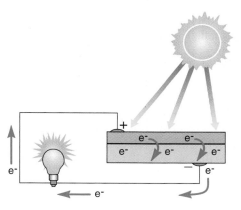

Figure 12.29 The operation of a photovoltaic cell. Boron impurities incorporated into the upper silicon crystal layers cause electrons (e-) to be released when solar radiation hits the cell. The released electrons move into the lower layer of the cell, thus creating a shortage of electrons, or a positive charge, in the upper layer and an oversupply of electrons, or negative charge, in the lower layer. The difference in charge creates an electric current in a wire connecting the two layers.

Figure 12.30 Solar roof tiles (*shiny area*) can generate enough electricity for a house full of efficient appliances. On sunny days, this array can produce a surplus to sell back to the utility company, making it even more cost-efficient.

semiconductor material (fig. 12.29). The photovoltaic effect, which is the basis of these devices, was first observed in 1839 by French physicist Alexandre-Edmond Becquerel, who also discovered radioactivity. His discovery didn't lead to any useful applications until 1954, when researchers at Bell Laboratories in New Jersey learned how to carefully introduce impurities into single crystals of silicon.

These handcrafted, single-crystal cells were much too expensive for any practical use until the advent of the U.S. space program. In 1958, when *Vanguard I* went into orbit, its radio was powered by six palm-sized photovoltaic cells that cost $2,000 per peak watt of output, more than 2,000 times as much as conventional energy at the time. Since then, prices have fallen dramatically. In 1970 they cost $100 per watt; in 2006 they were about $5 per watt. This makes solar energy cost-competitive with other sources in some areas. A photovoltaic array of about 30–40 m² will generate enough electricity for an efficient house (fig. 12.30).

By 2020 photovoltaic cells could be less than $1 per watt of generating capacity. With 15 percent efficiency and a 30-year life, they should be able to produce electricity for around 6¢ per kilowatt-hour. Commercial energy now costs around 10¢ per kWh in many places, which makes photovoltaic cells increasingly attractive.

During the past 25 years, the efficiency of energy captured by photovoltaic cells has increased from less than 1 percent of incident light to more than 10 percent under field conditions and over 75 percent in the laboratory. Promising experiments are underway using exotic metal alloys, such as gallium arsenide, and semiconducting polymers of polyvinyl alcohol, which are more efficient in energy conversion than silicon crystals. Every year college students from all over North America race solar-powered cars across the country to test solar technology and publicize its potential. Perhaps you and your classmates would like to build a solar car and enter the race.

One of the most promising developments in photovoltaic cell technology in recent years is the invention of **amorphous silicon collectors**. First described in 1968 by Stanford Ovshinky, a self-taught inventor from Detroit, these noncrystalline silicon semiconductors can be made into lightweight, paper-thin sheets that require much less material than conventional photovoltaic cells. They also are vastly cheaper to manufacture and can be made in a variety of shapes and sizes, permitting ingenious applications. Roof tiles with photovoltaic collectors layered on their surface already are available. Even flexible films can be coated with amorphous silicon collectors. Silicon collectors already are providing power to places where conventional power is unavailable, such as lighthouses, mountaintop microwave repeater stations, villages on remote islands, and ranches in the Australian outback.

You probably already use amorphous silicon photovoltaic cells. They are being built into solar-powered calculators, watches, toys, photosensitive switches, and a variety of other consumer products. Japanese electronic companies presently lead in this field, having foreseen the opportunity for developing a market for photovoltaic cells. This market is already more than $100 million per year. Japanese companies now have home-roof arrays capable of providing all the electricity needed for a typical home at prices in some areas competitive with power purchased from a utility. And Shanghai, China, recently announced a plan to install photovoltaic collectors on 100,000 roofs. This is expected to generate 430 million kWh annually and replace 20,000 tons of coal per year.

12.7 Water Power

Falling water is one of our oldest souces of power. In early American settlements, water-powered gristmills and sawmills were essential, and most early industrial cities were built where falling water could run mills. The invention of water turbines in the

Figure 12.31 Hydropower dams produce clean renewable energy but can be socially and ecologically damaging.

nineteenth century greatly increased the efficiency of hydropower dams (fig. 12.31). By 1925 falling water generated 40 percent of the world's electric power. Since then, hydroelectric production capacity has grown 15-fold, but fossil fuel use has risen so rapidly that water power is now only one-quarter of total electrical generation. Still, many countries produce most of their electricity from falling water. Norway, for instance, depends on hydropower for 99 percent of its electricity; Brazil, New Zealand, and Switzerland all produce at least three-quarters of their electricity with water power. Canada is the world's leading producer of hydroelectricity, running 400 power stations with a combined capacity exceeding 60,000 MW. First Nations people protest that their rivers are being diverted and lands flooded to generate electricity, most of which is sold to the United States.

The total world potential for hydropower is estimated to be about 3 million MW. If all of this capacity were put to use, the available water supply could provide between 8 and 10 terawatt hours (10^{12} watt-hours) of electrical energy. Currently, we use only about 10 percent of the potential hydropower supply. The energy derived from this source in 1994 was equivalent to about 500 million tons of oil, or 8 percent of the total world commercial energy consumption.

Most hydropower comes from large dams

Much of the hydropower development since the 1930s has focused on enormous dams. There is a certain efficiency of scale in giant dams, and they bring pride and prestige to the countries that build them, but they can have unwanted social and environmental effects that spark protests in many countries. China's Three Gorges Dam

on the Yangtze River, for instance, spans 2.0 km (1.2 mi) and is 185 m (600 ft) tall. The reservoir it creates is 644 km (400 mi) long and has displaced more than 1 million people (see related story "Three Gorges Dam" at www.mhhe.com/cunningham5e).

In tropical climates, large reservoirs often suffer enormous water losses. Lake Nasser, formed by the Aswan High Dam in Egypt, loses 15 billion m^3 each year to evaporation and seepage. Unlined canals lose another 1.5 billion m^3. Together, these losses represent one-half of the Nile River flow, or enough water to irrigate 2 million ha of land. The silt trapped by the Aswan High Dam formerly fertilized farmland during seasonal flooding and provided nutrients that supported a rich fishery in the delta region. Farmers now must buy expensive chemical fertilizers, and the fish catch has dropped almost to zero. Schistosomiasis, spread by snails that flourish in the reservoir, is an increasingly serious problem.

Unconventional hydropower comes from tides and waves

Ocean tides and waves contain enormous amounts of energy that can be harnessed to do useful work. A tidal station works like a hydropower dam, with its turbines spinning as the tide flows through them. A high-tide/low-tide differential of several meters is required to spin the turbines. Unfortunately, variable tidal periods often cause problems in integrating this energy source into the electric utility grid. Nevertheless, demand has kept some plants running for many decades.

Ocean wave energy can easily be seen and felt on any seashore. The energy that waves expend as millions of tons of water are picked up and hurled against the land, over and over, day after day, can far exceed the combined energy budget for both insolation (solar energy) and wind power in localized areas. Captured and turned into useful forms, that energy could make a substantial contribution to meeting local energy needs.

Dutch researchers estimate that 20,000 km of ocean coastline are suitable for harnessing wave power. Among the best places in the world for doing this are the west coasts of Scotland, Canada, the United States (including Hawaii), South Africa, and Australia. Wave energy specialists rate these areas at 40 to 70 kW per meter of shoreline. Altogether, it's calculated, if the technologies being studied today become widely used, wave power could amount to as much as 16 percent of the world's current electrical output.

Some of the designs being explored include oscillating water columns that push or pull air through a turbine, and a variety of floating buoys, barges, and cylinders that bob up and down as waves pass, using a generator to convert mechanical motion into electricity. It's difficult to design a mechanism that can survive the worst storms.

An interesting new development in this field is the Pelamis wave-power generator developed by the Scottish company Ocean Power Delivery (fig. 12.32). The first application of this technology is now in operation 5 km off the coast of Portugal, with three units producing 2.25 MW of electricity, or enough to supply 1,500 Portuguese households. Another 28 units are being installed. Each of the units consists of four cylindrical steel sections linked by hinged joints. Anchored to the seafloor at its nose, the snakelike machine points into the waves and undulates up and down and side to side as swells move along its 125 m length. This motion pumps fluid to hydraulic motors that drive electrical generators to produce power, which is carried to shore by underwater cables. Portugal considers wave energy one of its most promising sources of renewable energy.

Pelamis's inventor, Richard Yemm, says that survivability is the most important feature of a wave-power device. Being off-shore, the Pelamis isn't exposed to the pounding breakers that destroy shore-based wave-power devices. If waves get too steep, the Pelamis simply dives under them, much as a surfer dives under a breaker. These wave converters lie flat in the water and are positioned far offshore, so they are unlikely to stir up as much opposition as do the tall towers of wind generators.

Geothermal heat, tides, and waves could supply substantial amounts of energy in some places

The earth's internal temperature can provide a useful source of energy in some places. High-pressure, high-temperature steam fields exist below the earth's surface. Around the edges of continental plates or where the earth's crust overlays magma (molten rock) pools close to the surface, this **geothermal energy** is expressed in the form of hot springs, geysers, and fumaroles. Yellowstone National Park is the largest geothermal region in the United States. Iceland, Japan, and New Zealand also have high concentrations of geothermal springs and vents. Depending on the shape, heat content, and access to groundwater, these sources produce wet steam, dry steam, or hot water. Iceland, which sits on a midocean ridge (chapter 1), has abundant geothermal energy. Iceland has ambitious plans to be the first carbon-neutral country, largely because the earth's heat provides steam for heat and electric energy.

While few places have geothermal steam, the earth's warmth can help reduce energy costs nearly everywhere. Pumping water

Figure 12.32 The Pelamis wave converter (named after a sea snake) is a 125 m long and 3.5 m diameter tube, hinged so it undulates as ocean swells pass along it. This motion drives pistons that turn electrical generators. Energy experts calculate that capturing just 1 to 2 percent of global wave power could supply at least 16 percent of the world's electrical demand.

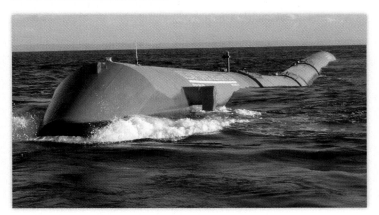

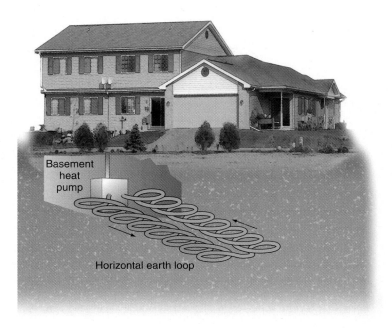

Figure 12.33 Geothermal energy can cut heating and cooling costs by half in many areas. In summer (*shown here*), warm water is pumped through buried tubing (*earth loops*), where it is cooled by constant underground temperatures. In winter, the system reverses and the relatively warm soil helps heat the house. Where space is limited, earth loops can be vertical. If more space is available, the tubing can be laid in shallow horizontal trenches, as shown here.

through deeply buried pipes can extract enough heat so that a heat pump will operate more efficiently. Similarly, the relatively uniform temperature of the ground can be used to augment air conditioning in the summer (fig. 12.33).

12.8 Fuel Cells

Rather than store and transport energy, another alternative is to generate it locally, on demand. **Fuel cells** are devices that use ongoing electrochemical reactions to produce an electrical current. They are very similar to batteries except that, rather than recharging them with an electrical current, you add more fuel for the chemical reaction. Depending on the environmental costs of input fuels, fuel cells can be a clean energy source for office buildings, hospitals, or even homes.

All fuel cells consist of a positive electrode (the cathode) and a negative electrode (the anode) separated by an electrolyte, a material that allows the passage of charged atoms, called ions, but is impermeable to electrons (fig. 12.34). In the most common systems, hydrogen or a hydrogen-containing fuel is passed over the anode, while oxygen is passed over the cathode. At the anode, a reactive catalyst, such as platinum, strips an electron from each hydrogen atom, creating a positively charged hydrogen ion (a proton). The hydrogen ion can migrate through the electrolyte to the cathode, but the electron is excluded. Electrons flow through an external circuit, and the electrical current generated by their passage can be

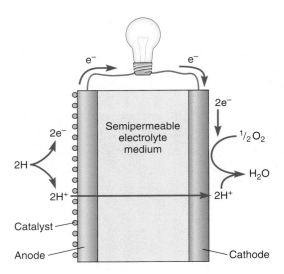

Figure 12.34 Fuel cell operation. Electrons are removed from hydrogen atoms at the anode to produce hydrogen ions (protons) that migrate through a semipermeable electrolyte medium to the cathode, where they reunite with electrons from an external circuit and oxygen atoms to make water. Electrons flowing through the circuit connecting the electrodes create useful electrical current.

used to do useful work. At the cathode, the electrons and protons are reunited and combined with oxygen to make water.

Fuel cells provide direct-current electricity as long as they are supplied with hydrogen and oxygen. For most uses, oxygen is provided by ambient air. Hydrogen can be supplied as a pure gas, but storing hydrogen gas is difficult and dangerous because of its explosive nature. Liquid hydrogen takes far less space than the gas but must be kept below –250°C (–400°F), not a trivial task for most mobile applications. The alternative is a device called a **reformer** or converter that strips hydrogen from fuels such as natural gas, methanol, ammonia, gasoline, ethanol, or even vegetable oil. Many of these fuels can be derived from biofuels, such as ethanol. Even methane effluents from landfills and wastewater treatment plants can be used as a fuel source. Where a fuel cell can be hooked permanently to a gas line, hydrogen can be provided by solar, wind, or geothermal facilities that use electricity to hydrolyze water.

A fuel cell that runs on pure oxygen and hydrogen produces no waste products except drinkable water and radiant heat. When a reformer is coupled to the fuel cell, some pollutants are released (most commonly carbon dioxide), but the levels are typically far less than conventional fossil fuel combustion in a power plant or an automobile engine (fig. 12.35). Although the theoretical efficiency of electrical generation of a fuel cell can be as high as 70 percent, the actual yield is closer to 40 or 45 percent. This is not much better than a very good fossil fuel power plant or a gas turbine electrical generator. On the other hand, the quiet, clean operation and variable size of fuel cells make them useful in buildings where waste heat can be captured for water heating or space heating. A 45-story office building at 4 Times Square, for example, has two 200 kW fuel cells on its fourth floor that provide both electricity and heat. The same building has photovoltaic panels on its façade, natural lighting, fresh-air intakes to reduce air conditioning, and a number of other energy conservation features.

Figure 12.35 The Long Island Power Authority has installed 75 stationary fuel cells to provide reliable backup power.

Utilities are promoting renewable energy

Utility restructuring currently being planned in the United States could include policies to encourage conservation and alternative energy sources. Among the proposed policies are (1) "distributional surcharges" in which a small per kilowatt-hour charge is levied on all utility customers to help finance renewable energy research and development, (2) "renewables portfolio" standards to require power suppliers to obtain a minimum percentage of their energy from sustainable sources, and (3) **green pricing** that allows utilities to profit from conservation programs and charge premium prices for energy from renewable sources.

Some states already are pursuing these policies. For example, Iowa has a Revolving Loan Fund supported by a surcharge on investor-owned gas and electric utilities. This fund provides low-interest loans for renewable energy and conservation. Several states have initiated green pricing programs as a way to encourage a transition to sustainable energy. One of the first was in Colorado, where 1,000 customers agreed to pay $2.50 per month above their regular electric rates to help finance a 10 MW wind farm being built on the Colorado–Wyoming border. Buying a 100 kW "block" of wind power provides the same environmental benefits as planting a half acre of trees or not driving an automobile 4,000 km (2,500 mi) per year. Not all green pricing plans are as straightforward as this, however. Some utilities collect the premium rates for facilities that already exist or for renewable energy bought from other utilities at much lower prices.

12.9 What Is Our Energy Future?

Global energy demand continues to rise, as cars, appliances, and factories become more abundant in both rich countries and developing countries. Growing use in China and India, the world's two most populous countries, is especially important in our energy future. Per capita use in wealthy countries, however, remains far above that in developing areas, and our rate of growth will also dictate the shape of our energy future. This growing demand is raising energy costs. High prices are also pushing innovation in alternative sources and better efficiency. None of the renewable energy sources discussed in this chapter will replace fossil fuels and nuclear power in the near future, because most of our energy infrastructure uses conventional energy sources and because our investments in alternatives has mostly been relatively small. Renewable alternatives can contribute substantially to sustainability, though. Since we import 35 million barrels of oil per day, a one percent savings through better efficiency or sustainable biofuels would represent a lot of oil.

The World Energy Council (WEC) projects that global energy demand will double by 2050, but they also project that we can meet energy needs while managing climate change, through greater cooperation, strategic investments, and clear rules for energy trading. The WEC calculates that renewables could provide about 40 percent of the world's cumulative energy consumption if political leaders take global climate change seriously. Curbing climate change is likely to involve taxes on carbon emissions—effectively making polluters pay society for the widespread environmental costs of their emissions. Carbon trading markets are already changing the ways we calculate the costs of energy. For some companies, it is cheaper to buy credits for carbon capture, for example, by funding tree planting or land conservation schemes, than to reduce carbon emissions. Carbon trading can increase the payback of carbon-capturing algae biofuels, or other carbon-free energy sources. These new calculations, together with policy changes in where we invest funds for research and development, could lead to alternative energy sources that provide more than 40 percent of our energy needs, if we take appropriate actions.

The United States, the world's largest energy consumer, has emphasized oil, coal, and nuclear power in energy policies. President George W. Bush and his father, President George H.W. Bush,

both have close family ties to oil producers in the United States and Saudi Arabia, and Vice President Dick Cheney (CEO of an oil services corporation) oversaw energy policies that did little to encourage conservation or alternative energy sources. These strategies helped to maintain U.S. control of critical energy resources. They also maintained our dependence on imported energy and allowed other countries to establish leadership in alternative strategies, such as automobile efficiency, better building practices, and wind power, that will drive much of our energy future. The decisions of future presidents, congresses, voters, and consumers will determine which strategies the United States will take in the future. Your choices, in where you live and what you drive, eat, and buy, will influence these energy futures (fig. 12.36).

Conclusion

Conventional resources, especially oil, coal, and natural gas, remain our dominant energy sources. Coal is extremely abundant, especially in North America, but extracting and burning coal have been major causes of environmental damage and air pollution, because coal contains many impurities. Clean-burning coal plants that use integrated gasification combined cycle (IGCC) technology exist but are not widely used in the United States. Coal provides most of our electricity. Oil (petroleum) provides most of our energy for transportation and heating. The United States has consumed over half of its recoverable oil resources, and the remaining resources (including controversial sources in the Arctic) would last about three years if we stopped all imports. Our biggest oil supplier is Canada, and much of this oil derives from tar sands, an unconventional oil source that has become economically

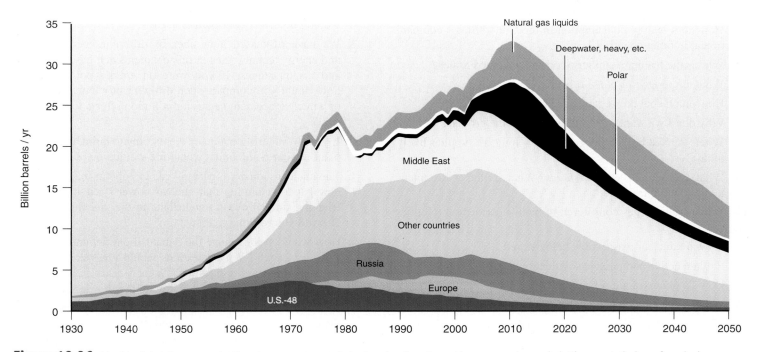

Figure 12.36 Liquid oil has been our dominant energy source, but planning for alternative sources is needed. The exact timing of peaks is uncertain, but experts agree on the shape of the curve. What strategies would you recommend?
Source: Association for the Study of Peak Oil and Gas, 2007.

recoverable since world oil prices have risen to nearly $150 per barrel. Extracting these resources, like coal extraction, is extremely damaging to land and water resources.

Natural gas is more abundant than oil and cleaner than coal, so its importance has been increasing. The main drawback of gas is the risk of shipping and storing this highly-combustible gas. Nuclear power creates heat using chain reactions in highly radioactive uranium pellets. This heat creates steam that drives electric turbines. The world has abundant uranium, but storing the highly dangerous waste from mining, processing, electricity production, and dismantling old reactors remains an expensive and unresolved problem. The American public has been unenthusiastic about nuclear power, but there has been renewed interest recently because nuclear power emits no greenhouse gases. The cost of building and decommissioning, as well as the risk of accidents or attacks, remains an important uncertainty.

Conservation is a key factor in a sustainable energy future. New designs in housing, office buildings, industrial production, and transportation can all save huge amounts of energy. Transportation consumes over 40 percent of the U.S. energy budget, so more efficient vehicles can have a substantial effect on our overseas oil dependence.

Biofuels include ethanol and oil (biodiesel) from plants. Sources vary greatly in their net energy yield and environmental effects. Thus far, U.S. biofuels have used corn and soybeans, but sources should diversify as production quadruples in the coming decade or so. Hydropower is our principal renewable resource. Hydro is clean and reliable, but a focus on huge dams has led to many environmental and social problems. Rapid innovations in solar, wind, wave, and other alternative energy forms have made these types of energy increasingly popular. In 2008, former Vice President Al Gore issued a bold and inspiring challenge to the United States. Currently, he said, "We're borrowing money from China to buy oil from the Persian Gulf to burn in ways that destroy the planet." However, renewable, environmentally friendly sources, such as wind, solar, and geothermal, could supply all the energy we need. He urged the country to move toward producing 100 percent of its energy from sustainable sources by 2018. Doing so, he proposed, would solve the three biggest crises we face—environmental, economic, and security—simultaneously. This ambitious project could create hundreds of thousands of jobs, spur economic development, and eliminate our addiction to imported oil. As you can see, the energy choices we make have profound effects on many aspects of our lives and our environment.

Practice Quiz

1. Where are Samsø and Ærø islands, and how do they supply their energy needs?
2. Define *energy*, *power*, and *kilowatt-hour* (kWh).
3. What are the major sources of global commercial energy?
4. How does energy consumption in the United States compare to that in other countries?
5. Who is the leading supplier of oil to the United States?
6. What are *proven-in-place reserves*?
7. How much coal do we have, and how long will it last?
8. Why don't we want to use all the coal in the ground?
9. Where is most liquid oil located? How long are supplies likely to last?
10. What are *tar sands* and *oil shales*? What are the environmental costs of their extraction?
11. Why is natural gas considered to be a superior fuel to either coal or oil?
12. How are nuclear wastes now being stored?
13. Explain active and passive solar energy.
14. How do photovoltaic cells work?
15. What's a *fuel cell*, and how does it work?
16. What are *biofuels*, and how could they contribute to sustainability?

Critical Thinking and Discussion Questions

Apply the principles you have learned in this chapter to discuss these questions with other students.

1. If you were the energy czar of your state or country, where would you invest your budget? Why?
2. We have discussed a number of different energy sources and energy technologies in this chapter. Each has advantages and disadvantages. If you were an energy policy analyst, how would you compare such different problems as the risk of a nuclear accident versus air pollution effects from burning coal?
3. If your local utility company were going to build a new power plant in your community, what kind would you prefer? Why?
4. The nuclear industry is placing ads in popular magazines and newspapers, claiming that nuclear power is environmentally friendly, since it doesn't contribute to the greenhouse effect. How do you respond to that claim?
5. How would you evaluate the debate about net energy loss or gain in biofuels? What questions would you ask the experts on each side of this question? What worldviews or hidden agendas do you think might be implicit in this argument?
6. Although we have used vast amounts of energy resources in the process of industrialization and development, some would say that it was a necessary investment to get to a point at which we can use energy more efficiently and sustainably. Do you agree? Might we have followed a different path?

Data Analysis: Energy Calculations

Most college students either already own or are likely to buy an automobile and a computer sometime soon. How do these items compare in energy usage? Suppose that you were debating between a high-mileage car, such as the Honda Insight, or a sport utility vehicle, such as a Ford Excursion. How do the energy requirements of these two purchases measure up? To put it another way, how long could you run a computer on the energy you would save by buying an Insight rather than an Excursion?

Here are some numbers you need to know. The Insight gets about 75 mpg, while the Excursion gets about 12 mpg. A typical American drives about 15,000 mi per year. A gallon of regular, unleaded gasoline contains about 115,000 Btu on average. Most computers use about 100 watts of electricity. One kilowatt-hour (kWh) = 3,413 Btu.

1. How much energy does the computer use if it is left on continuously? (You really should turn it off at night or when it isn't in use, but we'll simplify the calculations.)

   100 watt/h × 24 h/day × 365 days/yr = _____ kWh/yr

2. How much gasoline would you save in an Insight, compared with an Excursion?

   a. Excursion:

      15,000 mi/yr ÷ 12 mpg = _____ gal/yr

   b. Insight:

      15,000 mi/yr ÷ 75 mpg = _____ gal/yr

   c. Gasoline savings (a − b) = _____ gal/yr

   d. Energy savings:

      (gal × 115,000 Btu) = _____ Btu/yr

   e. Converting Btu to kWh:

      (Btu × 0.00029 Btu/kWh) = _____ kWh/yr saved

3. How long would the energy saved run your computer?

   kWh/yr saved by Insight ÷ kWh/yr consumed by computer = _____

For Additional Help in Studying This Chapter, please visit our website at www.mhhe.com/cunningham5e. You will find practice quizzes, key terms, answers to end of chapter questions, additional case studies, an extensive reading list, and Google Earth™ mapping quizzes.

 A crane unloads a garbage barge at Fresh Kills on Staten Island, the world's largest landfill before it closed in 2001.

13 Solid and Hazardous Waste

Learning Outcomes

After studying this chapter, you should be able to answer the following questions:

- What are the major components of the waste stream? How do we dispose of most of our waste?
- How does a sanitary landfill operate? Why are we searching for alternatives to landfills?
- Why is ocean dumping a problem?
- What are the benefits of recycling?
- What are the "three Rs" of waste reduction, and which is most important?
- What are some steps *you* can take to reduce waste production?
- How can biomass waste be converted to natural gas?
- What are toxic and hazardous waste? How do we dispose of them?
- What is bioremediation? Is it a promising solution in waste management?
- What is the Superfund, and has it shown progress?

> *We are living in a false economy where the price of goods and services does not include the cost of waste and pollution.*
>
> **–Lynn Landes, founder and director of Zero Waste America**

The New Alchemy: Creating Gold from Garbage

Most people think of recycling in terms of newspapers, plastic bottles, and other household goods. Your daily household recycling is the bedrock of recycling programs, but another growing and exciting area of recycling is done at commercial and industrial scales. The 230 million tons of garbage the United States produces each year includes some knotty problems: old furniture and carpeting, appliances and computers, painted wood, food waste. It's no wonder that we've simply dumped it all in landfills as long as we could. But landfills are becoming more scarce and more difficult to site (fig. 13.1).

Incinerators are a common alternative, but they are expensive to build and operate, and they can produce dangerous air contaminants, including dioxins from burned plastics, and heavy metals.

One of our largest sources of waste is construction and demolition debris—the rubble left over when a building is torn down, remodeled, or built. Construction and demolition account for over 140 million tons of waste per year, about 1.5 kg per person per day—on top of the 230 million tons per year of municipal solid waste. All this mixed debris is normally trucked to landfills, but alternatives have emerged in recent years.

A slowly growing number of cities and companies are sending their waste, including construction debris, to commercial recyclers. One example is Taylor Recycling, based in Montgomery, New York. Starting as a tree removal business, Taylor has expanded to construction and demolition waste and now operates in four states. The company recycles and sells 97 percent of the mixed debris it receives, well above the industry average of 30 to 50 percent. Trees are ground and converted to mulch

for landscaping. Dirt from stumps is screened and sold as clean garden soil. Mixed materials are sorted into recyclable glass, metals, and plastics. Construction debris is sorted and ground: broken drywall is ground to fresh gypsum, which is sold to drywall producers; wood is composted or burned; bricks are crushed for fill and construction material. Organic waste that can't be separated, such as food-soaked paper, is sent to a gassifier. The gassifier is like an enclosed, oxygen-free pressure cooker, which converts biomass to natural gas. The gas runs electric generators for the plant, and any extra gas can be sold. Waste heat warms the recycling facility. The 3 percent of incoming waste that doesn't get recycled is mainly mixed plastics, which are currently landfilled.

From their base outside of New York City, recycling is clearly a good idea. New York has used up most of its landfill space and now ships garbage to Virginia, Ohio, Pennsylvania, and South Carolina. Fuel and trucking costs alone drive up disposal costs, and with landfill capacity shrinking, tipping fees are climbing. According to Jim Taylor, the 1,000 garbage trucks leaving New York City each day travel an average of 300 miles round-trip, at less than 4 miles per gallon of fuel.

The story of garbage processing is changing globally. Garbage has long been one of the United States' largest exports, but increasingly the country is exporting sorted recycled materials, as well. Chinese manufacturers are finding valuable material sources in American waste. In Western Europe, where environmental regulation and landfill space are both tight, recycling, composting, and conversion of biomass to gas are booming businesses. The Swiss company Kompogas, one of many companies processing garbage, ferments organic waste in giant tanks, producing methane, compost, and fertilizer.

New technologies are providing innovative strategies for recycling waste. One method that is getting attention is Changing World Technologies' thermal depolymerization process ("thermal" means it involves heating; "depolymerization" essentially means breaking down organic molecules) that converts almost any kind of organic waste into clean, usable diesel oil. Animal by-products, sewage sludge, shredded tires, and other waste can be converted with 85 percent energy efficiency. Pilot plants have been built for specialized sources, such as a turkey processing plant in Colorado, but the technology could accommodate mixed sources as well.

Recycling is a rapidly growing industry because it makes money coming and going. Recyclers are paid to haul away waste, which they turn into marketable products. The business is also exciting because these companies, like Taylor Recycling, see the huge social and economic benefits of environmental solutions. Often when we discuss environmental problems, businesses are part of the problem. But these examples show that business owners can be just as excited as anybody about environmental quality. With a good business model, being green can be very rewarding.

Garbage disposal may be one of the most exciting stories in environmental science, because it shows so much promise and innovation. So far, the examples discussed here make up a minority of our waste management strategies, but they are expanding. In this chapter we'll look at our other waste management methods, the composition of our waste, and some of the differences between solid waste and hazardous waste.

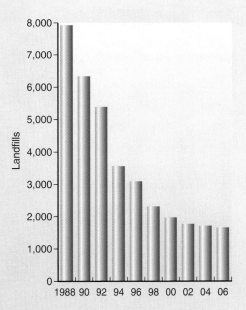

Figure 13.1 The number of landfills in the United States has fallen by nearly 80 percent over the past two decades, but the total capacity has remained relatively constant because landfills are now much larger than in the past.

Source: Data from Environmental Protection Agency, 2007.

13.1 Waste

Waste is everyone's business. We all produce unwanted by-products and residues in nearly everything we do. According to the Environmental Protection Agency (EPA), the United States produces 11 billion tons of solid waste each year. Nearly half of that amount consists of agricultural waste, such as crop residues and animal manure, which are generally recycled into the soil on the farms where they are produced. They represent a valuable resource as groundcover to reduce erosion and as fertilizer to nourish new crops, but they also constitute the single largest source of nonpoint air and water pollution in the country. About one-third of all solid wastes are mine tailings, overburden from strip mines, smelter slag, and other residues produced by mining and primary metal processing. Much of this material is stored in or near its source of production and isn't mixed with other kinds of wastes. Improper disposal practices, however, can result in serious and widespread pollution.

Industrial waste—other than mining and mineral production—amounts to some 400 million metric tons per year in the United States. Most of this material is recycled, converted to other forms, destroyed, or disposed of in private landfills or deep injection wells. About 60 million metric tons of industrial waste fall in a special category of hazardous and toxic waste, which we will discuss later in this chapter.

Municipal waste—a combination of household and commercial refuse—amounts to about 250 million metric tons per year in the United States. That's just over 2 kg (4.6 lbs) per person per day—twice as much per capita as Europe or Japan, and five to ten times as much as most developing countries (fig. 13.2).

The waste stream is everything we throw away

Does it surprise you to learn that you generate that much garbage? Think for a moment about how much we discard every year. There are organic materials, such as yard and garden wastes, food wastes, and sewage sludge from treatment plants; junked cars; worn-out furniture; and consumer products of all types. Newspapers, magazines, advertisements, and office refuse make paper one of our major wastes (fig. 13.3). In spite of recent progress in recycling, many of the 200 billion metal, glass, and plastic food and beverage containers used every year in the United States end up in the trash. Although plastic makes up only about 10 percent of our waste by weight, it comprises about 20 percent by volume. Wood, concrete, bricks, and glass come from construction and demolition sites, dust and rubble from landscaping and road building. All of this varied and voluminous waste has to arrive at a final resting place somewhere.

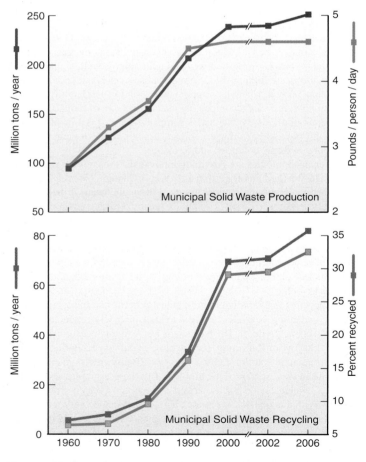

Figure 13.2 Solid waste production has leveled off in recent years on a per capita basis, but so has the recycling rate as a percentage of the total. The recycling rate includes composting of yard wastes.
Source: Data from Environmental Protection Agency, 2007.

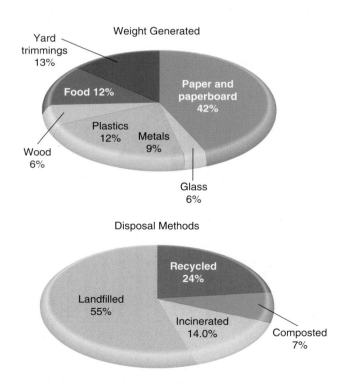

Figure 13.3 Composition of municipal solid waste in the United States by weight, before recycling, and disposal methods.
Source: Data from U.S. Environmental Protection Agency Office of Solid Waste Management, 2006.

The **waste stream** is a term that describes the steady flow of varied wastes that we all produce, from domestic garbage and yard wastes to industrial, commercial, and construction refuse. Many of the materials in our waste stream would be valuable resources if they were not mixed with other garbage. Unfortunately, our collecting and dumping processes mix and crush everything together, making separation an expensive and sometimes impossible task. In a dump or incinerator, much of the value of recyclable materials is lost.

Another problem with refuse mixing is that hazardous materials in the waste stream get dispersed through thousands of tons of miscellaneous garbage. This mixing makes the disposal or burning of what might have been rather innocuous stuff a difficult, expensive, and risky business. Spray-paint cans, pesticides, batteries (zinc, lead, or mercury), cleaning solvents, smoke detectors containing radioactive material, and plastics that produce dioxins and PCBs (polychlorinated biphenyls) when burned are mixed with paper, table scraps, and other nontoxic materials. The best thing to do with household toxic and hazardous materials is to separate them for safe disposal or recycling, as we will see later in this chapter.

13.2 Waste Disposal Methods

Where do our wastes go now? In this section, we will examine some historic methods of waste disposal, as well as some future options. Notice that our presentation begins with the least desirable—but most commonly used—measures and proceeds to discuss some preferable options. Keep in mind as you read this that modern waste management reverses this order and stresses the "three Rs" of reduction, reuse, and recycling before destruction or, finally, secure storage of wastes.

Figure 13.4 Trash disposal has become a crisis in the developing world, where people have adopted cheap plastic goods and packaging but lack good recycling or disposal options.

Open dumps release hazardous materials into the air and water

Often, the way people dispose of waste is to simply drop it someplace. Open, unregulated dumps are still the predominant method of waste disposal in most developing countries. The giant developing world megacities have enormous garbage problems (fig. 13.4). Mexico City, one of the largest cities in the world, generates some 10,000 tons of trash each day. Until recently, most of this torrent of waste was left in giant piles, exposed to the wind and rain, as well as rats, flies, and other vermin. Manila, in the Philippines, has at least ten huge open dumps. The most notorious is called "Smoky Mountain" because of its constant smoldering fires. Thousands of people live and work on this 30 m high heap of refuse. They spend their days sorting through the garbage for edible or recyclable materials. Health conditions are abysmal, but these people have nowhere else to go. The government would like to close these dumps, but how will the residents be housed and fed? Where else will the city put its garbage?

Most developed countries forbid open dumping, at least in metropolitan areas, but illegal dumping is still a problem. You have undoubtedly seen trash accumulating along roadsides and in vacant, weedy lots in the poorer sections of cities. Is this just a question of aesthetics? Consider the problem of waste oil and solvents. An estimated 200 million liters of waste motor oil are poured into the sewers or allowed to soak into the ground every year in the United States. This is about five times as much as was spilled by the *Exxon Valdez* in Alaska in 1989! No one knows the volume of solvents and other chemicals disposed of by similar methods.

Increasingly, these toxic chemicals are showing up in the groundwater supplies on which nearly half the people in America depend for drinking (chapter 10). An alarmingly small amount of oil or other solvents can pollute large quantities of drinking or irrigation water. One liter of gasoline, for instance, can make a million liters of water undrinkable. The problem of illegal dumping is likely to become worse as acceptable sites for waste disposal become more scarce and costs for legal dumping escalate. We clearly need better enforcement of antilittering laws, as well as a change in our attitudes and behavior.

Ocean dumping is nearly uncontrollable

The oceans are vast, but not so large that we can continue to treat them as carelessly as has been our habit. Every year some 25,000 metric tons (55 million lbs) of packaging, including half a million bottles, cans, and plastic containers, are dumped at sea. Beaches, even in remote regions, are littered with the nondegradable flotsam and jetsam of industrial society (fig. 13.5). About 150,000 tons (330 million lbs) of fishing gear—including more than 1,000 km (660 mi) of nets—are lost or discarded at sea each year. Environmental groups estimate that 50,000 northern fur seals are entangled in this refuse and drown or starve to death every year in the North Pacific alone.

Until recently, many cities in the United States dumped municipal refuse, industrial waste, sewage, and sewage sludge into

Figure 13.5 Dumping of trash at sea is a global problem. Even on the most remote islands, beaches are covered with plastic flotsam and jetsam.

landfills now have many of the safeguards of hazardous waste repositories described later in this chapter.

More careful attention is now paid to the siting of new landfills. Sites located on highly permeable or faulted rock formations are passed over in favor of sites with less leaky geologic foundations. Landfills are being built away from rivers, lakes, floodplains, and aquifer recharge zones, rather than near them, as was often done in the past. More care is being given to a landfill's long-term effects, so that costly cleanups and rehabilitation can be avoided.

Historically, landfills have been a convenient and relatively inexpensive waste-disposal option in most places, but this situation is changing rapidly. Rising land prices and shipping costs, as well as increasingly demanding landfill construction and maintenance requirements, are making this a more expensive disposal method. Over the past 30 years, many U.S. cities have had trash disposal rates increase five- to tenfold. The United States now spends about $10 billion per year to dispose of trash. A decade from now, it may cost us $100 billion per year to dispose of our trash and garbage.

Currently, 55 percent of all municipal solid waste in the United States is landfilled, 30 percent is recycled, and 15 percent is incinerated. Suitable places for waste disposal are becoming scarce in many areas. Other uses compete for open space. Citizens have become more concerned and vocal about health hazards, as well as aesthetics. It is difficult to find a neighborhood or community willing to accept a new landfill. Since 1984, when stricter financial and environmental protection requirements for landfills took effect, roughly 90 percent of all existing landfills in the United States have closed. In many cases, this means that old, small, uneconomical landfills closed, while larger, more modern ones replaced them. Nevertheless, many major cities are running out of local landfill space. They export their trash, at enormous expense, to neighboring communities and even other states. More than half the solid waste from New Jersey goes out of state, some of it up to 800 km (500 mi) away.

the ocean. Federal legislation now prohibits this dumping. New York City, the last to stop offshore sewage sludge disposal, finally ended this practice in 1992. Still, 60 million to 80 million m³ of dredge spoil—much of it highly contaminated—are disposed of at sea. Some people claim that the deep abyssal ocean plain is the most remote, stable, and innocuous place to dump our wastes. Others argue that we know too little about the values of these remote places or the rare species that live there to smother them with sludge and debris.

Landfills receive most of our waste

Over the past 50 years, most American and European cities have recognized the health and environmental hazards of open dumps. Increasingly, cities have turned to **sanitary landfills**, where solid waste disposal is regulated and controlled. To decrease smells and litter and to discourage insect and rodent populations, landfill operators are required to compact the refuse and cover it every day with a layer of dirt (fig. 13.6). This method helps control pollution, but the dirt fill also takes up as much as 20 percent of landfill space. Since 1994, all operating landfills in the United States have been required to control such hazardous substances as oil, chemical compounds, toxic metals, and contaminated rainwater that seep through piles of waste. An impermeable clay and/or plastic lining underlies and encloses the storage area. Drainage systems are installed in and around the liner to catch drainage and to help monitor chemicals that leak out. Modern municipal solid waste

Figure 13.6 In a sanitary landfill, trash and garbage are crushed and covered each day to prevent accumulation of vermin and spread of disease. A waterproof lining is now required to prevent leaching of chemicals into underground aquifers.

Life Cycle Analysis

One step toward understanding your place in the waste stream is to look at the life cycle of the materials you buy. Here is a rough approximation of the process. With another student, choose one item that you use regularly. On paper, list your best guess for the following: (1) a list of the major materials in it; (2) the original sources (geographic locations and source materials) of those materials; (3) the energy needed to extract/convert the materials; (4) the distances the materials traveled; (5) the number of businesses involved in getting the item to you; (6) where the item will go when you dispose of it; (7) what kinds of reused/recycled products could be made from the materials in it.

Figure 13.7 A Chinese woman smashes a cathode ray tube from a computer monitor in order to remove valuable metals. This kind of unprotected demanufacturing is highly hazardous to both workers and the environment.

When organic materials decompose in the anerobic conditions inside a landfill, they create methane gas. Globally, landfills are estimated to produce more than 700 million metric tons of methane annually. Since methane is 20-times as potent at absorbing heat as CO_2, this represents about 12 percent of all greenhouse gas emissions. Landfills are the single largest anthropogenic source of methane in the United States. Until recently, almost all this landfill methane was simply vented into the air. Now about half of all landfill gas in the United States is either flared (burned) on site or is collected and used as fuel for electrical generation. Methane recovery in the United States produces 440 trillion BTU per year, and is equivalent to removing 25 million vehicles from the highway. Some landfill operators are deliberately pumping water to their waste as a way of speeding up production of this valuable fuel.

We often export waste to countries ill-equipped to handle it

Although most industrialized nations agreed to stop shipping hazardous and toxic waste to less-developed countries in 1989, the practice still continues. In 2006, for example, 400 tons of toxic waste were illegally dumped at 14 open dumps in Abidjan, the capital of the Ivory Coast. The black sludge—petroleum wastes containing hydrogen sulfide and volatile hydrocarbons—killed ten people and injured many others. At least 100,000 city residents sought medical treatment for vomiting, stomach pains, nausea, breathing difficulties, nosebleeds, and headaches. The sludge—which had been refused entry at European ports—was transported by an Amsterdam-based multinational company on a Panamanian-registered ship and handed over to an Ivorian firm (thought to be connected to corrupt government officials) to be dumped in the Ivory Coast. The Dutch company agreed to clean up the waste and pay the equivalent of (U.S.) $198 million to settle claims.

One of the greatest sources of toxic material currently going to developing countries is outdated electronic devices. There are at least 2 billion television sets and personal computers in use globally. Televisions often are discarded after only about five years, while computers, play-stations, cellular telephones, and other electronics become obsolete even faster. It's estimated that 50 million tons of **electronic waste** (or **e-waste**) are discarded every year worldwide. Only about 20 percent of the components are currently recycled. The rest generally goes to open dumps or landfills. This waste stream contains at least 2.5 billion kg of lead (as well as mercury, gallium, germanium, nickel, palladium, beryllium, selenium, arsenic), and valuable metals, such as gold, silver, copper, and steel.

Until recently, most of this e-waste went to China, where villagers, including young children, would break it apart to retrieve valuable metals. Often, this scrap recovery was done under primitive conditions where workers had little or no protective gear (fig. 13.7). Health risks in this work are severe, especially for growing children. Soil, groundwater, and surface water contamination at these sites is severe. One Chinese village was reported to have the highest dioxin levels ever found anywhere in the world.

Shipping e-waste to China is now officially banned, but illegal smuggling continues. Meanwhile, other poor countries in Asia and Africa are importing more e-waste. The Basel Action Network, an international network of activists seeking to prevent the globalization of the toxic chemical trade, tracks international e-waste shipments and working conditions. The organization is named after the Swiss town where the agreement to ban international shipping of hazardous wastes was reached. Exporting waste to poor communities also occurs within rich countries (see What Do You Think? p. 314).

Environmental Justice

When a new landfill, petrochemical factory, incinerator, or other unwanted industrial facility is proposed for a minority neighborhood, charges of environmental racism often are raised by those who oppose this siting. Everyday experiences tell us that minority neighborhoods are much more likely to have high pollution levels and facilities that you wouldn't want to live near than are middle- or upper-class white neighborhoods. But does this prove that land-use decisions are racist or just that minorities are less politically powerful than middle- or upper-class residents? Could it be that land prices are simply cheaper and public resistance to locating a polluting facility in a place that's already polluted is less than putting it in a cleaner environment? Or does this distinction matter? Perhaps showing that a disproportionate number of minorities live in dirtier places is evidence enough of racism. How would you decide?

One of the first systematic studies showing this inequitable distribution of environmental hazards based on race in the United States was conducted by Robert D. Bullard in 1978. Asked for help by a predominantly black community in Houston that was slated for a waste incinerator, Bullard discovered that all five of the city's existing landfills and six of eight incinerators were located in African-American neighborhoods. In a book entitled *Dumping on Dixie*, Bullard showed that this pattern of risk exposure in minority communities is common throughout the United States. Among his findings are:

- Three of the five largest commercial hazardous waste landfills, accounting for about 40 percent of all hazardous waste disposal in the United States, are located in predominantly African-American or Hispanic communities.

- Sixty percent of African Americans and Latinos and nearly half of all Asians, Pacific Islanders, and Native Americans live in communities with uncontrolled toxic waste sites.

- The average percentage of the population made up by minorities in communities without a hazardous waste facility is 12 percent. By contrast, communities with one hazardous waste facility have, on average, twice as high (24 percent) a minority population, while those with two or more such facilities average three times as high a minority population (38 percent) as those without one.

But does this prove that race, not class or income, is the strongest determinant of who is exposed to environmental hazards? What additional information might you look for to make this distinction? One of the lines of evidence Dr. Bullard raises is the fact that the discrepancy between the pollution exposure of middle-class blacks and that of middle-class whites is even greater than the difference between poorer whites and blacks. While upper-class whites can "vote with their feet" and move out

Native Americans march in protest of toxic waste dumping on tribal lands.

of polluted and dangerous neighborhoods, Bullard argues, minorities are restricted by color barriers and prejudice to less desirable locations.

Some additional evidence uncovered by this research is variation in the way toxic waste sites are cleaned up and how polluters are punished in different neighborhoods. White communities see faster responses and get better results once toxic wastes are discovered than do minority communities. For instance, the EPA takes 20 percent longer to place a hazardous waste site in a minority community on the Superfund National Priority List than it does for one in a white community. Penalties assessed against polluters of white communities average six times higher than those against polluters of minority communities. Cleanup is more thorough in white communities as well. Most toxic wastes in white communities are treated—that is, removed or destroyed. By contrast, waste sites in minority neighborhoods are generally only "contained" by putting a cap over them, leaving contaminants in place to potentially resurface or leak into groundwater at a later date.

How would you evaluate these findings? Do they convince you that racism is at work, or do you think that other explanations might be equally likely? Which of these arguments do you find most persuasive, or what other evidence would you need to make a reasoned judgment about whether or not environmental racism is a factor in determining who gets exposed to pollution and who enjoys a cleaner, more pleasant environment?

Most of the world's obsolete ships are now dismantled and recycled in poor countries. The work is dangerous, and old ships often are full of toxic and hazardous materials, such as oil, diesel fuel, asbestos, and heavy metals. On India's Anlang Beach, for example, more than 40,000 workers tear apart outdated vessels using crowbars, cutting torches, and even their bare hands. Metal is dragged away and sold for recycling. Organic waste is often simply burned on the beach, where ashes and oily residue wash back into the water.

Incineration produces energy but causes pollution

Faced with growing piles of garbage and a lack of available landfills at any price, many cities have built waste incinerators to burn municipal waste. Another term commonly used for this technology is **energy recovery**, or waste-to-energy, because the heat derived from incinerated refuse is a useful resource. Burning garbage can produce steam used directly for heating buildings or generating electricity. Internationally, well over 1,000 waste-to-energy plants in Brazil, Japan, and Western Europe generate much-needed energy while reducing the amount that needs to be landfilled. In the United States more than 110 waste incinerators burn 45,000 metric tons of garbage daily. Some of these are simple incinerators; others produce steam and/or electricity.

Types of incinerators

Municipal incinerators are specially designed burning plants capable of burning thousands of tons of waste per day. In some plants, refuse is sorted as it comes in to remove unburnable or recyclable materials before combustion. This is called **refuse-derived fuel** because the enriched burnable fraction has a higher energy content than the raw trash. Another approach, called **mass burn**, is to dump everything smaller than sofas and refrigerators into a giant furnace and burn as much as possible (fig. 13.8). This technique avoids the expensive and unpleasant job of sorting through the garbage for nonburnable materials, but it often causes greater problems with air pollution and corrosion of burner grates and chimneys.

Incinerators have serious negative consequences. Residual ash and unburnable residues representing 10 to 20 percent of the original volume are usually taken to a landfill for disposal. Because the volume of burned garbage is reduced by 80 to 90 percent, disposal is a smaller task. However, the residual ash usually contains a variety of toxic components that make it an environmental hazard if not disposed of properly. Ironically, one worry about incinerators is whether enough garbage will be available to feed them. Often, incinerators compete with recycling programs for paper, plastics, and other materials. Some communities in which recycling has been really successful have had to buy garbage from neighbors to meet contractual obligations to waste-to-energy facilities. In other places, fears that this might happen have discouraged recycling efforts.

Incinerator cost and safety

The cost-effectiveness of garbage incinerators is the subject of heated debates. Initial construction costs are high—usually between $100 million and $300 million for a typical municipal facility. Tipping fees at an incinerator (the fee charged to haulers for each ton of garbage dumped) are often much higher than those at a landfill. As landfill space near metropolitan areas becomes more scarce and more expensive, however, landfill rates are certain to rise. It may pay in the long run to incinerate refuse so that the lifetime of existing landfills will be extended.

Environmental safety of incinerators is another point of concern. The EPA has found alarmingly high levels of dioxins, furans, lead, and cadmium in incinerator ash. These toxic materials are more concentrated in the fly ash (lighter, airborne particles capable of penetrating deep into the lungs) than in heavy bottom ash. Dioxin levels can be as high as 780 ppb (parts per billion). One part per billion of TCDD, the most toxic dioxin, is considered a health concern. All of the incinerators studied exceeded cadmium standards, and 80 percent exceeded lead standards. Proponents of incineration argue that, if they are run properly and equipped with appropriate pollution-control devices, incinerators are safe for the general public. Opponents counter that neither public officials nor pollution-control equipment can be trusted to keep the air clean. They argue that recycling and source reduction efforts are better ways to deal with waste problems.

The EPA, which generally supports incineration, acknowledges the health threat of incinerator emissions but holds that the danger is very slight. The EPA estimates that dioxin emissions from a typical municipal incinerator may cause one death per million people in 70 years of operation. Critics of incineration claim that a more accurate estimate is 250 deaths per million in 70 years.

One way to reduce these dangerous emissions is to remove batteries containing heavy metals and plastics containing chlorine before wastes are burned. Bremen, Germany, is one of several European cities now trying to control dioxin emissions by keeping all plastics out of incinerator waste. Bremen is requiring households to separate plastics from other garbage. This is expected to eliminate nearly all dioxins and other combustion by-products and prevent the expense of installing costly pollution-control equipment that otherwise would be necessary to keep the burners operating. Minneapolis has initiated a recycling program for the small "button" batteries used in hearing aids, watches, and calculators in an attempt to lower mercury emissions from its incinerator.

13.3 Shrinking the Waste Stream

Having less waste to discard is obviously better than struggling with disposal methods, all of which have disadvantages and drawbacks. In this section we will explore some of our options for recycling, reuse, and reduction of the wastes we produce.

Figure 13.8 A diagram of a municipal "mass burn" garbage incinerator. Steam produced in the boiler can be used to generate electricity or to heat nearby buildings.

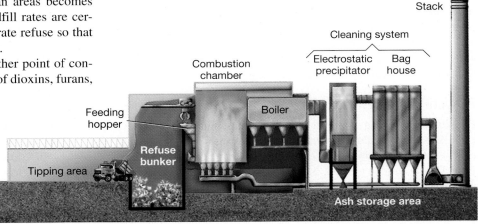

Figure 13.9 In some systems, residents sort recyclables, which are then transported in trucks with multiple compartments. In other systems, everything is mixed together for transport and then sorted by workers or complex machines at a recycling center.

Figure 13.10 Creating a stable, economically viable market for recycled products is essential for recycling success. Consumers can help by buying recycled products.

Recycling captures resources from garbage

The term *recycling* has two meanings in common usage. Sometimes we say we are recycling when we really are reusing something, such as refillable beverage containers. In terms of solid waste management, however, **recycling** is the reprocessing of discarded materials into new, useful products (fig. 13.9). Some recycling processes reuse materials for the same purposes; for instance, old aluminum cans and glass bottles are usually melted and recast into new cans and bottles. Other recycling processes turn old materials into entirely new products. Old tires, for instance, are shredded and turned into rubberized playground or road surfacing. Newspapers become cellulose insulation, kitchen wastes become a valuable soil amendment, and steel cans become new automobiles and construction materials.

There have been some dramatic successes in recycling in recent years. Minneapolis and Seattle, for instance, now claim a 60 percent recycling rate, something thought unattainable a decade ago. The high value of aluminum scrap (as much as $1,200 per ton in recent years) has spurred a large percentage of aluminum recycling nearly everywhere. About two-thirds of all aluminum cans are now recycled, up from only 15 percent 20 years ago. This recycling is so rapid that half of all the aluminum cans now on grocery shelves will be made into another can within two months.

Recycling challenges

Despite encouraging gains in recycling rates, major challenges exist. Despite the value of aluminum, Americans still throw away nearly 350,000 metric tons of aluminum beverage containers each year. That is enough to make 3,800 Boeing 747 airplanes. Plastics recyclers have developed innovative methods and products, but the low price of virgin plastic, made from oil, is usually less than the cost of transporting and storing used plastics. Consequently,

less than 5 percent of the United States' 24 million tons of plastic waste is recycled each year.

Wild fluctuations in commodity prices make it still harder to develop a market for recycled materials. Newsprint, for example, cost $160 a ton in 1995; by 1999 it dropped to just $42 per ton and then climbed to $650 per ton in 2006 (fig. 13.10).

Contamination is a major obstacle in plastics recycling. Most of the 24 billion plastic soft drink bottles sold every year in the United States are made of PET (polyethylene terphthalate), which can be remanufactured into carpet, fleece clothing, plastic strapping, and nonfood packaging. However, even a trace of vinyl—a single PVC (polyvinyl chloride) bottle in a truckload, for example—can make PET useless. Although most bottles are now marked with a recycling number, it's hard for consumers to remember which is which. Because single-use beverage containers are so costly to recycle, they have been outlawed in Denmark and Finland.

The growing popularity of bottled water is a becoming a serious waste disposal problem. Of the 300 billion bottles of water consumed each year globally, less than 20 percent are recycled even though they are generally made of easily recyclable plastic. It takes around 75 billion liters (500 million barrels) of oil to manufacture and ship these bottles. In most American cities, tap water is both safer and better tasting than bottled water. The best way to control this problem is through bottle deposits. States with deposit laws recover about 78 percent of all beverage containers, while those without generally have recycling rates of 20 percent or less.

The value of recycled plastics also tends to undergo wide swings. In 2006, for example, a slowing economy created a glut of both recycled and virgin plastic resins. Prices fell by 20 percent or more, and recyclers reported huge piles sitting in their warehouses. This high variability is a problem for recycling programs. Processors need both a relatively stable source of material and a

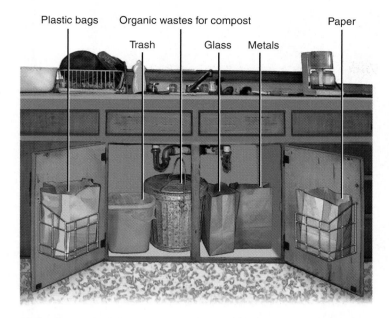

Figure 13.11 Source separation in the kitchen—the first step in a strong recycling program.

Recycling saves money, energy, and space

Recycling is usually a better alternative to either dumping or burning wastes. It saves money, energy, raw materials, and land space, while also reducing pollution. Recycling also encourages individual awareness and responsibility for the refuse produced (fig. 13.11). Curbside pickup of recyclables costs around $35 per ton, as opposed to the $80 paid to dispose of them at an average metropolitan landfill. Many recycling programs cover their own expenses with materials sales and may even bring revenue to the community.

Recycling drastically reduces pressure on landfills and incinerators. Philadelphia is investing in neighborhood collection centers that will recycle 600 tons a day, enough to eliminate the need for a previously planned, high-priced incinerator. New York City, down to one available landfill but still producing 27,000 tons of garbage a day, set a target of 50 percent waste reduction to be accomplished by recycling office paper and household and commercial waste. In 2002 Mayor Michael Bloomberg discontinued most recycling, arguing that the program was too expensive. The city quickly found that disposing of waste was more expensive than recycling, and most programs were reinstated.

Japan probably has the most successful recycling program in the world. Half of all household and commercial wastes in Japan are recycled, while the rest are about equally incinerated or landfilled. The country has begun a push to increase recycling, because

incineration costs almost as much. Some communities have raised recycling rates to 80 percent, and others aim to reduce waste altogether by 2020. This level of recycling takes a high level of participation and commitment. In Yokohama, a city of 3.5 million, there are now 10 categories of recyclables, including used clothing and sorted plastics. Some communities have 30 or 40 categories for sorting recyclables.

Recycling lowers demands for raw resources (fig. 13.12). The United States cuts down 2 million trees every day to produce newsprint and paper products, a heavy drain on its forests. Recycling the print run of a single Sunday issue of the *New York Times* would spare 75,000 trees. Every piece of plastic made in the United States reduces the reserve supply of petroleum and makes the country more dependent on foreign oil. Recycling 1 ton of aluminum saves 4 tons of bauxite (aluminum ore) and 700 kg of petroleum coke and pitch, as well as keeping 35 kg of aluminum fluoride out of the air.

Recycling also reduces energy consumption and air pollution. Plastic bottle recycling could save 50 to 60 percent of the energy needed to make new ones. Making new steel from old scrap offers up to 75 percent energy savings. Producing aluminum from scrap instead of bauxite ore cuts energy use by 95 percent, yet the United States still throws away more than a million tons of aluminum every year. If aluminum recovery were doubled worldwide, more than a million tons of air pollutants would be eliminated every year.

Reducing litter is an important benefit of recycling. Ever since disposable paper, glass, metal, foam, and plastic packaging began to accompany nearly everything we buy, these discarded

reliable market for their product in order to operate efficiently. Often, because of tax policies and economies of scale, it's cheaper to use virgin resin than recycled plastic. Governments can help create incentives for consumers to recycle products and for producers to use those materials.

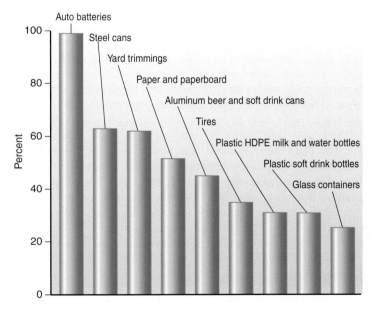

Figure 13.12 Recycling rates for selected materials in the United States. Battery recycling, which is required by law, is very successful. Other materials, even though valuable for reuse, have mixed recycling success.

Source: Data from Environmental Protection Agency, 2007.

Figure 13.13 Composting is a good way to convert yard waste, vegetable scraps, and other organic materials into useful garden mulch. Mix everything together, keep it moist and well aerated, and in a few weeks, you will have a rich, odor-free mulch. This three-bin system allows you to have batches in different stages.

wrappings have collected on our roadsides and in our lakes, rivers, and oceans. Litter is a costly as well as unsightly problem. Americans pay an estimated 32¢ for each piece of litter picked up by crews along state highways, which adds up to $500 million every year. "Bottle bills" requiring deposits on bottles and cans have reduced littering in many states.

Composting recycles organic waste

Pressed for landfill space, many cities have banned yard waste from municipal garbage. Rather than bury this valuable organic material, they are turning it into a useful product through **composting**: biological degradation or breakdown of organic matter under aerobic (oxygen-rich) conditions. The organic compost resulting from this process makes a nutrient-rich soil amendment that aids water retention, slows soil erosion, and improves crop yields. A home compost pile is an easy and inexpensive way to dispose of organic waste in an interesting and environmentally friendly way. All you need to do is to pile up lawn clippings, vegetable waste, fallen leaves, wood chips, or other organic matter in an out-of-the way place, keep it moist, and turn it over every week or so (fig. 13.13). Within a few months, naturally occurring microorganisms will decompose the organic material into a rich, pleasant-smelling compost that you can use as a soil amendment.

Some cities and counties have developed successful composting programs. Residents pay a small amount to drop off yard waste, and they can buy back rich, inexpensive soil amendment.

Energy from waste

Every year we throw away the energy equivalent of 80 million barrels of oil in organic waste in the United States. In developing countries up to 85 percent of the waste stream is food, textiles, vegetable matter, and other biodegradable materials. Worldwide, at least one-fifth of municipal waste is organic kitchen and garden refuse.

Figure 13.14 Reusing discarded products is a creative and efficient way to reduce wastes. This recycling center in Berkeley, California, is a valuable source of used building supplies and a money saver for the whole community.

Methane capture from landfills is relatively inefficient. A better strategy is converting organic waste to methane in a contained, anaerobic digester. While composting breaks down organic waste in oxygen-rich environments, an oxygen-free environment produces methane. Many new technologies are being developed or are in use: most use steam and pressure to "cook" waste and release clean methane.

Anaerobic digestion also can be done on a small scale. Millions of household methane generators provide fuel for cooking and lighting for homes in China and India (chapter 12). In the United States some farmers produce all the fuel they need to run their farms—both for heating and for running trucks and tractors—by generating methane from animal manure.

Reuse is even more efficient than recycling

Even better than recycling or composting is cleaning and reusing materials in their present form, thus saving the cost and energy of remaking them into something else. We do this already with some specialized items. Auto parts are regularly sold from junkyards, especially for older car models. In some areas stained-glass windows, brass fittings, fine woodwork, and bricks salvaged from old houses bring high prices. Some communities sort and reuse a variety of materials received in their dumps (fig. 13.14).

In many cities, glass and plastic bottles are routinely returned to beverage producers for washing and refilling. The reusable, refillable bottle is the most efficient beverage container we have. It is better for the environment than remelting and more profitable for local communities. A reusable glass container makes an average of 15 round-trips between factory and customer before it becomes so scratched and chipped that it has to be recycled. Reusable containers also favor local bottling companies and help preserve regional differences.

Since the advent of cheap, lightweight, disposable food and beverage containers, many small, local breweries, canneries, and bottling companies have been forced out of business by huge national conglomerates. These big companies can afford to ship food and beverages great distances, as long as it is a one-way trip. If they had to collect their containers and reuse them, canning and bottling factories serving large regions would be uneconomical. Consequently, the national companies favor recycling rather than refilling because they prefer fewer, larger plants and don't want to be responsible for collecting and reusing containers. In some circumstances, life-cycle assessment shows that washing and decontaminating containers takes as much energy and produces as much air and water pollution as manufacturing new ones.

In many less-affluent nations, reuse of all sorts of manufactured goods is an established tradition. Where most manufactured products are expensive and labor is cheap, it pays to salvage, clean, and repair products. Cairo, Manila, Mexico City, and many other cities have large populations of poor people who make a living by scavenging. Entire ethnic populations may survive on scavenging, sorting, and reprocessing scraps from city dumps. Eliminating waste shipments can reduce pollution but may threaten livelihoods. How can we resolve this dilemma?

Reducing waste is often the cheapest option

Most of our attention in waste management focuses on recycling. But slowing the consumption of throw-away products is by far the most effective way to save energy, materials, and money. The "three Rs" waste hierarchy—reduce, reuse, recycle—lists the most important strategy first. Industries are increasingly finding that reducing saves money. Soft-drink makers use less aluminum per can than they did 20 years ago, and plastic bottles use less plastic. 3M has saved over $500 million in the past 30 years by reducing its use of raw materials, reusing waste products, and increasing efficiency. Individual action is essential, too (see What Can You Do?).

Excess packaging of food and consumer products is one of our greatest sources of unnecessary waste. Paper, plastic, glass, and metal packaging material makes up 50 percent of our domestic trash by volume (see related story "South Africa's National Flower?" at www.mhhe.com/cunningham5e). Much of that packaging is primarily for marketing and has little to do with product protection (fig. 13.15). Manufacturers and retailers might be persuaded to reduce these wasteful practices if consumers ask for products without excess packaging. Canada's National Packaging Protocol (NPP) recommends that packaging minimize depletion of virgin resources and production of toxins in manufacturing. The preferred hierarchy is (1) no packaging, (2) minimal packaging, (3) reusable packaging, and (4) recyclable packaging. This plan sets an ambitious target of 50 percent reduction in excess packaging.

In 2008, China banned ultrathin (less than 0.025 mm) plastic bags and called for a return to reusable cloth bags for shopping. This could eliminate up to 3 billion plastic bags used every day in China. Japan, Ireland, South Africa, and Taiwan also have discouraged single-use plastic bags through taxes or prohibitions. In 2007, San Francisco became the first American city to outlaw petroleum-based plastic grocery bags.

Figure 13.15 How much more do we need? Where will we put what we already have?
Reprinted with special permission of King Features Syndicate.

What Can You Do?

Reducing Waste

1. Buy foods that come with less packaging; shop at farmers' markets or co-ops, using your own containers.
2. Take your own washable, refillable beverage container to meetings or convenience stores.
3. When you have a choice at the grocery store among plastic, glass, or metal containers for the same food, buy the reusable or easier-to-recycle glass or metal.
4. Separate your cans, bottles, papers, and plastics for recycling.
5. Wash and reuse bottles, aluminum foil, plastic bags, and so on for your personal use.
6. Compost yard and garden wastes, leaves, and grass clippings.
7. Help your school develop responsible systems for disposing of electronics and other waste.
8. Write to your senators and representatives, and urge them to vote for container deposits, recycling, and safe incinerators or landfills.

Source: Data from Minnesota Pollution Control Agency.

Figure 13.16 Modern society produces large amounts of toxic and hazardous waste.

Where disposable packaging is necessary, we still can reduce the volume of waste in our landfills by using materials that are compostable or degradable. **Photodegradable plastics** break down when exposed to ultraviolet radiation. **Biodegradable plastics** incorporate such materials as cornstarch that microorganisms can decompose. Several states have introduced legislation requiring biodegradable or photodegradable six-pack beverage yokes, fast-food packaging, and disposable diapers. These degradable plastics often don't decompose completely; they only break down to small particles that remain in the environment. In doing so, they can release toxic chemicals. And in modern, lined landfills, they don't decompose at all. Furthermore, they make recycling less feasible and may lead people to believe that littering is okay.

13.4 Hazardous and Toxic Wastes

The most dangerous aspect of the waste stream is that it often contains highly toxic and hazardous materials that are injurious to both human health and environmental quality (fig. 13.16). We

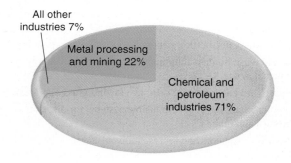

Figure 13.17 Producers of hazardous wastes in the United States.
Source: Data from the U.S. Environmental Protection Agency, 2006.

now produce and use a vast array of flammable, explosive, caustic, acidic, and highly toxic chemical substances for industrial, agricultural, and domestic purposes. According to the EPA, U.S. industries generate about 900 million metric tons of officially classified hazardous wastes each year, about 3 metric tons for each person in the country. In addition, considerably more toxic and hazardous waste material is generated by industries or processes not regulated by the EPA. Shockingly, at least 40 million metric tons (22 billion lbs) of toxic and hazardous wastes are released into the air, water, and land in the United States each year. The biggest sources of these toxins are the chemical and petroleum industries (fig. 13.17).

Hazardous waste includes many dangerous substances

Legally, a **hazardous waste** is any discarded material, liquid or solid, that contains substances known to be (1) fatal to humans or laboratory animals in low doses; (2) toxic, carcinogenic, mutagenic, or teratogenic to humans or other life-forms; (3) ignitable with a flash point less than 60°C; (4) corrosive; or (5) explosive or highly reactive (undergoes violent chemical reactions either by itself or when mixed with other materials). Notice that this definition includes both toxic and hazardous materials, as defined in chapter 8. Certain compounds are exempt from regulation as hazardous waste if they are accumulated in less than 1 kg (2.2 lbs) of commercial chemicals or 100 kg of contaminated soil, water, or debris. Even larger amounts (up to 1,000 kg) are exempt when stored at an approved waste treatment facility for the purpose of being beneficially used, recycled, reclaimed, detoxified, or destroyed.

Most hazardous waste is recycled, converted to nonhazardous forms, stored, or otherwise disposed of on-site by the generators—chemical companies, petroleum refiners, and other large industrial facilities—so that it doesn't become a public problem. Still, the hazardous waste that does enter the waste stream or the environment represents a serious environmental problem. And orphan wastes left behind by abandoned industries remain a serious threat to both environmental quality and human health. For years little attention was paid to this material. Wastes stored on private property, buried, or allowed to soak into the ground were considered of little concern to the public. An estimated 5 billion metric tons of highly poisonous chemicals were improperly disposed of in the United States between 1950 and 1975 before regulatory controls became more stringent.

Federal legislation regulates hazardous waste

Two important federal laws regulate hazardous waste management and disposal in the United States. The Resource Conservation and Recovery Act (RCRA, pronounced "rickra") of 1976 is a comprehensive program that requires rigorous testing and management of toxic and hazardous substances. A complex set of rules requires generators, shippers, users, and disposers of these materials to keep meticulous account of everything they handle and what happens to it from generation (cradle) to ultimate disposal (grave) (fig. 13.18).

The Comprehensive Environmental Response, Compensation, and Liability Act (CERCLA or Superfund Act), passed in 1980 and modified in 1984 by the Superfund Amendments and Reauthorization Act (SARA), is aimed at rapid containment, cleanup, or remediation of abandoned toxic waste sites. This statute authorizes the EPA to undertake emergency actions when a threat exists that toxic material will leak into the environment. The EPA is empowered to bring suit for the recovery of its costs from potentially responsible parties, such as site owners, operators, waste generators, or transporters.

SARA also established (under title III) a community right to know and state emergency response plans that give citizens access to information about what is present in their communities. One of the most useful tools in this respect is the **Toxic Release Inventory**, which requires 20,000 manufacturing facilities to report annually on releases of more than 300 toxic materials. You can find specific information in the inventory about what is in your neighborhood.

The government does not have to prove that anyone violated a law or what role he or she played in a Superfund site. Rather, liability under CERCLA is "strict, joint, and several," meaning that anyone associated with a site can be held responsible for the entire cost of cleaning it up, no matter how much of the mess they made. In some cases property owners have been assessed millions of dollars for removal of wastes left there years earlier by previous owners. This strict liability has been a headache for the real estate and insurance businesses.

CERCLA was amended in 1995 to make some of its provisions less onerous. In cases where treatment is unavailable or too costly and it is likely that a less costly remedy will become available within a reasonable time, interim containment is now allowed. The EPA also now has the discretion to set site-specific cleanup levels, rather than adhere to rigid national standards.

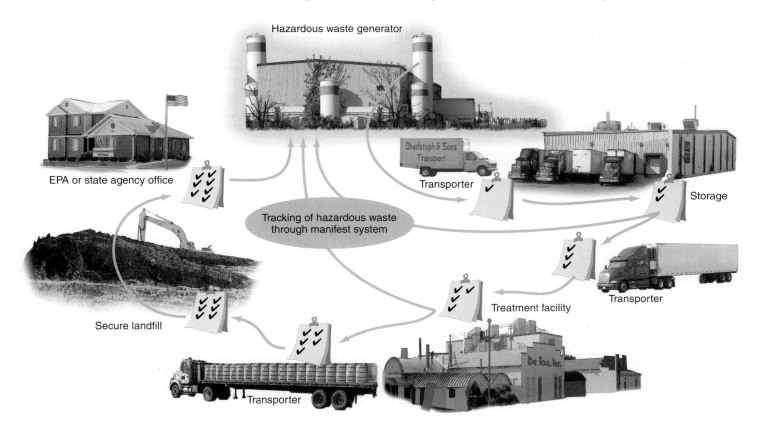

Figure 13.18 Toxic and hazardous wastes must be tracked from "cradle to grave" by detailed shipping manifests.

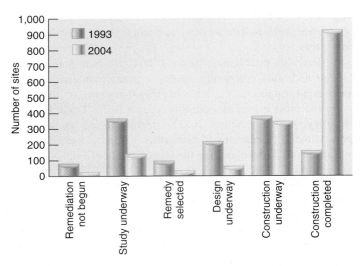

Figure 13.19 Progress on Superfund National Priority List (NPL) sites. After years of little progress, the number of completed sites jumped from 155 in 1993 to 926 in 2004. Over 90 percent of the 1,529 currently listed NPL sites are under construction or completed.
Source: Data from Environmental Protection Agency, 2004.

Figure 13.20 Some of the hazardous waste sites on the EPA priority cleanup list. Sites located on aquifer recharge zones represent an especially serious threat. Once groundwater is contaminated, cleanup is difficult and expensive. In some cases, it may not be possible.
Source: Environmental Protection Agency.

Superfund sites are those listed for federal cleanup

The EPA estimates that there are at least 36,000 seriously contaminated sites in the United States. The General Accounting Office (GAO) places the number much higher, perhaps more than 400,000 when all are identified. Originally, about 1,671 sites were placed on the National Priority List (NPL) for cleanup with financing from the federal Superfund program. The **Superfund** is a revolving pool designed to (1) provide an immediate response to emergency situations that pose imminent hazards and (2) to clean up or remediate abandoned or inactive sites. Without this fund, sites would languish for years or decades while the courts decided who was responsible for paying for the cleanup. Originally a $1.6 billion pool, the fund peaked at $3.6 billion. From its inception, the fund was financed by taxes on producers of toxic and hazardous wastes. Industries opposed this "polluter pays" tax, because current manufacturers are often not the ones responsible for the original contamination. In 1995 Congress agreed to let the tax expire. Since then the Superfund has dwindled, and the public has picked up an increasing share of the bill. In the 1980s the public covered less than 20 percent of the Superfund. In 2004, however, general revenues (public funds) paid the entire cost off a greatly reduced program, and the industry share was zero.

Total costs for hazardous waste cleanup in the United States are estimated to be between $370 billion and $1.7 trillion, depending on how clean sites must be and what methods are used. For years, Superfund money was spent mostly on lawyers and consultants, and cleanup efforts were often bogged down in disputes over liability and best cleanup methods. During the 1990s, however, progress improved substantially, with a combination of rule adjustments and administrative commitment to cleanup. By 2004, more than half (926) of the original NPL sites were listed as completed in cleanup or containment (fig. 13.19). In recent years, though, progress has slowed because of underfunding and a lower priority among administrators.

What qualifies a site for the NPL? These sites are considered to be especially hazardous to human health and environmental quality because they are known to be leaking or have a potential for leaking supertoxic, carcinogenic, teratogenic, or mutagenic materials (chapter 8). The ten substances of greatest concern or most commonly detected at Superfund sites are lead, trichloroethylene, toluene, benzene, PCBs, chloroform, phenol, arsenic, cadmium, and chromium. These and other hazardous or toxic materials are known to have contaminated groundwater at 75 percent of the sites now on the NPL. In addition, 56 percent of these sites have contaminated surface waters, and airborne materials are found at 20 percent of the sites. Seventy million Americans, including 10 million children, live within 6 km of a Superfund site.

Where are these thousands of hazardous waste sites, and how did they get contaminated? Old industrial facilities, such as smelters, mills, petroleum refineries, and chemical manufacturing plants, are highly likely to have been sources of toxic wastes. Regions of the country with high concentrations of aging factories, such as the "rust belt" around the Great Lakes or the Gulf Coast petrochemical centers, have large numbers of Superfund sites (fig. 13.20). Mining districts also are prime sources of toxic and hazardous waste. Within cities, factories and places such as railroad yards, bus repair barns, and filling stations, where solvents, gasoline, oil, and other petrochemicals were spilled or dumped on the ground, often are highly contaminated.

Some of the most infamous toxic waste sites were old dumps where many different materials were mixed together indiscriminately. For instance, Love Canal in Niagara Falls, New York, was an open dump that both the city and nearby chemical factories used as a disposal site. More than 20,000 tons of toxic chemical waste were buried under what later became a housing development. Another infamous example occurred in Hardeman County, Tennessee, where about a quarter of a million barrels of chemical waste were buried in shallow pits that subsequently leaked toxins into the groundwater.

Brownfields present both liability and opportunity

Among the biggest problems in cleaning up hazardous waste sites are questions of liability and the degree of purity required. In many cities, these problems have created large areas of contaminated properties, known as **brownfields**, that have been abandoned or are not being used to their potential because of real or suspected pollution. Up to one-third of all commercial and industrial sites in the urban core of many big cities fall in this category. In heavy industrial corridors the percentage typically is higher.

For years no one was interested in redeveloping brownfields because of liability risks. Who would buy a property, knowing that they might be forced to spend years in litigation and negotiations and be forced to pay millions of dollars for pollution they didn't create? Even if a site has been cleaned to current standards, there is a worry that additional pollution might be found in the future or that more stringent standards might be applied.

In many cases, property owners complain that unreasonably high levels of purity are demanded in remediation programs. Consider the case of Columbia, Mississippi. For many years a 35-ha (81-acre) site in Columbia was used for turpentine and pine tar manufacturing. Soil tests showed concentrations of phenols and other toxic organic compounds exceeding federal safety standards. The site was added to the Superfund NPL, and remediation was ordered. Some experts recommended that the best solution was to simply cover the surface with clean soil and enclose the property with a fence to keep people out. The total costs would have been about $1 million. Instead, the EPA ordered Reichhold Chemical, the last known property owner, to excavate more than 12,500 tons of soil and haul it to a commercial hazardous waste dump in Louisiana at a cost of some $4 million. The intention is to make the site safe enough to be used for any purpose, including housing—even though no one has proposed building anything there. According to the EPA, the dirt must be clean enough for children to play in— even eat—without risk.

Similarly, in places where contaminants have seeped into groundwater, the EPA generally demands that cleanup be carried to drinking-water standards. Many critics believe that these pristine standards are unreasonable. Former Congressman Jim Florio, a principal author of the original Superfund Act, says, "It doesn't make any sense to clean up a rail yard in downtown Newark so it can be used as a drinking water reservoir." Depending on where the site is, what else is around it, and what its intended uses are, much less stringent standards may be perfectly acceptable.

Brownfield redevelopment is increasingly seen as an opportunity for rebuilding cities, creating jobs, increasing the tax base, and preventing needless destruction of open space at urban mar-

gins. In 2002 the EPA established a new brownfields revitalization fund designed to encourage restoration of more sites, as well as more kinds of sites. In some communities former brownfields are being turned into "eco-industrial parks" that feature environmentally friendly businesses and bring in much-needed jobs to inner-city neighborhoods (chapter 14).

Hazardous waste must be processed or stored permanently

What shall we do with toxic and hazardous wastes? In our homes, we can reduce waste generation and choose less toxic materials. Buy only what you need for the job at hand. Use up the last little bit, or share leftovers with a friend or neighbor. Many common materials that you probably already have make excellent alternatives to commercial products.

Produce less waste

As with other wastes, the safest and least expensive way to avoid hazardous waste problems is to avoid creating the wastes in the first place. Manufacturing processes can be modified to reduce or eliminate waste production. In Minnesota, the 3M Company reformulated products and redesigned manufacturing processes to eliminate more than 140,000 metric tons of solid and hazardous wastes, 4 billion liters (1 billion gal) of wastewater, and 80,000 metric tons of air pollution each year. It frequently found that these new processes not only spared the environment but also saved money by using less energy and fewer raw materials.

Recycling and reusing materials also eliminates hazardous wastes and pollution. Many waste products of one process or industry are valuable commodities in another. Already, about 10 percent of the wastes that would otherwise enter the waste stream in the United States are sent to surplus material exchanges, where they are sold as raw materials for use by other industries. This figure could probably be raised substantially with better waste management. In Europe at least one-third of all industrial wastes are exchanged through clearinghouses, where beneficial uses are found. This represents a double savings: the generator doesn't have to pay for disposal, and the recipient pays little, if anything, for raw materials.

Convert to less hazardous substances

Several processes are available to make hazardous materials less toxic. *Physical treatments* tie up or isolate substances. Charcoal or resin filters absorb toxins. Distillation separates hazardous components from aqueous solutions. Precipitation and immobilization in ceramics, glass, or cement isolate toxins from the environment, so that they become essentially nonhazardous. One of the few ways to dispose of metals and radioactive substances is to fuse them in silica at high temperatures to make a stable, impermeable glass that is suitable for long-term storage. Plants, bacteria, and fungi can also concentrate or detoxify contaminants (see Exploring Science p. 324).

Incineration is applicable to mixtures of wastes. A permanent solution to many problems, it is quick and relatively easy, but

Exploring
SCIENCE:

Bioremediation

Cleaning up the thousands of hazardous waste sites at factories, farms, and gas stations is an expensive project. In the United States alone, waste cleanup is projected to cost at least $700 billion. Usually hazardous waste remediation (cleanup) involves digging up soil and incinerating it, potentially releasing toxins into the air, or trucking it to a secure landfill. Contaminated groundwater is frequently pumped out of the ground; hopefully, contaminants are retrieved at the same time.

How do plants, bacteria, and fungi do all this? Many of the biophysical details are poorly understood, but in general, plant roots are designed to efficiently extract nutrients, water, and trace minerals from soil and groundwater. The mechanisms involved may aid extraction of metallic and organic contaminants. Some plants also use toxic elements as a defense against herbivores: locoweed, for example, selectively absorbs elements such as selenium, concentrat-

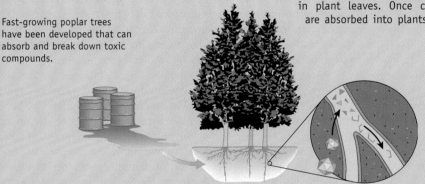

Fast-growing poplar trees have been developed that can absorb and break down toxic compounds.

ing toxic levels in its leaves. Absorption can be extremely effective. Bracken fern growing in Florida has been found to contain arsenic at concentrations more than 200 times higher than in the soil in which it was growing.

Genetically modified plants are also being developed to process toxins. Poplars have been developed to process toxins, using a gene borrowed from bacteria that transforms a toxic compound of mercury into a safer form. In another experiment, a gene for producing mammalian liver enzymes, which specialize in breaking down toxic organic compounds, was inserted into tobacco plants. The plants succeeded in producing the liver enzymes and breaking down toxins absorbed through their roots.

These remediation methods are not without risks. Insects could consume leaves containing concentrated substances, allowing contaminants to enter the food chain. Some absorbed contaminants are volatilized, or emitted in gaseous form, through pores in plant leaves. Once contaminants are absorbed into plants, the plants

themselves are usually toxic and must be landfilled. But the cost of phytoremediation can be less than half the cost of landfilling or treating toxic soil, and the volume of plant material requiring secure storage is a fraction of the volume of the contaminated dirt.

Cleaning up hazardous and toxic waste sites will be a big business for the foreseeable future, in North America and around the world. Innovations such as bioremediation offer promising prospects for business development, as well as for environmental health and saving taxpayer money.

A promising alternative to these methods involves **bioremediation**, or biological waste treatment. Microscopic bacteria and fungi can absorb, accumulate, and detoxify a remarkable variety of toxic compounds. They can also accumulate heavy metals, and some have been developed that can metabolize (break down) PCBs. Aquatic plants such as water hyacinths and cattails can also be used to purify contaminated effluent.

Recently, an increasing variety of plants have been used in phytoremediation (cleanup using plants). Some types of mustard can extract lead, arsenic, zinc, and other metals from contaminated soil. Radioactive strontium and cesium have been extracted from soil near the Chernobyl nuclear power plant using common sunflowers. Poplar trees can absorb and break down toxic organic chemicals. Natural bacteria in groundwater, when provided with plenty of oxygen, can neutralize contaminants in aquifers. Experiments have shown that pumping air *into* groundwater can be a more effective cleanup method than pumping water *out*.

---

not necessarily cheap—nor always clean—unless done correctly. Wastes must be heated to over 1,000°C (2,000°F) for a sufficient period of time to complete destruction. The ash resulting from thorough incineration is reduced in volume up to 90 percent and often is safer to store in a landfill or another disposal site than the original wastes. Nevertheless, incineration remains highly controversial (fig. 13.21).

Chemical processing can transform materials to make them nontoxic. Included in this category are neutralization, removal of metals or halogens (chlorine, bromine, etc.), and oxidation. The Sunohio Corporation of Canton, Ohio, for instance, has developed a process called PCBx, in which chlorine in such molecules as PCBs is replaced with other ions that render the compounds less toxic. A portable unit can be moved to the location of the hazardous wastes, eliminating the need for shipping them.

Store permanently

Inevitably, there will be some materials that we can't destroy, make into something else, or otherwise cause to vanish. We will have to store them out of harm's way. There are differing opinions about how best to do this.

Retrievable Storage   Dumping wastes in the ocean or burying them in the ground generally means that we have lost control of them. If we learn later that our disposal technique was a mistake, it is difficult, if not impossible, to go back and recover the wastes. For many supertoxic materials, the best way to store them may be in **permanent retrievable storage**. This means placing waste storage containers in a secure building, salt mine, or bedrock cavern, where they can be inspected periodically and retrieved, if

Figure 13.21 Actor Martin Sheen joins local activists in a protest in East Liverpool, Ohio, site of the largest hazardous waste incinerator in the United States. About 1,000 people marched to the plant to pray, sing, and express their opposition. Involving celebrities draws attention to your cause. A peaceful, well-planned rally builds support and acceptance in the broader community.

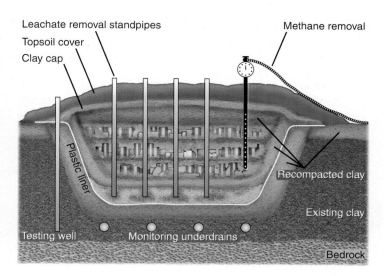

Figure 13.22 A secure landfill for toxic waste. A thick plastic liner and two or more layers of impervious compacted clay enclose the landfill. A gravel bed between the clay layers collects any leachate, which can then be pumped out and treated. Well samples are tested for escaping contaminants and methane is collected for combustion.

necessary, for repacking or for transfer if a better means of disposal is developed. This technique is more expensive than burial in a landfill because the storage area must be guarded and monitored continuously to prevent leakage, vandalism, or other dispersal of toxic materials. Remedial measures are much cheaper with this technique, however, and it may be the best system in the long run.

Secure Landfills One of the most popular solutions for hazardous waste disposal has been landfilling. Although, as we saw earlier in this chapter, many such landfills have been environmental disasters, newer techniques make it possible to create safe, modern **secure landfills** that are acceptable for disposing of many hazardous wastes. The first line of defense in a secure landfill is a thick bottom cushion of compacted clay that surrounds the pit like a bathtub (fig. 13.22). Moist clay is flexible and resists cracking if the ground shifts. It is impermeable to groundwater and will safely contain wastes. A layer of gravel is spread over the clay liner, and perforated drainpipes are laid in a grid to collect any seepage that escapes from the stored material. A thick polyethylene liner, protected from punctures by soft padding materials, covers the gravel bed. A layer of soil or absorbent sand cushions the inner liner, and the wastes are packed in drums, which then are placed into the pit, separated into small units by thick berms of soil or packing material.

When the landfill has reached its maximum capacity, a cover much like the bottom sandwich of clay, plastic, and soil—in that order—caps the site. Vegetation stabilizes the surface and improves its appearance. Sump pumps collect any liquids that filter through the landfill, either from rainwater or leaking drums. This leachate is treated and purified before being released. Monitoring wells check groundwater around the site to ensure that no toxins have escaped.

Most landfills are buried below ground level to be less conspicuous; however, in areas where the groundwater table is close to the surface, it is safer to build above-ground storage. The same protective construction techniques are used as in a buried pit. An advantage to such a facility is that leakage is easier to monitor because the bottom is at ground level.

Transportation of hazardous wastes to disposal sites is of concern because of the risk of accidents. Emergency-preparedness officials conclude that the greatest risk in most urban areas is not nuclear war or natural disaster but crashes involving trucks or trains carrying hazardous chemicals through densely packed urban corridors. Another worry is who will bear financial responsibility for abandoned waste sites. Hazardous wastes remain toxic long after the businesses that created them are gone. As is the case with nuclear wastes (chapter 12), we may need new institutions for perpetual care of these wastes.

Conclusion

In many traditional societies, people reuse nearly everything because they can't afford to discard useful resources. Modern society, however, produces a prodigious amount of waste. Government policies and economies of scale make it cheaper and more convenient to extract virgin raw materials to make new consumer products rather than to reuse or recycle items that still have useful life. We're now beginning to recognize the impacts of this wasteful lifestyle. We see the problems associated with waste disposal as well as the impacts of energy and material resource extraction. The increasing toxicity of modern products makes waste reduction even more urgent. The mantra of reduction, reuse, and recycle is becoming more widely accepted.

There are increasing opportunities to exchange materials with others who can use them, or to recycle them into other products. A big market for used construction supplies and surplus chemicals allows salvage of stuff that would otherwise go to landfills. Vehicles, electronics, and other complex products are demanufactured to reclaim valuable metals. Paint, used carpet, food and beverage containers, and many other unwanted consumer products are transformed into new merchandise. Organic matter can be composted into beneficial soil amendments. Some pioneers in sustainability find they can live comfortably while producing no waste at all if they practice reduction, reuse, and recycling faithfully.

How much waste do you produce, and where does it go after you toss it into the garbage can? What can you do to reduce your personal waste flow? Is recycling and reuse widely accepted in your community? If not, what could you do to change attitudes toward trash?

Practice Quiz

1. List some items that can be recycled from construction and demolition waste.

2. What are *solid wastes* and *hazardous wastes*? What is the difference between them?

3. Describe the difference between an open dump, a sanitary landfill, and a modern, secure, hazardous waste disposal site.

4. Describe some concerns about waste incineration.

5. List some benefits and drawbacks of recycling wastes. What are the major types of materials recycled from municipal waste, and how are they used?

6. What is *e-waste*? How is most of it disposed of, and what are some strategies for improving recycling rates?

7. What is *composting*, and how does it fit into solid waste disposal?

8. What materials are most recycled in the United States?

9. What are *brownfields*, and why do cities want to redevelop them?

10. What are *bioremediation* and *phytoremediation*? What are some advantages to these methods?

Critical Thinking and Discussion Questions

Apply the principles you have learned in this chapter to discuss these questions with other students.

1. A toxic waste disposal site has been proposed for the Pine Ridge Indian Reservation in South Dakota. Many tribal members oppose this plan, but some favor it because of the jobs and income it will bring to an area with 70 percent unemployment. If local people choose immediate survival over long-term health, should we object or intervene?

2. Should industry officials be held responsible for dumping chemicals that were legal when they did it but are now known to be extremely dangerous? At what point can we argue that they should have known about the hazards involved?

3. Suppose that your brother or sister has decided to buy a house next to a toxic waste dump because it costs $20,000 less than a comparable house elsewhere. What do you say to him or her?

4. Is there a fundamental difference between incinerating municipal, medical, or toxic industrial waste? Would you oppose an incinerator in your neighborhood for one type of waste but not others? Why or why not?

5. Some scientists argue that permanent retrievable storage of toxic and hazardous wastes is preferable to burial. How can we be sure that material that will be dangerous for thousands of years will remain secure? If you were designing such a repository, how would you address this question?

 **Data Analysis:** How Much Waste Do You Produce, and How Much Do You Know How to Manage?

As people become aware of waste disposal problems in their communities, more people are recycling more materials. Some things are easy to recycle, such as newsprint, office paper, or aluminum drink cans. Other things are harder to classify. Most of us give up pretty quickly and throw things in the trash if we have to think too hard about how to recycle them.

1. Take a poll to find out how many people in your class know how to recycle the items in the table on page 327. Once you have taken your poll, convert the numbers to percentages: divide the number who know how to recycle each item by the number of students in your class, and then multiply by 100.

2. Now find someone on your campus who works on waste management. This might be someone in your university/college administration, or it might be someone who actually empties trash containers. (You might get more interesting and straightforward answers from the latter.) Ask the following questions: (1) Can this person fill in the items your class didn't know about? (2) Is there a college/university policy about recycling? What are some of the points on that policy? (3) How much does the college spend each year on waste disposal? How many tuition payments does that total? (4) What are the biggest parts of the waste stream? (5) Does the school have a plan for reducing that largest component?

Item	Percentage Who Know How to Recycle
Newspapers	
Paperboard (cereal boxes)	
Cardboard boxes	
Cardboard boxes with tape	
Plastic drink bottles	
Other plastic bottles	
Styrofoam food containers	
Food waste	
Plastic shopping bags	
Plastic packaging materials	
Furniture	
Last year's course books	
Left-over paint	

Like mega cities in many developing countries, New Delhi, India, suffers from traffic congestion and air pollution.

14 Economics and Urbanization

Learning Outcomes

After studying this chapter, you should be able to answer the following questions:

- How have the size and location of the world's largest cities changed over the past century?
- Define slum and shantytown, and describe the conditions you might find in them.
- What is urban sprawl? How have automobiles contributed to sprawl?
- What are some principles of smart growth and new urbanism?
- Describe sustainable development and why it's important.
- What value do we get from free ecological services?
- What's the difference between GNP and GPI?
- What do we mean by internalizing external costs?

What kind of world do you want to live in? Demand that your teachers teach you what you need to know to build it.

–Peter Kropotkin

Curitiba: A Model Sustainable City

A few decades ago, Curitiba, Brazil, like many cities in the developing world, faced rapid population growth, air pollution, congested streets, and inadequate waste disposal systems. In 1969, Jamie Lerner, a landscape architect and former student activist, ran for mayor on a platform of urban renewal, social equity, and environmental protection. For the next 30 years, first as mayor, and then governor of Parana State, Lerner instituted social reforms and urban planning that have made the city an outstanding model of sustainable urban development. Today, with a population of about 1.8 million, Curitiba has an international reputation for progressive social programs, environmental protection, cleanliness, and livability. An international conference of mayors and city planners called Curitiba the most innovative city in the world.

The master plan instituted by Lerner called for rational municipal zoning and land-use regulations—rare provisions for a city in the developing world. An extensive network of parks and open space protected from future development extends throughout the city and provides access to nature as well as recreation opportunities. It also helps protect the watershed and prevent floods that once afflicted the city. Tanguá Park, for example, Curitiba's newest, is built on land once destined to be a garbage dump. The dump was moved elsewhere, however, and the park now encompasses about 250 ha of open space, including a rare stand of Brazilian pine (*Araucaria agustafolia*), the symbol of Parana State. Together, the many parks, bicycle trails, city gardens, and public riverbanks make the city a beautiful and pleasant place to live. Curitiba has more open space per capita than most American cities.

The heart of Curitiba's environmental and social plan is education for everyone. The first federal university in Brazil was built here in 1912.

Figure 14.1 Much of Curitiba's city center has been converted to pedestrian shopping streets, while historic buildings have been converted to new uses. Black and white tile mosaics are featured throughout the city.

Free municipal schools give the city the highest literacy rate in the country. Environmental education reaches to everyone as well. School children study ecology along with Portuguese and math. With the help of children, who encourage their parents, the city has instituted a complex program that separates organic waste, trash, plastic, glass, and metal for recycling. Everything reusable is salvaged from the waste stream and sold as raw material to local industries. The city calculates that 1,200 trees per day are saved by paper recycled in this program. "Imagine if the whole of Brazil did this," Lerner exclaims. "We could save 26 million trees per year!" More than 70 percent of the city residents now participate in recycling.

Environmental protection and resource conservation also contribute to social welfare programs. Low-income Curitibanos are employed to pick up and sort recyclables, as well as to work in civic gardens and a city-run farm. Along with the many street sweepers in their distinctive orange suits, these programs provide jobs and pride while also keeping the city clean and beautiful. More than 1,100 municipal facilities provide services to the entire population. Free day care is available to all working mothers with children under 6 years old. Eldercare programs provide psychological services as well as food and housing to seniors. New job-training and small-business incubators run by the city help Curitibanos learn technical skills and launch new enterprises. The city has also built a technology park to attract new-economy businesses.

One of the crises that originally motivated Mayor Lerner was preservation of the historic central business district of his city. Traffic choked narrow streets and air pollution drove wealthy residents and businesses out of the central city, leaving beautiful old buildings to decay and eventually be torn down. In one of his boldest moves, Lerner banned most vehicles from the central business district and turned the streets into pedestrian malls. Rather than depend on private automobiles to move people around, Curitiba built a remarkably successful and economical bus-based rapid transit system. More than 340 feeder routes throughout the metropolitan area link to high-speed articulated buses that travel on dedicated busways. Everyone in the city lives within walking distance of frequent, economical, rapid public transportation. Today more than three-quarters of all trips made each day in the city utilize this innovative system, a percentage beyond the fondest imagination of most American city planners.

Benches, fountains, landscaping, distinctive lighting, and other amenities along with European-style shops and sidewalk cafes now make Curitiba's central district a pleasant place to stroll, shop, or simply gather with friends. Historic buildings have been refurbished, and many have been recycled for new purposes (fig. 14.1). Throughout the city center, walking streets are paved with characteristic black and white ceramic mosaics that draw on the city's colonial past.

Curitiba illustrates a number of ways that we can live sustainably with our environment and with each other. In this chapter, we'll look at other aspects of city planning and urban environments as well as some principles of ecological economics that help us understand the nature of resources and the choices we face both as individuals and communities.

14.1 Cities Are Places of Crisis and Opportunity

More than half of humans now live in cities, and in the next quarter century that number will approach three-quarters of us. This is a dramatic change from all previous human history, in which most humans lived by hunting and gathering, farming, or fishing. Since the beginning of the industrial revolution about 300 years ago, cities have grown rapidly in both size and power (fig. 14.2). By 1950, 38 percent of the world's population lived in cities; by 2030 that proportion will nearly double (table 14.1).

The vast majority of urban growth will occur in less-developed countries (fig. 14.3). Populations in these cities are expanding far faster than infrastructure, including roads and transportation, housing, water supplies, sewage treatment, and schools, can possibly grow. Building new infrastructure is especially hard in a poor country, where incomes are low and tax collection is insufficient to support public services. All these challenges make Curitiba's success that much more astonishing. However, even Curitiba is straining to keep up, as the city has more than doubled in size in less than two decades.

Despite these challenges, cities are also places where innovation occurs. Ideas mix and experimentation happens in urban areas. Diverse employment opportunities and new economies arise in cities, as well as concentrations of poverty. Huge **urban agglomerations** (mergers of multiple municipalities) are forming throughout the world. Some have become **megacities** (with populations over 10 million people). While these cities pave over vast landscapes and consume inconceivable amounts of resources, they are also relatively efficient in resource use. Environmental degradation would probably be much worse if that many people were spread across the countryside. Cities, for all their ills, are one of the places where we can learn new ways to live sustainably. New York City, one of the largest in the world, has established new codes for "green" building, for water conservation, and for recycling. More New Yorkers use public transportation and walk to work than in any other major American city.

Cities can be engines of economic progress and social reform. Some of the greatest promise for innovation comes from cities like Curitiba, where innovative leaders can focus knowledge and

Figure 14.2 In less than 20 years, Shanghai, China, has built Pudong, a new city of 1.5 million residents and 500 skyscrapers on former marshy farmland across the Huang Pu River from the historic city center. This kind of rapid urban growth is occurring in many developing countries.

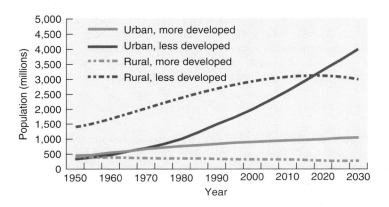

Figure 14.3 Growth of urban and rural populations in more-developed regions and in less-developed regions.
Source: United Nations Population Division. *World Urbanization Prospects*, 2004.

Table 14.1	Urban Share of Total Population (Percentage)		
	1950	**2000**	**2030***
Africa	18.4	40.6	57.0
Asia	19.3	43.8	59.3
Europe	56.0	75.0	81.5
Latin America	40.0	70.3	79.7
North America	63.9	77.4	84.5
Oceania	32.0	49.5	60.7
World	38.3	59.4	70.5

*Projected.
Source: Data from United Nations Population Division, 2003.

resources on common problems. Cities can be efficient places to live, where mass transportation can move people around and goods and services are more readily available than in the country. Concentrating people in urban areas leaves open space available for farming and biodiversity. But cities can also be dumping grounds for poverty, pollution, and unwanted members of society. Providing food, housing, transportation, jobs, clean water, and sanitation to the 2 or 3 billion new urban residents expected to crowd into cities—especially those in the developing world—in this century may be one of the preeminent challenges of this century.

As Curitiba shows, there is much we can do to make our cities more livable. It's especially encouraging that a Brazilian city, where per capita income is only one-fifth that of the United States, can make so much progress. But what of countries where personal wealth is only one-tenth that of Brazil? What hope is there for them?

Large cities are expanding rapidly

You can already see the dramatic shift in size and location of big cities. In 1900 only 13 cities in the world had populations over 1 million (table 14.2). All of those cities except Tokyo and Peking were in Europe or North America. London was the only city in the world with more than 5 million residents. By 2007, there were at least 300 cities—100 of them in China alone—with more than 1 million residents. Of the 13 largest of these metropolitan areas, none are in Europe. Only New York City and Los Angeles are in a developed country. By 2025, it's expected that at least 93 cities will have populations over 5 million, and three-fourths of those cities will be in developing countries (fig. 14.4). In just the next

Table 14.2		The World's Largest Urban Areas (Populations in Millions)	
1900		**2015****	
London, England	6.6	Tokyo, Japan	31.0
New York, USA	4.2	New York, USA	29.9
Paris, France	3.3	Mexico City	21.0
Berlin, Germany	2.4	Seoul, Korea	19.8
Chicago, USA	1.7	São Paolo, Brazil	18.5
Vienna, Austria	1.6	Osaka, Japan	17.6
Tokyo, Japan	1.5	Jakarta, Indonesia	17.4
St. Petersburg, Russia	1.4	Delhi, India	16.7
Philadelphia, USA	1.4	Los Angeles, USA	16.6
Manchester, England	1.3	Beijing, China	16.0
Birmingham, England	1.2	Cairo, Egypt	15.5
Moscow, Russia	1.1	Manila, Philippines	13.5
Peking, China*	1.1	Buenos Aires, Brazil	12.9

*Now spelled Beijing.
**Projected.
Source: Data from T. Chandler, *Three Thousand Years of Urban Growth*, 1974, Academic Press; and *World Gazetter*, 2003.

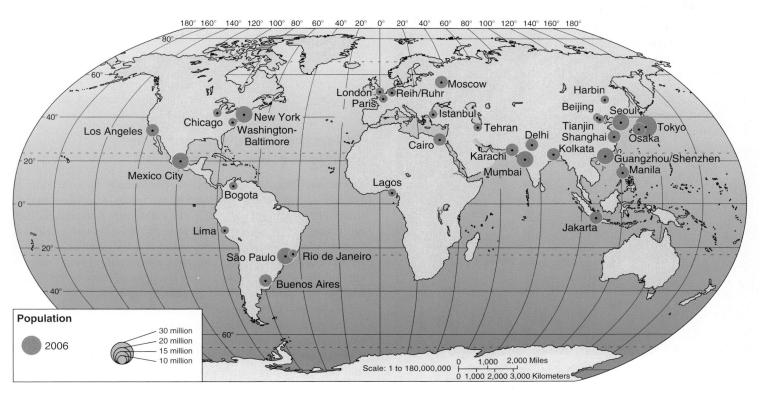

Figure 14.4 Some of the world's largest urban agglomerations in 2006. Sizes of many of these metropolitan areas are only educated guesses. As much as half the population of some cities in developing countries can be undocumented migrant workers or shantytown residents who are difficult to count. Most population growth in this century is expected to be in the megacities of the developing world.

25 years, Mumbai, India; Delhi, India; Karachi, Pakistan; Manila, Philippines; and Jakarta, Indonesia, all are expected to grow by at least 50 percent.

China represents the largest demographic shift in human history. Since the end of Chinese collectivized farming and factory work in 1986, around 250 million people have moved from rural areas to cities. And in the next 25 years an equal number is expected to join this vast exodus. In addition to expanding existing cities, China plans to build 400 new urban centers with populations of at least 500,000 over the next 20 years. Already at least half of the concrete and one-third of the steel used in construction around the world each year is consumed in China.

Consider Shanghai, for example. In 1985, the city had a population of about 10 million. It's now about 19 million—including at least 4 million migrant laborers. In the past decade, Shanghai has built 4,000 skyscrapers (buildings with more than 25 floors). The city already has twice as many tall buildings as Manhattan, and proposals have been made for 1,000 more. The problem is that most of this growth has taken place in a swampy area called Pudong, across the Huang Pu River from the historic city center (see fig. 14.2). Pudong is now sinking about 1.5 cm per year due to groundwater drainage and the weight of so many buildings.

Other Chinese cities have plans for similar massive building projects to revitalize blighted urban areas. Harbin, a city of about 9 million people and the capital of Heilongjiang Province, for example, recently announced plans to relocate across the Songhua River on 740 km² (285 mi², or roughly the size of New York City) of former farmland. Residents hope these new towns will be both

more livable for their residents and more ecologically sustainable than the old cities they're replacing. In 2005, the Chinese government signed a long-term contract with a British engineering firm to build at least five "eco-cities," each the size of a large Western capital. Plans calls for these cities to be self-sufficient in energy, water, and most food products, with the aim of zero emissions of greenhouse gases from transportation.

Immigration is driven by push and pull factors

People migrate to cities for many reasons. In China over the past 20 years—or in America during the twentieth century—mechanization eliminated jobs and drove people off the land. Where cropland is owned by a minority of wealthy landlords, as is the case in many developing countries, subsistence farmers are often ejected when new cash crops or cattle grazing become economically viable. Many people also move to the city because of the opportunities and independence offered there. Cities offer jobs, better housing, entertainment, and freedom from the constraints of village traditions. Possibilities exist in the city for upward social mobility, prestige, and power not ordinarily available in the country. Cities support specialization in arts, crafts, and professions for which markets don't exist elsewhere.

Government policies often favor urban over rural areas in ways that both push and pull people into cities. Developing countries commonly spend most of their budgets on improving urban areas (especially around the capital city, where leaders live). This gives the major cities a virtual monopoly on new jobs, housing, education, and finance, all of which bring in rural people searching for a better life. Lima, for example, has only 20 percent of Peru's population, but has 50 percent of the national wealth, 60 percent of the manufacturing, 65 percent of the retail trade, 73 percent of the industrial wages, and 90 percent of all banking in the country. Similar statistics pertain to many national capitals.

Congestion, pollution, and water shortages plague many cities

First-time visitors to a supercity—particularly in a developing country—often are overwhelmed by the immense crush of pedestrians and vehicles of all sorts jostling for space in the streets. The noise, congestion, and confusion of traffic make it seem suicidal to venture onto the street. Jakarta, Indonesia, for instance, is one of the most densely populated cities in the world (fig. 14.5). Traffic is chaotic almost all the time. People often spend three or four hours each way commuting to work from outlying areas.

Pollution from burgeoning traffic and from unregulated factories degrades air quality in many urban areas. China's spectacular economic growth has resulted in an flood of private automobiles mainly in cities. Beijing, for instance, has doubled the number of cars on its streets in just the past five years to 2.5 million. China is the world's second-largest producer of greenhouse gases, and the World Bank warns that it is home to 16 of the world's 20 cities with the worst air pollution. Chinese health authorities say that a third of the country's urban residents are exposed to

Figure 14.5 Motorized rickshaws, motor scooters, bicycles, street vendors, and pedestrians all vie for space on the crowded streets of Jakarta. The heat, noise, smells, and sights are overpowering. In spite of the difficulties of living here, people work hard and have hope for the future.

harmful air pollution levels. They blame this pollution for more than 400,000 premature deaths each year.

Few cities in developing countries can afford to build modern waste treatment systems for their rapidly growing populations. The World Bank estimates that only one-third of urban residents in developing countries have satisfactory sanitation services. In Latin America, only 2 percent of urban sewage receives any treatment. In Egypt, Cairo's sewer system was built about 50 years ago to serve a population of 2 million people. It is now being overwhelmed by more than five times that many residents. Less than 1 percent of India's 500,000 towns and villages have even partial sewage systems or water treatment facilities.

It's often difficult to find clean drinking water for urban areas. According to Qiu Baoxing, Chinese minister of construction, 70 percent of his country's surface water is so polluted by industrial toxins, human waste, and agricultural chemicals that it is unsuited for human consumption. One hundred of China's 660 cities face severe water shortages, he reported. Worldwide, according to the United Nations, at least 1.1 billion people don't have safe drinking water, and twice that many don't have adequate sanitation. It may not be so bad to have primitive sanitation if you live in a sparsely populated rural area, but imagine what it's like to live in a densely crowded megacity, such as Jakarta, with 10 million people, fewer than half of whom are served by a municipal sanitary system.

Many cities lack sufficient housing

The United Nations estimates that at least 1 billion people live in crowded, unsanitary slums of the central cities or in the vast shantytowns and squatter settlements that ring the outskirts of most major cities in the developing world. Around 100 million people have no home at all. In Mumbai (formerly Bombay), India, for example, it's estimated that half a million people sleep on the streets, sidewalks, and traffic circles because they can find no other place to live (fig. 14.6).

Slums are generally legal but inadequate multifamily tenements or rooming houses, often converted from some other use. Families live crowded in small rooms with inadequate ventilation and sanitation. Often these structures are rickety and unsafe. In 1999, for example, a 7.4 magnitude earthquake hit eastern Turkey, killing more than 14,000 people when shoddy, poorly built apartments collapsed.

Shantytowns, with shacks built of corrugated metal, discarded packing crates, brush, plastic sheets, and other scavenged materials, grow on the outskirts of many cities in the developing world. They can house millions of people but generally lack clean water, sanitation, or safe electrical power. Shantytowns are usually illegal, but they quickly fill in the empty space in towns where squatters can build shelters close to jobs. With little or no public services, shantytowns often fill with trash and debris. Many governments try to clean out illegal settlements by torching or bulldozing the huts and sending riot police to drive out residents, but people either move back in or relocate to another shantytown. In 2005, the government of Zimbabwe destroyed the homes of some 700,000 people in shantytowns around the capital of Harare. Families were evicted in the middle of the night during the cold-

Figure 14.6 In Mumbai, India, as many as half a million people sleep on the streets because they have no other place to live. Ten times as many live in crowded, dangerous slums and shantytowns throughout the city.

est weather of the year, often with only minutes to gather their belongings. President Robert Mugabe justified this blitzkreig as necessary to control crime, but critics claimed it was mainly to remove political opponents.

Two-thirds of the population of Kolkata are thought to live in unplanned squatter settlements, and nearly half the 25 million residents of Mexico City occupy the unauthorized *colonias* around the city. Often, shantytowns occupy the most polluted, dangerous parts of cities where no one else wants to live. In Bhopal, India, and Mexico City, for example, squatter settlements were built next to deadly industrial sites. In Brazil, shantytowns called *favelas* perch on steep hillsides unwanted for other building (fig. 14.7). As desperate and inhumane as conditions are in these slums and shantytowns, many people do more than merely survive there. They work hard, raise families, educate their children, and often improve their living standard little by little as they make some money.

Many countries are recognizing that the only way they can house all their citizens is to cooperate with shantytown dwellers. Recognizing land rights, providing financing for home improvements, and

Figure 14.7 Shantytowns called *favelas* perch on hillsides above Rio de Janeiro.

supporting community efforts to provide water, sewers, and power can greatly improve living conditions for many poor people.

14.2 Urban Planning

How can we live together in cities in ways that are environmentally sound, socially just, and economically sustainable? Starting with Greek cities thousands of years ago, planners have debated the best ways for us to organize ourselves.

Transportation is crucial in city development

Most of the world's major cities grew up around a port, a river crossing, a railroad hub, or some other focal point for transportation. Often, those original reasons for city placement no longer apply, and a location that made sense for a small, backwoods community no longer works for a major metropolitan center.

Getting people around within a large urban area has become one of the most difficult problems that many city officials face. A century ago, most American cities were organized around transportation corridors. First horse-drawn carriages, then electric streetcars provided a way for people to get to work, school, and shops. Everyone, rich or poor, wanted to live as close to the city center as possible, and within easy walking distance of the trolley or streetcar line. When Henry Ford introduced the first affordable,

Table 14.3	Characteristics of Urban Sprawl
1.	Unlimited outward extension
2.	Low-density residential and commercial development
3.	Leapfrog development that consumes farmland and natural areas
4.	Fragmentation of power among many small units of government
5.	Dominance of freeways and private automobiles
6.	No centralized planning or control of land uses
7.	Widespread strip-malls and "big-box" shopping centers
8.	Great fiscal disparities among localities
9.	Reliance on deteriorating older neighborhoods for low-income housing
10.	Decaying city centers as new development occurs in previously rural areas

Source: Data from PlannersWeb, Burlington, Vermont, 2001.

Figure 14.8 Huge houses on sprawling lots consume land, alienate us from our neighbors, and make us ever more dependent on automobiles.

© 2003 Regents of the University of Minnesota. All rights reserved. Used with permission of the Design Center for American Urban Landscape.

mass-produced automobile, it allowed people to build houses on larger lots in areas served only by streets. Freeway construction, which began in America in the 1950s, allowed people to move even further out into the country. Cities that were once compact began to spread over the landscape, consuming space and wasting resources. This pattern of development is known as **sprawl**. While there is no universally accepted definition of the term, sprawl generally includes the characteristics outlined in table 14.3.

In most American metropolitan areas, the bulk of new housing is in large, tract developments that leapfrog out beyond the city edge in a search for inexpensive rural land with few restrictions on land use or building practices (fig. 14.8). The U.S. Department of Housing and Urban Development estimates that urban sprawl

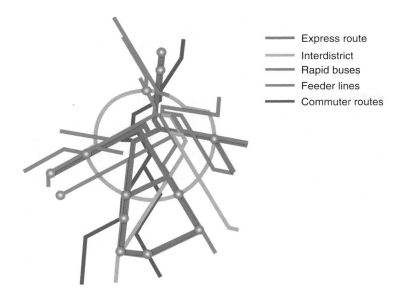

Figure 14.9 Curitiba's transit system has more than 340 interlinked bus routes. Neighborhood feeder lines connect to high-volume, express buses running on dedicated busways. Interdistrict routes carry passengers between suburbs, while specialized commuter routes carry workers directly from distant neighborhoods to the city center. Altogether this system moves more than 1.9 million people per day, or more than 75 percent of all personal transportation in the city.

consumes some 200,000 ha (roughly 500,000 acres) of farmland every year. Although the price of tract homes often is less than comparable urban property, there are external costs in the form of new roads, sewers, water mains, power lines, schools, shopping centers, and other infrastructure required by this low-density development. Ironically, people who move to the country to escape from urban problems such as congestion, crime, and pollution often find that they have simply brought those problems with them.

Because many Americans live far from where they work, shop, or recreate, they consider it essential to own a private automobile. The average U.S. driver spends 443 hours per year behind a steering wheel, or the equivalent of one full 8-hour day per week in an automobile. The freeway system was designed to allow drivers to travel at high speeds from source to destination without ever having to stop. As more and more vehicles clog highways, however, the reality is far different. In Los Angeles, for example, which has the worst congestion in the United States, the average speed in 1982 was 58 mph (93 kph), and the average driver spent less than 4 hours per year in traffic jams. In 2004, the average speed was only 35.6 mph (57.3 kph), and the typical driver spent 97 hours in bumper-to-bumper traffic.

Altogether, it's estimated that traffic congestion costs the United States $78 billion per year in wasted time and fuel. Some people argue that the existence of traffic jams in cities shows that more highways are needed. Often, however, building more traffic lanes simply encourages more people to drive farther and put more cars on the road. Meanwhile, about one-third of Americans are too young, too old, or too poor to drive. For these people, car-oriented

development causes isolation and makes daily tasks like grocery shopping difficult. Parents spend long hours transporting young children. Teenagers and aging grandparents are forced to drive, often presenting a hazard on public roads.

As the opening case study for this chapter shows, Curitiba provides an excellent example of successful mass transit and environmentally sound, socially just, and economically sustainable urban planning. Curitiba's bus rapid transit system focuses around five transportation corridors radiating from the city center (fig. 14.9). Within these corridors, high-speed, articulated buses, each of which can carry 270 passengers, travel on dedicated roadways closed to all other vehicles. These bus-trains make limited stops at transfer terminals that link to 340 feeder routes extending throughout the city. Everyone in the city is within walking distance of a bus stop that has frequent, convenient, affordable service.

Each bus station along the dedicated busway is made up of tubular structures elevated to be at the same height as the bus floor, providing access for disabled persons and making entry and exit quicker and easier for everyone (fig. 14.10). Passengers pay as they go through a turnstile to enter the terminal. When a bus pulls up, it opens multiple doors through which passengers can enter or exit. Dozens of passengers might disembark and an equal number get on the bus during the 60-second stop, making a trip over many kilometers remarkably fast. The system charges a single fee for a trip in one direction, regardless of the number of transfers involved. This makes the system equitable for those who can't afford to live in expensive neighborhoods close to the city center.

Curitiba's buses make more than 21,000 trips per day, traveling more than 440,000 km (275,000 mi) and carrying 1.9 million passengers, or about three-quarters of all personal trips within the city. One of the best things about this system is its economy. Working with existing roadways for the most part, the city was able to construct this system for one-tenth the cost of a light rail system or freeway system, and one-hundredth the cost of a subway. The success of bus rapid transit has allowed Curitiba to remain relatively compact and to avoid the sprawl engendered by an American-style freeway system.

Figure 14.10 High-speed, bi-articulated buses travel on dedicated transit ways in Curitiba and make limited stops at elevated tubular bus terminals that allow many passengers to disembark and reload during 60-second stops.

Exploring
SCIENCE:

Traditionally, most ecologists have studied pristine, natural ecosystems, trying to understand how nature works without human disturbance. In recent years, however, we have come to recognize that cities are ecological systems, too. If ecology is the relationship among organisms and their environment, what could be more ecological than studying the organism (humans) with the greatest impact on other species and our environment? As the human population becomes increasingly urbanized, perhaps the most important place to study human ecology is in cities. Rather than merely seeing humans as disturbing factors, some scientists are beginning to gather data on how urban ecosystems work.

Urban areas share many characteristics with natural ecosystems. Energy flows into and out of the city, and is degraded and dispersed as it is used to do work. Materials are used to create structures, some living and some non-living. Species compete for shelter, food, habitat, and other resources. The most successful species flourish and proliferate; less fit ones dwindle and eventually vanish. Every day, a typical American city of a million people consumes roughly 1 million metric tons (roughly 2 billion lbs) of raw materials. This urban material flow includes 500,000 m^3 (1.3 million gal) of water, the energy equivalent of about 20,000 metric tons of fossil fuels, 13,000 metric tons of other minerals (including food packaging and construction materials), 12,000 metric tons of farm products, and 10,000 metric tons of wood and paper products.

To remove municipal waste from this city of a million residents, about 1,000 fully loaded garbage trucks travel every day to rural areas carrying the discarded remnants of this urban metabolism. In addition, the city's municipal wastewater treatment plants discharge somewhat more than 500,000 m^3 of water (that lost to evaporation is offset by precipitation and solids added to the wastewater stream). You may remember from chapter 2 that most natural ecosystems recycle materials. Recycling and reuse also occur in urban ecology, although rarely with the speed and efficiency of natural systems.

Urban biodiversity and wildlife also can be important. While many biologists have tended to assume that cities are devoid of life other than humans and our domesticated companion animals and plants, we are now beginning to appreciate that urban open space can be essential refuges for wild creatures as well. As cities grow, it's not uncommon for consolidated urban areas to be more than 100 miles (160 km) across. This would present an insurmountable barrier to many migrating animals if the city consists of nothing but concrete, glass, and metal. Urban parks, forests, corridors along rivers, and even ordinary yards, if managed properly, can provide a refuge for migrating wildlife and can sustain local populations that contribute significantly to overall biodiversity. Planting native vegetation and designing landscapes to provide maximum food and shelter are important steps in protecting and promoting urban biodiversity.

Long-term ecological research (LTER) sites are ecologically significant sites identified and funded by the National Science Foundation to allow long-term, in-depth research on ecological problems and questions. Although most of the LTER sites are in remote locations, where nature has minimal human interference, two LTER sites were set up in Phoenix and Baltimore. Funded by the National Science Foundation for an initial period of six years at nearly a million dollars per year, but with an understanding that the research will go on much beyond that, these projects use modern technology to study every aspect of urban ecology. How does the city shape its own weather? What plants and animals live in the city, and where, and how? What ecological processes cycle materials and energy? How do human activities influence those processes?

One of the first things that scientists are discovering is that cities are not homogenous; like other ecosystems, they have shifting patches of contrasting conditions. Using computer models and sophisticated mapping techniques, ecologists can explore how patches change in space and time. Elsewhere, environmental scientists have investigated important problems of pollution and environmental health in cities. Where are toxic and hazardous materials generated, stored, and released in the city? How do they move around and where do they accumulate?

In Detroit, a group of students worked with experts to map data on more than 5,000 children with elevated blood lead levels. Not surprisingly, they found a correlation between low incomes, old housing, incidence of poisoning, and concentration of special-education students. Sometimes public awareness of a problem is one of the best outcomes of this research. Another large group of students in the Detroit area used geographic information systems (digital maps) together with chemical and biological analysis to prepare a detailed study of water quality and ecosystem health in the Rouge River. Perhaps students at your school could organize a similar study of the urban ecology in your area.

We can make our cities more livable

Are there alternatives to unplanned sprawl and wasteful resource use? One option proposed by many urban planners is **smart growth**, which makes effective use of land resources and existing infrastructure by encouraging in-fill development that avoids costly duplication of services and inefficient land use (table 14.4). Smart growth aims to provide a mix of land uses to create a variety of affordable housing choices and opportunities. It also attempts to provide a variety of transportation choices, including pedestrian-friendly neighborhoods. This approach to planning also seeks to maintain a unique sense of place by respecting local cultural and natural features.

By making land-use planning open and democratic, smart growth makes urban expansion fair, predictable, and cost-effective. All stakeholders are encouraged to participate in creating a vision for the city and to collaborate with rather than confront each other. Goals are established for staged and managed growth in urban transition areas with compact development patterns. This approach is not opposed to growth. It recognizes that the goal is not to block growth but to channel it to areas where it can be sustained over the long term. Smart growth strives to enhance access to equitable public and private resources for everyone and to promote the safety, livability, and revitalization of existing urban and rural communities.

Smart growth protects environmental quality. It tries to reduce traffic and to conserve farmlands, wetlands, and open space. As cities grow and transportation and communications enable more community interaction, the need for regional planning becomes greater and more pressing. Community and business leaders must

Table 14.4	Goals for Smart Growth
1.	Create a positive self-image for the community.
2.	Make the downtown vital and livable.
3.	Alleviate substandard housing.
4.	Solve problems with air, water, toxic waste, and noise pollution.
5.	Improve communication between groups.
6.	Improve community member access to the arts.

Source: Data from Vision 2000, Chattanooga, Tennessee.

make decisions based on a clear understanding of regional growth needs and how infrastructure can be built most efficiently and for the greatest good.

One of the best examples of successful urban land-use planning in the United States is Portland, Oregon, which has rigorously enforced a boundary on its outward expansion, requiring instead that development be focused on in-filling unused space within the city limits. Because of its many urban amenities, Portland is considered one of the best cities in America. Between 1970 and 1990 the Portland population grew by 50 percent, but its total land area grew only 2 percent. During this time, Portland property taxes decreased 29 percent and vehicle miles traveled increased only 2 percent. By contrast, Atlanta, which had similar population growth, experienced an explosion of urban sprawl that increased its land area three-fold, drove up property taxes 22 percent, and increased traffic miles by 17 percent. A result of this expanding traffic and increasing congestion was that Atlanta's air pollution increased by 5 percent, while Portland, which has one of the best public transit systems in the nation, saw a decrease of 86 percent.

New urbanism incorporates smart growth

Rather than abandon the cultural history and infrastructure investment in existing cities, a group of architects and urban planners is attempting to redesign metropolitan areas to make them more appealing, efficient, and livable. European cities such as Stockholm, Sweden; Helsinki, Finland; Leichester, England; and Neerlands, the Netherlands, have a long history of innovative urban planning. In the United States, Andres Duany, Elizabeth Plater-Zyberk, Peter Calthorpe, and Sym Van Der Ryn have been leaders in this movement. Using what is sometimes called a neo-traditionalist approach, these designers attempt to recapture some of the best features of small towns and livable cities of the past. They are designing urban neighborhoods that integrate houses, offices, shops, and civic buildings. Ideally, no house should be more than a five-minute walk from a neighborhood center with a convenience store, a coffee shop, a bus stop, and other amenities. A mix of apartments, townhouses, and detached houses in a variety of price ranges ensures that neighborhoods will include a diversity of ages and income levels. Some design principles of this movement include

- Limit city size or organize cities in modules of 30,000 to 50,000 people—large enough to be a complete city but small enough to be a community.

Figure 14.11 This walking street in Queenstown, New Zealand, provides opportunities for shopping, dining, and socializing in a pleasant outdoor setting.

- Maintain greenbelts in and around cities. These provide recreational space and promote efficient land use, as well as help ameliorate air and water pollution.

- Determine in advance where development will take place. This protects property values and prevents chaotic development. Planning can also protect historical sites, agricultural resources, and ecological services of wetlands, clean rivers, and groundwater replenishment.

- Locate everyday shopping and services so people can meet daily needs with greater convenience, less stress, less automobile dependency, and less use of time and energy (fig. 14.11). This might be accomplished by encouraging small-scale commercial development in or close to residential areas.

- Encourage walking or the use of small, low-speed, energy-efficient vehicles (microcars, motorized tricycles, bicycles, etc.) for many local trips now performed in full-size automobiles. Creating special traffic lanes, reducing the number or size of parking spaces, and closing shopping streets to big cars might encourage such alternatives.

- Promote more diverse, flexible housing as an alternative to conventional detached, single-family houses. In-fill building between existing houses saves energy, reduces land costs, and might help provide a variety of living arrangements. Allowing single-parent families or groups of unrelated adults to share housing and to use facilities cooperatively also provides alternatives to those not living in a traditional nuclear family.

- Make cities more self-sustainable by growing food locally, recycling wastes and water, using renewable energy sources, reducing noise and pollution, and creating a cleaner, safer environment. Encourage community gardening (fig. 14.12).

Figure 14.12 Many cities have large amounts of unused open space that could be used to grow food. Residents often need help decontaminating soil and gaining access to the land.

Reclaimed inner-city space or a greenbelt of agricultural and forestland around the city provides food and open space, and also contributes valuable ecological services, such as purifying air, supplying clean water, and protecting wildlife habitat and recreation land.

- Equip buildings with "green roofs" or rooftop gardens that improve air quality, conserve energy, reduce stormwater runoff, reduce noise, and help reduce urban heat island effects. Intensive gardens can include large trees, shrubs, flowers, and may require regular maintenance (fig. 14.13). Extensive gardens require less soil, add less weight to the

building, and usually have simple plantings of prairie plants or drought-resistant species, such as sedum, that require minimum care. They can last twice as long as conventional roofs. In Europe more than 1 million m^2 of green roofs are installed every year. Urban roofs are also a good place for solar collectors or wind turbines.

- Plan cluster housing, or open-space zoning, which preserves at least half of a subdivision as natural areas, farmland, or other forms of open space. Studies have shown that people who move to the country don't necessarily want to live miles from the nearest neighbor; what most desire is long views across an interesting landscape and an opportunity to see wildlife. By carefully clustering houses on smaller lots, a conservation subdivision can provide the same number of buildable lots as a conventional subdivision and still preserve 50 to 70 percent of the land as open space (fig. 14.14). This not only reduces development costs (less distance to build roads, lay telephone lines, sewers, power cables, etc.) but also helps to foster a greater sense of community among new residents.

- Preserve urban habitat. It can make a significant contribution toward saving biodiversity as well as improving mental health and giving us access to nature.

These planning principles aren't just a matter of aesthetics. Dr. Richard Jackson, former director of the National Center for Environmental Health in Atlanta, points out a strong association between urban design and our mental and physical health. As our cities have become ever more spread out and impersonal, we have fewer opportunities for healthful exercise and socializing. Chronic diseases, such as cardiovascular diseases, asthma, diabetes, obesity, and depression, are becoming the predominant health concerns in the United States.

Figure 14.13 This award-winning green roof on the Chicago City Hall is functional as well as beautiful. It reduces rain runoff by about 50 percent, and keeps the surface as much as 30°F cooler than a conventional roof on hot summer days.

Figure 14.14 This conservation development clusters houses on one-third of its property, and the rest is preserved as native prairie and oak woodland. Being close together, neighbors develop a sense of community, yet everyone has expansive views and access to open space.

"Despite common knowledge that exercise is healthful," Dr. Jackson says, "fewer than 40% of adults are regularly active, and 25% do no physical activity at all. The way we design our communities makes us increasingly dependent on automobiles for the shortest trip, and recreation has become not physical but observational." Long commutes and a lack of reliable mass transit and walkable neighborhoods mean that we spend more and more time in stressful road congestion. "Road rage" isn't imaginary. Every commuter can describe unpleasant encounters with rude drivers. Urban design that offers the benefits of more walking, more social contact, and surroundings that include water and vegetation can provide healthful physical exercise and psychic respite.

WHAT DO YOU THINK?

The Architecture of Hope

How sustainable and self-sufficient can urban areas be? An exciting experiment in minimal impact in London gives us an image of what our future may be. BedZED, short for the Beddington Zero Energy Development, is an integrated urban project built on the grounds of an old sewage plant in South London. BedZED's green strategies begin with recycling the ground on which it stands. Designed by architect Bill Dunster and his colleagues, the complex demonstrates dozens of energy-saving and water-saving ideas. BedZED has been occupied, and winning awards, since it was completed in 2003.

Most of BedZED's innovations involve the clever combination of simple, even conventional ideas. Expansive, south-facing, triple-glazed windows provide abundant light, minimize the use of electric lamps, and provide passive solar heat in the winter. Thick, superinsulated walls keep interiors warm in winter and cool in summer. Rotating "wind cowls" on roofs turn to catch fresh breezes, which cool spaces in summer. In winter, heat exchangers use the heat of stale, outgoing air to warm fresh, incoming air. Energy used in space heating is nearly eliminated. Building materials are recycled, reclaimed, or renewable, which reduces the "embodied energy" invested in producing and transporting them.

BedZED does use energy, but the complex generates its own heat and electricity with a small, on-site, superefficient plant that uses local tree trimmings for fuel. Thus BedZED uses no *fossil* fuels, and it is "carbon-neutral" because the carbon dioxide released by burning wood was recently captured from the air by trees. In addition, photovoltaic cells on roofs provide enough free energy to power 40 solar cars. Fuel bills for BedZED residents can be as little as 10 percent of what other Londoners pay for similar-sized homes.

Water-efficient appliances and toilets reduce water use. Rainwater collection systems provide "green water" for watering gardens, flushing toilets, and other nonconsumptive uses. Reed-bed filtration systems purify used water without chemicals. Water meters allow residents to see how much water they use. Just knowing about consumption rates helps encourage conservation. Residents use about half as much water per person as other Londoners.

BedZED residents can save money and time by not using or even owning a car. Office space is available on-site, so some residents can work where they live, and the commuter rail station is just a ten-minute walk away. The site is also linked to bicycle trails that facilitate bicycle commuting. Car pools and rent-by-the-hour auto memberships allow many residents to avoid owning (and parking) a vehicle altogether.

Building interiors are flooded with natural light, ceilings are high, and most residences have rooftop gardens. Community events and common spaces encourage humane, healthy lifestyles and community ties. Child-care services, shops, entertainment, and sports facilities are built into the project. The approximately 100 housing units are designed for a range of income levels ensuring a racially, ethnically, and age-diverse

South-facing windows heat homes, and colorful, rotating "wind cowls" ventilate rooms at BedZED, an ecological housing complex in South London, U.K.

community. Prices are lower than many similar-sized London homes, and few in this price range or inner-city location have abundant sunlight or gardens.

Similar projects are being built across Europe and even in some developing countries, such as China. Architect Dunster says that BedZED-like developments on cleaned-up brownfields could provide all the 3 million homes that the U.K. expects to need in the next decade with no sacrifice of open space. And as green building techniques, designs, and materials become standard, he argues, they will cost no more than conventional, energy-wasting structures.

What do you think? Would you enjoy living in a dense, urban setting such as BedZED? Would it involve a lower standard of living than you now have? How much would it be worth to avoid spending 8 to 10 hours per week not fighting bumper-to-bumper traffic while commuting to school or work? The average cost of owning and driving a car in the United States is about $9,000 per year. What might you do with that money if owning a vehicle were unnecessary? Try to imagine what urban life might be like if most private automobiles were to vanish. If you live in a typical American city, how much time do you have to enjoy the open space that the suburbs and freeways were supposed to provide? Perhaps, most importantly, how will we provide enough water, energy, and space for the 3 billion people expected to crowd into cities worldwide over the next few decades if we don't adopt some of the sustainable practices and approaches represented by BedZED?

14.3 Economics and Sustainable Development

Like many of our environmental issues, improving urban conditions will ultimately be decided by economics and policy decisions. We'll discuss policy in chapter 15. In the next half of this chapter, we'll review some of the principles of environmental economics.

Can development be sustainable?

By now it is clear that security and living standards for the world's poorest people are inextricably linked to environmental protection. One of the most important questions in environmental science is how we can continue improvements in human welfare within the limits of the earth's natural resources. *Development* means improving people's lives. *Sustainability* means living on the earth's renewable resources without damaging the ecological processes that support us all. **Sustainable development** is an effort to marry these two ideas. A popular definition describes this goal as "meeting the needs of the present without compromising the ability of future generations to meet their own needs."

But is this possible? As you've learned elsewhere in this book, many people argue that our present population and economic levels are exhausting the world's resources. There's no way, they insist, that more people can live at a higher standard without irreversibly degrading our environment. Others claim that there's enough for everyone if we just share equitably and live modestly. Let's look a little deeper into this important debate.

Our definitions of resources shape how we use them

To understand the problems and promise of sustainability, you need to understand the different kinds of resources we use. The way we treat resources depends largely on how we view and define them. **Classical economics**, developed in the 1700s by philosophers such as Adam Smith (1723–1790) and Thomas Malthus (1766–1834), assumes that natural resources are finite—that resources such as iron, gold, water, and land exist in fixed amounts. According to this view, as populations grow, scarcity of these resources reduces quality of life, increases competition, and ultimately causes populations to fall again. In a free market, where fully informed buyers and sellers make free, independent decisions to buy and sell, the price of a pound of butter depends on the supply of available butter (it is cheap when plenty is available) and the demand for butter (buyers pay more when they must compete to get the resource, fig. 14.15).

This price mechanism has been modified by the idea of **marginal costs**. Buyers and sellers evaluate the added return for buying (or making and selling) slightly more of a product, or the marginal cost of slightly greater production or purchase. If the return is greater than the marginal cost, then a sale is made.

The nineteenth-century economist John Stuart Mill assumed that most resources are finite, but he developed the idea of a **steady-state economy**. Rather than boom-and-bust cycles of population and resource use, as envisioned by Malthus, Mill proposed that economies can achieve an equilibrium of resource use and production. Intellectual and moral development continues, he argued, once this stable, secure state is achieved.

Neoclassical economics, developed in the nineteenth century, expanded the idea of resources to include labor, knowledge, and capital. Labor and knowledge are resources because they are necessary to create goods and services; they are not finite because every new person can add more labor and energy to an economy. **Capital** is any form of wealth that contributes to the production of more wealth. Money can be invested to produce more money. Mineral resources can be developed and manufactured into goods that return more money. Economists distinguish several kinds of capital:

1. Natural capital: goods and services provided by nature
2. Human capital: knowledge, experience, human enterprise
3. Manufactured (built) capital: tools, buildings, roads, technology

To this list some social theorists would add social capital, the shared values, trust, cooperation, and organization that can develop in a group of people but cannot exist in one individual alone.

Because the point of capital is the production of more capital (that is, wealth), neoclassical economics emphasizes the idea of growth. Growth results from the flow of resources, goods, and services (fig. 14.16). Continued growth is always necessary for continued prosperity, according to this view. Natural resources contribute to production and growth, but they are not critical supplies that limit growth. They are not limiting because resources are considered to be interchangeable and substitutable. As one resource becomes scarce, neoclassical economics predicts that a substitute will be found.

Since production of wealth is central to neoclassical economics, an important measure of growth and wealth is consumption.

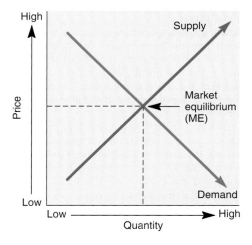

Figure 14.15 Classic supply/demand curves. When price is low, supply is low and demand is high. As prices rise, supply increases but demand falls. Market equilibrium is the price at which supply and demand are equal.

If a society consumes more oil, more minerals, and more food, it is presumably becoming wealthier. This idea has extended to the idea of *throughput*, the amount of resources a society uses and discards. More throughput is a measure of greater consumption and greater wealth, according to this view. Throughput is commonly measured in terms of **gross national product (GNP)**, the sum of all products and services bought and sold in an economy. Because GNP includes activities of offshore companies, economists sometimes prefer **gross domestic product (GDP)**, which more accurately reflects the local economy by accounting for only those goods and services bought and sold locally.

Natural resource economics extends the neoclassical viewpoint to treat natural resources as important waste sinks (absorbers), as well as sources of raw materials. Natural capital (resources) is considered more abundant, and therefore cheaper, than built or human-made capital.

Ecological economics incorporates principles of ecology

Ecological economics applies ecological ideas of system functions and recycling to the definition of resources. This school of thought also recognizes efficiency in nature, and it acknowledges the importance of ecosystem functions for the continuation of human economies and cultures. In nature, one species' waste is another's food, so that nothing is wasted. We need an economy that recycles materials and uses energy efficiently, much as a biological community does. Ecological economics also treats the natural environment as part of our economy, so that natural capital becomes a key consideration in economic calculations. Ecological functions, such as absorbing and purifying wastewater, processing air pollution, providing clean water, carrying out photosynthesis, and creating soil, are known as **ecological services** (table 14.5).

Table 14.5 Important Ecological Services
We depend on our environment to continually provide
1. A regulated global energy balance and climate; chemical composition of the atmosphere and oceans; water catchment and groundwater recharge; production and recycling of organic and inorganic materials; maintenance of biological diversity.
2. Space and suitable substrates for human habitation, crop cultivation, energy conversion, recreation, and nature protection.
3. Oxygen, fresh water, food, medicine, fuel, fodder, fertilizer, building materials, and industrial inputs.
4. Aesthetic, spiritual, historic, cultural, artistic, scientific, and educational opportunities and information.

Source: Data from R. S. de Groot, *Investing in Natural Capital,* 1994.

These services are free: we don't pay for them directly (although we often pay indirectly when we suffer from their absence). Therefore, they are often excluded from conventional economic accounting, a situation that ecological economists attempt to rectify (fig. 14.17).

Many ecological economists also promote the idea of a steady-state economy. As with John Stuart Mill's original conception of steady states, these economists argue that economic health can be maintained without constantly growing consumption and throughput. Instead, efficiency and recycling of resources can allow steady prosperity where there is little or no population growth. Low birth rates and death rates (like *K*-adapted species, see chapter 3), political and social stability, and reliance on renewable energy would characterize such a steady-state economy. Like Mill, these economists argue that human and social capital—knowledge, happiness, art, life expectancies, and cooperation—can continue to grow even without constant expansion of resource use.

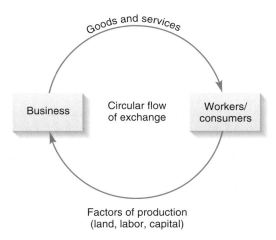

Figure 14.16 The neoclassical model of the economy focuses on the flow of goods, services, and factors of production (land, labor, capital) between business and individual workers/consumers. The social and environmental consequences of these relationships are irrelevant in this view.

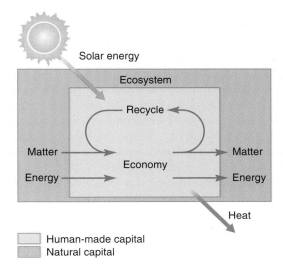

Figure 14.17 An ecological economics view considers natural capital and recycling integral to the economy. Human-made capital is created using limited supplies of natural capital.

Source: Data from Herman Daly in A. M. Jansson et al., *Investing in Natural Capital,* ISEE.

Figure 14.18 Biological resources are renewable in that they replace themselves by reproduction, but if overused or misused, populations die. If a species is lost, it cannot be re-created. It is permanently lost as a component of its ecosystem and as a resource to humans.

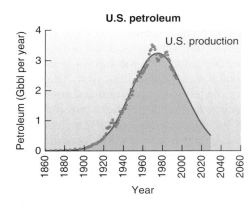

Figure 14.19 U.S. petroleum production. Dots indicate actual production. The bell-shaped curve is a theoretical Hubbert curve for a nonrenewable resource. The shaded area under the curve, representing 220 Gbbl (Gbbl = Gigabarrels or billions of standard 42-gallon barrels), is an estimate of the total economically recoverable resource.

Both ecological economics and neoclassical economics distinguish between renewable and nonrenewable resources. **Nonrenewable resources** exist in finite amounts: minerals, fossil fuels, and also groundwater that recharges extremely slowly are all fixed, at least on a human timescale. **Renewable resources** are naturally replenished and recycled at a fairly steady rate. Fresh water, living organisms, air, and food resources are all renewable (fig. 14.18).

These categories are important, but they are not as deterministic as you might think. Nonrenewable resources, such as iron and gold, can be extended through more efficient use: cars now use less steel than they once did, and gold is mixed in alloys to extend its use. Substitution also reduces demand for these resources: car parts once made of iron are now made of plastic and ceramics; copper wire, once stockpiled to provide phone lines, is now being replaced with cheap, lightweight fiber-optic cables made from silica (sand). Recycling also extends supplies of nonrenewable resources. Aluminum, platinum, gold, silver, and many other valuable metals are routinely recycled now, further reducing the demand for extracting new sources. The only limit to recycling is usually the relative costs of extracting new resources compared with collecting used materials. Recoverable sources of nonrenewable resources are also expanded by technological improvements. New methods make it possible to mine very dilute metal ores, for example. Gold ore of extremely low concentrations is now economically recoverable—that is, you can make money on it—even though the price of gold has fallen because of greater efficiency, more discoveries, and resource substitution. Scarcity of resources, seen by classical economists as the trigger for conflict and suffering, can actually provide the impetus for much of the innovation that leads to substitution, recycling, and efficiency.

On the other hand, renewable resources can become exhausted if they are managed badly. This is especially apparent in biological resources, such as the passenger pigeon, American bison, and Atlantic cod. All these species once existed in extraordinary numbers, but within a few years, each was brought to the brink of extinction (or eliminated entirely) by overharvesting.

Scarcity can lead to innovation

Are we about to run out of essential natural resources? It stands to reason that, if we consume a fixed supply of nonrenewable resources at a constant rate, we'll eventually use up all the economically recoverable reserves. There are many warnings in the environmental literature that our extravagant depletion of nonrenewable resources sooner or later will result in catastrophe, misery, and social decay. Models for exploitation rates of nonrenewable resources—called Hubbert curves after Stanley Hubbert, who developed them—often closely match historic experience for natural resource depletion (fig. 14.19).

Many economists, however, contend that human ingenuity and enterprise often allow us to respond to scarcity in ways that postpone or alleviate dire effects of resource use. The question of whether this view is right has to do with the important theme of **limits to growth**.

In the early 1970s an influential study of resource limitations was funded by the Club of Rome, an organization of wealthy business owners and influential politicians. The study was carried out by a team of scientists, from the Massachusetts Institute of Technology, headed by Donnela Meadows. The results of this study were published in the 1972 book *Limits to Growth*. A complex computer model of the world economy was used to examine various scenarios of different resource depletion rates, growing population, pollution, and industrial output. Given the Malthusian assumptions built into this model, catastrophic social and environmental collapse seemed inescapable.

Figure 14.20 shows one example of the world model. Food supplies and industrial output rise as population grows and resources are consumed. Once past the carrying capacity of the environment, however, a crash occurs as population, food production, and industrial output all decline precipitously. Pollution continues to grow as society decays and people die, but eventually it also falls. Notice the similarity between this set of curves and the "boom-and-bust" population cycles described in chapter 3.

Many economists criticized this model because it underestimated technological development and factors that might mitigate the effects of scarcity. In 1992 the Meadows group published updated computer models in *Beyond the Limits* that include technological progress, pollution abatement, population stabilization, and new public policies that work for a sustainable future. If we adopt these changes sooner rather than later, the models show an outcome like that in figure 14.21, in which all factors stabilize sometime in this century at an improved standard of living for everyone. Of course, none of these computer models shows what will happen, only what some possible outcomes might be, depending on the choices we make.

Communal property resources are a classic problem in economics

One of the difficulties of economics and resource management is that there are many resources we all share but nobody owns. Clean air, fish in the ocean, clean water, wildlife, and open space are all natural amenities that we exploit but that nobody clearly controls.

In 1968 biologist Garret Hardin wrote **"The Tragedy of the Commons,"** an article describing how commonly held resources are degraded and destroyed by self-interest. Using the metaphor of the "commons," or community pastures in colonial New England villages, Hardin theorized that it behooves each villager to put more cows on the pasture. Each cow brings more wealth to the individual farmer, but the costs of overgrazing are shared by the entire community. The individual farmer, then, suffers only part of the cost, but gets to keep all the profits from the extra cows s/he put on the pasture. Consequently, the commons becomes

overgrazed, exhausted, and depleted. This dilemma is also known as the "free rider problem." The best solution, Hardin argued, is to give coercive power to the government or to privatize resources so that a single owner controls resource use.

This metaphor has been applied to many resources, especially to human population growth. It benefits every poor villager to produce a few more children, but collectively these children consume all the resources available, making us all poorer in the end. The same argument has been applied to many resource overuse problems, such as depletion of ocean fisheries, pollution, African famines, and urban crime.

Recent critics have pointed out that what Hardin was really describing was not a commons, or collectively owned and managed resource, but an **open access system**, in which there are no rules to manage resource use. In fact, many common resources have been managed successfully for centuries by cooperative agreements among users. Native American management of wild rice beds, Swiss village-owned mountain forests and pastures, Maine lobster fisheries, and communal irrigation systems in Spain, Bali, Laos, and many other countries have all remained viable for centuries under communal management.

Each of these "commons," or **communal resource management systems**, shares a number of features: (1) community members have lived on the land or used the resource for a long time and anticipate that their children and grandchildren will as well, thus giving them a strong interest in sustaining the resource and maintaining bonds with their neighbors; (2) the resource has clearly defined boundaries; (3) the community group size is known and enforced; (4) the resource is relatively scarce and highly variable, so that the community is forced to be interdependent; (5) the management strategies appropriate for local conditions have evolved over time and are collectively enforced; that is, those affected by the rules have a say in them; (6) the resource and its use are actively monitored, discouraging anyone from cheating or taking too much; (7) conflict resolution mechanisms reduce discord; and

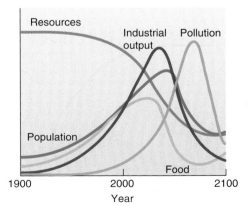

Figure 14.20 A run of one of the world models in *Limits to Growth*. This model assumes business-as-usual for as long as possible until Malthusian limits cause industrial society to crash. Notice that pollution continues to increase well after industrial output, food supplies, and population have all plummeted.

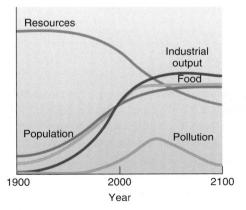

Figure 14.21 A run of the world model from *Beyond the Limits*. This model assumes that population and consumption are curbed, new technologies are introduced, and sustainable environmental policies are embraced immediately, rather than after resources are exhausted.

(8) incentives encourage compliance with rules, while sanctions for noncompliance keep community members in line.

Rather than being the only workable solution to problems in common pool resources, privatization and increasing external controls often prove to be disastrous. Where small villages have owned and operated local jointly held forests and fishing grounds for generations, nationalization and commodification of resources generally have led to rapid destruction of both society and ecosystems. Where communal systems once enforced restraint over harvesting, privatization encouraged narrow self-interest and allowed outsiders to take advantage of the weakest members of the community.

14.4 Natural Resource Accounting

Decision making about sustainable resource use often entails **cost-benefit analysis (CBA)**, the process of accounting and comparing the costs of a project and its benefits. Ideally, this process assigns values to social and environmental effects of a given undertaking, as well as the value of the resources consumed or produced. However, the results of CBA often depend on how resources are accounted for and measured in the first place. CBA is one of the main conceptual frameworks of resource economics, and it is used by decision makers around the world as a way of justifying the building of dams, roads, and airports, as well as in considering

what to do about biodiversity loss, air pollution, and global climate change. CBA is a useful way of rational decision making about these projects. It is also widely disputed because it tends to discount the value of natural resources, ecological services, and human communities, and it is used to justify projects that jeopardize all these resources.

In CBA the monetary value of all benefits of a project are counted up and compared with the monetary costs of the project. Usually, the direct expenses of a project are easy to ascertain: how much will you have to pay for land, materials, and labor? The monetary worth of lost opportunities—to swim or fish in a river or to see birds in a forest—on the other hand, is much harder to appraise, as are inherent values of the existence of wild species or wild rivers. What is a bug or a bird worth, for instance, or the opportunity for solitude or inspiration? Eventually, the decision maker compares all the costs and benefits to see whether the project is justified or whether an alternative action might bring greater benefit at less cost.

Critics of CBA point out its absence of standards, inadequate attention to alternatives, and the placing of monetary values on intangible and diffuse or distant costs and benefits. Who judges how costs and benefits will be estimated? How can we compare things as different as the economic gain from cheap power with loss of biodiversity or the beauty of a free-flowing river? Critics claim that placing monetary values on everything could lead to a belief that only money and profits count and that any behavior is acceptable as long as you can pay for it. Sometimes speculative or even hypothetical results are given specific numerical values in CBA and then treated as if they were hard facts.

Figure 14.22 shows an example of a cost-benefit analysis for reducing particulate air pollution (soot) in Poland. As you can see, removing the highest 40 percent of particulates is highly cost effective. Approaching 70 percent particulate removal, however, the costs may exceed benefits. Data such as these can be useful

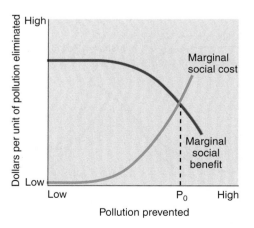

Figure 14.22 To achieve maximum economic efficiency, regulations should require pollution prevention up to the optimum point (P_0) at which the costs of eliminating pollution just equal the social benefits of doing so.

Table 14.6	Estimated Annual Value of Ecological Services	
Ecosystem Services	Value (trillion $ U.S.)
Soil formation	17.1
Recreation	3.0
Nutrient cycling	2.3
Water regulation and supply	2.3
Climate regulation (temperature and precipitation)	1.8
Habitat	1.4
Flood and storm protection	1.1
Food and raw materials production	0.8
Genetic resources	0.8
Atmospheric gas balance	0.7
Pollination	0.4
All other services	1.6
Total value of ecosystem services	33.3

Source: Adapted from R. Costanza et al., "The Value of the World's Ecosystem Services and Natural Capital," in *Nature*, vol. 387, 1997.

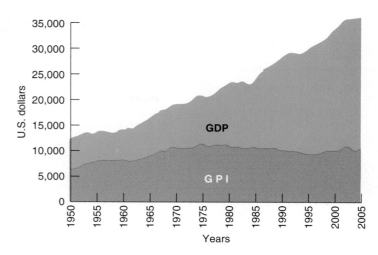

Figure 14.23 Although per capita GDP in the United States nearly doubled between 1970 and 2000 in inflation-adjusted dollars, a genuine progress index that takes into account natural resource depletion, environmental damage, and options for future generations hardly increased at all.
Source: Data from *Redefining Progress*, 2006.

to decision makers. On the other hand, this same study showed that controlling sulfur emissions had high costs and negligible benefits. Might this conclusion result from the ways "benefits" were evaluated?

Values such as wildlife, nonhuman ecological systems, and ecological services can be incorporated with natural resource accounting. In theory this accounting contributes to sustainable resource use because it can put a value on long-term or intangible goods that are necessary but often disregarded in economic decision making. One important part of natural resource accounting is assigning a value to ecological services (table 14.6). The total value of nature's services of $33.3 trillion per year is about twice the entire world GNP. Another is using alternative measures of wealth and development. As mentioned earlier, GNP is a widely used measure of wealth that is based on rates of consumption and throughput. GNP doesn't account for natural resource depletion or ecosystem damage, however. The World Resources Institute, for example, estimates that soil erosion in Indonesia reduces the value of crop production about 40 percent per year. If natural capital were taken into account, Indonesian GNP would be reduced by at least 20 percent annually. Similarly, Costa Rica experienced impressive increases in timber, beef, and banana production between 1970 and 1990. But decreased natural capital during this period, represented by soil erosion, forest destruction, biodiversity losses, and accelerated water runoff, added up to at least $4 billion, or about 25 percent of annual GNP. A number of countries, including Canada and China, now use a "green GDP" that measures environmental costs as part of economic accounting.

New approaches measure real progress

A number of systems have been proposed as alternatives to GNP that reflect genuine progress and social welfare. In their 1989 book, Herman Daly and John Cobb proposed a **genuine progress index (GPI)**, which takes into account real per capita income, quality of life, distributional equity, natural resource depletion, environmental damage, and the value of unpaid labor. They point out that, while per capita GDP in the United States nearly doubled between 1970 and 2000, per capita GPI increased only 4 percent (fig. 14.23). Some social service organizations would add to this index the costs of social breakdown and crime, which would decrease real progress even further over this time span.

The United Nations Development Program (UNDP) uses a benchmark called the **human development index (HDI)** to track social progress. HDI incorporates life expectancy, educational attainment, and standard of living as critical measures of development. Gender issues are accounted for in the gender development index (GDI), which is simply HDI adjusted or discounted for inequality or achievement between men and women.

In its annual Human Development Report, the UNDP compares country-by-country progress. As you might expect, the highest development levels are generally found in North America, Europe, and Japan. In 2003 Norway ranked first in the world in both HDI and GDI. The United States ranked seventh and Canada was eighth, although subtracting the HDI rank from GDP index placed Canada first in the world on an equity-adjusted basis. The 25 countries with the lowest HDI in 2003 were all in Africa. Haiti ranked the lowest in the Western Hemisphere.

Although poverty remains widespread in many places, encouraging news also can be found in development statistics. Poverty

What Can You Do?

Personally Responsible Consumerism

Each of us can do many things to lower our ecological impacts and support "green" businesses through responsible consumerism and ecological economics.

- Practice living simply. Ask yourself if you really need more material goods to make your life happy and fulfilled.

- Rent, borrow, or barter when you can. Can you reduce the amount of stuff you consume by renting, instead of buying, machines and equipment you actually use only rarely?

- Recycle or reuse building materials: doors, windows, cabinets, appliances. Shop at salvage yards, thrift stores, yard sales, or other sources of used clothes, dishes, appliances, etc.

- Consult the *National Green Pages* from Co-Op America for a list of eco-friendly businesses. Write to companies from which you buy goods or services, and ask them what they are doing about environmental protection and human rights.

- Buy green products. Look for efficient, high-quality materials that will last and that are produced in the most environmentally friendly manner possible. Subscribe to clean-energy programs if they are available in your area. Contact your local utility and ask that it provide this option if it doesn't now.

- Buy locally grown or locally made products made under humane conditions by workers who receive a fair wage.

- Think about the total life-cycle costs of the things you buy, especially big purchases, such as cars. Try to account for the environmental impacts, energy use, and disposal costs, as well as initial purchase price.

- Stop junk mail. Demand that your name be removed from mass-mailing lists.

- Invest in socially and environmentally responsible mutual funds or "green" businesses when you have money for investment.

Internalizing external costs

One of the factors that can make resource-exploiting enterprises look good in cost-benefit analysis is externalizing costs. **Externalizing costs** is the act of disregarding or discounting resources or goods that contribute to producing something but for which the producer does not actually pay. Usually, external costs are diffuse and difficult to quantify. Generally, they belong to society at large, not to the individual user. When a farmer harvests a crop in the fall, for example, the value of seeds, fertilizer, and the sale of the crop is tabulated; the value of soil lost to erosion, water quality lost to nonpoint-source pollution, and depleted fish populations is not accounted for. These are most often costs shared by the whole society, rather than borne by the resource user. They are external to the accounting system, and they are generally ignored in cost-benefit analysis—or when the farmer evaluates whether the year was profitable. Larger enterprises, such as dam building, logging, and road building, generally externalize the cost of ecological services lost along the way.

One way to use the market system to optimize resource use is to make sure that those who reap the benefits of resource use also bear all the external costs. This is referred to as **internalizing costs**. Calculating the value of ecological services or diffuse pollution is not easy, but it is an important step in sustainable resource accounting.

14.5 Trade, Development, and Jobs

A sustainable society requires some degree of equitable resource distribution: if most wealth is held by just a few people, the misery and poverty of the majority eventually lead to social instability and instability of resource supplies. Accordingly, wealthy industrial nations have worked harder in recent decades to assist in developing the economies in poorer countries.

International trade can stimulate growth but externalize costs

Expanding trade relations has been promoted as a way to distribute wealth, stimulate economies around the world, and at the same time satisfy the desires of consumers in wealthy countries. According to the economic theory of *comparative advantage*, each place has some sort of goods or services it can make and sell cheaper, or better, than others can. International trade allows us to take advantage of all the best or cheapest products from around the world. If Egypt can produce cotton cheaper than Texas, then we should buy our cotton from Egypt. If Malaysia can manufacture athletic shoes with labor that costs a few cents an hour, then we should buy our shoes from Malaysia. Building factories in Malaysia also stimulates the Malaysian economy, even though workers earn only a small fraction of what a worker in a wealthy country would earn.

has fallen more in the past 50 years, the UNDP reports, than in the previous 500 years. Child death rates in developing countries as a whole have been more than halved. Average life expectancy has increased by 30 percent, while malnutrition rates have declined by almost a third. The proportion of children who lack primary school has fallen from more than half to less than a quarter. And the share of rural families without access to safe water has fallen from nine-tenths to about one-quarter.

Some of the greatest progress has been made in Asia. China and a dozen other countries with populations that add up to more than 1.6 billion have decreased the proportion of their people living below the poverty line by half. Still, in the 1990s the number of people with incomes less than $1 (U.S.) per day increased by almost 100 million to 1.3 billion—and the number appears to be growing in every region except Southeast Asia and the Pacific. Even in industrial countries, more than 100 million people live below the poverty line and 37 million are chronically unemployed.

A problem with international trade is that it externalizes costs on a grand scale. Tropical hardwood products can be sold extremely cheaply in the United States—as lumber, plywood, shipping pallets, and so on. The environmental costs of producing those hardwood products occur far from the consumer who buys a piece of cheap Brazilian plywood. Making matters worse, the environmental costs of the plywood are usually exported to places where there are few legal controls on pollution and resource extraction. A factory in the United States, for example, is legally bound to minimize its production of air and water pollution. Pollution control can be expensive, and internalizing this cost makes a factory less profitable. A similar factory in Mexico might have far less responsibility for pollution control, making it cheaper, at least in the short term, to produce goods there than in the United States. Ongoing protests in the United States and elsewhere around the world against the World Trade Organization (WTO) and other forces of globalization have been largely about such exporting and externalizing of environmental and social costs of production.

Another criticism of international trade is that the international banking systems that finance it are set up by and for the wealthy countries. The WTO and the General Agreement on Tariffs and Trade (GATT), for example, regulate 90 percent of all international trade. The WTO and GATT are both made up of relationships and agreements between corporations in a few very wealthy countries. Representatives of less powerful countries often charge that these agreements trap poorer regions into the role of suppliers of natural resources—timber, mineral ores, fruit, and cheap labor. These countries are forced to mine their natural capital for only small returns in wealth.

Socially responsible development can help people and protect their environment

The World Bank has more influence on the financing and policies of developing countries than any other institution. Of some $25 billion loaned each year for Third World projects by multinational development banks, about two-thirds comes from the World Bank. This institution was founded in 1945 to provide aid to war-torn Europe and Japan. In the 1950s its emphasis shifted to development aid for Third World countries. This aid was justified on humanitarian grounds, but it also conveniently provided markets and political support for growing American and European multinational corporations.

The World Bank is jointly owned by 150 countries, but one-third of its support comes from the United States. Its president has always been an American. The bulk of its $66.8 billion capital comes from private investors. The bank makes loans for development projects, and the investors hope to make a profit on the investment, usually from interest paid on the loans.

Many World Bank projects have been environmentally destructive and highly controversial. In Botswana, for example, $18 million was provided to increase beef production for export by 20 percent, despite already severe overgrazing on fragile grasslands. The project failed, as had two previous beef production projects in the same area. In Ethiopia, rich floodplains in the Awash River Valley were flooded to provide electric power and irrigation water for cash export crops. More than 150,000 subsistence farmers were displaced, and food production was seriously reduced.

Figure 14.24 A small loan can help start a business like this Balinese woman's food stand.

Recently, the World Bank has begun to attempt some environmental and social review of its loans. In part this change comes from demands from the U.S. Congress that projects be environmentally and socially benign. Whether this will improve the World Bank's track record remains to be seen.

The World Bank deals in huge loans for massive projects. These are impressive to investors and to the countries that borrow the money. They are also a huge economic gamble, and on average the economic return on large loans has been quite low.

Recently, smaller, local development programs, often called **microlending**, have begun to develop. These are aimed at small-scale, widespread development, and their results have been very promising. The first of these was Bangladesh's Grammeen (village) Bank network. These banks make small loans, averaging just $67, to help poor people buy a sewing machine, a bicycle, a loom, a cow, or some other commodity that will help them start, or improve, a home business (fig. 14.24). Ninety percent of the customers are women, usually with no collateral or steady income. Still, loan repayment rates are 98 percent—compared with only 30 percent at a conventional bank. This program enhances dignity, respect, and cooperation in a village community, and it teaches individual responsibility and enterprise.

Comparable programs have now sprung up around the world. In the United States, more than a hundred organizations have begun providing microloans and small grants for training. The Women's Self-Employment Project in Chicago, for instance, teaches job skills to single mothers in housing projects. Similarly, "tribal circle" banks on Native American reservations successfully finance microscale economic development projects.

Try Your Hand at Microlending

The best way to observe microlending at work is to try it out. Collect donations of $1–$5 from people in your class, until the total is $25. Go to www.kiva.org, a microlending organization that pools small loans, and select a business to support. You can use Paypal to send the money, and for the next year, you will receive periodic reports on how the business is going. To evaluate whether you're getting a good rate of return, consider that for most stock market investments, about 5–10 percent annual return is reasonable.

1. What is 5–10 percent of $25?

Also poll the class to get these averages:

2. What percentage interest do you earn from your bank accounts?

3. What percentage do you pay to credit card companies?

Answers: 1. $1.25–$2.50; 2., 3. Answers will vary, but probably <2% and <15%.

Table 14.7   Goals for an Eco-Efficient Economy

- Introduce no hazardous materials into the air, water, or soil.
- Measure prosperity by how much natural capital we can accrue in productive ways.
- Measure productivity by how many people are gainfully and meaningfully employed.
- Measure progress by how many buildings have no smokestacks or dangerous effluents.
- Make the thousands of complex governmental rules that now regulate toxic or hazardous materials unnecessary.
- Produce nothing that will require constant vigilance from future generations.
- Celebrate the abundance of biological and cultural diversity.
- Live on renewable solar income rather than fossil fuels.

14.6 Green Business and Green Design

Businesspeople and consumers are increasingly aware of the unsustainability of producing the goods we use every day. Recently, a number of business innovators have tried to develop green businesses, which produce environmentally and socially sound products. Environmentally conscious, or "green," companies, such as the Body Shop, Patagonia, Aveda, Malden Mills, Johnson and Johnson, and others, have shown that operating according to the principles of sustainable development and environmental protection can be good for public relations, employee morale, and sales (table 14.7).

Green business works because consumers are becoming aware of the ecological consequences of their purchases. Increasing interest in environmental and social sustainability has caused an explosive growth of green products. The *National Green Pages* published by Co-Op America currently lists more than 2,000 green companies. You can find eco-travel agencies, telephone companies that donate profits to environmental groups, entrepreneurs selling organic foods, shade-grown coffee, straw-bale houses, paint thinner made from orange peels, sandals made from recycled auto tires, and a plethora of hemp products, including burgers, ale, clothing, shoes, rugs, and shampoo. Although these eco-entrepreneurs represent a tiny sliver of the $7 trillion per year U.S. economy, they often are pioneers in developing new technologies and offering innovative services. Markets also grow over time: organic food marketing has grown from a few funky local co-ops to a $7 billion market segment. Most supermarket chains now carry some organic food choices. Similarly, natural-care health and beauty products reached $2.8 billion in sales in 1999 out of a $33 billion industry. By supporting these products, you can ensure that they will continue to be available and, perhaps, even help expand their penetration into the market.

Corporations committed to eco-efficiency and clean production include such big names as Monsanto, 3M, DuPont, and Duracell. Applying the famous three *R*s—reduce, reuse, recycle—these firms have saved money and gotten welcome publicity (see related story "Eco-Efficient Carpeting" at www.mhhe.com/cunningham5e). Savings can be substantial. Pollution-prevention programs at 3M, for example, have saved $857 million over the past 25 years. In a major public relations achievement, DuPont has cut its emissions of airborne cancer-causing chemicals almost 75 percent since 1987. Small operations can benefit as well. Stanley Selengut, owner of three eco-tourist resorts in the U.S. Virgin Islands, attributes $5 million worth of business to free press coverage about the resorts' green building features and sustainable operating practices.

Green design is good for business and the environment

Architects are starting to get on board the green bandwagon, too. Acknowledging that heating, cooling, lighting, and operating buildings is one of our biggest uses of energy and resources, architects such as William McDonough are designing "green office" projects. Among McDonough's projects are the Environmental Defense Fund headquarters in New York City; the Environmental Studies Center at Oberlin College in Ohio; the European headquarters for Nike in Hilversum, the Netherlands; and the Gap corporate offices in San Bruno, California (fig. 14.25). Each uses a combination of energy-efficient designs and technologies, including natural lighting and efficient water systems. The Gap office building, for example, is intended to promote employee well-being and productivity, as well as efficiency. It has high ceilings, abundant skylights, windows that open, a full-service fitness center (including pool), and a landscaped atrium for each office bay that brings the outside in. The

Figure 14.25 The award-winning Gap, Inc. corporate offices in San Bruno, California, demonstrate some of the best features of environmental design. A roof covered with native grasses provides insulation and reduces runoff. Natural lighting, an open design, and careful relation to its surroundings make this a pleasant place to work.

roof is covered with native grasses. Warm interior tones and natural wood surfaces (all wood used in the building was harvested by certified sustainable methods) give a friendly feeling. Paints, adhesives, and floor coverings are low-toxicity, and the building is one-third more energy-efficient than strict California laws require. The pleasant environment helps improve employee effectiveness and retention. Gap, Inc., estimates that the increased energy and operational efficiency will have a four- to eight-year payback (table 14.8).

Environmental protection creates jobs

For years business leaders and politicians have portrayed environmental protection and jobs as mutually exclusive. They claim that pollution control, protection of natural areas and endangered species, and limits on use of nonrenewable resources will strangle the economy and throw people out of work. Ecological economists dispute this claim, however. Their studies show that only 0.1 percent of all large-scale layoffs in the United States in recent years were due to government regulations (fig. 14.26). Environmental protection, they argue, is not only necessary for a healthy economic system; it actually creates jobs and stimulates business.

Green businesses often create far more jobs and stimulate local economies far more than environmentally destructive ones. Wind energy, for example, provides about five times as many jobs per kilowatt-hour of electricity than does coal-fired power (chapter 12).

Japan, already a leader in efficiency and environmental technology, has recognized the multibillion-dollar economic potential of "green" business. The Japanese government is investing $4 billion per year on research and development, and now it is selling about $12 billion worth of equipment and services per year worldwide. It is already marketing advanced waste incinerators, pollution-control equipment, alternative energy sources, and water treatment systems. Superefficient "hybrid" gas-electric cars are helping Japanese automakers flourish, while U.S. corporations that once dominated the world slide toward bankruptcy. Unfortunately, the United States has been resisting international pollution-control conventions, rather than recognizing the potential for economic growth and environmental protection in the field of "green" business.

Table 14.8   McDonough Design Principles

Inspired by the way living systems actually work, Bill McDonough offers three simple principles for redesigning processes and products:

- *Waste equals food.* This principle encourages elimination of the concept of waste in industrial design. Every process should be designed so that the products themselves, as well as leftover chemicals, materials, and effluents, can become "food" for other processes.

- *Rely on current solar income.* This principle has two benefits: First, it diminishes, and may eventually eliminate, our reliance on hydrocarbon fuels. Second, it means designing systems that sip energy rather than gulping it down.

- *Respect diversity.* Evaluate every design for its impact on plant, animal, and human life. What effects do products and processes have on identity, independence, and integrity of humans and natural systems? Every project should respect the regional, cultural, and material uniqueness of its particular place.

Conclusion

More than half of us now live in cities, and in a generation three-quarters of us will live in cities. Most of this growth will occur in the large cities of developing countries, where resources are already strained. Cities draw immigrants from the countryside by offering jobs, social mobility, education, and other opportunities unavailable in rural areas. Weak rural economies and a lack of access to land also push people to cities. Shortages of water and housing are

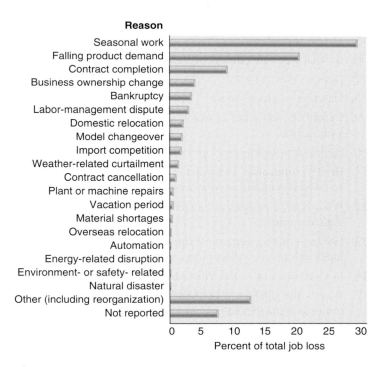

Figure 14.26 Although opponents of environmental regulation often claim that protecting the environment costs jobs, studies by economist E. S. Goodstein show that only 0.1 percent of all large-scale layoffs in the United States were the result of environmental laws.
Source: Data from E. S. Goodstein, Economic Policy Institute, Washington, D.C.

especially urgent problems in fast-growing cities of the developing world. Illegal shantytowns often develop as people seek a place to live on the outskirts of cities. Crime and pollution plague residents of these developments, who often have nowhere else to go.

Urban planning tries to minimize the strains of urbanization. Good planning saves money, because providing roads and services to sprawling suburbs is very expensive. Transportation, which is key to our economy, is a critical part of planning because it determines how far-flung urban development will be. Smart growth, cluster development, and improved standards for environmentally conscious building are also important in urban planning.

Economic policies are often at the root of the success or failure of cities. These policies build from some basic sets of assumptions about the nature of resources. Classical economics assumes that resources are finite, so that we compete to control them. Neoclassical economics assumes that resources are based on capital, which can include knowledge and social capital as well as

resources, and that constant growth is both possible and essential. Natural resource economics extends neoclassical ideas to internalize the value of ecological services in economic accounting.

The "tragedy of the commons" is a classic description of our inability to take care of public resources. Subsequent explanations have pointed that collective rules of ownership are necessary for the survival of shared resources. Our ability to agree on these rules appears to depend on a number of factors, including the scarcity of the resource and our ability to monitor its use.

Because classic productivity indices count many social and environmental ills as positive growth, alternatives have been proposed, including the genuine progress index (GPI). Measures like the GPI can be used to help ensure fair and responsible growth in developing areas. Microlending is another innovative strategy that promotes equity in economic growth. Green business and green design are fast-growing parts of many economies. These approaches save money by minimizing consumption and waste.

Practice Quiz

1. How many people now live in urban areas?
2. How many cities were over 1 million in 1900? How many are now?
3. Why do people move to urban areas?
4. What is the difference between a shantytown and a slum?
5. Define *sprawl*.
6. In what ways are cities ecosystems?
7. Define *smart growth*.
8. Describe a "green" roof.
9. Describe a few ways in which BedZED is self-sufficient and sustainable.
10. Define *sustainable development*.
11. Briefly summarize the differences in how neoclassical and ecological economics view natural resources.
12. What is the estimated economic value of all the world's ecological services?
13. How is it that nonrenewable resources can be extended indefinitely, while renewable resources are exhaustible?
14. In your own words, describe what is shown in figure 14.22.
15. What's the difference between open access and communal resource management?
16. Describe the genuine progress index (GPI).
17. What is *microlending*?

Critical Thinking and Discussion Questions

Apply the principles you have learned in this chapter to discuss these questions with other students.

1. This chapter presents a number of proposals for suburban redesign. Which of them would be appropriate or useful for your community? Try drawing up a plan for the ideal design of your neighborhood.
2. A city could be considered an ecosystem. Using what you learned in chapters 2 and 3, describe the structure and function of a city in ecological terms.
3. If you were doing a cost-benefit study, how would you assign a value to the opportunity for good health or the existence of rare and endangered species in faraway places? Is there a danger or cost in simply saying some things are immeasurable and priceless and therefore off limits to discussion?
4. What would be the effect on the developing countries of the world if we were to change to a steady-state economic system? How could we achieve a just distribution of resource benefits while still protecting environmental quality and future resource use?
5. When an ecologist warns that we are using up irreplaceable natural resources and an economist rejoins that ingenuity and enterprise will find substitutes for most resources, what underlying premises and definitions shape their arguments?

Data Analysis: Using a Logarithmic Scale

We've often used very large numbers in this book. Millions of people suffer from common diseases. Hundreds of millions are moving from the country to the city. Billions of people will probably be added to the world population in the next half century. Cities that didn't exist a few decades ago now have millions of residents. How can we plot such rapid growth and such huge numbers? If you use ordinary graph paper, making a scale that goes to millions or billions will run off the edge of the page unless you make the units very large.

Figure 1, for example, shows the growth of Mumbai, India, over the past 150 years plotted with an **arithmetic scale** (showing constant intervals) for the Y-axis. It looks as if there is very little growth in the first third of this series and then explosive growth during the last few decades, yet we know that the *rate* of growth was actually greater at the beginning than at the end of this time. How could we display this differently? One way to make the graph easier to interpret is to use a **logarithmic scale.** A logarithmic scale, or "log scale," progresses by factor of 10. So the Y-axis would be numbered 0, 1, 10, 100, 1,000. . . . The effect on a graph is to spread out the smaller values and compress the larger values. In figure 2, the same data are plotted using a log scale for the Y-axis, which makes it much easier to see what happened throughout this time period.

Do these two graphing techniques give you a different impression of what's happening in Mumbai? How might researchers use one or the other of these scales to convey a particular message or illustrate details in a specific part of the growth curve?

For Additional Help in Studying This Chapter, please visit our website at www.mhhe.com/cunningham5e. You will find practice quizzes, key terms, answers to end of chapter questions, additional case studies, an extensive reading list, and Google Earth™ mapping quizzes.

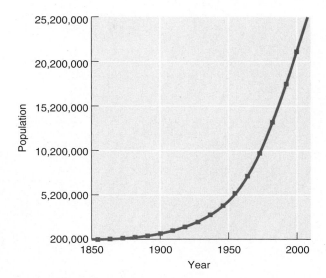

Figure 1 The Growth of Mumbai.

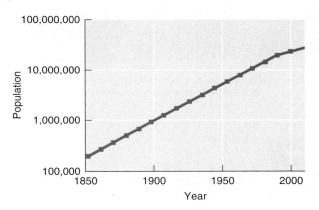

Figure 2 The Growth of Mumbai.

China is planting trees to stabilize soil, reduce erosion and flooding, and hold back the dust storms that plague northern China. In 2006, some 500 million Chinese planted more than 2 billion trees.

Environmental Policy and Sustainability

Learning Outcomes

After studying this chapter, you should be able to answer the following questions:

- What is environmental policy, and how is it formed?
- Why are public policies sometimes unfair or irrational?
- Why do some people want to modify or do away with the NEPA?
- What are some of the most important U.S. environmental laws, and what do they do?
- What are adaptive management and ecosystem management? How do they work?
- Why are international environmental laws and conventions sometimes ineffective?
- What is citizen science, and what opportunities does it offer?
- What can individuals do to contribute to environmental protection?
- How can we work together for these same ends?
- What is sustainability, and why is it important?

You must be the change you wish to see in the world.

–Mahatma Gandhi

CASE STUDY

Is China Greening?

With 1.3 billion people and one of the world's fastest growing econo-mies, China has both tremendous potential and enormous challenges. Since 1980, China's GDP growth has averaged 9 percent per year, and its economy is now ten times larger than it was three decades ago. This phenomenal increase has lifted hundreds of million people out of poverty and accounts for a large proportion of global social progress in recent years.

But rapid growth has also created serious environmental problems. Energy consumption has been rising faster than the economy, and two-thirds of that energy comes from coal—the dirtiest of all sources. Approx-imately one-half of global cement production and one-third of global steel production currently occur in China. Within a decade, China's new vehicle sales are expected to pass those in the United States (fig. 15.1).

According the United Nations Environment Program, 16 of the 20 smoggiest cities in the world are in China, and one-third of the country is affected by acid precipitation. Seventy percent of all Chinese cities don't meet national air quality standards. It's estimated that 400,000 Chinese die each year from the effects of air pollution. China is now the world's largest producer of sulfur dioxide, chlorofluo-rocarbons, and carbon dioxide. Three-quarters of Chinese rivers and lakes fail to meet water quality standards. About half the surface water is so contaminated that it isn't useful even for agriculture. Two-thirds of Chinese cities don't have enough water to meet demands, and at least 300 million people live in areas with severe water shortages. Unsustain-able farming, grazing, and logging have degraded one-third of China's land, and 400 million people are threatened by expanding deserts.

But there also is positive news among this litany of woes. Chinese leaders have recently pledged to take steps to improve environmental quality. In 2006, the government raised the sulfur pollution tax on busi-nesses and began requiring flue gas desulfurization equipment on power plants. And in 2008, the government carried out the first-ever national pollu-tion survey. China plans to spend (U.S.) $125 billion over the next five years to reduce water pollution and bring clean drinking water to everyone in the country by 2015. Officials also promise that 16 percent of China's energy will come from renewable sources within a decade. Logging on steep hillsides has been officially banned, and more than 50 billion trees have been planted on 500,000 km² of marginal land to hold back deserts and reduce blowing dust. Construction has already begun on five new cities designed to be largely self-sufficient in energy, water, and food, and to be climate neutral with respect to greenhouse gases.

Although there are doubts about whether the government can actu-ally deliver on all these promises, it's encouraging that they're mov-ing toward sustainability. A generation ago, environmental conditions weren't even mentioned by Chinese officials. Mere survival dominated the political agenda. Providing enough food, jobs, and housing for a rap-idly growing population and stabilizing a chaotic political system preoc-cupied public attention. Now, at least environmental quality and resource use is something that leaders feel they need to address.

But why should you be interested in public policies half a world away? Part of the answer is that we all share a single planet. What hap-pens in one place affects us all. You may have noticed that rising Chi-nese demand for commodities, such as oil, copper, steel, and seafood, are driving up prices worldwide. Less well known is the fact that on some days, three-fourths of the air pollution in some North American west coast cities can be traced to China. And if China meets its rapidly growing power demand by burning more coal, the effects on our global climate will be disastrous. It has been said that what happens in China in the next few decades will determine the future of the modern world.

There are huge opportunities for progress in China. Currently, it takes about 20 times more energy to create one dollar of goods in China than it does in the United States or Western Europe. Increasing energy efficiency in Chinese factories should be relatively easy using existing technology. It may be possible for China (and other developing coun-tries) to learn from Western experience to bypass some of the most envi-ronmentally damaging stages of industrialization and to move directly to sustainable ways of living. China already has higher vehicle efficiency requirements than the United States.

Although China currently has the world's largest population, and, therefore, a huge impact on our global resources and environment, its problems aren't unique. Many of the countries in which the poorer four-fifths of all humans live face similar environmental challenges. Finding ways that all of us can live sustainably within the limits of our resource base and without damaging nature's life-support systems is the preemi-nent challenge of environmental science. And, as is the case with China, while the world faces many serious environmental problems, there are also signs of progress in many places that give us hope for the future. In this chapter, we'll look both at public policies that hurt or benefit our environment, as well as actions we can take individually and collectively to live sustainably.

Figure 15.1 Between 1994 and 2004, Chinese GDP more than doubled, while the number of private automobiles grew about four-fold. Chemical oxygen demand (COD) fell by more than half as citizens demanded effluent controls on industry, but sulfur dioxide (SO_2) emis-sions increased as more coal was burned.
Source: Shao, M., et al., 2006.

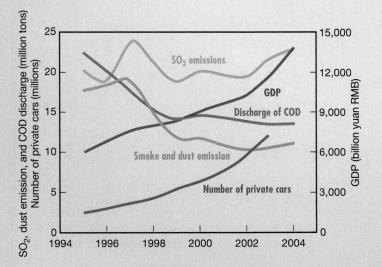

15.1 Environmental Policy and Law

A policy is a rule or decision about how to act or deal with problems. On a personal, informal level, you might have a policy always to get your homework in on time. On a national, formal level, we have policies such as the Clean Air Act, a set of rules, agreed upon by a majority of the U.S. Congress, restricting air pollutants and setting fines for those who exceed legal limits. There are international policies as well: the 1987 Montreal Protocol set agreed-upon limits to the production of ozone-depleting chemicals (chapter 9), and the Convention on International Trade in Endangered Species (CITES) is an international agreement to restrict trade in endangered species (chapter 5).

Environmental policy is both the official rules and regulations concerning the environment that are adopted, implemented, and enforced by government agencies, as well as the general public opinion about environmental issues. National policies are established through negotiation and compromise in a democratic society. Sometimes this wrangling can take decades. Theoretically, it allows all voices to be heard, and the resulting policy serves the interest of the majority.

Making policy isn't always fair or just. Rules can seem harsh and unfair to minorities. The unfair influence of money and power in government also can be discouraging to voters. But sound policy for the public good can be formed—especially when the public is active and vocal in defending its interests. Events don't always follow principles, however. The best way to make sure they do is to develop an informed and involved populace.

How is policy created?

How do policy issues and options make their way onto the stage of public debate? Problems are identified and acted upon in a **policy cycle** that acts to continually define and improve the public agenda (fig. 15.2). The first stage in this process is problem identification. Sometimes the government identifies issues for groups that have no voice or don't recognize problems themselves. In other cases, the public identifies a problem, such as loss of biodiversity or health effects of exposure to toxic waste, and demands redress by the government. In either case, proponents describe the issue—either privately or publicly—and characterize the risks and benefits of their preferred course of action.

Seizing the initiative in issue identification often allows a group to define terms, set the agenda, organize stakeholders, choose tactics, aggregate related issues, and legitimate (or delegitimate) issues and actors. It can be a great advantage to set the format or choose the location of a debate. Next, stakeholders develop proposals for preferred policy options, often in the form of legislative proposals or administrative rules. Proponents build support for their position through media campaigns, public education, and personal lobbying of decision makers. By following the legislative or administrative process through its many steps, interest groups ensure that their proposals finally get enacted into law or established as a rule or regulation.

The next step is implementation. Ideally, government agencies faithfully carry out policy directives as they organize bureaucracies, provide services, and enforce rules and regulations, but often it takes continued monitoring to make sure the system works as it should. Evaluating the results of policy decisions is as important as establishing them in the first place. Measuring impacts on target and nontarget populations shows us whether the intended goals, principles, and course of action are being attained. Finally, suggested changes or adjustments are considered that will make the policy fairer or more effective.

There are different routes by which this cycle is carried out. Special economic interest groups, such as industry associations, labor unions, or wealthy and powerful individuals, don't need (or often want) much public attention or support for their policy initiatives. They generally carry out the steps of issue identification, agenda setting, and proposal development privately because they can influence legislative or administrative processes directly through their contacts with decision makers. Public interest groups, on the other hand, often lack direct access to corridors of power and need to rally broad general support to legitimate their proposals.

An important method for getting public interest on the table is to attract media attention. Organizing a dramatic protest or media event can generate a lot of free publicity (fig. 15.3). Often, officials ignore public opinion, but when protests get strong enough, decision makers are forced to listen. Some of the newfound governmental concern about the Chinese environment described in the opening case study for this chapter comes from public demand. In 2005 there were at least 87,000 public protests in China, many of which were about pollution, environmental health, property rights, and similar issues. A campaign led by artists, students, and writers forced the government to cancel plans for a series of 13 large dams on the Nu River (the Chinese portion of the Salween) in an area of high biological and cultural diversity. There are now more than 2,000 nongovernmental organizations (NGOs) in China working on social and environmental issues, and for the first time, they have officially recognized status.

There appears to be a dramatic worldwide shift in public attitudes toward environmental protection. In a 2007 BBC poll of 22,000 residents of 21 countries, 83 percent agreed that individuals would definitely or probably have to make lifestyle changes

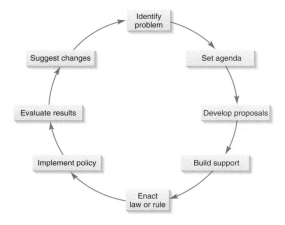

Figure 15.2 The policy cycle.

to reduce the amount of climate-changing gases they produce (fig. 15.4). Overall, 70 percent said they were personally ready to make sacrifices to protect the environment. Concern about environmental quality varied by country, however, with just over 40 percent of Russians willing to change their lifestyle to prevent global warming, compared to nearly 90 percent of Canadians. The Chinese are the most enthusiastic when it comes to energy taxes to prevent climate change. Eighty-five percent of the Chinese polled agreed that such taxes are necessary, over 24 percent more than the next most-supportive countries.

Policy formation can be complicated

The policies we establish depend to a great extent on the system within which they operate. For many of us, the ideal political system is one that is open, honest, transparent, and reaches the best possible decisions to maximize benefits to everyone. In a pluralistic, democratic society, we aim to give everyone an equal voice in policy making. Ideally, many separate interests put forward their solutions to public problems that are discussed, debated, and evaluated fairly and equally. Facts and access are open to everyone. Policy choices are made democratically but compassionately; implementation is reasonable, fair, and productive. Unfortunately, this isn't always the way political systems work. Those with the most power and money usually have the greatest voice in public affairs. Still, there have been altruistic movements in the past to ensure human rights and environmental protection that reveal an impulse for generosity and goodwill among the general public.

Another model for public decision making is rational choice and science-based management. In this utilitarian approach, no policy should have greater total costs than benefits. In choosing among policy alternatives, we should always prefer those with the greatest cumulative welfare and the least negative impacts. Professional administrators would weigh various options and make an objective, methodical decision that would bring maximum social gain. There are many reasons, however, that rational choice doesn't always work in the public arena.

- Many conflicting values and needs cannot be compared because they aren't comparable or we don't have perfect information.

- There are few generally agreed-upon broad societal goals, but rather, benefits to specific groups and individuals, many of which are in conflict.

- Policymakers generally aren't motivated to make decisions on the basis of societal goals, but rather to maximize their own rewards: power, status, money, or reelection.

- Large investments in existing programs and policies create "path dependence" and "sunken costs" that prevent policymakers from considering good alternatives foreclosed by previous decisions.

- Uncertainty about the consequences of various policy options compels decision makers to stick as closely as possible to previous policies to reduce the likelihood of adverse, calamitous, unanticipated consequences.

- Policymakers, even if well meaning, don't have sufficient intelligence or adequate data or models to calculate accurate costs and benefits when large numbers of diverse political, social, economic, and cultural values are at stake.

- The segmented nature of policy making in large bureaucracies makes coordinating decision making difficult.

National policies play a critical role in environmental protection

Signed into law by President Nixon in 1970, the **National Environmental Policy Act (NEPA)** is the cornerstone of U.S. environmental policy and a model for many other countries. Conservationists

Figure 15.3 Protestors demonstrate against monetary policies and international institutions that threaten livelihoods and environmental quality.

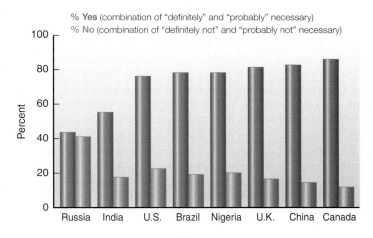

Figure 15.4 About 70 percent of the 22,000 people in 21 countries polled by the BBC in 2007 agreed with the statement, "I am ready to make significant changes to the way I live to help prevent global warming or climate change."

Figure 15.5 Every major federal project in the United States must be preceded by an environmental impact statement (EIS).

see this act as a powerful tool for environmental protection, but commercial interests blame it for gridlock and consider it an impediment to business. Conservatives call for streamlining NEPA to make it less burdensome for industry. Conservationists worry this is really an effort to weaken the law and return to laissez-faire resource management.

NEPA does three important things: (1) it authorizes the Council on Environmental Quality (CEQ), the oversight board for general environmental conditions; (2) it directs federal agencies to take environmental consequences into account in decision making; and (3) it requires an **environmental impact statement (EIS)** be published for every major federal project likely to have an important impact on environmental quality. NEPA doesn't forbid environmentally destructive activities if they comply otherwise with relevant laws, but it demands that agencies admit publicly what they plan to do. Once embarrassing information is revealed, however, few agencies will bulldoze ahead, ignoring public opinion. And an EIS can provide valuable information about government actions to public interest groups that wouldn't otherwise have access to these resources.

What kinds of projects require an EIS? The activity must be federal and it must be major, with a significant environmental impact (fig. 15.5). Evaluations are always subjective as to whether specific activities meet these characteristics. Each case is unique and depends on context, geography, the balance of beneficial versus harmful effects, and whether any areas of special cultural, scientific, or historical importance might be affected. A complete EIS for a project is usually time-consuming and costly. The final document is often hundreds of pages long and generally takes six to nine months to prepare. Sometimes just requesting an EIS is enough to sideline a questionable project. In other cases, the EIS process gives adversaries time to rally public

opposition and information with which to criticize what's being proposed. If agencies don't agree to prepare an EIS voluntarily, citizens can petition the courts to force them to do so.

Every EIS must contain the following elements: (1) purpose and need for the project, (2) alternatives to the proposed action (including taking no action), (3) a statement of positive and negative environmental impacts of the proposed activities. In addition, an EIS should make clear the relationship between short-term resources and long-term productivity, as well as any irreversible commitment of resources resulting from project implementation.

Among the areas in which lawmakers have tried recently to ignore or limit NEPA include forest policy, energy exploration, and marine wildlife protection. The "Healthy Forest Initiative," for example, called for bypassing EIS reviews for logging or thinning projects, and prohibited citizen appeals of forest management plans. Similarly, when the Bureau of Land Management proposed 77,000 coal-bed methane wells in Wyoming and Montana, promoters claimed that water pollution and aquifer depletion associated with this technology didn't require environmental review (chapter 12).

To be informed environmental citizens, Americans need to know something about how policies and laws like NEPA are created and applied. What do you think? Is NEPA an essential safeguard for our environmental quality, or merely an antiquated impediment for business and personal property rights? What information would you need to judge the effectiveness of this important law?

Laws affirm public policy

Laws are rules set by authority, society, or custom. Church laws, social mores, administrative regulations, and a variety of other codes of behavior can be considered laws if they are backed by some enforcement power. **Environmental law** includes official rules, decisions, and actions concerning environmental quality, natural resources, and ecological sustainability. In the United States a wide variety of environmental laws are promulgated at both the local and national levels. Because every country has different legislative and legal processes, this chapter will focus primarily on the U.S. system in the interest of simplicity and space. Environmental law can be established or modified in each of the three branches of government: legislative, judicial, and executive—statutory, case, and administrative law, respectively. Understanding how these systems work is an important step in becoming an environmentally literate person.

The legislative branch establishes statutory law

Establishing laws at either the state or the federal level is one of the most important ways of protecting our environment. Many environmental groups spend a good deal of their time and resources trying to influence the legislative process. In this section we'll look at how that system works.

Federal laws (statutes) are enacted by Congress and must be signed by the president. They originate as legislative proposals called bills, which are usually drafted by the congressional staff, often in consultation with representatives of various interest groups. Thousands of bills are introduced every year in Congress. Some are very narrow, providing funds to build a specific section of road or to help a particular person, for instance. Others are

extremely broad, perhaps overhauling the Social Security system or changing the entire tax code. Similarly, environmental legislation might deal with a very specific local problem or a national or international issue. Often a number of competing bills on a single issue may be introduced as proponents from different sides attempt to incorporate their views into law. A bill may have a single sponsor if it is the pet project of a particular legislator, or it may have 100 or more co-authors if it is an issue of national importance. Table 15.1 lists some of the most important U.S. environmental legislation.

Citizens can be involved in this process either by writing or calling their elected representatives, or by appearing at public hearings (fig. 15.6). A personal letter or statement is always more persuasive than simply signing a petition. Some politicians treat all the copies of a similar form letter or email as a single statement, because they all contain the same information. Still, a petition with a million signatures will probably catch the attention of a legislator—especially if they are all potential voters.

Being involved in local election campaigns can greatly increase your access to legislators. Writing letters or making telephone calls also are highly effective ways to get your message across. All legislators now have email addresses, but letters

Table 15.1 Major U.S. Environmental Laws

Legislation	Provisions
National Environmental Policy Act of 1969	Declared national environmental policy, required environmental impact statements, created Council on Environmental Quality.
Clean Air Act of 1970	Established national primary and secondary air quality standards. Required states to develop implementation plans. Major amendments in 1977 and 1990.
Clean Water Act of 1972	Set national water quality goals and created pollutant discharge permits. Major amendments in 1977 and 1996.
Federal Pesticides Control Act of 1972	Required registration of all pesticides in U.S. commerce. Major modifications in 1996.
Marine Protection Act of 1972	Regulated dumping of waste into oceans and coastal waters.
Coastal Zone Management Act of 1972	Provided funds for state planning and management of coastal areas.
Endangered Species Act of 1973	Protected threatened and endangered species. Directed FWS to prepare recovery plans.
Safe Drinking-Water Act of 1974	Set standards for safety of public drinking-water supplies and to safeguard ground water. Major changes made in 1986 and 1996.
Toxic Substances Control Act of 1976	Authorized EPA to ban or regulate chemicals deemed a risk to health or the environment.
Federal Land Policy and Management Act of 1976	Charged the BLM with long-term management of public lands. Ended homesteading and most sales of public lands.
Resource Conservation and Recovery Act of 1976	Regulated hazardous-waste storage, treatment, transportation, and disposal. Major amendments in 1984.
National Forest Management Act of 1976	Gave statutory permanence to national forests. Directed USFS to manage forests for "multiple use."
Surface Mining Control and Reclamation Act of 1977	Limited strip-mining on farmland and steep slopes. Required restoration of land to original contours.
Alaska National Interest Lands Act of 1980	Protected 40 million ha (100 million acres) of parks, wilderness, and wildlife refuges.
Comprehensive Environmental Response, Compensation and Liability Act of 1980	Created $1.6 billion "Superfund" for emergency response, spill prevention, and site remediation for toxic wastes. Established liability for cleanup costs.
Superfund Amendments and Reauthorization Act of 1994	Increased Superfund to $8.5 billion. Shared responsibility for cleanup among potentially responsible parties. Emphasized remediation and public "right to know."

Source: Data from N. Vig and M. Kraft, *Environmental Policy in the 1990s*, 3rd Congressional Quarterly Press.

Figure 15.6 Citizens line up to testify at a legislative hearing. By getting involved in the legislative process, you can be informed and have an impact on government policy.

and phone calls are usually taken more seriously. Getting media attention can sway the opinions of decision makers. Organizing protests, marches, demonstrations, or other kinds of public events can call attention to your issue. Public education campaigns, press conferences, TV ads, and a host of other activities can be helpful. Joining together with other like-minded groups can greatly increase your clout and ability to get things done. It's hard for a single individual or even a small group to have much impact, but if you can organize a mass movement, you may be very effective in influencing public policy.

The judicial branch shapes case law

The judicial branch of government establishes environmental law by ruling on the constitutionality of statutes and interpreting their meaning. We describe the body of legal opinions built up by many court cases as **case law**. Often legislation is written in vague and general terms so as to make it widely enough accepted to gain passage. Congress, especially in the environmental area, often leaves it to the courts to "fill in the gaps." When trying to interpret a law, the courts depend on the legislative record from hearings and debates to determine congressional intent. What was a particular statute meant to do by those who wrote and passed it? In ruling on cases, the court builds up a body of precedent that influences subsequent deliberations. Judges usually respect the rulings of their predecessors unless some new information becomes available.

Criminal law derives from those federal and state statutes that prohibit wrongs against the state or society. Serious crimes, like murder or rape, that are punishable by long jail terms or heavy fines are called felonies. Lesser crimes, such as shoplifting or vandalism, that result in smaller fines or shorter sentences in a county or city jail are labeled misdemeanors. A criminal case is always initiated by a government prosecutor. Guilt or innocence of the defendant is determined by a jury of peers, but the sentence is imposed, often in consultation with the jury, by the judge. The judge is responsible

for keeping order in the hearings and for determining points of law (fig. 15.7). The jury acts as a fact-finding body that weighs the truth and reliability of the witnesses and evidence.

Violation of many environmental statutes constitutes a criminal offense. In 1975 the U.S. Supreme Court ruled that corporate officers can be held criminally liable for violations of environmental laws if they were grossly negligent, or the illegal actions can be considered willful and knowing violations. In one of the toughest criminal sentences imposed so far, the president of a Colorado company was sentenced in 1999 to 14 years in prison for knowingly dumping chlorinated solvents that contaminated the water table.

Civil law is a body of laws regulating relations between individuals or between individuals and corporations. Issues such as property rights and personal dignity and freedom are protected by civil law. The defendant in a civil case has a right to be tried by a jury, but in highly technical issues, this right often is waived and the case is heard only by the judge. Being found guilty of a civil offense can result in financial penalties but not jail time.

In contrast to a criminal case, where the burden of proof lies with the prosecution and defendants are considered innocent until proven guilty, civil cases can be decided on a "preponderance of evidence." This makes civil cases considerably easier to win than criminal cases when evidence is ambiguous. A number of mitigating factors also are taken into account in determining guilt and assigning penalties in civil cases. Guilt or innocence is based on whether the defendant could reasonably have anticipated and avoided the offense. A "good faith effort" to comply or solve the problem can be a factor. The compliance history is important. Is this a first-time offender or a habitual repeater? Finally, is there evidence of economic benefit to the perpetrator? That is, did the violator gain personally from the action? If so, it is more likely that willful intent was involved.

Civil cases can be brought in both state and federal court. In 2000, the Koch Oil Company, one of the largest pipeline and refinery operators in the United States, agreed to pay $35 million

Figure 15.7 In a trial court, the judge presides over trials, rules on motions, and interprets the law.

in fines and penalties to state and federal authorities for negligence in more than 300 oil spills in Texas, Oklahoma, Kansas, Alabama, Louisiana, and Missouri over the previous decade. Koch also agreed to spend more than $1 billion on cleanup and improved operations.

Sometimes the purpose of a civil suit is to seek an injunction or some other protection from the actions of an individual, a corporation, or a governmental agency. You might ask the courts, for example, to order the government to cease and desist from activities that are in violation of either the spirit or the letter of the law. This sort of civil action is heard only by a judge; no jury is present. Environmental groups have been very successful in asking courts to stop logging and mining operations, to enforce implementation of the Endangered Species Act, to require agencies to enforce air and water pollution laws, and a host of other efforts to protect the environment and conserve natural resources.

The executive branch enforces administrative law

More than 100 federal agencies and thousands of state and local boards and commissions have environmental oversight. They usually have power to set rules, adjudicate disputes, and investigate misconduct.

Agency rule-making and standard-setting can be an important way for environmentally concerned citizens and public interest groups to have an impact on environmental policy. Rule-making is often a complex, highly technical process that is difficult for citizen groups to understand and monitor. The proceedings are usually less dramatic and colorful than criminal trials, and yet can be very important for environmental protection. The Bush administration has made profound changes in U.S. environmental policy through rule-making that avoids open debate or public scrutiny.

The EPA is the primary agency with responsibility for protecting environmental quality in the United States. Created in 1970 at the same time as NEPA, the EPA has more than 18,000 employees and ten regional offices. Often in conflict with Congress, other agencies of the executive branch, and environmental groups, the EPA has to balance many competing interests and conflicting opinions. Greatly influenced by politics, the agency changes dramatically, depending on which party is in power and what attitudes toward the environment prevail at any given time.

The Departments of the Interior and Agriculture are to natural resources what the EPA is to pollution. Interior is home to the National Park Service, which is responsible for more than 376 national parks, monuments, historic sites, and recreational areas. It also houses the Bureau of Land Management (BLM), which administers some 140 million ha (350 million acres) of land, mostly in the western United States. In addition, Interior is home to the U.S. Fish and Wildlife Service, which operates more than 500 national wildlife refuges and administers endangered species protection.

The Department of Agriculture is home to the U.S. Forest Service, which manages about 175 national forests and grasslands, totaling some 78 million ha (193 million acres). With 39,000 employees, the Forest Service is nearly twice as large as the EPA (fig. 15.8). The Department of Labor houses the Occupational Safety and Health Administration (OSHA), which oversees workplace safety. Research that forms the basis for OSHA standards is carried out by the National Institute for Occupational Safety and Health (NIOSH). In addition, several independent agencies that are not tied to any specific department also play a role in environmental protection and public health. The Consumer Products Safety Commission passes and enforces regulations to protect consumers, and the Food and Drug Administration is responsible for the purity and wholesomeness of food and drugs.

How can we manage complex systems?

Managing ecosystems is a complicated and controversial affair. A classic example of the difficulty brought about by clashing human and natural interconnections can be seen in the Florida Everglades. L. Gunderson, C. Holling, and S. Light discuss ways in which policymakers find solutions for difficult problems in their 1995 book *Barriers and Bridges*.

Over the past century, the Everglades ecosystem has been transformed from an immense subtropical wetland into a highly managed, multiple-use system as the result of one of the world's largest public works projects. This huge hydraulic system stretches from just south of Orlando through the many springs and lakes that feed the Kissimmee River, through Lake Okeechobee to the Everglades National Park at the southern tip of Florida. In 1947, Marjorie Stoneman Douglas aptly named this the "river of grass" to describe the broad, slowly moving sheet of water that trickles through the vast wetlands of South Florida.

Management policies have lurched from one approach to another as environmental surprises and human intervention have transformed the area. Natural crises include floods, droughts, fires, and storms. Human effects, including agricultural pollution, invasive species, and the increasing water demands of rapidly growing coastal cities, further complicate the picture.

Figure 15.8 Smokey Bear symbolizes the Forest Service's role in extinguishing forest fires.

Historic flow | Current flow | Planned flow

Figure 15.9 Planned results of the Comprehensive Everglades Restoration Plan. Red dashed outline shows the national park boundary.

Table 15.2	Principles of Ecosystem Management
•	Plan at multiple scales and time periods
•	Use ecological boundaries rather than political ones
•	Protect essential ecological processes
•	Monitor the system and use adaptive management
•	Allow for organizational change
•	Make human dimensions part of the plan
•	Identify values of nature and stakeholders

The earliest settlers were intent on "reclaiming" wetlands in order to farm the rich muck soils. They dug canals to drain the land as fast as possible. This prevented the wetland from absorbing and storing excess rainfall. The result was severe flooding in abnormally wet years, such as 1947, when more than 90 percent of the crop was destroyed. The U.S. Army Corps of Engineers responded by creating a massive water control plan, including 2,200 km of canals, 1,000 km of levees, and pumping stations with 3.8 billion liters/day capacity (fig. 15.9).

In 1971, however, the worst drought in 40 years hit the region. This occurred just as the population of southern Florida topped the 2 million mark, and sugarcane production had more than tripled. Competition for water between the Everglades National Park, agricultural interests, and rapidly growing cities demanded new approaches to water management. Stakeholders debated for years about how the system should be managed. In 2000, the state of Florida and the Army Corps agreed to begin a restoration project that would remove many levees and canals, build new reservoirs

to store water, dechannelize rivers, and construct 14,000 ha of new filtration wetlands. Despite billions of dollars already spent, however, this project is behind schedule and criticized by all sides as being inadequate (see related story on "The World's Biggest Restoration Project" at www.mhhe.com/cunningham5e).

Some preservationists suggest that we should simply leave ecosystems, such as the Everglades, alone to let nature take its own course. Technological optimists, on the other hand, claim that we need to apply even more engineering to make nature do whatever we want. The fact is that we have intervened for so long and to such a great extent in the Everglades that we can't just walk away now and hope that things will work out for the best. At the same time, we need to recognize that the Everglades are so complex and highly variable that we will probably never understand or control them completely.

Ecologists applaud the attempt to restore natural functions of the ecosystem and to manage it as an entire system rather than to tackle problems as if they were each unique and unconnected to any others. They suggest **adaptive management**, in which action plans are continuously monitored, evaluated, and adjusted to fit changing realities and understandings of the system (fig. 15.10). We also need adaptive models, they propose, for social and institutional change. In some ways, human behavior may be more difficult to understand and predict than that of nature.

Ecosystem management, recognizes that we often need to take action in the face of uncertainty. Decision makers who work at multiple scales should see the interconnections between different levels of organization. If we hope to preserve biodiversity, defend ecological integrity, and maintain ecological processes, we have to recognize ecological boundaries. Humans can't be separated from nature. We inevitably affect ecological processes and are in turn affected by them. We also need to manage over time periods much longer than the normal planning cycle or term-in-office for most human institutions (table 15.2).

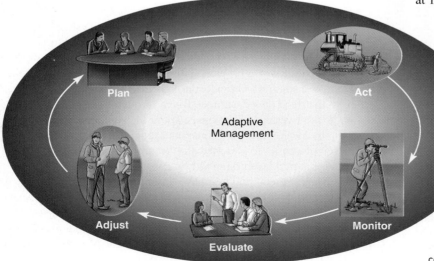

Adaptive Management

Plan · Act · Monitor · Evaluate · Adjust

Figure 15.10 Adaptive management recognizes that we need to treat management plans for ecosystems as a scientific experiment in which we monitor, evaluate, and adjust our policies to fit changing conditions and knowledge.

Table 15.3	Some Major International Treaties and Conventions	
Date	**Title**	**Covers**
1971	Convention on Wetlands of International Importance (Ramsar)	Protects wetlands, especially as waterfowl habitat
1972	Convention to Protect World Cultural and Natural Heritage	Protects cultural sites and natural resources
1973	Convention on International Trade in Endangered Species (CITES)	Restricts trade in endangered plants and animals
1979	Convention on Migratory Species (CMS)	Protects migratory species, especially birds
1982	United Nations Law of the Sea (UNCLOS)	Declares the oceans international commons
1985	Protocol on Substances That Deplete the Ozone Layer (Ozone)	Initiates phaseout of chlorofluorocarbons
1989	Convention on the Transboundary Movements of Hazardous Waste (Basel)	Bans shipment of hazardous waste
1992	Convention on Biodiversity (CBD)	Protects biodiversity as national resources
1992	United Nations Framework Convention on Climate Change (UNFCCC)	Rolls back carbon dioxide production in industrialized countries
1994	Convention to Combat Desertification (CCD)	Provides assistance to fight desertification, especially in Africa

Of course it takes good science to understand and manage complex ecosystems, but what is good science? Advocates from across the political spectrum tend to promote evidence that supports their own agenda, but to label data that contradicts their ideology as "junk science." Does that mean that all research is equally valid? As we discussed in chapter 1, there are objective standards to identify good science. In the first place, it needs to be impartial and open to public scrutiny. Secret, proprietary research may be useful in business, but it shouldn't be used to formulate public policy. Research in which some special interest can determine what questions will be asked, or how data are collected, analyzed, and published, doesn't usually yield sound science. For evidence to be considered valid, it must be reproducible. Materials, methods, and results should be published in peer-reviewed journals in enough detail so that others can independently replicate them. Scientists have a right—as does everyone—to their own personal opinions and to participate in public debates. Sometimes issues, such as global climate change, are so important that leading scientists feel compelled to speak out. But they have to be careful to distinguish when they are speaking as scientific experts, and when they're speaking as interested citizens.

15.2 International Treaties and Conventions

As recognition of the interconnections in our global environment has advanced, the willingness of nations to enter into protective **international treaties and conventions** has grown (table 15.3). The earliest of these conventions had no nations as participants; they were negotiated entirely by panels of experts. Not only the number of parties taking part in these negotiations has grown, but the rate at which parties are signing on and the speed at which agreements take force also have increased rapidly (fig. 15.11). The Convention on International Trade in Endangered Species (CITES), for example, was not enforced until 14 years after ratification, but the Convention on Biological Diversity was enforceable after just one year and had 160 signatories only four years after introduction. Over the past 25 years, more than 170 treaties and conventions have been negotiated to protect our global environment. Designed to regulate activities ranging from intercontinental shipping of hazardous waste, to deforestation, overfishing, trade in endangered species, global warming, and wetland

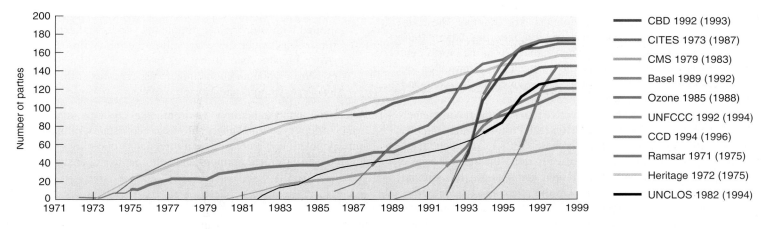

Figure 15.11 Additions of participating parties to some major international environmental treaties. The thick portion of each line shows when the agreement went into effect (*date in parentheses*). See table 15.3 for complete treaty names.
Source: Data from United Nations Environment Programme from Global Environment Outlook–2000.

protection, these agreements theoretically cover almost every aspect of human impacts on the environment.

Unfortunately, many of these environmental treaties constitute little more than vague good intentions. Even though we often call them laws, there is no body that can legislate or enforce international environmental protection. The United Nations and a variety of regional organizations bring stakeholders together to negotiate solutions to a variety of problems, but the agreed-upon solutions generally rely on moral persuasion and public embarrassment for compliance. Most nations are unwilling to give up sovereignty. There is an international court, but it has no enforcement power. Nevertheless, there are creative ways to strengthen international environmental protection.

One of the problems with most international treaties is the tradition that they must be passed by unanimous consent. A single recalcitrant nation effectively has veto power over the wishes of the vast majority. For instance, more than 100 countries at the UN Conference on Environment and Development (UNCED), held in Rio de Janeiro in 1992, agreed to restrictions on the release of greenhouse gases. At the insistence of U.S. negotiators, however, the climate convention was reworded so that it only urged—but did not require—nations to stabilize their emissions.

As a way of avoiding this problem, some treaties incorporate innovative voting mechanisms. When a consensus cannot be reached, they allow a qualified majority to add stronger measures in the form of amendments that do not need ratification. All members are legally bound to the whole document unless they expressly object. This approach was used in the Montreal Protocol, passed in 1987 to halt the destruction of stratospheric ozone by chlorofluorocarbons (CFCs). The agreement allowed a vote of two-thirds of the 140 participating nations to amend the protocol. Although initially the protocol called for only a 50 percent reduction in CFC production, subsequent research showed that ozone was being depleted faster than previously thought (chapter 9). The protocol was strengthened by amendment to an outright ban on CFC production, in spite of the objection of a few countries.

When strong accords with meaningful sanctions cannot be passed, sometimes the pressure of world opinion generated by revealing the sources of pollution can be effective. Activists can use this information to expose violators. For example, the environmental group Greenpeace discovered monitoring data in 1990 showing that Britain was disposing of coal ash in the North Sea. Although not explicitly forbidden by the Oslo Convention on ocean dumping, this evidence proved to be an embarrassment, and the practice was halted.

Trade sanctions can be an effective tool to compel compliance with international treaties. The Montreal Protocol, for example, bound signatory nations not to purchase CFCs or products made using them from countries that refused to ratify the treaty. Because many products employed CFCs in their manufacture, this stipulation proved to be very effective. On the other hand, trade agreements also can work against environmental protection. The World Trade Organization (WTO) was established to make international trade more fair and to encourage development. It has been used, however, to expand trade globalization and to weaken national

environmental laws. In 1990, after decades of public protests and boycotts, the United States banned the import of tuna caught using methods that killed thousands of dolphins each year. Shrimp caught with nets that kill endangered sea turtles were also banned. Mexico filed a complaint with the WTO, contending that dolphin-safe tuna laws represented an illegal barrier to trade. Thailand, Malaysia, India, and Pakistan filed a similar suit over turtle-friendly shrimp laws. The WTO ordered the United States to allow the import of both tuna and shrimp from countries that allow fisheries to kill dolphins and turtles. Environmentalists point out that the WTO has never ruled against a corporate suit, in part because rules are decided by a committee composed mainly of appointed industry leaders, rather than by elected officials representing a spectrum of political and economic interests.

15.3 What Can Individuals Do?

Whatever your skills and interests, you can participate in policy formation and contribute to understanding and protecting our common environment. If you enjoy science, there are many disciplines that contribute to environmental science. As you know by now, biology, chemistry, geology, ecology, climatology, geography, demography, and other sciences all provide essential ideas and data to environmental science. Environmental scientists usually focus on one of these disciplines, but their work serves the others. An environmental chemist, for example, might study contaminants in a stream system, and this work might help an aquatic ecologist understand changes in a stream's food web.

You can also help seek environmental solutions if you prefer writing, art, working with children, history, politics, economics, or other areas of study. As you have read, environmental science depends on communication, education, good policies, and economics as well as on science.

Believe it or not, you can also help understand our environment just by enjoying it (fig. 15.12). As author Edward Abbey wrote,

> It is not enough to fight for the land; it is even more important to enjoy it. While it is still there. So get out there and mess around with your friends, ramble out yonder and explore the forests, encounter the grizz, climb the mountains. Run the rivers, breathe deep of that yet sweet and lucid air, sit quietly for a while and contemplate the precious stillness, that lovely mysterious and awesome space.

When you understand how environmental systems function—from nutrient cycles and energy flows to ecosystems, climate systems, population dynamics, agriculture, and economies—you can develop well-informed opinions and help find useful answers.

Environmental education is an important tool

In 1990 the United State Congress recognized the importance of environmental education by passing the National Environmental Education Act. The act established two broad goals: (1) to improve

Figure 15.12 Enjoying nature helps you understand what needs to be saved, and it gives you motivation to keep working.

Figure 15.13 Environmental education helps develop awareness and appreciation of ecological systems and how they work.

understanding among the general public of the natural and built environment and the relationships between humans and their environment, including global aspects of environmental problems, and (2) to encourage postsecondary students to pursue careers related to the environment. Specific objectives proposed to meet these goals include developing an awareness and appreciation of our natural and social/cultural environment, knowledge of basic ecological concepts, acquaintance with a broad range of current environmental issues, and experience in using investigative, critical-thinking, and problem-solving skills in solving environmental problems (fig. 15.13). Several states, including Arizona, Florida, Maryland, Minnesota, Pennsylvania, and Wisconsin, have successfully incorporated these goals and objectives into their curricula (table 15.4).

Speaking in support of the National Environmental Education Act, former Environmental Protection Agency administrator William K. Reilly called for broad **environmental literacy** in which every citizen is fluent in the principles of ecology and has a "working knowledge of the basic grammar and underlying syntax of environmental wisdom." Environmental literacy, according to Reilly, can help establish a stewardship ethic—a sense of duty to care for and manage wisely our natural endowment and our productive resources for the long haul. It isn't enough, he says, "for a few specialists to know what is going on while the rest of us wander about in ignorance."

You have made a great start toward learning about your environment by reading this book and taking a class in environmental science. Pursuing your own environmental literacy is a life-long process. Some of the most influential environmental books of all time examine environmental problems and suggest solutions (table 15.5). To this list we'd add some personal favorites: *The Singing Wilderness* by Sigurd F. Olson, *My First Summer in the Sierra* by John Muir, and *Remaking Society: Pathways to a Green Future* by Murray Bookchin.

Citizen science encourages everyone to participate

Many students are discovering they can make authentic contributions to scientific knowledge through active learning and undergraduate research programs. Internships in agencies or environmental organizations are one way of doing this. Another is to get involved in organized **citizen science** projects in which ordinary people join with established scientists to answer real scientific questions. Community-based research was pioneered in the Netherlands, where several dozen research centers now study environmental issues ranging from water quality in the Rhine

Table 15.4	Outcomes from Environmental Education
The natural context: An environmentally educated person understands the scientific concepts and facts that underlie environmental issues and the interrelationships that shape nature.	
The social context: An environmentally educated person understands how human society is influencing the environment, as well as the economic, legal, and political mechanisms that provide avenues for addressing issues and situations.	
The valuing context: An environmentally educated person explores his or her values in relation to environmental issues; from an understanding of the natural and social contexts, the person decides whether to keep or change those values.	
The action context: An environmentally educated person becomes involved in activities to improve, maintain, or restore natural resources and environmental quality for all.	

Source: Data from *A Greenprint for Minnesota*, Minnesota Office of Environmental Education, 1993.

Table 15.5   The Environmentalist's Bookshelf

What are some of the most influential and popular environmental books? In a survey[1] of environmental experts and leaders around the world on the best books on nature and the environment, the top ten were

A Sand County Almanac by Aldo Leopold (100)[2]

Silent Spring by Rachel Carson (81)

State of the World by Lester Brown and the Worldwatch Institute (31)

The Population Bomb by Paul Ehrlich (28)

Walden by Henry David Thoreau (28)

Wilderness and the American Mind by Roderick Nash (21)

Small Is Beautiful: Economics as If People Mattered by E. F. Schumacher (21)

Desert Solitaire: A Season in the Wilderness by Edward Abbey (20)

The Closing Circle: Nature, Man, and Technology by Barry Commoner (18)

The Limits to Growth: A Report for the Club of Rome's Project on the Predicament of Mankind by Donella H. Meadows, et al. (17)

[1]Robert Merideth, 1992, G. K. Hall/Macmillan, Inc.

[2]Indicates number of votes for each book. Because the preponderance of respondents were from the United States (82 percent), American books are probably overrepresented.

Figure 15.14 Many interesting, well-paid jobs are opening up in environmental fields. Here, a technician takes a well sample to test for water contamination.

River, cancer rates by geographic area, and substitutes for harmful organic solvents. In each project, students and neighborhood groups team with scientists and university personnel to collect data. Their results have been incorporated into official government policies.

Similar research opportunities exist in the United States and Canada. The Audubon Christmas Bird Count, (see Exploring Science, p. 365), is a good example. Earthwatch offers a much smaller but more intense opportunity to take part in research. Every year hundreds of Earthwatch projects each field a team of a dozen or so volunteers who spend a week or two working on issues ranging from loon nesting behavior to archaeological digs. The American River Watch organizes teams of students to measure water quality. You might be able to get academic credit as well as helpful practical experience in one of these research experiences.

Environmental careers range from engineering to education

The need for both environmental educators and environmental professionals opens up many job opportunities in environmental fields. The World Wildlife Fund estimates, for example, that 750,000 new jobs will be created over the next decade in the renewable energy field alone. Scientists are needed to understand the natural world and the effects of human activity on the environment (fig. 15.14). Lawyers and other specialists are needed to develop government and industry policy, laws, and regulations to protect the environment. Engineers are needed to develop technologies and products to clean up pollution and to prevent its production in the first place. Economists, geographers, and social scientists are

needed to evaluate the costs of pollution and resource depletion and to develop solutions that are socially, culturally, politically, and economically appropriate for different parts of the world. In addition, business will be looking for a new class of environmentally literate and responsible leaders who appreciate how products sold and services rendered affect our environment.

Trained people are essential in these professions at every level, from technical and clerical support staff to top managers. Perhaps the biggest national demand over the next few years will be for environmental educators to help train an environmentally literate populace. We urgently need many more teachers at every level who are trained in environmental education. Outdoor activities and natural sciences are important components of this mission, but environmental topics such as responsible consumerism, waste disposal, and respect for nature can and should be incorporated into reading, writing, arithmetic, and every other part of education.

How much is enough?

Our consumption of resources and disposal of wastes often have destructive impacts on the earth. Technology has made consumer goods and services cheap and readily available in the richer countries of the world. As you already know, we in the industrialized world use resources at a rate out of proportion to our percentage of the population. If everyone in the world were to attempt to live at our level of consumption, given current methods of production, the results would surely be disastrous. In this section we will look at some options for consuming less and reducing our environmental impacts. Perhaps no other issue in this book represents so clear an ethical question as the topic of responsible consumerism.

A century ago, economist and social critic, Thorstein Veblen, in his book, *The Theory of the Leisure Class*, coined the term

Exploring
SCIENCE:

The Christmas Bird Count

Figure 1 Participating in citizen-science projects is a good way to learn about science and your local environment.

Black-capped chickadee
Christmas Bird Count
2002–03

Total counted at a CBC circle
- 0
- <10
- 10-50
- 50-100
- 100-500
- >500

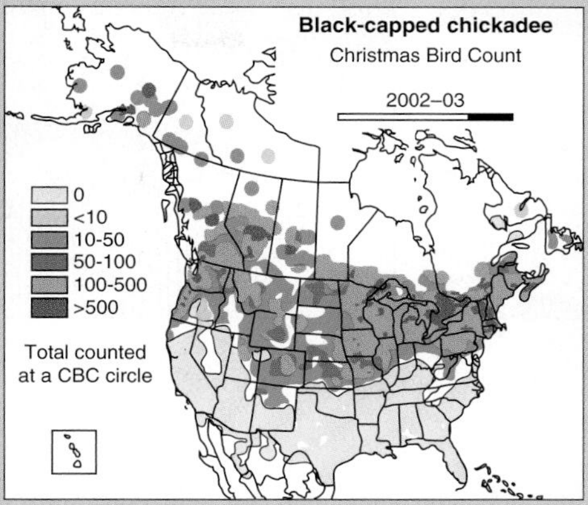

Figure 2 Volunteer data collection can produce a huge, valuable data set. Christmas Bird Count data, such as this map, are available online.
Data from Audubon Society.

Every Christmas since 1900, dedicated volunteers have counted and recorded all the birds they can find within their team's designated study site (fig. 1). This effort has become the largest, longest-running citizen-science project in the world. For the 100th count, nearly 50,000 participants in about 1,800 teams observed 58 million birds belonging to 2,309 species. Although about 70 percent of the counts in 2000 were made in the United States or Canada, 650 teams in the Caribbean, Pacific Islands, and Central and South America also participated. Participants enter their bird counts on standardized data sheets, or submit their observations over the Internet. Compiled data can be viewed and investigated online, almost as soon as they are submitted.

Frank Chapman, the editor of *Bird-Lore* magazine and an officer in the newly formed Audubon Society, started the Christmas Bird Count in 1900. For years, hunters had gathered on Christmas Day for a competitive hunt, often killing hundreds of birds and mammals as teams tried to outshoot each other. Chapman suggested an alternative contest: to see which team could observe and identify the most birds, and the most species, in a day. The competition has grown and spread. In the 100th annual count, the winning team was in Monte Verde, Costa Rica, with an amazing 343 species tallied in a single day.

The tens of thousands of bird-watchers participating in the count gather vastly more information about the abundance and distribution of birds than biologists could gather alone. These data provide important information for scientific research on bird migrations, populations, and habitat change. Now that the entire record for a century of bird data is available on the BirdSource website (www.birdsource.org), both professional ornithologists and amateur bird-watchers can study the geographical distribution of a single species over time, or they can examine how all species vary at a single site through the years. Those concerned about changing climate can look for variation in long-term distribution of species. Climatologists can analyze the effects of weather patterns such as El Niño or La Niña on where birds occur. One of the most intriguing phenomena revealed by this continent-wide data collection is irruptive behavior: that is, appearance of massive numbers of a particular species in a given area in one year, and then their move to other places in subsequent years following weather patterns, food availability, and other factors.

In 2005 the 105th Christmas Bird Count collected data on nearly 70 million birds from 2,019 volunteer groups. This citizen-science effort has produced a rich, geographically broad data set far larger than any that could be produced by professional scientists (fig. 2). Following the success of the Christmas Bird Count, other citizen-science projects have been initiated. Project Feeder Watch, which began in the 1970s, has more than 15,000 participants, from school-children and backyard bird-watchers to dedicated birders. The Great Backyard Bird Count of 2005 collected records on over 600 species and more than 6 million individual birds. In other areas, farmers have been enlisted to monitor pasture and stream health; volunteers monitor water quality in local streams and rivers; and nature reserves solicit volunteers to help gather ecological data. You can learn more about your local environment, and contribute to scientific research, by participating in a citizen-science project. Contact your local Audubon chapter or your state's department of natural resources to find out what you can do.

How does counting birds contribute to sustainability? Citizen-science projects are one way individuals can learn more about the scientific process, become familiar with their local environment, and become more interested in community issues. In this chapter, we'll look at other ways individuals and groups can help protect nature and move toward a sustainable society.

conspicuous consumption to describe buying things we don't want or need just to impress others. How much more shocked he would be to see current trends. The average American now consumes twice as many goods and services as in 1950. The average house is now more than twice as big as it was 50 years ago, even though the typical family has half as many people. We need more space to hold all the stuff we buy. Shopping has become the way many people define themselves. As Marx predicted, everything has become commodified; getting and spending have eclipsed family, ethnicity, even religion as the defining matrix of our lives. But the futility and irrelevance of much American consumerism leaves a psychological void. Once we possess things, we find they don't make us young, beautiful, smart, and interesting as they promised. With so much attention on earning and spending money, we don't have time to have real friends, to cook real food, to have creative hobbies, or to do work that makes us feel we have accomplished

Figure 15.15 Is shopping our highest purpose?

something with our lives. Some social critics call this drive to possess stuff **"affluenza."**

A growing number of people find themselves stuck in a vicious circle: They work frantically at a job they hate, to buy things they don't need, so they can save time to work even longer hours (fig. 15.15). Seeking a measure of balance in their lives, some opt out of the rat race and adopt simpler, less-consumptive lifestyles. As Thoreau wrote in *Walden*, "Our life is frittered away by detail . . . simplify, simplify."

The United Nations Environment Programme (UNEP) has held workshops on sustainable consumption in Paris and Tokyo. Recognizing that making people feel guilty about their lifestyles and purchasing habits isn't working, UNEP is attempting to find ways to make sustainable living something consumers will adopt willingly. The goal is economically, socially, and environmentally viable solutions that allow people to enjoy a good quality of life while consuming fewer natural resources and polluting less. A good example of this approach is a British automaker that provides a mountain bike with every car it sells, urging buyers to use the bike for short journeys. Another example cited by UNEP is European detergent makers who encourage customers to switch to low-temperature washing liquids and powders, not just to save energy but because it's good for their clothes.

Although each of our individual choices may make a small impact, collectively they can be important. The What Can You Do? box on p. 367 offers some suggestions for reducing waste and pollution.

Green consumerism has its limits

To quote Kermit the Frog, "It's not easy being green." Even with the help of endorsement programs, doing the right thing from an environmental perspective may not be obvious. Often we are faced with complicated choices. Do the social benefits of buying rainforest nuts justify the energy expended in transporting them here, or would it be better to eat only locally grown products? In switching from Freon propellants to hydrocarbons, we spare the stratospheric ozone but increase hydrocarbon-caused smog. By choosing reusable diapers over disposable ones, we decrease the amount of material going to the landfill, but we also increase water pollution, energy consumption, and pesticide use (cotton is one of the most pesticide-intensive crops grown in the United States).

China, Ireland, South Africa, Australia, Japan, and Taiwan have either banned ultrathin plastic bags entirely or established taxes to control litter. Obviously, paper from naturally growing trees is a better environmental choice, isn't it? Well, not necessarily. Paper making consumes water and causes much more water pollution than does plastic manufacturing. Paper mills also release air pollutants, including foul-smelling sulfides and captans as well as highly toxic dioxins.

Growing, harvesting, and transporting logs from agroforestry plantations can be as environmentally disruptive as oil production. It takes a great deal of energy to pulp wood and dry newly made paper. Paper is also heavier and bulkier to ship than plastic. Although the resins used to make a plastic bag contain many calories, in the end, paper bags are generally more energy-intensive to produce and market than plastic ones.

If both paper and plastic go to a landfill in your community, the plastic bag takes up less space. It doesn't decompose in the landfill, but neither does the paper in an air-tight, water-tight landfill. If paper is recycled but plastic is not, then the paper bag may be the better choice. If you are lucky enough to have both paper and plastic recycling, the plastic bag is probably a better choice since it recycles more easily and produces less pollution in the process. The best choice of all is to bring your own reusable cloth bag.

Complicated, isn't it? We often must make decisions without complete information, but it's important to make the best choices we can. Don't assume that your neighbors are wrong if they reach conclusions different from yours. They may have valid considerations of which you are unaware. The truth is that simple black and white answers often don't exist.

Taking personal responsibility for your environmental impact can have many benefits. Recycling, buying green products, and other environmental actions not only set good examples for your friends and neighbors, they also strengthen your sense of involvement and commitment in valuable ways. There are limits, however, to how much we can do individually through our buying habits and

Figure 15.16 Student volunteers plant native trees and bushes in a watershed protection project. Working together on a practical problem can be enjoyable and productive.

personal actions to bring about the fundamental changes needed to save the earth. Green consumerism generally can do little about larger issues of global equity, chronic poverty, oppression, and the suffering of millions of people in the developing world. There is a danger that exclusive focus on such problems as whether to choose paper or plastic bags, or to sort recyclables for which there are no markets, will divert our attention from the greater need to change basic institutions.

15.4 How Can We Work Together?

While a few exceptional individuals can be effective working alone to bring about change, most of us find it more productive and more satisfying to work with others.

Collective action multiplies individual power (fig. 15.16). You get encouragement and useful information from meeting regularly with others who share your interests. It's easy to get discouraged by the slow pace of change; having a support group helps maintain your enthusiasm. You should realize, however, that there is a broad spectrum of environmental and social action groups. Some will suit your particular interests, preferences, or beliefs more than others. In this section, we will look at some environmental organizations as well as options for getting involved.

Student environmental groups can have lasting effects

A number of organizations have been established to teach ecology and environmental ethics to elementary and secondary school students, as well as to get them involved in active projects to clean up their local community. Groups such as Kids Saving the Earth or Eco-Kids Corps are an important way to reach this vital audience. Family education results from these efforts as well. In a

What Can You Do?

Reducing Your Impact

Purchase Less

Ask yourself whether you really need more stuff.

Avoid buying things you don't need or won't use.

Use items as long as possible (and don't replace them just because a new product becomes available).

Use the library instead of purchasing books you read.

Make gifts from materials already on hand, or give nonmaterial gifts.

Reduce Excess Packaging

Carry reusable bags when shopping and refuse bags for small purchases.

Buy items in bulk or with minimal packaging; avoid single-serving foods.

Choose packaging that can be recycled or reused.

Avoid Disposable Items

Use cloth napkins, handkerchiefs, and towels.

Bring a washable cup to meetings; use washable plates and utensils rather than single-use items.

Buy pens, razors, flashlights, and cameras with replaceable parts.

Choose items built to last and have them repaired; you will save materials and energy while providing jobs in your community.

Conserve Energy

Walk, bicycle, or use public transportation.

Turn off (or avoid turning on) lights, water, heat, and air conditioning when possible.

Put up clotheslines or racks in the backyard, carport, or basement to avoid using a clothes dryer.

Carpool and combine trips to reduce car mileage.

Save Water

Water lawns and gardens only when necessary.

Use water-saving devices and fewer flushes with toilets.

Don't leave water running when washing hands, food, dishes, and teeth.

Based on material by Karen Oberhauser, Bell Museum Imprint, University of Minnesota, 1992. Used by permission.

World Wildlife Fund survey, 63 percent of young people said they "lobby" their parents about recycling and buying environmentally responsible products.

Organizations for secondary and college students often are among our most active and effective groups for environmental change. The largest student environmental group in North America is the **Student Environmental Action Coalition (SEAC)**. Formed in 1988 by students at the University of North Carolina at Chapel Hill, SEAC has grown rapidly to more than 30,000 members in some 500 campus environmental groups. SEAC is both

Figure 15.17 The Adam Joseph Lewis Center for Environmental Studies at Oberlin College is designed to be self-sustaining even in northern Ohio's cool, cloudy climate. Large, south-facing windows let in sunlight, while 370 m² of solar panels on the roof generate electricity. A constructed wetland outside and a living machine inside purify wastewater.

an umbrella organization and grassroots network that functions as an information clearinghouse and a training center for student leaders. Member groups undertake a diverse spectrum of activities ranging from politically neutral recycling promotion to confrontational protests of government or industrial projects. National conferences bring together thousands of activists who share tactics and inspiration while also having fun. If there isn't a group on your campus, why not look into organizing one?

Another important student organizing group is the network of Public Interest Research Groups active on most campuses in the United States. While not focused exclusively on the environment, the PIRGs usually include environmental issues in their priorities for research. By becoming active, you could probably introduce environmental concerns to your local group if they are not already working on problems of importance to you. Remember that you are not alone. Others share your concerns and want to work with you to bring about change; you just have to find them. There is power in numbers. As Margaret Mead once said, "Never doubt that a small, highly committed group of individuals can change the world; indeed, it is the only thing that ever has."

Schools can be sources of information and experimentation in sustainable living. They have knowledge and expertise to figure out how to do new things, and they have students who have the energy and enthusiasm to do much of the research, and for whom that discovery will be a valuable learning experience. At many colleges and universities, students have undertaken campus audits to examine water and energy use, waste production and disposal, paper consumption, recycling, buying locally produced food, and

many other examples of sustainable resource consumption. At more than 100 universities and colleges across America, graduating students have taken a pledge that reads:

> "I pledge to explore and take into account the social and environmental consequences of any job I consider and will try to improve these aspects of any organization for which I work."

Could you introduce something similar at your school?

Campuses often have building projects that can be models for sustainability research and development. Some recent examples of prize-winning sustainable design can be found at Stanford University, Oberlin College in Ohio, and the University of California at Santa Barbara. Stanford's Jasper Ridge building will provide classroom, laboratory, and office space for its biological research station. Stanford students worked with the administration to develop *Guidelines for Sustainable Buildings*, a booklet that covers everything from energy-efficient lighting to native landscaping. With 275 photovoltaic panels to catch sunlight, there should be no need to buy electricity for the building. In fact, it's expected that surplus energy will be sold back to local utility companies to help pay for building operation.

Oberlin's Environmental Studies Center, designed by architect Bill McDonough, features 370 m² of photovoltaic panels on its roof, a geothermal well to help heat and cool the building, large south-facing windows for passive solar gain, and a "living machine" for water treatment, including plant-filled tanks in an indoor solarium and a constructed wetland outside (fig. 15.17).

The Campus Climate Challenge, recently launched by a coalition of nonprofit groups, seeks to engage students, faculty, and staff at 500 college campuses in the United States and Canada in a long-term campaign to eliminate global warming pollution. Many campuses have invested in clean energy, set strict green building standards for new construction, purchased fuel-efficient vehicles, and adopted other policies to save energy and reduce their greenhouse gas emissions. Some examples include Concordia University in Austin, Texas, the first college or university in the country to purchase all of its energy from renewable sources. The 5.5 million kilowatt-hours of "green power" it uses each year will eliminate about 8 million pounds of CO_2 emissions annually, the equivalent of planting 1,000 acres of trees or taking 700 cars off the roads. Emory University in Atlanta, Georgia, has taken the lead in green building standards, with 11 buildings that are or could become LEED certified. Emory's Whitehead Biomedical Research Building was the first facility in the Southeast to be LEED certified. Like a number of other colleges, Carleton College in Northfield, Minnesota, has built its own windmill, which is expected to provide about 40 percent of the school's electrical needs. The $1.8 million wind turbine is expected to pay for itself in about ten years. The Campus Climate Challenge website at http://www.energyaction.net contains valuable resources, including strategies and case studies, an energy action packet, a campus organizing guide, and more.

At many schools, students have persuaded the administration to buy locally produced food and to provide organic, vegetarian,

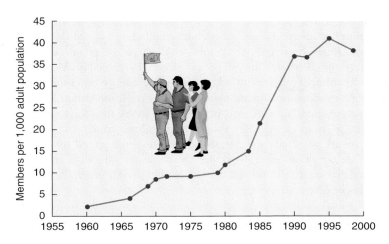

Figure 15.18 Growth of national environmental organizations in the United States.

and fair trade options in campus cafeterias. This not only benefits your health and the environment, but can also serve as a powerful teaching tool and everyday reminder that individuals can make a difference. Could you do something similar at your school? See the Data Analysis box p. 374 for other suggestions.

National organizations are influential but sometimes complacent

Among the oldest, largest, and most influential environmental groups in the United States are the National Wildlife Federation, the World Wildlife Fund, the Audubon Society, the Sierra Club, the Izaak Walton League, Friends of the Earth, Greenpeace, Ducks Unlimited, the Natural Resources Defense Council, and The Wilderness Society. Sometimes known as the "group of 10," these organizations are criticized by other environmentalists for their tendency to compromise and cooperate with the establishment. Although many of these groups were militant—even extremist—in their formative stages, they now tend to be more staid and conservative. Members are mostly passive and know little about the inner workings of the organization, joining as much for publications or social aspects as for their stands on environmental issues. Collectively, these groups grew rapidly during the 1980s (fig. 15.18), but many of their new members had little contact with them beyond making a one-time donation.

Still, these groups are powerful and important forces in environmental protection. Their mass membership, large professional staffs, and long history give them a degree of respectability and influence not found in newer, smaller groups. The Sierra Club, for instance, with about half a million members and chapters in almost every state, has a national staff of about 400, an annual budget over $20 million, and 20 full-time professional lobbyists

in Washington, D.C. These national groups have become a potent force in Congress, especially when they band together to pass specific legislation, such as the Alaska National Interest Lands Act or the Clean Air Act.

Although much of the focus of the big environmental groups is in Washington, Audubon, Sierra Club, and Izaak Walton have local chapters, outings, and conservation projects. This can be a good way to get involved. Go to some meetings, volunteer, offer to help. You may have to start out stuffing envelopes or some other unglamorous job, but if you persevere, you may have a chance to do something important and fun. It's a good way to learn and meet people.

Some environmental groups, such as the Environmental Defense Fund (EDF), The Nature Conservancy (TNC), the National Resources Defense Council (NRDC), and the Wilderness Society (WS), have limited contact with ordinary members except through their publications. They depend on a professional staff to carry out the goals of the organization through litigation (EDF and NRDC), land acquisition (TNC), or lobbying (WS). Although not often in the public eye, these groups can be very effective because of their unique focus. TNC buys land of high ecological value that is threatened by development. With more than 3,200 employees and assets around $3 billion, TNC manages 7 million acres in what it describes as the world's largest private sanctuary system (fig. 15.19). Still, the Conservancy is controversial for some of its management decisions, such as gas and oil drilling in some reserves, and including executives from some questionable companies on its governing board and advisory council. The

Figure 15.19 The Nature Conservancy buys land with high biodiversity or unique natural values to protect it from misuse and development.

Figure 15.20 Protests, marches, and public demonstrations can be an effective way to get your message out and to influence legislators.

Conservancy replies that it is trying to work with these companies to bring about change rather than just criticize them.

New players have brought energy to the environmental movement

There's a new energy today in environmental groups. In 2004, Michael Shellenberger and Ted Nordhaus, consultants for the Environmental Grant-making Foundation, proclaimed the "Death of Environmentalism." The major environmental organizations, they claimed, had become so embedded in Washington politics and concerned about their own jobs, that they had become largely irrelevant. The greatest evidence of this impotence, they charged, was the failure to influence policy on global climate change despite years of work and hundreds of millions of dollars spent on the issue. Even political candidates, such as John Kerry and Al Gore, who had stellar records in Congress as environmental advocates, barely mentioned it when running for office.

But just when the prospects for environmental progress seemed darkest, we seemed to reach a tipping point. The evidence of global climate change became impossible to ignore. But several emerging leaders deserve credit and thanks as well. Al Gore, who lost the U.S. presidency by just one vote in the Supreme Court, reinvented himself as an environmental champion. In a single year, he won a Grammy, an Emmy, and an Oscar, wrote a best-selling book, and shared the Nobel Peace Prize with the IPCC. Once reviled and rejected, he became a global environmental hero.

Now, a broader, diverse, savvy, and passionate movement is taking shape. Author and teacher Bill McKibben, together with a few students from Middlebury College, created the "Step It Up" network that inspired tens of thousands of citizens in 2007 to participate in more than 1,500 climate protests across the United States (fig. 15.20). The Power Shift conference, later that year, brought together more than 5,500 youth activists to work on climate and energy issues. New leaders, including Majora Carter and Van Jones, brought poverty, jobs, and justice groups into the conversation. Business and religious leaders also joined in. More than 450 colleges pledged to adopt renewable energy and other measures to become climate neutral. A new coalition called the Climate Action Network was organized to lobby for climate policy, and a group called 1Sky sought to mobilize ordinary Americans to work for green jobs, transform energy policy, and freeze climate pollution levels. Together these groups hope to bring about a major change in human history. Wouldn't you like to be part of this effort?

International nongovernmental organizations

International **nongovernmental organizations (NGOs)** can be vital in the struggle to protect areas of outstanding biological value. Without this help, local groups could never mobilize the public interest or financial support for projects, such as preventing dams on China's Nu River.

The rise in international NGOs in recent years has been phenomenal. At the Stockholm Conference in 1972, only a handful of environmental groups attended, almost all from fully developed countries. Twenty years later, at the Rio Earth Summit, more than 30,000 individuals representing several thousand environmental groups, many from developing countries, held a global Ecoforum to debate issues and form alliances for a better world.

Some NGOs are located primarily in the more highly developed countries of the north and work mainly on local issues. Others are headquartered in the north but focus their attention on the problems of developing countries in the south. Still others are truly global, with active groups in many different countries. A few are highly professional, combining private individuals with representatives of government agencies on quasi-government boards or standing committees with considerable power. Others are on the fringes of society, sometimes literally voices crying in the wilderness. Many work for political change, more specialize in gathering and disseminating information, and some undertake direct action to protect a specific resource (see related case study "Saving a Gray Whale Nursery" at www.mhhe.com/cunningham5e).

Public education and consciousness-raising using protest marches, demonstrations, civil disobedience, and other participatory public actions and media events are generally important tactics for these groups. Greenpeace, for instance, carries out well-publicized confrontations with whalers, seal hunters, toxic waste dumpers, and others who threaten very specific and visible resources. Greenpeace may well be the largest environmental organization in the world, claiming some 2.5 million contributing members.

In contrast to these highly visible groups, others choose to work behind the scenes, but their impact may be equally important. Conservation International has been a leader in debt-for-nature swaps to protect areas particularly rich in biodiversity. It also has some interesting initiatives in economic development, seeking products made by local people that will provide income along with environmental protection (fig. 15.21).

Figure 15.21 International conservation groups often initiate economic development projects that provide local alternatives to natural resource destruction.

Figure 15.22 A rapidly growing economy has brought increasing affluence to China that has improved standards of living for many Chinese people, but it also brings environmental and social problems associated with Western life styles.

15.5 Sustainability Is a Global Challenge

As developing countries become more affluent, they are adopting many of the wasteful and destructive lifestyle patterns of the West. Take the example of China from the opening case study for this chapter. In the early 1960s, it's estimated that 300 million Chinese suffered from chronic hunger, and at least 30 million starved to death in the worst famine in world history. Now chronic hunger has decreased from about 30 percent of the population 40 years ago to less than 10 percent today. There now are 300 million people in China who consider themselves middle class (a number equal to the entire population of the United States). Many of those Chinese are adopting a fast-food diet, and sedentary lifestyle (fig. 15.22). Obesity—especially among children—has more than doubled in the past decade. Now twice as many Chinese are overweight or obese as are undernourished. As chapter 8 points out, diseases associated with affluent lifestyles, such as diabetes, cardiovascular disease, depression, and traffic accidents, are becoming the leading causes of morbidity and mortality worldwide.

Still, most Chinese live at a very low level of material consumption by European or American standards. Despite rapid eco-

nomic growth, the average Chinese uses one-tenth as much energy as the average American. As we discussed in chapter 4, one way of comparing impacts is by our environmental footprint. It now takes about 9.7 global hectares to support the average American. By contrast, the average Chinese citizen has a footprint of only 1.6 global hectares. If all the 1.3 billion Chinese were to try to match the American level of consumption, it would take about four extra planets using the same technology and mind-set we now have.

On the other hand, the economic growth now taking place in China and other developing countries is alleviating a huge amount of suffering. Many scholars and social activists believe that poverty is at the core of the world's most serious human problems: hunger, child deaths, migrations, insurrections, and environmental degradation. It's in all our best interests to try to reduce these problems. Aside from moral and humanitarian concerns, bringing people out of poverty means more productive workers and customers in the global economy.

Can development be truly sustainable?

Can the lives of the world's poor be improved without destroying our shared environment? A possible solution to this dilemma is **sustainable development**, a term popularized by *Our Common Future*, the 1987 report of the World Commission on Environment and Development, chaired by then Norwegian Prime Minister Gro Harlem Brundtland. In the words of this report, sustainable

Table 15.6   Millennium Development Goals

Goals

1. Eradicate extreme poverty and hunger.
2. Achieve universal primary education.
3. Promote gender equality and empower women.
4. Reduce child mortality.
5. Improve maternal health.
6. Combat HIV/AIDS, malaria, and other diseases.
7. Ensure environmental sustainability.
8. Develop a global partnership for development.

Specific Objectives

1a.	Reduce by half the proportion of people living on less than a dollar a day.
1b.	Reduce by half the proportion of people who suffer from hunger.
2a.	Ensure that all boys and girls complete a full course of primary schooling.
3a.	Eliminate gender disparity in primary and secondary education by 2015.
4a.	Reduce by two-thirds the mortality rate among children under five.
5a.	Reduce by three-quarters the maternal mortality ratio.
6a.	Halt and begin to reverse the spread of HIV/AIDS.
6b.	Halt and begin to reverse the spread of malaria and other major diseases.
7a.	Integrate the principles of sustainable development into policies and programs; reverse the loss of environmental resources.
7b.	Reduce by half the proportion of people without sustainable access to safe drinking water.
7c.	Achieve significant improvement in the lives of 100 million slum dwellers by 2020.
8a.	Develop further an open trading and financial system that is rule-based, predictable, and nondiscriminatory, including a commitment to good governance, development, and poverty reduction.
8b.	Address the least-developed countries' special needs. This includes tariff- and quota-free access for their exports; enhanced debt relief for heavily indebted poor countries.

Figure 15.23 A model for integrating ecosystem health, human needs, and sustainable economic growth.
Modified from Raymond Grizzle and Christopher Barrett, personal communication.

development means "meeting the needs of the present without compromising the ability of future generations to meet their own needs."

Another way of saying this is that we are dependent on nature for food, water, energy, fiber, waste disposal, and other life-support services. We can't deplete resources or create wastes faster than nature can recycle them if we hope to be here for the long term. Development means improving people's lives. Sustainable development, then, means progress in human well-being that can be extended or prolonged over many generations rather than just a few years. To be truly enduring, the benefits of sustainable development must be available to all humans rather than to just the members of a privileged group (fig. 15.23).

But economic growth is not sufficient in itself to meet all essential needs. As the Brundtland Commission pointed out, political stability, democracy, and human rights are needed to ensure that the poor will gain a fair share of benefits. A 2008 study by researchers at Yale and Columbia Universities found a significant correlation between environmental stability, open political systems, and good government. Of the 133 countries in this study, New Zealand, Sweden, Finland, Czech Republic, and the United Kingdom held the top five places (in that order). The United States ranked 28th, behind countries such as Japan, and most of Western Europe.

The millennium assessment sets development goals

In 2000 United Nations Secretary-General Kofi Annan called for a **millennium assessment** of the consequences of ecosystem change on human well-being as well as the scientific basis for actions to enhance the conservation and sustainable use of those systems. More than 1,360 experts from around the world worked on technical reports about the conditions and trends of ecosystems, scenarios for the future, and possible responses. Each part of this study was scrutinized by governments, independent scientists, and other experts to ensure the robustness of its findings. As a result of this assessment, the United Nations has developed a set of goals and objectives for sustainable development (table 15.6). From what you've learned in this book, how do you think we could work—individually and collectively—to accomplish these goals?

Conclusion

All through this book you've seen evidence of environmental degradation and resource depletion, but there are also many cases in which individuals and organizations are finding ways to stop pollution, use renewable rather than irreplaceable resources, and even restore biodiversity and habitat. Sometimes all it takes is the catalyst of a pilot project to show people how things can be done differently to change attitudes and habits. In this chapter, you've learned some practical approaches to living more lightly on the world individually as well as working collectively to create a better world.

The findings from the millennium assessment serve as a good summary of what you've learned. Among the key conclusions are:

- All of us depend on nature and ecosystem services to provide the conditions for a decent, healthy, and secure life.
- We have made unprecedented changes to ecosystems in recent decades to meet growing demands for food, fresh water, fiber, and energy.
- These changes have helped improve the lives of billions, but at the same time they weakened nature's ability to deliver other key services, such as purification of air and water, protection from disasters, and the provision of medicine.

- Among the outstanding problems we face are the dire state of many of the world's fish stocks, the intense vulnerability of the 2 billion people living in dry regions, and the growing threat to ecosystems from climate change and pollution.
- Human actions have taken the planet to the edge of a massive wave of species extinctions, further threatening our own well-being.
- The loss of services derived from ecosystems is a significant barrier to reducing poverty, hunger, and disease.
- The pressures on ecosystems will increase globally unless human attitudes and actions change.
- Measures to conserve natural resources are more likely to succeed if local communities are given ownership of them, share the benefits, and are involved in decisions.
- Even today's technology and knowledge can reduce considerably the human impact on ecosystems. They are unlikely to be deployed fully, however, until ecosystem services cease to be perceived as free and limitless.
- Better protection of natural assets will require coordinated efforts across all sections of governments, businesses, and international institutions.

Practice Quiz

1. What is a *policy*? How are policies formed?
2. Describe three important provisions of NEPA.
3. List four important U.S. environmental laws (besides NEPA), and briefly describe what each does.
4. Why are international environmental conventions and treaties often ineffective? What can make them more successful?
5. Why is the World Trade Organization controversial?
6. List two broad goals of environmental education identified by the National Environmental Education Act.
7. What is *citizen science*, and what are some of its benefits? Describe one such important project.
8. List five things each of us could do to help preserve our common environment.
9. Describe some things schools and students have done to promote sustainable living.
10. Define *sustainability* and describe some of its principal tenets.

Critical Thinking and Discussion Questions

Apply the principles you have learned in this chapter to discuss these questions with other students.

1. In your opinion, how much environmental protection is too much? Think of a practical example in which some stakeholders may feel oppressed by governmental regulations. How would you justify or criticize these regulations?
2. Which is a better choice at the grocery store, paper or plastic? How would you evaluate the trade-offs between packaging choices? What evidence would you look for to make this decision in your life?
3. Do you agree with Margaret Mead that a small group of committed individuals is the only thing that can change the world? What do you think she meant? Think of some examples of groups of individuals who have changed the world. How did they do it?
4. Suppose that you were going to make a presentation on sustainability to your school administrators. What suggestions would you make for changes on your campus? What information would you need to support these proposals?
5. Is sustainability an achievable goal? What do you think are the main barriers to attaining this objective? What have you learned in this book that gives you hope for the future?

Data Analysis: Campus Environmental Audit

How sustainable is your school? What could you, your fellow students, the faculty, staff, and administration do to make your campus more environmentally friendly? Perhaps you and your classmates could carry out an environmental audit of your school. Some of the following items are things you could observe for yourself; other information you'd need to get from the campus administrators.

1. *Energy.* How much total energy does your campus use each year? Is any of it from renewable sources? How does your school energy use compare to that of a city with the same population? Could you switch to renewable sources? How much would that cost? How long would the payback time be for various renewable sources? Is there a campus policy about energy conservation? What would it take to launch a campaign for using resources efficiently?

2. *Buildings.* Are any campus buildings now LEED certified? Do any campus buildings now have compact fluorescent bulbs, high-efficiency fans, or other energy-saving devices? Do you have single-pane or double-pane windows? Are lights turned off when rooms aren't in use? At what temperatures (winter and summer) are classrooms, offices, and dorms maintained? Who makes this decision? Could you open a window in hot weather? Are new buildings being planned? Will they be LEED certified? If not, why?

3. *Transportation.* Does your school own any fuel-efficient vehicles (hybrids or other high-mileage models)? If you were making a presentation to an administrator to encourage him or her to purchase efficient vehicles, what arguments would you use? How many students commute to campus? Are they encouraged to carpool or use public transportation? How might you promote efficient transportation? How much total space on your campus is devoted to parking? What's the cost per vehicle to build and maintain parking? How else might that money be spent to facilitate efficient transportation? Where does runoff from parking lots and streets go? What are the environmental impacts of this storm runoff?

4. *Water use.* What's the source of your drinking water? How much does your campus use? Where does wastewater go? How many toilets are on the campus? How much water does each use for every flush? How much would it cost to change to low-flow appliances? How much would it save in terms of water use and cost?

5. *Food.* What's the source of food served in campus dining rooms? Is any of it locally grown or organic? How much junk food is consumed annually? What are the barriers to buying locally grown, fair-trade, organic, free-range food? Does the campus grow any of its own food? Would that be possible?

6. *Ecosystem restoration.* Are there opportunities for reforestation, stream restoration, wetland improvements, or other ecological repair projects on your campus. What percentage of the vegetation on campus is native? What might be the benefits of replacing non-native species with indigenous varieties? Have gardeners considered planting species that provide food and shelter for wildlife?

What other aspects of your campus life could you study to improve sustainability? How could you organize a group project to promote beneficial changes in your school's environmental impacts?

For Additional Help in Studying This Chapter, please visit our website at www.mhhe.com/cunningham5e. You will find practice quizzes, key terms, answers to end of chapter questions, additional case studies, an extensive reading list, and Google Earth™ mapping quizzes.

Appendixes

Vegetation

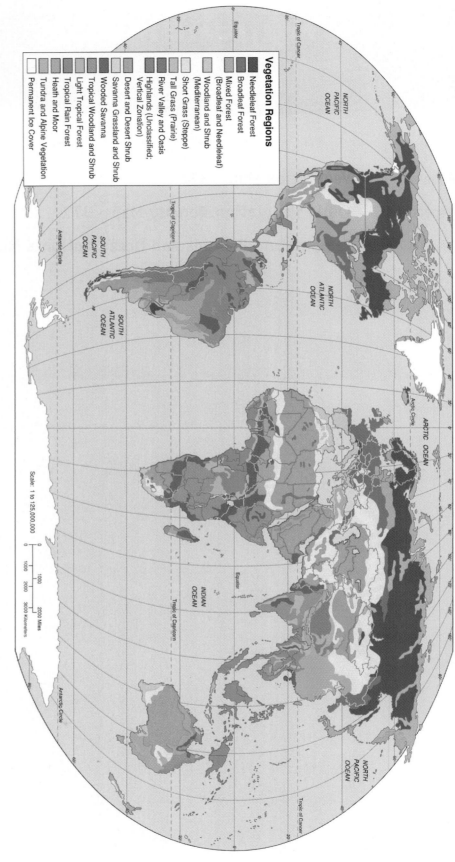

Vegetation Regions

- Needleleaf Forest
- Broadleaf Forest
- Mixed Forest (Broadleaf and Needleleaf)
- Woodland and Shrub (Mediterranean)
- Short Grass (Steppe)
- Tall Grass (Prairie)
- River Valley and Oasis
- Highlands (Unclassified; Vertical Zonation)
- Desert and Desert Shrub
- Savanna Grassland and Shrub
- Wooded Savanna
- Tropical Woodland and Shrub
- Light Tropical Forest
- Tropical Rain Forest
- Heath and Moor
- Tundra and Alpine Vegetation
- Permanent Ice Cover

Scale: 1 to 125,000,000

Vegetation is the most visible consequence of the distribution of temperature and precipitation. The global distribution of vegetation types and the global distribution of climate are closely related, but not all vegetation types are the consequence of temperature and precipitation or other climatic variables. Many types of vegetation, in many areas of the world, are the consequence of human activities, particularly the grazing of domesticated livestock, burning, and forest clearance.

World Population Density

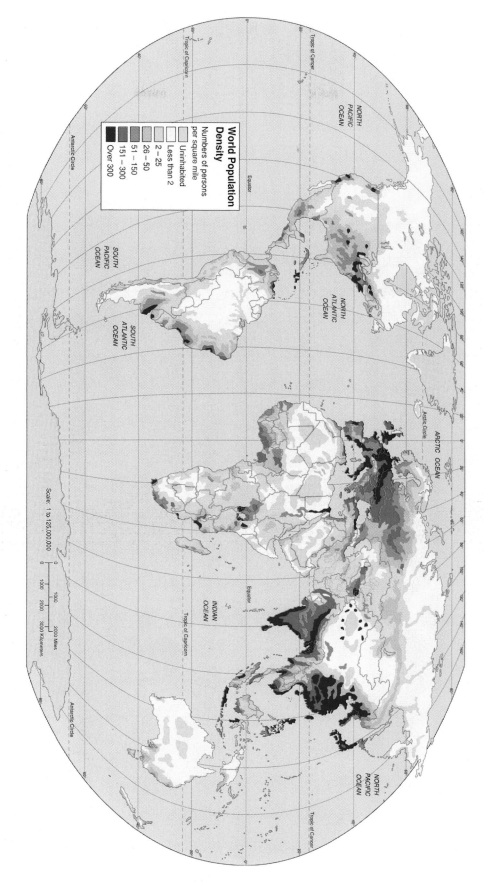

World Population Density

Numbers of persons
per square mile

- Uninhabited
- Less than 2
- 2 – 25
- 26 – 50
- 51 – 150
- 151 – 300
- Over 300

Scale: 1 to 125,000,000

0   1000   2000   2000 Miles

0   1000   2000   3000 Kilometers

No feature of human activity is more reflective of environmental conditions than where people live. In the areas of densest population, a mixture of natural and human factors have combined to allow maximum food production, maximum urbanization, and especially concentrated economic activity. Three such great concentrations appear on the map—East Asia, South Asia, and Europe—with a fourth lesser concentration in eastern North America (the "Megalopolis" region of the United States and Canada). The areas of future high density (in addition to those already existing) are likely to be in Middle and South America and Africa, where population growth rates are well above the world average. Population that is extremely dense or growing at an excessive rate when measured against a region's habitability is one of the greatest indicators of environmental deterioration.

Temperature Regions and Ocean Currents

Along with precipitation, temperature is one of the two most important environmental variables, defining the climatic conditions so essential for the distribution of human activities and the human population. Ocean currents exert a significant influence over the climate of adjacent continents and are the most important mechanism for redistributing surplus heat from the equatorial region into middle and high latitudes.

Surface Temperature Regions

- Always cold: polar regions and high altitudes
- Cold winter and cool summer; always cool in tropical higher altitudes
- Cold winter and mild summer
- Cool winter and mild summer
- Hot summer and cold winter
- Hot summer and cool winter
- Hot summer and mild winter
- Always hot
- Always mild

Hot = above 68°F (20°C)
Mild = 50° – 68°F (10° – 20°C)
Cool = 32° – 50°F (0° – 10°C)
Cold = below 32°F (0°C)

→ Cool/cold current
→ Warm current

Scale: 1 to 125,000,000

0   1000   2000   3000 Miles
0   1000   2000   3000 Kilometers

NORTH PACIFIC OCEAN

SOUTH PACIFIC OCEAN

NORTH ATLANTIC OCEAN

SOUTH ATLANTIC OCEAN

INDIAN OCEAN

ARCTIC OCEAN

Tropic of Cancer

Equator

Tropic of Capricorn

Arctic Circle

Antarctic Circle

Glossary

A

abundance The number of individuals of a species in an area.

acid precipitation Acidic rain, snow, or dry particles deposited from the air due to increased acids released by anthropogenic or natural resources.

acids Substances that release hydrogen atoms in water.

active solar systems Mechanical systems that use moving substances to collect and transfer solar energy.

acute effects A sudden onset of symptoms or effects of exposure to some factor.

acute poverty Insufficient income or access to resources needed to provide the basic necessities for life, such as food, shelter, sanitation, clean water, medical care, and education.

adaptation Physical changes that allow organisms to survive in a given environment.

adaptive management A management plan designed from the outset to "learn by doing" and to actively test hypotheses and adjust treatments as new information becomes available.

administrative law Executive orders, administrative rules and regulations, and enforcement decisions by administrative agencies and special administrative courts.

aerosols Minute particles or liquid droplets suspended in the air.

affluenza An addiction to spending and consuming beyond one's needs.

agency rule-making The formal process of establishing rules and standards by administrative agencies.

albedo A description of a surface's reflective properties.

allergens Substances that activate the immune system and cause an allergic response; may not be directly antigenic themselves but may make other materials antigenic.

allopatric speciation Species that arise from a common ancestor through geographic isolation or some other barrier to reproduction.

ambient air The air immediately around us.

amorphous silicon collectors Photovoltaic cells made from randomly assembled silicon molecules rather than silicon crystals. Amorphous collectors are less efficient but far cheaper than crystalline collectors.

analytical thinking A way of systematic analysis that asks, "How can I break this problem down into its constituent parts?"

anemia Low levels of hemoglobin due to iron deficiency or lack of red blood cells.

anthropocentric Believing that humans hold a special place in nature; being centered primarily on humans and human affairs.

B

antigens Substances that stimulate the production of, and react with, specific antibodies.

aquifers Porous, water-bearing layers of sand, gravel, and rock below the earth's surface; reservoirs for groundwater.

arbitration A formal process of dispute resolution resulting in a legally binding decision that all parties must obey.

arithmetic growth A pattern of growth that increases at a constant amount per unit time, such as 1, 2, 3, 4 or 1, 3, 5, 7.

atmospheric deposition Sedimentation of solids, liquids, or gaseous materials from the air.

atom The smallest particle that exhibits the characteristics of an element.

atomic number The characteristic number of protons per atom of an element.

autotroph An organism that synthesizes food molecules from inorganic molecules by using an external energy source, such as light energy.

barrier islands Low, narrow, sandy islands that form offshore from a coastline.

bases Substances that readily bond with hydrogen ions in an aqueous solution.

Batesian mimicry Evolution by one species to resemble another species that is protected from predators by a venomous stinger, bad taste, or some other defensive adaptation.

benthic The bottom of a sea or lake.

binomials Scientific or Latin names that combine the genus and species, e.g., *Zea mays.*

bioaccumulation The selective absorption and concentration of molecules by cells.

biocentrism The belief that all creatures have rights and values; being centered on nature rather than humans.

biochemical oxygen demand (BOD) A standard test for measuring the amount of dissolved oxygen utilized by aquatic microorganisms.

biodegradable plastics Plastics that can be decomposed by microorganisms.

biodiversity The genetic, species, and ecological diversity of the organisms in a given area.

biofuel Fuel made from biomass.

biogeochemical cycles Movement of matter within or between ecosystems; caused by living organisms, geologic forces, or chemical reactions. The cycling of nitrogen, carbon, sulfur, oxygen, phosphorus, and water are examples.

biological community The populations of plants, animals, and microorganisms living and interacting in a certain area at a given time.

biological controls Use of natural predators, pathogens, or competitors to regulate pest populations.

biomagnification Increase in concentration of certain stable chemicals (for example, heavy metals or fat-soluble pesticides) in successively higher trophic levels of a food chain or web.

biomass The accumulated biological material produced by living organisms.

biomass fuel Organic material produced by plants, animals, or microorganisms that can be burned directly as a heat source or converted into a gaseous or liquid fuel.

biomass pyramid A metaphor or diagram that explains the relationship between the amounts of biomass at different trophic levels.

biomes Broad, regional types of ecosystems characterized by distinctive climate and soil conditions and distinctive kinds of biological community adapted to those conditions.

bioremediation Use of biological organisms to remove pollution or restore environmental quality.

biosphere The zone of air, land, and water at the surface of the earth that is occupied by organisms.

biosphere reserves World heritage sites identified by the IUCN as worthy for national park or wildlife refuge status because of high biological diversity or unique ecological features.

biotic potential The maximum reproductive rate of an organism, given unlimited resources and ideal environmental conditions.

birth control Any method used to reduce births, including celibacy, delayed marriage, contraception; devices or medications that prevent implantation of fertilized zygotes and induced abortions.

blind experiments A design in which researchers don't know which subjects were given experimental treatment until after data have been gathered and analyzed.

bogs Areas of waterlogged soil that tend to be peaty; fed mainly by precipitation; low productivity; some bogs are acidic.

boreal forest A broad band of mixed coniferous and deciduous trees that stretches across northern North America (and Europe and Asia); its northernmost edge, the taiga, intergrades with the arctic tundra.

brownfields Abandoned or underused urban areas in which redevelopment is blocked by liability or financing issues related to toxic contamination.

C

cancer Invasive, out-of-control cell growth that results in malignant tumors.

capital Any form of wealth, resources, or knowledge available for use in the production of more wealth.

carbohydrate An organic compound consisting of a ring or chain of carbon atoms with hydrogen and oxygen attached; examples are sugars, starches, cellulose, and glycogen.

carbon cycle The circulation and reutilization of carbon atoms, especially via the processes of photosynthesis and respiration.

carbon management Projects to reduce carbon dioxide emissions from fossil fuel or to ameliorate their effects.

carbon monoxide Colorless, odorless, nonirritating but highly toxic gas produced by incomplete combustion of fuel, incineration of biomass or solid waste, or partially anaerobic decomposition of organic material.

carbon neutral Producing no net carbon dioxide emissions.

carbon sink Places of carbon accumulation, such as in large forests (organic compounds) or ocean sediments (calcium carbonate).

carcinogens Substances that cause cancer.

carnivores Organisms that mainly prey upon animals.

carrying capacity The maximum number of individuals of any species that can be supported by a particular ecosystem on a long-term basis.

case law Precedents from both civil and criminal court cases.

cell Minute compartments surrounded by semipermeable membranes within which the processes of life are carried out by all living organisms.

cellular respiration The process in which a cell breaks down sugar or other organic compounds to release energy used for cellular work; may be anaerobic or aerobic, depending on the availability of oxygen.

chain reaction A self-sustaining reaction in which the fission of nuclei produces subatomic particles that cause the fission of other nuclei.

chaparral A biological community characterized by thick growth of thorny, evergreen shrubs typical of a Mediterranean climate.

chemical bond The force that holds molecules together.

chemical compounds Molecules made up of two or more kinds of atoms held together by chemical bonds.

chemical energy Potential energy stored in chemical bonds of molecules.

chemosynthesis Extracting energy for life from inorganic chemicals, such as hydrogen sulfide, rather than from sunlight.

chlorofluorocarbons Chemical compounds with a carbon skeleton and one or more attached chlorine and fluorine atoms. Commonly used as refrigerants, solvents, fire retardants, and blowing agents.

chloroplasts Chlorophyll-containing organelles in eukaryotic organisms; sites of photosynthesis.

chronic effects Long-lasting results of exposure to a toxin; can be a permanent change caused by a single, acute exposure or a continuous, low-level exposure.

citizen science Projects in which trained volunteers work with scientific researchers to answer real-world questions.

civil law A body of laws regulating relations between individuals or between individuals and corporations concerning property rights, personal dignity and freedom, and personal injury.

classical economics Modern, Western economic theories of the effects of resource scarcity, monetary policy, and competition on supply and demand of goods and services in the market-place. This is the basis for the capitalist market system.

clear-cutting Cutting every tree in a given area, regardless of species or size; an appropriate harvest method for some species; can be destructive if not carefully controlled.

climate A description of the long-term pattern of weather in a particular area.

climax community A long-lasting, self-sustaining community resulting from ecological succession that is resistant to disturbance.

closed-canopy A forest where tree crowns spread over 20 percent of the ground; has the potential for commercial timber harvests.

closed system A system in which there is no exchange of energy or matter with its surroundings.

cloud forests High mountain forests where temperatures are uniformly cool and fog or mist keeps vegetation wet all the time.

coevolution The process in which species exert selective pressure on each other and gradually evolve new features or behaviors as a result of those pressures.

cogeneration The simultaneous production of electricity and steam or hot water in the same plant.

coliform bacteria Bacteria that live in the intestines (including the colon) of humans and other animals; used as a measure of the presence of feces in water or soil.

commensalism A symbiotic relationship in which one member is benefited and the second is neither harmed nor benefited.

common law The body of court decisions that constitutes a working definition of individual rights and responsibilities where no formal statutes define these issues.

communal resource management systems Resources managed by a community for long-term sustainability.

community-based planning Involving community stakeholders in pluralistic, adaptive, inclusive, proactive planning.

community ecology The study of interactions of all populations living in the ecosystem of a given area.

community (ecological) structure The patterns of spatial distribution of individuals, species, and communities.

competitive exclusion A theory that no two populations of different species will occupy the same niche and compete for exactly the same resources in the same habitat for very long.

complexity The number of species at each trophic level and the number of trophic levels in a community.

composting The biological degradation of organic material under aerobic (oxygen-rich) conditions to produce compost, a nutrient-rich soil amendment and conditioner.

compound Substances composed of different kinds of atoms.

confined animal-feeding operation Feeding large numbers of livestock at a high density in pens or barns.

conifer A needle-bearing tree that produces seeds in cones.

conservation medicine Attempts to understand how changes we make in our environment threaten our health as well as that of natural communities on which we depend.

conservation of matter In any chemical reaction, matter changes form; it is neither created nor destroyed.

conspicuous consumption A term coined by economist and social critic Thorstein Veblen to describe buying things we don't want or need to impress others.

constructed wetlands Artificially constructed wetlands.

consumers Organisms that obtain energy and nutrients by feeding on other organisms or their remains. See also *heterotroph*.

consumption The fraction of withdrawn water that is lost in transmission or that is evaporated, absorbed, chemically transformed, or otherwise made unavailable for other purposes as a result of human use.

contour plowing Plowing along hill contours; reduces erosion.

controlled studies Comparisons made between two populations that are identical (as far as possible) in every factor except the one being studied.

control rods Neutron-absorbing material inserted into spaces between fuel assemblies in nuclear reactors to regulate fission reaction.

convection currents Rising or sinking air currents that stir the atmosphere and transport heat from one area to another. Convection currents also occur in water.

conventional (criteria) pollutants The seven substances (sulfur dioxide, carbon monoxide, particulates, hydrocarbons, nitrogen oxides, photochemical oxidants, and lead) identified by the Clean Air Act as the most serious threat of all pollutants to human health and welfare.

convergent evolution Species evolve from different origins but under similar environmental conditions to have similar traits.

coral reefs Prominent oceanic features composed of hard, limy skeletons produced by coral animals; usually formed along edges of shallow, submerged ocean banks or along shelves in warm, shallow, tropical seas.

core The dense, intensely hot mass of molten metal, mostly iron and nickel, thousands of kilometers in diameter at the earth's center.

core habitat A habitat patch large enough and with ecological characteristics suitable to support a critical mass of the species that make up a particular community.

Coriolis effect The tendency for air above the earth to appear to be deflected to the right (in the Northern Hemisphere) or the left (in the South) because of the earth's rotation.

corridors Strips of natural habitat that connect two adjacent nature preserves to allow migration of organisms from one place to another.

cost-benefit analysis (CBA) An evaluation of large-scale public projects by comparing the costs and benefits that accrue from them.

cover crops Plants, such as rye, alfalfa, or clover, that can be planted immediately after harvest to hold and protect the soil.

creative thinking Original, independent thinking that asks, "How might I approach this problem in new and inventive ways?"

criminal law A body of court decisions based on federal and state statutes concerning wrongs against persons or society.

criteria pollutants See *conventional pollutants*.

critical factor The single environmental factor closest to a tolerance limit for a given species at a given time.

critical thinking An ability to evaluate information and opinions in a systematic, purposeful, efficient manner.

crude birth rate The number of births in a year per 1000 (using the midyear population).

crude death rate The number of deaths per thousand persons in a given year; also called crude mortality rate.

crust The cool, lightweight, outermost layer of the earth's surface that floats on the soft, pliable underlying layers; similar to the "skin" on a bowl of warm pudding.

cultural eutrophication An increase in biological productivity and ecosystem succession caused by human activities.

D

debt-for-nature swaps Forgiveness of international debt in exchange for nature protection in developing countries.

deciduous Trees and shrubs that shed their leaves at the end of the growing season.

decomposer Fungus or bacterium that breaks complex organic material into smaller molecules.

deductive reasoning "Top down" reasoning in which we start with a general principle and derive a testable prediction about a specific case.

deforestation Removing trees from a forest.

delta Fan-shaped sediment deposit found at the mouth of a river.

demanufacturing Disassembly of products so components can be reused or recycled.

demographic transition A pattern of falling death rates and birth rates in response to improved living conditions; typically leads to rapid then stabilizing population growth.

demography The statistical study of human populations relating to growth rate, age structure, geographic distribution, etc., and their effects on social, economic, and environmental conditions.

density-dependent factors Either internal or external factors that affect growth rates of a population depending on the density of the organisms in the population.

dependency ratio The number of nonworking members compared with working members for a given population.

dependent variable Also known as the response variable; is one affected by other variables.

desalinization (or desalination) Removal of salt from water by distillation, freezing, or ultrafiltration.

desertification Denuding and degrading a once fertile land, initiating a desert-producing cycle that feeds on itself and causes long-term changes in soil, climate, and biota of an area.

deserts Biomes characterized by low moisture levels and infrequent and unpredictable precipitation. Daily and seasonal temperatures fluctuate widely.

detritivore Organisms that consume organic litter, debris, and dung.

dieback A sudden population decline; also called a population crash.

disability-adjusted life years (DALYs) A health measure that assesses the total burden of disease by combining premature deaths and loss of a healthy life that result from illness or disability.

discharge The amount of water that passes a fixed point in a given amount of time; usually expressed as liters or cubic feet of water per second.

discount rate The amount we discount or reduce the value of a future payment. When you bor-

row money from the bank at 10 percent annual interest, you are in effect saying that having the money now is worth 10 percent more to you than having the same amount one year from now.

disease A deleterious change in the body's condition in response to destabilizing factors, such as nutrition, chemicals, or biological agents.

dissolved oxygen (DO) content Amount of oxygen dissolved in a given volume of water at a given temperature and atmospheric pressure; usually expressed in parts per million (ppm).

disturbance Any force that disrupts the established patterns and processes, such as species diversity and abundance, community structure, community properties, or species relationships.

disturbance-adapted species Species that depend on repeated disturbance for their survival and propagation.

divergent evolution Separation of a species into new types.

diversity The number of species present in a community (species richness), as well as the relative abundance of each species.

DNA Deoxyribonucleic acid; the long, double-helix molecule in the nucleus of cells that contains the genetic code and directs the development and functioning of all cells.

double-blind design Neither the subject (participant) nor the experimenter knows which participants are receiving the experimental or the control treatments until after data have been gathered and analyzed.

drip irrigation Uses pipe or tubing perforated with very small holes to deliver water one drop at a time directly to the soil around each plant.

dust domes High concentrations of dust and aerosols in the air over cities.

E

earthquakes Sudden, violent movement of the earth's crust.

ecological diseases Sudden, wide-spread epidemics among livestock and wild species.

ecological economics Application of ecological insights to economic analysis; incorporating ecological principles and priorities into economic accounting systems.

ecological footprint An estimate of our individual and collective environmental impacts. It is usually calculated and expressed as the area of bioproductive land required to support a particular lifestyle.

ecological niche The functional role and position of a species in its ecosystem, including what resources it uses, how and when it uses the resources, and how it interacts with other species.

ecological services Processes or materials, such as clean water, energy, climate regulation, and nutrient cycling, provided by ecosystems.

ecological succession The process by which organisms gradually occupy a site, alter its ecological conditions, and are eventually replaced by other organisms.

ecology The scientific study of relationships between organisms and their environment. It is concerned with the life histories, distribution, and behavior of individual species as well as the structure and function of natural systems at the level of populations, communities, and ecosystems.

economic development A rise in real income per person; usually associated with new technology that increases productivity or resources.

ecosystem A specific biological community and its physical environment interacting in an exchange of matter and energy.

ecosystem management An integration of ecological, economic, and social goals in a unified systems approach to resource management.

ecosystem restoration To reinstate an entire community of organisms to as near its natural condition as possible.

ecotones Boundaries between two types of ecological communities.

ecotourism A combination of adventure travel, cultural exploration, and nature appreciation in wild settings.

edge effects A change in species composition, physical conditions, or other ecological factors at the boundary between two ecosystems.

electron A negatively charged subatomic particle that orbits around the nucleus of an atom.

electronic waste (e-waste) Discarded electronic equipment, including TVs, cell phones, computers, etc.

element A substance that cannot be broken into simpler units by chemical means.

El Niño A climatic change marked by shifting of a large warm water pool from the western Pacific Ocean toward the east. Wind direction and precipitation patterns are changed over much of the Pacific and perhaps around the world.

emergent disease A new disease or one that has been absent for at least 20 years.

emergent properties Properties that make a system more than the sum of its parts.

emigration The movement of members from a population.

emission standards Regulations for restricting the amounts of air pollutants that can be released from specific point sources.

endangered species A species considered to be in imminent danger of extinction.

endemism A species that is restricted to a single region, country, or other area.

endocrine hormone disrupters Chemicals that interfere with the function of endocrine hormones such as estrogen, testosterone, thyroxine, adrenaline, or cortisone.

energy The capacity to do work, such as moving matter over a distance.

energy intensity The amount of energy needed to provide the goods and services consumed in an economy.

energy recovery The incineration of solid waste to produce useful energy.

entropy A measure of disorder and usefulness of energy in a system.

environment The circumstances or conditions that surround an organism or a group of organisms as well as the complex of social or cultural conditions that affect an individual or a community.

environmental ethics A search for moral values and ethical principles in human relations with the natural world.

environmental health The science of external factors that cause disease, including elements of the natural, social, cultural, and technological worlds in which we live.

environmental impact statement (EIS) An analysis of the effects of any major program or project planned by a federal agency; required by provisions in the National Environmental Policy Act of 1970.

environmental justice Fair access to a clean, healthy environment, regardless of class, race, income level, or other status.

environmental law Legal rules, decisions, and actions concerning environmental quality, natural resources, and ecological sustainability.

environmental literacy A basic understanding of ecological principles and the ways society affects, or responds to, environmental conditions.

environmental policy The official rules or regulations concerning the environment adopted, implemented, and enforced by some government agency.

environmental science The systematic, scientific study of our environment as well as our role in it.

epidemiology The study of the distribution and causes of disease and injuries in human populations.

epigenetics Effects (both positive and negative) expressed in future generations that are not caused by nuclear mutations and are not inherited by normal Mendelian genetics.

epiphyte A plant that grows on a substrate other than the soil, such as the surface of another organism.

estuaries Bays or drowned valleys where a river empties into the sea.

eutrophic Rivers and lakes rich in organic material (*eu* = well; *trophic* = nourished).

evolution A theory that explains how random changes in genetic material and competition for scarce resources cause species to change gradually.

evolutionary species concept A definition of species that depends on evolutionary relationships.

exotic organisms Alien species introduced by human agency into biological communities where they would not naturally occur.

explanatory variable An independent variable that helps explain the relationship to a dependent variable.

exponential growth Growth at a constant rate of increase per unit of time; can be expressed as a constant fraction or exponent. See also *geometric growth.*

externalizing costs Shifting expenses, monetary or otherwise, to someone other than the individuals or groups who use a resource.

extinction The irrevocable elimination of species; can be a normal process of the natural world as species outcompete or kill off others or as environmental conditions change.

F

family planning Controlling reproduction; planning the timing of birth and having only as many babies as are wanted and can be supported.

famines Acute food shortages characterized by large-scale loss of life, social disruption, and economic chaos.

fauna All of the animals present in a given region.

fecundity The physical ability to reproduce.

federal laws (statutes) Laws passed by the federal legislature and signed by the chief executive.

fens Wetlands fed mainly by groundwater.

feral A domestic animal that has taken up a wild existence.

fetal alcohol syndrome A tragic set of permanent physical, mental, and behavioral birth defects that result when mothers drink alcohol during pregnancy.

first law of thermodynamics States that energy is conserved; that is, it is neither created nor destroyed under normal conditions.

flood An overflow of water onto land that normally is dry.

floodplains Low lands along riverbanks, lakes, and coastlines subjected to periodic inundation.

flora All of the plants present in a given region.

food security The ability of individuals to obtain sufficient food on a day-to-day basis.

food web A complex, interlocking series of individual food chains in an ecosystem.

fossil fuels Petroleum, natural gas, and coal created by geologic forces from organic wastes and dead bodies of formerly living biological organisms.

fragmentation Disruption of habitat into small, isolated fragments.

freshwater ecosystems Ecosystems in which the fresh (nonsalty) water of streams, rivers, ponds, or lakes plays a defining role.

fuel assembly A bundle of hollow metal rods containing uranium oxide pellets; used to fuel a nuclear reactor.

fuel cells Mechanical devices that use hydrogen or hydrogen-containing fuel, such as methane, to produce an electric current. Fuel cells are clean, quiet, and highly efficient sources of electricity.

fugitive emissions Substances that enter the air without going through a smokestack, such as dust from soil erosion, strip mining, rock crushing, construction, and building demolition.

fungi Nonphotosynthetic, eukaryotic organisms with cell walls, filamentous bodies, and absorptive nutrition.

fungicide A chemical that kills fungi.

G

gamma rays Very short wavelength forms of the electromagnetic spectrum.

gap analysis A biogeographical technique of mapping biological diversity and endemic species to find gaps between protected areas that leave endangered habitats vulnerable to disruption.

gasohol A mixture of gasoline and ethanol.

gene A unit of heredity; a segment of DNA nucleus of the cell that contains information for the synthesis of a specific protein, such as an enzyme.

genetically modified organisms (GMOs) Organisms created by combining natural or synthetic genes using the techniques of molecular biology.

genetic engineering Laboratory manipulation of genetic material using molecular biology.

genuine progress index (GPI) An alternative to GNP or GDP for economic accounting that measures real progress in quality of life and sustainability.

geoengineering Attempts to deliberately alter global-scale ecological or geochemical processes to create more desirable environmental conditions.

geographic isolation Geographical changes that isolate populations of a species and prevent reproduction or gene exchange for a long enough time so that genetic drift changes the populations into distinct species.

geometric growth Growth that follows a geometric pattern of increase, such as 2, 4, 8, 16, etc. See also *exponential growth.*

geothermal energy Energy drawn from the internal heat of the earth, either through geysers, fumaroles, hot springs, or other natural geothermal features or through deep wells that pump heated groundwater.

GIS Geographical information systems that use computers to combine and analyze geographical data.

global environmentalism The extension of modern environmental concerns to global issues.

grasslands Biomes dominated by grasses and associated herbaceous plants.

greenhouse effect Trapping of heat by the earth's atmosphere, which is transparent to incoming visible light waves but absorbs outgoing long-wave infrared radiation.

greenhouse gas A gas that traps heat in the atmosphere.

green plans Integrated national environmental plans for reducing pollution and resource consumption while achieving sustainable development and environmental restoration.

green political parties Political organizations based on environmental protection, participatory democracy, grassroots organization, and sustainable development.

green pricing Plans in which consumers can voluntarily pay premium prices for renewable energy.

green revolution Dramatically increased agricultural production brought about by "miracle" strains of grain; usually requires high inputs of water, plant nutrients, and pesticides.

gross domestic product (GDP) The total economic activity within national boundaries.

gross national product (GNP) The sum total of all goods and services produced in a national economy. Gross domestic product (GDP) is used to distinguish economic activity within a country from that of offshore corporations.

groundwater Water held in gravel deposits or porous rock below the earth's surface; does not include water or crystallization held by chemical bonds in rocks or moisture in upper soil layers.

gully erosion Removal of layers of soil, creating channels or ravines too large to be removed by normal tillage operations.

H

habitat The place or set of environmental conditions in which a particular organism lives.

half-life The time required for one-half of a sample to decay or change into some other form.

hazardous waste Any discarded material containing substances known to be toxic, mutagenic, carcinogenic, or teratogenic to humans or other life-forms; ignitable, corrosive, explosive, or highly reactive alone or with other materials.

health A state of physical and emotional well-being; the absence of disease or ailment.

heap-leach extraction A technique for separating gold from extremely low-grade ores. Crushed ore is piled in huge heaps and sprayed with a dilute alkaline-cyanide solution, which percolates through the pile to extract the gold.

heat Total kinetic energy of atoms or molecules in a substance not associated with the bulk motion of the substance.

heat islands Areas of higher temperatures around cities.

herbicide A chemical that kills plants.

herbivores Organisms that eat only plants.

heterotroph An organism that is incapable of synthesizing its own food and, therefore, must feed upon organic compounds produced by other organisms.

high-level waste repository A place where intensely radioactive wastes can be buried and remain unexposed to groundwater and earthquakes for tens of thousands of years.

high-quality energy Intense, concentrated, and high-temperature energy that is considered high-quality because of its usefulness in carrying out work.

HIPPO Habitat destruction, Invasive species, Pollution, Population (human), and Overharvesting, the leading causes of extinction.

histogram A graph of frequency distributions.

holistic science The study of entire, integrated systems rather than isolated parts. Often takes a descriptive or an interpretive approach.

homeostasis A dynamic, steady state in a living system maintained through opposing, compensating adjustments.

hormesis Nonlinear effects of toxic materials.

human development index (HDI) A measure of quality of life using data for life expectancy, child survival, adult literacy, education, gender equity, access to clean water and sanitation, and income.

hydrologic cycle The natural process by which water is purified and made fresh through evaporation and precipitation. This cycle provides all the freshwater available for biological life.

hypothesis A conditional explanation that can be verified or falsified by observation or experimentation.

I

igneous rocks Crystalline minerals solidified from molten magma from deep in the earth's interior; basalt, rhyolite, andesite, lava, and granite are examples.

independent variable One that does not respond to other variables in a particular test.

indicators Species that have very specific environmental requirements and tolerance levels that make them good indicators of pollution or other environmental conditions.

indigenous people Natives or original inhabitants of an area, those who have lived in a particular place for a very long time.

inductive reasoning "Bottom-up" reasoning in which we study specific examples and try to discover patterns and derive general explanations from collected observations.

infiltration The process of water percolation into the soil and pores and hollows of permeable rocks.

inholdings Private lands within public parks, forests, or wildlife refuges.

insecticide A chemical that kills insects.

insolation Incoming solar radiation.

integrated gasification combined cycle (IGCC) A process in which a fuel (coal or biomass) is heated in the presence of high oxygen levels to produce a variety of gases, mostly hydrogen and carbon dioxide. Impurities, including CO_2, can easily be removed and the synthetic hydrogen gas, or syngas, is burned in a turbine to produce electricity. Superheated gas from the turbine is used to generate steam that produces more electricity, raising the efficiency of the system.

integrated pest management (IPM) An ecologically based pest-control strategy that relies on natural mortality factors, such as natural enemies, weather, cultural control methods, and carefully applied doses of pesticides.

Intergovernmental Panel on Climate Change (IPCC) A large group of scientists from many nations and a wide variety of fields assembled by the United Nations Environment Program and World Meteorological Organization to assess the current state of knowledge about climate change.

internalizing costs Planning so that those who reap the benefits of resource use also bear all the external costs.

international treaties and conventions Agreements between nations on important issues.

interspecific competition In a community, competition for resources between members of different species.

intraspecific competition In a community, competition for resources among members of the same species.

invasive species Organisms that thrive in new territory where they are free of predators, diseases, or resource limitations that may have controlled their population in their native habitat.

ionizing radiation High-energy electromagnetic radiation or energetic subatomic particles released by nuclear decay.

ionosphere The lower part of the thermosphere.

ions Electrically charged atoms that have gained or lost electrons.

irruptive growth See *Malthusian growth.*

island biogeography The study of rates of colonization and extinction of species on islands or other isolated areas based on size, shape, and distance from other inhabited regions.

isotopes Forms of a single element that differ in atomic mass due to a different number of neutrons in the nucleus.

J

J curve A growth curve that depicts exponential growth; called a J curve because of its shape.

joule A unit of energy. One joule is the energy expended in 1 second by a current of 1 amp flowing through a resistance of 1 ohm.

K

K-selected species Organisms whose population growth is regulated by internal (or intrinsic) as well as external factors. Large animals, such as whales and elephants, as well as top predators, generally fall in this category. They have relatively few offspring and often stabilize their population size near the carrying capacity of their environment.

keystone species A species whose impacts on its community or ecosystem are much larger and more influential than would be expected from mere abundance. This could be a top predator, a plant that shelters or feeds other organisms, or an organism that plays a critical ecological role.

kinetic energy Energy contained in moving objects, such as a rock rolling down a hill, the wind blowing through the trees, or water flowing over a dam.

Kyoto Protocol An international treaty adopted in Kyoto, Japan, in 1997, in which 160 nations agreed to roll back CO_2, methane, and nitrous oxide emissions to reduce the threat of global climate change.

L

landscape ecology The study of the reciprocal effects of spatial pattern on ecological processes.

landslides Mass wasting or mass movement of rock or soil downhill. Often triggered by seismic events or heavy rainfall.

La Niña The opposite of El Niño.

latent heat Stored energy in a form that is not sensible (detectable by ordinary senses).

LD50 A chemical dose lethal to 50 percent of a test population.

legal standing The right to take part in legal proceedings.

life expectancy The average age that a newborn infant can expect to attain in a particular time and place.

life span The longest period of life reached by a type of organism.

limiting factors Chemical or physical factors that limit the existence, growth, abundance, or distribution of an organism.

limits to growth A belief that the world has a fixed carrying capacity for humans.

lobbying Using personal contacts, public pressure, or political action to persuade legislators to vote in a particular manner.

logarithmic scale One that uses logarithms as units in a sequence that progresses by a factor of 10 in each step.

logical thinking A rational way of thought that asks, "How can orderly, deductive reasoning help me think clearly?"

logistic growth Growth rates regulated by internal and external factors that establish an equilibrium with environmental resources. See also *S curve.*

LULUs Locally Unwanted Land Uses, such as toxic waste dumps, incinerators, smelters, airports, freeways, and other sources of environmental, economic, or social degradation.

M

magma Molten rock from deep in the earth's interior; called lava when it spews from volcanic vents.

malnourishment A nutritional imbalance caused by lack of specific dietary components or inability to absorb or utilize essential nutrients.

Malthusian growth A population explosion followed by a population crash; also called irruptive growth.

Man and Biosphere (MAB) program A design for nature preserves that divides protected areas into zones with different purposes. A highly protected core is surrounded by a buffer zone and peripheral regions in which multiple-use resource harvesting is permitted.

mangrove forests Diverse groups of salt-tolerant trees and other plants that grow in intertidal zones of tropical coastlines.

manipulative experiment Altering a particular factor for a test or experiment while holding all others (as much as possible) constant.

mantle A hot, pliable layer of rock that surrounds the earth's core and underlies the cool outer crust.

marasmus A widespread human protein deficiency disease caused by a diet low in calories and protein or imbalanced in essential amino acids.

marginal costs The cost to produce one additional unit of a good or service.

marshes Wetlands without trees; in North America, this type of land is characterized by cattails and rushes.

mass burn The incineration of unsorted solid waste.

matter Anything that takes up space and has mass.

mean Average.

megacities See *megalopolis*.

megalopolis Also known as a megacity or supercity; megalopolis indicates an urban area with more than 10 million inhabitants.

megawatt (MW) Unit of electrical power equal to 1,000 kilowatts or 1 million watts.

mesosphere The atmospheric layer above the stratosphere and below the thermosphere; the middle layer; temperatures are usually very low.

metamorphic rocks Igneous and sedimentary rocks modified by heat, pressure, and chemical reactions.

methane hydrate Small bubbles or individual molecules of methane (natural gas) trapped in a crystalline matrix of frozen water.

microlending Small loans made to poor people who otherwise don't have access to capital.

midocean ridges Mountain ranges on the ocean floor where magma wells up through cracks and creates new crust.

Milankovitch cycles Periodic variations in tilt, eccentricity, and wobble in the earth's orbit; Milutin Milankovitch suggested these are responsible for cyclic weather changes.

millennium assessment A set of ambitious environmental and human development goals established by the United Nations in 2000.

mineral A naturally occurring, inorganic, crystalline solid with definite chemical composition, a specific internal crystal structure, and characteristic physical properties.

minimills Mills that use scrap metal as their starting material.

minimum viable population The number of individuals needed for long-term survival of rare and endangered species.

modern environmentalism A fusion of conservation of natural resources and preservation of nature with concerns about pollution, environmental health, and social justice.

molecules Combinations of two or more atoms.

monitored, retrievable storage Holding wastes in underground mines or secure surface facilities, such as dry casks, where they can be watched and repackaged, if necessary.

monoculture forestry Intensive planting of a single species; an efficient wood production approach, but one that encourages pests and disease infestations and conflicts with wildlife habitat or recreation uses.

morals A set of ethical principles that guides our actions and relationships.

morbidity Illness or disease.

more-developed countries (MDC) Industrialized nations characterized by high per capita incomes, low birth and death rates, low population growth rates, and high levels of industrialization and urbanization.

mortality Death rate in a population, such as number of deaths per thousand people per year.

Müllerian (or Muellerian) mimicry Evolution of two species, both of which are unpalatable and have poisonous stingers or some other defense mechanism, to resemble each other.

mutagens Agents, such as chemicals or radiation, that damage or alter genetic material (DNA) in cells.

mutation A change, either spontaneous or by external factors, in the genetic material of a cell; mutations in the gametes (sex cells) can be inherited by future generations of organisms.

mutualism A symbiotic relationship between individuals of two different species in which both species benefit from the association.

N

National Environmental Policy Act (NEPA) The law that established the Council on Environmental Quality and that requires environmental impact statements for all federal projects with significant environmental impacts.

natural experiment Observation of natural events to deduce causal relationships.

natural increase Crude death rate subtracted from crude birth rate.

natural resource economics Economics that takes natural resources into account as valuable assets.

natural resources Goods and services supplied by the environment.

natural selection The mechanism for evolutionary change in which environmental pressures cause certain genetic combinations in a population to become more abundant; genetic combinations best adapted for present environmental conditions tend to become predominant.

negative feedbacks Factors that result from a process and, in turn, reduce that same process.

neo-classical economics The branch of economics that attempts to apply the principles of modern science to economic analysis in a mathematically rigorous, noncontextual, abstract, predictive manner.

net energy yield Total useful energy produced during the lifetime of an entire energy system minus the energy used, lost, or wasted in making useful energy available.

net primary productivity The amount of biomass produced by photosynthesis and stored in a community after respiration, emigration, and other factors that reduce biomass.

neurotoxins Toxic substances, such as lead or mercury, that specifically poison nerve cells.

neutron A subatomic particle, found in the nucleus of the atom, that has no electromagnetic charge.

NIMBY Not-In-My-Back-Yard: the position of those opposed to LULUs.

nitrogen cycle The circulation and reutilization of nitrogen in both inorganic and organic phases.

nitrogen-fixing bacteria Bacteria that convert nitrogen from the atmosphere or soil solution into ammonia that can then be converted to plant nutrients by nitrite- and nitrate-forming bacteria.

nitrogen oxides Highly reactive gases formed when nitrogen in fuel or combustion air is heated to over 650°C (1,200°F) in the presence of oxygen or when bacteria in soil or water oxidize nitrogen-containing compounds.

noncriteria pollutants See *unconventional pollutants*.

nongovernmental organizations (NGOs) Pressure and research groups, advisory agencies, political parties, professional societies, and other groups concerned about environmental quality, resource use, and many other issues.

nonpoint sources Scattered, diffuse sources of pollutants, such as runoff from farm fields, golf courses, and construction sites.

nonrenewable resources Minerals, fossil fuels, and other materials present in essentially fixed amounts (within human time scales) in our environment.

normal distribution A bell-shaped curve or Gaussian distribution of measurements or data.

nuclear fission The radioactive decay process in which isotopes split apart to create two smaller atoms.

nuclear fusion A process in which two smaller atomic nuclei fuse into one larger nucleus and release energy; the source of power in a hydrogen bomb.

nucleic acids Large organic molecules made of nucleotides that function in the transmission of hereditary traits, in protein synthesis, and in control of cellular activities.

nucleus The center of the atom; occupied by protons and neutrons. In cells, the organelle that contains the chromosomes (DNA).

O

obese Pathologically overweight, having a body mass greater than 30 kg/m^2, or roughly 30 pounds above normal for an average person.

oil shales Fine-grained sedimentary rock rich in solid organic material called kerogen. When heated, the kerogen liquefies to produce a fluid petroleum fuel.

old-growth forests Forests free from disturbance for long enough (generally 150 to 200 years) to have mature trees, physical conditions, species diversity, and other characteristics of equilibrium ecosystems.

oligotrophic Condition of rivers and lakes that have clear water and low biological productivity (*oligo* = little; *trophic* = nourished); are usually clear, cold, infertile headwater lakes and streams.

omnivores Organisms that eat both plants and animals.

open access system A commonly held resource for which there are no management rules.

open canopy A forest where tree crowns cover less than 20 percent of the ground; also called woodland.

open system A system that exchanges energy and matter with its environment.

organic compounds Complex molecules organized around skeletons of carbon atoms arranged in rings or chains; includes biomolecules, molecules synthesized by living organisms.

overgrazing Allowing domestic livestock to eat so much plant material that it degrades the biological community.

overharvesting Harvesting so much of a resource that it threatens its existence.

overnutrition Receiving too many calories.

oxygen sag Oxygen decline downstream from a pollution source that introduces materials with high biological oxygen demands.

ozone A highly reactive molecule containing three oxygen atoms; a dangerous pollutant in ambient air. In the stratosphere, however, ozone forms an ultraviolet absorbing shield that protects us from mutagenic radiation.

P

paradigms Overarching models of the world that shape our worldviews and guide our interpretation of how things are.

parasite An organism that lives in or on another organism, deriving nourishment at the expense of its host, usually without killing it.

parasitism A relationship in which one organism feeds on another without immediately killing it.

parsimony A principle that says where two equally plausible explanations for a phenomenon are possible, we should choose the simpler one (also known as Ockham's razor).

particulate material Atmospheric aerosols, such as dust, ash, soot, lint, smoke, pollen, spores, algal cells, and other suspended materials; originally applied only to solid particles but now extended to droplets of liquid.

passive solar absorption The use of natural materials or absorptive structures without moving parts to gather and hold heat; the simplest and oldest use of solar energy.

pastoralists People who live by herding domestic animals.

pathogens Organisms that produce disease in host organisms, disease being an alteration of one or more metabolic functions in response to the presence of the organisms.

peat Deposits of moist, acidic, semidecayed organic matter.

pelagic Zones in the vertical water column of a water body.

perennial species Plants that grow for more than two years.

permafrost A permanently frozen layer of soil that underlies the arctic tundra.

permanent retrievable storage Placing waste storage containers in a secure location where they can be inspected periodically and retrieved, if necessary, for repacking or for transfer if a better means of disposal or reuse is developed.

persistent organic pollutants (POPs) Chemical compounds that persist in the environment and retain biological activity for a long time.

pest Any organism that reduces the availability, quality, or value of a useful resource.

pesticide Any chemical that kills, controls, drives away, or modifies the behavior of a pest.

pH A value that indicates the acidity or alkalinity of a solution on a scale of 0 to 14, based on the proportion of H^+ ions present.

phosphorus cycle The movement of phosphorus atoms from rocks through the biosphere and hydrosphere and back to rocks.

photochemical oxidants Products of secondary atmospheric reactions. See also *smog*.

photodegradable plastics Plastics that break down when exposed to sunlight or to a specific wavelength of light.

photosynthesis The biochemical process by which green plants and some bacteria capture light energy and use it to produce chemical bonds. Carbon dioxide and water are consumed while oxygen and simple sugars are produced.

photovoltaic cell An energy-conversion device that captures solar energy and directly converts it to electrical current.

phylogenetic species concept A definition of species that depends on genetic similarities (or differences).

phytoplankton Microscopic, free-floating, autotrophic organisms that function as producers in aquatic ecosystems.

pioneer species In primary succession on a terrestrial site, the plants, lichens, and microbes that first colonize the site.

plankton Primarily microscopic organisms that occupy the upper water layers in both freshwater and marine ecosystems.

plasma A hot, electrically neutral gas of ions and free electrons.

poaching Hunting wildlife illegally.

point sources Specific locations of highly concentrated pollution discharge, such as factories, power plants, sewage treatment plants, underground coal mines, and oil wells.

policy A societal plan or statement of intentions intended to accomplish some social or economic goal.

policy cycle The process by which problems are identified and acted upon in the public arena.

pollution To make foul, unclean, dirty; any physical, chemical, or biological change that adversely affects the health, survival, or activities of living organisms or that alters the environment in undesirable ways.

pollution charges Fees assessed per unit of pollution based on the "polluter pays" principle.

population All members of a species that live in the same area at the same time.

population crash A sudden population decline caused by predation, waste accumulation, or resource depletion; also called a dieback.

population explosion Growth of a population at exponential rates to a size that exceeds environmental carrying capacity; usually followed by a population crash.

population momentum A potential for increased population growth as young members reach reproductive age.

positive feedbacks Factors that result from a process and, in turn, increase that same process.

potential energy Stored energy that is latent but available for use. A rock poised at the top of a hill or water stored behind a dam are examples of potential energy.

power The rate of energy delivery; measured in horsepower or watts.

precautionary principle The rule that we should leave a margin of safety for unexpected developments. This principle implies that we should strive to prevent harm to human health and the environment even if risks are not fully understood.

predator An organism that feeds directly on other organisms in order to survive; live-feeders, such as herbivores and carnivores.

predator-mediated competition A situation in which the effects of a predator dominate population dynamics.

preservation A philosophy that emphasizes the fundamental right of living organisms to exist and to pursue their own ends.

primary pollutants Chemicals released directly into the air in a harmful form.

primary producers Photosynthesizing organisms.

primary productivity Synthesis of organic materials (biomass) by green plants using the energy captured in photosynthesis.

primary standards Regulations of the 1970 Clean Air Act; intended to protect human health.

primary succession Ecological succession that begins in an area where no biotic community previously existed.

primary treatment A process that removes solids from sewage before it is discharged or treated further.

principle of competitive exclusion A result of natural selection whereby two similar species in a community occupy different ecological niches, thereby reducing competition for food.

probability The likelihood that a situation, a condition, or an event will occur.

producer An organism that synthesizes food molecules from inorganic compounds by using an external energy source; most producers are photosynthetic.

productivity The amount of biomass (biological material) produced in a given area during a given period of time.

prokaryotic Cells that do not have a membrane-bounded nucleus or membrane-bounded organelles.

pronatalist pressures Influences that encourage people to have children.

prospective study A study in which experimental and control groups are identified before exposure to some factor. The groups are then monitored and compared for a specific time after the exposure to determine any effects the factor may have.

proteins Chains of amino acids linked by peptide bonds.

proton A positively charged subatomic particle found in the nucleus of an atom.

proven-in-place reserves Energy sources that have been thoroughly mapped and are likely to be economically recoverable with available technology.

pull factors Conditions that draw people from the country into the city.

push factors Conditions that force people out of the country and into the city.

R

r-selected species Organisms whose population growth is regulated mainly by external factors. They tend to have rapid reproduction and high mortality of offspring. Given optimum environmental conditions, they can grow exponentially. Many "weedy" or pioneer species fit in this category.

radioactive decay A change in the nuclei of radioactive isotopes that spontaneously emit high-energy electromagnetic radiation and/or subatomic particles while gradually changing into another isotope or different element.

rainforest A forest with high humidity, constant temperature, and abundant rainfall (generally over 380 cm [150 in.] per year); can be tropical or temperate.

rain shadow Dry area on the downwind side of a mountain.

random sample A subset of a collection of items or observations chosen at random.

rangeland Grasslands and open woodlands suitable for livestock grazing.

rational choice Public decision making based on reason, logic, and science-based management.

reasoned judgment Thoughtful decisions based on careful, logical examination of available evidence.

recharge zones Areas where water infiltrates into an aquifer.

reclamation Chemical, biological, or physical cleanup and reconstruction of severely contaminated or degraded sites to return them to something like their original topography and vegetation.

recycling   Reprocessing of discarded materials into new, useful products; not the same as reuse of materials for their original purpose, but the terms are often used interchangeably.

reduced tillage systems   Farming methods that preserve soil and save energy and water through reduced cultivation; includes minimum till, conserve-till, and no-till systems.

reflective thinking   A thoughtful, contemplative analysis that asks, "What does this all mean?"

reformer   A device that strips hydrogen from fuels such as natural gas, methanol, ammonia, gasoline, or vegetable oil so they can be used in a fuel cell.

refuse-derived fuel   Processing of solid waste to remove metal, glass, and other unburnable materials; organic residue is shredded, formed into pellets, and dried to make fuel for power plants.

regenerative farming   Farming techniques and land stewardship that restore the health and productivity of the soil by rotating crops, planting ground cover, protecting the surface with crop residue, and reducing synthetic chemical inputs and mechanical compaction.

regulations   Rules established by administrative agencies; regulations can be more important than statutory law in the day-to-day management of resources.

rehabilitation   Rebuilding basic structure or function in an ecological system without necessarily achieving complete restoration to its original condition.

relative humidity   At any given temperature, a comparison of the actual water content of the air with the amount of water that could be held at saturation.

relevé   A rapid assessment of vegetation types and biodiversity in an area.

remediation   Cleaning up chemical contaminants from a polluted area.

renewable resources   Resources normally replaced or replenished by natural processes; resources not depleted by moderate use; examples include solar energy, biological resources such as forests and fisheries, biological organisms, and some biogeochemical cycles.

renewable water supplies   Annual freshwater surface runoff plus annual infiltration into underground freshwater aquifers that are accessible for human use.

replacement rate   The number of children per couple needed to maintain a stable population. Because of early deaths, infertility, and nonreproducing individuals, this is usually about 2.1 children per couple.

replication   Repeating studies or tests.

reproducibility   Making an observation or obtaining a particular result consistently.

residence time   The length of time a component, such as an individual water molecule, spends in a particular compartment or location before it moves on through a particular process or cycle.

resilience   The ability of a community or ecosystem to recover from disturbances.

resource partitioning   In a biological community, various populations sharing environmental resources through specialization, thereby reducing direct competition. See also *ecological niche.*

resources   In economic terms, anything with potential use in creating wealth or giving satisfaction.

restoration ecology   Seeks to repair or reconstruct ecosystems damaged by human actions.

retrospective study   A study that looks back in history at a group of people (or other organisms) who suffer from some condition to try to identify something in their past life that the whole group shares but that is not found in the histories of a control group as near as possible to those being studied but who do not suffer from the same condition.

riders   Amendments attached to bills in conference committee, often completely unrelated to the bill to which they are added.

rill erosion   The removing of thin layers of soil as little rivulets of running water gather and cut small channels in the soil.

risk   The probability that something undesirable will happen as a consequence of exposure to a hazard.

risk assessment   Evaluation of the short-term and long-term risks associated with a particular activity or hazard; usually compared with benefits in a cost-benefit analysis.

rock   A solid, cohesive aggregate of one or more crystalline minerals.

rock cycle   The process whereby rocks are broken down by chemical and physical forces; sediments are moved by wind, water, and gravity; sedimented and reformed into rock; and then crushed, folded, melted, and recrystallized into new forms.

rotational grazing   Confining grazing animals in a small area for a short time to force them to eat weedy species as well as the more desirable grasses and forbes.

runoff   The excess of precipitation over evaporation; the main source of surface water and, in broad terms, the water available for human use.

S

salinity   The amount of dissolved salts (especially sodium chloride) in a given volume of water.

salinization   A process in which mineral salts accumulate in the soil, killing plants; occurs when soils in dry climates are irrigated profusely.

saltwater intrusion   The movement of saltwater into freshwater aquifers in coastal areas where groundwater is withdrawn faster than it is replenished.

sample   To analyze a small but representative portion of a population to estimate the characteristics of the entire class.

sanitary landfills   Landfills in which garbage and municipal waste are buried every day under enough soil or fill to eliminate odors, vermin, and litter.

savannas   An open prairie or grassland with scattered groves of trees.

scientific consensus   A general agreement among informed scholars.

scientific method   A systematic, precise, objective study of a problem. Generally this requires observation, hypothesis development and testing, data gathering, and interpretation.

scientific theory   An explanation or idea accepted by a substantial number of scientists.

S curve   A curve that depicts logistic growth; called an S curve because of its shape.

sea-grass beds   Large expanses of rooted, submerged, or emergent aquatic vegetation, such as eel grass or salt grass.

secondary pollutants   Chemicals modified to a hazardous form after entering the air or that are formed by chemical reactions as components of the air mix and interact.

secondary succession   Succession on a site where an existing community has been disrupted.

secondary treatment   Bacterial decomposition of suspended particulates and dissolved organic compounds that remain after primary sewage treatment.

second law of thermodynamics   States that, with each successive energy transfer or transformation in a system, less energy is available to do work.

secure landfills   Solid waste disposal sites lined and capped with an impermeable barrier to prevent leakage or leaching.

sedimentary rocks   Rocks composed of accumulated, compacted mineral fragments, such as sand or clay; examples include shale, sandstone, breccia, and conglomerates.

sedimentation   The deposition of organic materials or minerals by chemical, physical, or biological processes.

selection pressure   Limited resources or adverse environmental conditions that tend to favor certain adaptations in a population. Over many generations, this can lead to genetic change, or evolution.

selective cutting   Harvesting only mature trees of certain species and size; usually more expensive than clear-cutting but less disruptive for wildlife and often better for forest regeneration.

shade-grown coffee and cocoa   Plants grown under a canopy of taller trees, which provides habitat for birds and other wildlife.

sheet erosion   Peeling off thin layers of soil from the land surface; accomplished primarily by wind and water.

shelterwood harvesting   Mature trees are removed from the forest in a series of two or more cuts, leaving young trees and some mature trees as a seed source for future regeneration.

sick building syndrome   A cluster of allergies and other illnesses caused by sensitivity to molds, synthetic chemicals, or other harmful compounds trapped in insufficiently ventilated buildings.

significant numbers   Meaningful data that can be measured accurately and reproducibly.

sinkholes   A large surface crater caused by the collapse of an underground channel or cavern; often triggered by groundwater withdrawal.

sludge   A semisolid mixture of organic and inorganic materials that settles out of wastewater at a sewage treatment plant.

smart growth   The efficient use of land resources and existing urban infrastructure that encourages in-fill development, provides a variety of affordable housing and transportation choices, and seeks to maintain a unique sense of place by respecting local cultural and natural features.

smelting   Roasting ore to release metals from mineral compounds.

smog   The combination of smoke and fog in the stagnant air of London; now often applied to photochemical pollution.

social justice   Equitable access to resources and the benefits derived from them; a system that recognizes inalienable rights and adheres to what is fair, honest, and moral.

soil   A complex mixture of weathered rock material, partially decomposed organic molecules, and a host of living organisms.

soil creep The slow, downhill movement of soil due to erosion.

soil horizons Horizontal layers that reveal a soil's history, characteristics, and usefulness.

Southern Oscillation The combination of El Niño and La Niña cycles.

southern pine forest United States coniferous forest ecosystem characterized by a warm, moist climate.

speciation Evolution of new species.

species All the organisms genetically similar enough to breed and produce live, fertile offspring in nature.

species diversity The number and relative abundance of species present in a community.

specific heat The amount of heat energy needed to change the temperature of a body. Water has a specific heat of 1, which is higher than most substances.

sprawl Unlimited, unplanned growth of urban areas that consumes open space and wastes resources.

stability In ecological terms, a dynamic equilibrium among the physical and biological factors in an ecosystem or a community; relative homeostasis.

stable runoff The fraction of water available year-round; usually more important than total runoff when determining human uses.

state shift An abrupt response to a disturbance that causes a persistent change in a system to a new set of conditions and relationships.

statistics Mathematical analysis of the collection, organization, and interpretation of numerical data.

statutory law Rules passed by a state or national legislature.

steady-state economy Characterized by low birth and death rates, use of renewable energy sources, recycling of materials, and emphasis on durability, efficiency, and stability.

stewardship A philosophy that holds that humans have a unique responsibility to manage, care for, and improve nature.

Strategic Lawsuits against Public Participation (SLAPP) Lawsuits that have no merit but are brought merely to intimidate and harass private citizens who act in the public interest.

strategic metals and minerals Materials a country cannot produce itself but that it uses for essential materials or processes.

stratosphere The zone in the atmosphere extending from the tropopause to about 50 km (30 mi) above the earth's surface; temperatures are stable or rise slightly with altitude; has very little water vapor but is rich in ozone.

stress Physical, chemical, or emotional factors that place a strain on an animal. Plants also experience physiological stress under adverse environmental conditions.

strip-cutting Harvesting trees in strips narrow enough to minimize edge effects and to allow natural regeneration of the forest.

strip-farming Planting different kinds of crops in alternating strips along land contours; when one crop is harvested, the other crop remains to protect the soil and prevent water from running straight down a hill.

strip-mining Extracting shallow mineral deposits (especially coal) by scraping off surface layers with giant earth-moving equipment; creates a huge open pit; an alternative to underground or deep open-pit mines.

Student Environmental Action Coalition (SEAC) A grassroots coalition of student and youth environmental groups, working together to protect our planet and our future.

subduction (subducted) Where the edge of one tectonic plate dives beneath the edge of another.

subsidence Settling of the ground surface caused by the collapse of porous formations that result from withdrawal of large amounts of groundwater, oil, or other underground materials.

subsoil A layer of soil beneath the topsoil that has lower organic content and higher concentrations of fine mineral particles; often contains soluble compounds and clay particles carried down by percolating water.

sulfur cycle The chemical and physical reactions by which sulfur moves into or out of storage and through the environment.

sulfur dioxide A colorless, corrosive gas directly damaging to both plants and animals.

Superfund A fund established by Congress to pay for containment, cleanup, or remediation of abandoned toxic waste sites. The fund is financed by fees paid by toxic waste generators and by cost recovery from cleanup projects.

surface mining Some minerals are also mined from surface pits. See also *strip-mining*.

surface soil The A horizon in a soil profile; the soil just below the litter layer.

surface tension The tendency for a surface of water molecules to hold together, producing a surface that resists breaking.

sustainability Ecological, social, and economic systems that can last over the long term.

sustainable agriculture (regenerative farming) Ecologically sound, economically viable, socially just agricultural system. Stewardship, soil conservation, and integrated pest management are essential for sustainability.

sustainable development A real increase in well-being and standard of life for the average person that can be maintained over the long term without degrading the environment or compromising the ability of future generations to meet their own needs.

sustained yield Utilization of a renewable resource at a rate that does not impair or damage its ability to be fully renewed on a long-term basis.

swamps Wetlands with trees, such as the extensive swamp forests of the southern United States.

symbiosis The intimate living together of members of two species; includes mutualism, commensalism, and, in some classifications, parasitism.

sympatric speciation A gradual change (generally through genetic drift) so that offspring are genetically distinct from their ancestors even though they live in the same place.

synergism When an injury caused by exposure to two environmental factors together is greater than the sum of exposure to each factor individually.

synergistic effects The combination of several processes or factors is greater than the sum of their individual effects.

systems Networks of interdependent components and processes.

T

taiga The northernmost edge of the boreal forest, including species-poor woodland and peat deposits; intergrading with the arctic tundra.

tailings Mining waste left after mechanical or chemical separation of minerals from crushed ore.

taking The unconstitutional confiscation of private property.

tar sands Geologic deposits composed of sand and shale particles coated with bitumen, a viscous mixture of long-chain hydrocarbons.

tectonic plates Huge blocks of the earth's crust that slide around slowly, pulling apart to open new ocean basins or crashing ponderously into each other to create new, larger landmasses.

telemetry Locating or studying organisms at a distance using radio signals or other electronic media.

temperate rainforest The cool, dense, rainy forest of the northern Pacific coast; enshrouded in fog much of the time; dominated by large conifers.

temperature A measure of the speed of motion of a typical atom or molecule in a substance.

temperature inversions Atmospheric conditions in which a layer of warm air lies on top of cooler air and blocks normal convection currents. This can trap pollutants and degrade air quality.

teratogens Chemicals or other factors that specifically cause abnormalities during embryonic growth and development.

terracing Shaping the land to create level shelves of earth to hold water and soil; requires extensive hand labor or expensive machinery, but it enables farmers to farm very steep hillsides.

tertiary treatment The removal of inorganic minerals and plant nutrients after primary and secondary treatment of sewage.

thermal pollution Artificially raising or lowering of the temperature of a water body in a way that adversely affects the biota or water quality.

thermocline In water, a distinctive temperature transition zone that separates an upper layer that is mixed by the wind (the epilimnion) and a colder deep layer that is not mixed (the hypolimnion).

thermodynamics The branch of physics that deals with transfers and conversions of energy.

thermohaline ocean conveyor A large-scale oceanic circulation system in which warm water flows from equatorial zones to higher latitudes where it cools, evaporates, and becomes saltier and more dense, which causes it to sink and flow back toward the equator in deep ocean currents.

thorn scrub A dry, open woodland or shrubland characterized by sparse, spiny shrubs.

threatened species While still abundant in parts of its territorial range, this species has declined significantly in total numbers and may be on the verge of extinction in certain regions or localities.

throughput The flow of energy and/or matter into and out of a system.

tide pools Small pools of water left behind by falling tides.

tolerance limits See *limiting factors*.

topsoil The first true layer of soil; layer in which organic material is mixed with mineral particles; thickness ranges from a meter or more under virgin prairie to zero in some deserts.

total fertility rate The number of children born to an average woman in a population during her entire reproductive life.

total growth rate The net rate of population growth resulting from births, deaths, immigration, and emigration.

total maximum daily loads (TMDL) The amount of particular pollutant that a water body can receive from both point and nonpoint sources and still meet water quality standards.

Toxic Release Inventory A program created by the Superfund Amendments and Reauthorization Act of 1984 that requires manufacturing facilities and waste handling and disposal sites to report annually on releases of more than 300 toxic materials. You can find out from the EPA whether any of these sites are in your neighborhood and what toxins they release.

toxins Poisonous chemicals that react with specific cellular components to kill cells or to alter growth or development in undesirable ways; often harmful, even in dilute concentrations.

tradable permits Pollution quotas or variances that can be bought or sold.

"Tragedy of the Commons" An inexorable process of degradation of communal resources due to selfish self-interest of "free riders" who use or destroy more than their fair share of common property. See *open access system.*

transpiration The evaporation of water from plant surfaces, especially through stomates.

trophic level Step in the movement of energy through an ecosystem; an organism's feeding status in an ecosystem.

tropical rainforests Forests near the equator in which rainfall is abundant—more than 200 cm (80 in.) per year—and temperatures are warm to hot year-round.

tropical seasonal forests Semievergreen or partly deciduous forests tending toward open woodlands and grassy savannas dotted with scattered, drought-resistant trees.

tropopause The boundary between the troposphere and the stratosphere.

troposphere The layer of air nearest to the earth's surface; both temperature and pressure usually decrease with increasing altitude.

tsunami Far-reaching waves caused by earthquakes or undersea landslides.

tundra Treeless arctic or alpine biome characterized by cold, dark winters; a short growing season; and potential for frost any month of the year; vegetation includes low-growing perennial plants, mosses, and lichens.

U

unconventional pollutants Toxic or hazardous substances, such as asbestos, benzene, beryllium, mercury, polychlorinated biphenyls, and vinyl chloride, not listed in the original Clean Air Act because they were not released in large quantities; also called noncriteria pollutants.

urban agglomerations Urban areas where several cities or towns have coalesced.

utilitarian conservation The philosophy that resources should be used for the greatest good for the greatest number for the longest time.

V

values An estimation of the worth of things; a set of ethical beliefs and preferences that determines our sense of right and wrong.

vertical stratification The vertical distribution of specific subcommunities within a community.

vertical zonation Vegetation zones determined by climate changes brought about by altitude changes.

volatile organic compounds Organic chemicals that evaporate readily and exist as gases in the air.

volcanoes Vents in the earth's surface through which molten lava (magma), gases, and ash escape to create mountains.

vulnerable species Naturally rare organisms or species whose numbers have been so reduced by human activities that they are susceptible to actions that could push them into threatened or endangered status.

W

warm front A long, wedge-shaped boundary caused when a warmer advancing air mass slides over neighboring cooler air parcels.

waste stream The steady flow of varied wastes, from domestic garbage and yard wastes to industrial, commercial, and construction refuse.

water cycle The recycling and reutilization of water on earth, including atmospheric, surface, and underground phases and biological and nonbiological components.

waterlogging Water saturation of soil that fills all air spaces and causes plant roots to die from lack of oxygen; a result of overirrigation.

water scarcity Having less than 1,000 m^3 (264,000 gal) of clean fresh water available per person per year.

watershed The land surface and groundwater aquifers drained by a particular river system.

water stress Countries that consume more than 10 percent of renewable water supplies.

water table The top layer of the zone of saturation; undulates according to the surface topography and subsurface structure.

watt One joule per second.

weather The physical conditions of the atmosphere (moisture, temperature, pressure, and wind).

weathering Changes in rocks brought about by exposure to air, water, changing temperatures, and reactive chemical agents.

wetlands Ecosystems of several types in which rooted vegetation is surrounded by standing water during part of the year. See also *swamps, marshes, bogs, fens.*

wind farms Large numbers of windmills concentrated in a single area; usually owned by a utility or large-scale energy producer.

withdrawal A description of the total amount of water taken from a lake, a river, or an aquifer.

woodland A forest where tree crowns cover less than 20 percent of the ground; also called open canopy forest.

work The application of force through a distance; requires energy input.

world conservation strategy A proposal for maintaining essential ecological processes, preserving genetic diversity, and ensuring that utilization of species and ecosystems is sustainable.

X

X ray Very short wavelength in the electromagnetic spectrum; can penetrate soft tissue; although it is useful in medical diagnosis, it also damages tissue and causes mutations.

Z

zero population growth (ZPG) A condition in which births and immigration in a population just balance deaths and emigration.

zone of aeration Upper soil layers that hold both air and water.

zone of saturation Lower soil layers where all spaces are filled with water.

Credits

Photographs

Chapter 1

Opener: © William P. Cunningham; 1.1: © Vol. DV13/Getty Images; 1.2: © Stockgrek/age fotostock; 1.3: © Vol. 6/Corbis; 1.6: © Norbert Schiller/The Image Works; 1.9 & 1.10: © William P. Cunningham; 1.14: © The McGraw-Hill Companies, Inc./Barry Barker, photographer; 1.17: David L. Hansen, University of Minnesota Agricultural Experiment Station; 1.18: © Mary Ann Cunningham; p. 14: © Martin Strmko/istockphoto; 1.19: David L. Hansen, University of Minnesota Agricultural Experiment Station; 1.20: © The McGraw-Hill Companies, Inc./John A. Karachewski, photographer; 1.21: © Mary Ann Cunningham; 1.22a: Courtesy of the Bancroft Library at the University of California, Berkeley #Muir, John - POR 65; 1.22b: Courtesy of Grey Towers National Historic Landmark; 1.22c: © Bettmann/Corbis; 1.22d: © AP Wide World Photos; 1.23: © Joe Klune/istockphoto; 1.24a: AP/Wide World Photos; 1.24b: © Tom Turner/The Brower Fund/Earth Island Institute; 1.24c: Columbia University Archives; Columbia University in the City of New York; 1.24d: © AP/Wide World Photos

Chapter 2

Opener: © 2006 Peter Menzel/menzelphoto .com; 2.4: © Mary Ann Cunningham; p. 31: © G.I. Bernard/Animals Animals; 2.11: © William P. Cunningham; 2.12: NOAA; p. 39 Fig. 2: The SeaWiFS Project, NASA/Goddard Space Flight Center and ORBIMAGE; 2.24: © Nigel Cattlin/Visuals Unlimited

Chapter 3

Opener & 3.1: Courtesy of David Tilman; 3.3: © Vol. 6/Corbis RF; 3.4: © William P. Cunningham; 3.6: © visual safari/Alamy Images; 3.7 & 3.14: © William P. Cunningham; 3.15: © Corbis RF; 3.16: © D.P. Wilson/Photo Researchers; 3.17: © Creatas/PunchStock; 3.18a–b: © Edward Ross; 3.19: Courtesy Tom Finkle; 3.20a: © William P. Cunningham; 3.20b: © PhotoDisc/Getty Images; 3.19c: © William P. Cunningham; 3.21: © Vol. 6 PhotoDisc/Getty Images; 3.26a: © Digital

Vision/Getty Images; 3.26b: © Stockbyte Royalty Free; 3.26c: © PhotoDisc; 3.26d: © Corbis Royalty Free; 3.27a: © Corbis Royalty Free; 3.27b: © Eric and David Hosking/Corbis; 3.27c: © image100/PunchStock RF; 3.29: © Vol. 262/Corbis; 3.33 & 3.34: © William P. Cunningham

Chapter 4

Opener: © Pierre Tremblay/Masterfile; 4.1: © Apichart Weerawong/AP/Wide World Photos; 4.2: © Vol. EP39/PhotoDisc/Getty Images; 4.7: © Alain Le Garsmeur/Corbis; 4.11a: © William P. Cunningham; 4.11b: © Vol. 17/PhotoDisc/Getty Images

Chapter 5

Opener: © Digital Vision Royalty Free; p. 95: © Mary Ann Cunningham; 5.7: © Vol. 60/PhotoDisc; 5.8: © Vol. 6/Corbis; 5.9: © William P. Cunningham; 5.10: © Mary Ann Cunningham; 5.11: © William P. Cunningham; 5.12: © Vol. 90/Corbis; 5.13: © William P. Cunningham; 5.14: © Mary Ann Cunningham; 5.15: Courtesy of Sea WiFS/NASA; 5.17: NOAA; 5.18a: © Vol. 89/PhotoDisc/Getty Images; 5.18b: © Robert Garvey/Corbis; 5.18c: © Andrew Martinez/Photo Researchers; 5.18d: © Pat O'Hara/Corbis; 5.20a: © Vol. 16/PhotoDisc/Getty Images; 5.20b: © William P. Cunningham; 5.20c: © Raymond Gehman/Corbis; 5.21: © The McGraw-Hill Companies, Inc./Barry Barker, photographer; 5.23 & 5.24: © William P. Cunningham; p. 112 (fig. 2): © L. David Mech; 5.28: © Mary Ann Cunningham; 5.29: © Nature Picture Library/Alamy; 5.30: © Mary Ann Cunningham; 5.31: © William P. Cunningham; 5.32: © Lynn Funkhouser/Peter Arnold; 5.33: © Comstock/PunchStock

Chapter 6

Opener: © Ron Thiele; 6.5: © William P. Cunningham; 6.6a–c: Courtesy United Nations Environment Programme; 6.8: © William P. Cunningham; 6.9: © Digital Vision/Getty Images; 6.10: Courtesy U.S. Fish & Wildlife Service/J & K Hollingworth; 6.11: © Gary Braasch/Stone/Getty Images; 6.12: Courtesy of John McColgan, Alaska Fires Service/Bureau of Land Management; 6.13: © William P. Cun-

ningham; 6.15: Courtesy of Tom Finkle; 6.16, 6.17 & 6.19: © William P. Cunningham; 6.21: © American Heritage Center, University of Wyoming # 20475; 6.22: © William P. Cunningham; 6.23: © Richard Hamilton Smith/Corbis; 6.24: © Digital Vision/Getty Images RF; 6.25: © William P. Cunningham; 6.29: Courtesy of R.O. Bierregaard

Chapter 7

Opener: © John Maier/The New York Times/Redux Pictures; 7.5: © Norbert Schiller/The Image Works; 7.6: © Scott Daniel Peterson; 7.7: © Lester Bergman/Corbis; 7.11: © William P. Cunningham; 7.13a: Photo by Jeff Vanuga, USDA Natural Resources Conservation Service; 7.14: © Novastock/PhotoEdit; 7.15 & 7.18: © FAO photo/R. Faidutti; 7.21a: Photo by Lynn Betts, courtesy of USDA Natural Resources Conservation Service; 7.21b: Photo by Jeff Vanuga, courtesy of USDA Natural Resources Conservation Center; 7.21c: © Corbis Royalty Free; 7.22: © Corbis Royalty Free; 7.23 & 7.25: © William P. Cunningham; 7.27: © Michael Rosenfeld/Stone/Getty Images; 7.29: Photo by Lynn Betts, courtesy of USDA Natural Resources Conservation Service; 7.30 & p. 170: © William P. Cunningham; 7.24: © Vol. 120/Corbis; 7.31: © William P. Cunningham; 7.32: © 2008 Star Tribune/Minneapolis-St. Paul; 7.33: © William P. Cunningham

Chapter 8

Opener: © The Carter Center/Emily Staub; 8.1: Courtesy of Donald R. Hopkins; 8.3: © William P. Cunningham; 8.4a: © Vol. 40/Corbis; 8.4b: © Image Source/Getty RF; 8.4c: Courtesy of Stanley Erlandsen, University of Minnesota; 8.7: © Getty Images RF; 8.18: © The McGraw-Hill Companies, Inc./Sharon Farmer, photographer

Chapter 9

Opener: Courtesy of Dave Hansen, College of Agriculture Experiment Station, University of Minnesota; 9.1: Courtesy Jacques Descloitres, MODIS Rapid Response Team, NASA/GSFC; 9.5: Courtesy Candace Kohl, University of California, San Diego; 9.11a–b: Photographer Lisa McKeon, courtesy of Glacier National Park

Index